Applied Environmental Science and Engineering for a Sustainable Future

Series Editors
Jega V. Jegatheesan
Deakin University, Geelong, Australia
Li Shu
Deakin University, Geelong, Australia
Piet Lens
UNESCO-IHE Institute for Water Education, Delft, The Netherlands
Chart Chiemchaisri
Kasetsart University, Bangkok, Thailand

Applied Environmental Science and Engineering for a Sustainable Future (AESE) series covers a variety of environmental issues and how they could be solved through innovations in science and engineering. Our societies thrive on the advancements in science and technology which pave way for better and better standard of living. The adverse effects of such improvements are the deterioration of the environment. Thus, better catchment management in order to sustainably manage all types of resources (including water, minerals and others) becomes paramount important. Water and wastewater treatment and reuse, solid and hazardous waste management, industrial waste minimisation, soil and agriculture as well as myriad of other topics needs better understanding and application. The book series will aim at fulfilling such a task in coming years.

More information about this series at http://www.springer.com/series/13085

Ram Lakhan Singh
Editor

Principles and Applications of Environmental Biotechnology for a Sustainable Future

Editor
Ram Lakhan Singh
Professor of Biochemistry
Dr. Ram Manohar Lohia Avadh University
Faizabad, India

Applied Environmental Science and Engineering for a Sustainable Future
ISBN 978-981-10-9465-1 ISBN 978-981-10-1866-4 (eBook)
DOI 10.1007/978-981-10-1866-4

Softcover reprint of the hardcover 1st edition 2016

Printed on acid-free paper

This Springer imprint is published by Springer Nature
The registered company is Springer Science+Business Media Singapore Pte Ltd.

To Uma,
my best friend, companion, and devoted loving wife.

Preface

People need a good environment to be healthy and happy. Here, only the importance of environmental biotechnology begins. Wastes require proper disposal to prevent pollution, and water must be reconditioned for repeated use. Environmental biotechnology is about using living organisms or their products to manage the environment and environmental quality. Environmental biotechnology is a broad field, including resolution to pollution and environmental problems by usage of microorganisms and other living systems to solve them such as the detoxification of pollutants and clean-up of oil spillage. It is an environmentally focused approach, which removes the barrier of traditional rigid boundaries to provide a role in remediation of contaminated land, water and air, treatment of industrial wastes, biopesticide production and application.

It gives me immense pleasure to introduce this book *Principles and Applications of Environmental Biotechnology for a Sustainable Future*. The purpose of this book is to present a fair reflection of the practical biological approaches currently employed to address environmental problems and to provide the reader with a working knowledge of the science that underpins them. Starting with the fundamentals of biotechnology, the book provides a detailed discussion on global environmental problems, microbes and their interaction with the environment, xenobiotics and their remediation, solid waste management, waste water treatment, bioreactors, biosensors, biomining and biopesticides. This book also covers renewable and non-renewable bioenergy resources, biodiversity and its conservation, approaches to control biotechnological industries, bioassay for pollutants' analysis and genetically modified microorganism and foods to increase awareness for improvement in industrial productivity.

All chapters have been designed and prepared by the authors in such a way that presents the subject in depth following a student-friendly approach. A systematic reading of the text from the beginning will allow the students to gain a clear understanding of all the aspects of environmental biotechnology. The book will also serve the new entrants in research and professionals in biotechnology industry to understand the basics of environmental biotechnology which they can utilize for climbing further steps in their career and solve certain tricky problems coming in their way. Each chapter has a short bibliography which may serve as an entry point to the research students.

Key Features

The text of the book includes certain important features to facilitate better understanding of the topics discussed in the chapters.

Abstract at the beginning of each chapter highlights the important concepts discussed and enables recapitulation.

Contents of each chapter list the main topics included in that chapter.

Tables and figures interspersed throughout the chapters enable easy understanding of the concepts discussed.

Bibliography at the end of each chapter familiarizes the readers with important texts and articles cited in the text.

Suggested further readings are a ready reference for students, teachers and researchers to pursue their studies further on those aspects of environmental biotechnology that may be of interest to them.

Organization of the Book

The book is organized into 15 chapters.

Chapter 1 traces the brief introduction, scope and applications of environmental biotechnology.

Chapter 2 describes the current global issues and environmental problems.

Chapter 3 is devoted to microbes and their interaction with the environment. Microbial communication with the environment provides essential information required for understanding the form, function and systematic relationship of host–microorganism.

Chapter 4 starts with the introduction of xenobiotics, types, environmental factor and remediation by using different microorganisms. The discussion extends to details of variations in the structure and functions of xenobiotics and mechanisms of its bioremediation.

Chapter 5 presents the essential technologies involved in solid waste management and soil pollution control. This is one of the most necessary topics in environmental biotechnology. The chapter presents a detailed discussion on solid waste recycling and reuse of resources, quantitative physical measures and variables.

Chapter 6 is devoted to waste water treatment levels and processes. This chapter describes disposal, phase separation, oxidation and polishing of waste water. It also covers the parameters essential for treatment.

Chapter 7 deals with bioreactors, classification and its types. It describes biochemical processes which employ different microbes, plant cells or mammalian cell systems for production of biological products. Bioreactor provides a controlled environment for the production of metabolites which can help to achieve the optimal growth of microbes for efficient catalytic reactions.

Chapter 8 focuses on biopesticide types, resources and their scope in agriculture. This chapter also covers its industrial applications and worldwide market/product analysis.

Chapter 9 presents the renewable and non-renewable bioenergy resources including biofuels. Fossil fuels make up a large portion of today's energy market, and enough promising new renewable technologies are emerging. Each of them have benefits and challenges and relate to unique technologies that play a role in our current energy systems.

Chapter 10 addresses biodiversity and its conservation. Biodiversity is the degree of variation of life. It is a measure of variety of organisms present in different ecosystems. Conservation of biodiversity is about saving life on earth in all its forms and keeping natural ecosystems functioning and healthy. Biodiversity conservation relies on a number of disciplines working together, including ecology and other areas such as economics, law, public policy and psychology.

Chapter 11 describes biosensors, bioreceptors, types and its applications. Biosensor is an analytical device used for detection of an analyte that combines a biological component with a physico-chemical detector.

Chapter 12 covers environmental control of biotechnology industries primarily involving the use of organisms for remedial measures against pollution. The ever-increasing world population and industries have deteriorated the world environment to such an alarming extent that it has already started spelling doom for many species including man.

Chapter 13 highlights biochemistry of bioleaching and biomining. Bioleaching is a result of slow decomposition of ore, releasing sulphuric compounds that, in turn, also release metals into the environment. Bioleaching utilizes billions of rock-eating bacteria, acting as catalyst, to extract copper, iron, gold, silver, cobalt and other metals from centuries-old mining waste. This process converts the insoluble metal sulphides in mine wastes into water-soluble metal sulphates, thereby neutralizing the toxins like sulphur, arsenic, etc.

Chapter 14 covers the latest research happening in the field of genetically modified microorganisms (GMOs). GMOs are the sources of genetically modified foods and are widely used in scientific research and to produce goods other than food. The scope and applications of GMOs in agriculture and industries are discussed along with the assessment of probable harm caused to the environment.

Chapter 15 covers bioplastics, degradable biobased polymers, renewable resources, photodegradable polymers and the future of plastics.

Acknowledgements

I thank all the students in my environmental biochemistry and biotechnology classes as well as my research students over the period of my career as teacher who helped me a lot to learn the science of biotechnology and who really encouraged me to present this book for the useful reading and understanding of concepts of environmental biotechnology by the students, researchers and other stakeholders. I am deeply indebted to all contributing authors for their tremendous efforts to finish their work on time and for their gracious tolerance of my repeated haggling for revisions. I wish to express special appreciation to the editorial and production staffs of Springer for their excellent work. The team of Springer publishing group (Ms. Raman Shukla from Springer, India and Joseph Daniel and S. Koperundevi from SPi Technologies) has played a great role throughout, always helpful and supportive. Special thanks are due to Ms. Aakanksha Tyagi, Associate Editor-Life Sciences, Springer India with whom I started the project at proposal level and got constant critical advise throughout the project. I acknowledge the generosity of Jega V. Jegatheesan, Li Shu, Piet Lens and Chart Chiemchaisri, the series editors of Applied Environmental Science and Engineering for a Sustainable Future, for accepting this book in the series.

I want to thank all the persons who have helped me along the way for their support, valuable guidance and innumerable suggestions. Finally, I wish to extend appreciation to my family and friends for providing strong moral support and encouragement.

I am sure that this book will prove of equally high value to advanced undergraduate and graduate students, research scholars and industrial designers of equipment systems for waste management and pollution abatement activities.

I would like to receive your valuable feedback to improve the content of this book in the next edition.

August 2016 Ram Lakhan Singh

Contents

Contributors

Monika Asthana Department of Biotechnology, School of Life Sciences, Dr. Bhim Rao Ambedkar University, Agra, UP, India

Rajveer Singh Chauhan Phycology Unit, Department of Botany, Lucknow University, Lucknow, India

Minal Garg Department of Biochemistry, Lucknow University, Lucknow, UP, India

Rajeeva Gaur Department of Microbiology (Centre of Excellence), Dr. Ram Manohar Lohia Avadh University, Faizabad, UP, India

Ankit Gupta National Institute of Allergy and Infectious Disease, National Institute of Health, Rockville, MD, USA

Ankur Gupta Department of Microbiology, School of Life Sciences, Dr. Bhim Rao Ambedkar University, Agra, UP, India

Rasna Gupta Department of Biochemistry, Dr. Ram Manohar Lohia Avadh University, Faizabad, India

Avnish Kumar Department of Biotechnology, School of Life Sciences, Dr. Bhim Rao Ambedkar University, Agra, UP, India

Shiv Kumar Department of Biochemistry, Lucknow University, Lucknow, UP, India

Surabhi Mahajan Department of Microbiology, School of Life Sciences, Dr. Bhim Rao Ambedkar University, Agra, UP, India

Sudhir Mehrotra Department of Biochemistry, Lucknow University, Lucknow, UP, India

Shiv Shankar Department of Environmental Science, School for Environmental Sciences, Babasaheb Bhim Rao Ambedkar University (A Central University), Lucknow, India

B.S. Sharma Department of Environmental Sciences, School of Life Sciences, Dr. Bhim Rao Ambedkar University, Agra, UP, India

Rajendra Sharma Department of Botany, School of Life Sciences, Dr. Bhim Rao Ambedkar University, Agra, UP, India

Rajesh Sharma Department of Biotechnology, VBS Purvanchal University, Jaunpur, India

V.P. Sharma CSIR-Indian Institute of Toxicology Research, Lucknow, India

Shikha Department of Environmental Science, School for Environmental Sciences, Babasaheb Bhim Rao Ambedkar University (A Central University), Lucknow, India

Kartikeya Shukla Department of Environmental Sciences, VBS Purvanchal University, Jaunpur, India

Anurag Singh Department of Microbiology (Centre of Excellence), Dr. Ram Manohar Lohia Avadh University, Faizabad, UP, India

Pankaj Singh Department of Biochemistry, Dr. Ram Manohar Lohia Avadh University, Faizabad, India

Pradeep Kumar Singh Department of Biochemistry, Dr. Ram Manohar Lohia Avadh University, Faizabad, India

R.P. Singh Department of Biochemistry, Dr. Ram Manohar Lohia Avadh University, Faizabad, India

Ram Lakhan Singh Department of Biochemistry, Faculty of Science, Dr. Ram Manohar Lohia Avadh University, Faizabad, India

Ranjan Singh Department of Microbiology (Centre of Excellence), Dr. Ram Manohar Lohia Avadh University, Faizabad, UP, India

Rishi Srivastva Department of Microbiology, VBS Purvanchal University, Jaunpur, India

Soni Tiwari Department of Microbiology (Centre of Excellence), Dr. Ram Manohar Lohia Avadh University, Faizabad, UP, India

S.P. Tiwari Department of Microbiology, VBS Purvanchal University, Jaunpur, India

Ashutosh Tripathi Department of Microbiology (Centre of Excellence), Dr. Ram Manohar Lohia Avadh University, Faizabad, UP, India

Mohd Zahid Department of Biochemistry, Lucknow University, Lucknow, UP, India

Abbreviations

%	Percentage
°C	Degree Celsius
2,4,5-T	2,4,5-Trichloroacetic acid
2,4-D	2,4-Diphenoxy acetic acid
4-CP	4-Chlorophenol
A^0	Angstrom
AAO	Anaerobic ammonium oxidation
ABE	Acetone–butanol–ethanol
AChE	Acetyl cholinesterase
ACP	Anaerobic contact process
AFP	α-Fetoprotein
AM	Arbuscular mycorrhizae
AMD	Acid mine drainage
AOA	Ammonia-oxidizing archaea
APEs	Alkylphenolethoxylates
APs	Alkylphenols
ASP	Activated sludge process
AST	Advanced sludge treatment
ASTM	American Society of the International Association for Testing and Materials
ATP	Adenosine triphosphate
ATSDR	Agency for toxic substances and disease registry
BCR/ABL gene	Breakpoint cluster region/Abelson gene
BHC	Benzene hexachloride
BHC	Beta-hexachlorocyclohexane
BIOX	Biooxidation
BIS	Bureau of Indian Standards
BMW	Biomedical waste
BNP	B-type natriuretic peptide
BnzE	Benzene dihydrodiol dehydrogenase
BOD	Biological oxygen demand
BR	Biosphere reserve
BRCA1gene	Breast cancer 1 gene
BSCs	Biological safety cabinets

Bt	*Bacillus thuringiensis*
C 21	Carbon 21
CA 125	Cancer antigen 125
CA 15-3	Cancer antigen 15-3
CA 19-9	Cancer antigen 19-9
CBMWTDES	Common bio-medical waste treatment and disposal facilities
CBOD	Carbonaceous biological oxygen demand
CDM	Clean development mission
CEA	Carcinoembryonic antigen
CFCs	Chlorofluorocarbons
CFR	Code of Federal Regulations
CFU	Colony-forming unit
CIA	Capillary-based immunoassay
CNT	Carbon nanotubes
Co A	Coenzyme A
COD	Chemical oxygen demand
CPCB	Central Pollution Control Board
CRP	C-reactive protein
CSIRO	Commonwealth Scientific and Industrial Research Organisation
CSP	Concentrated solar power
CSTR	Continuously stirred tank reactor
CTnI	Cardiac troponin I
DAF	Dissolved air floatation
DBP	4,4′-Dichlorobenzophenone
DCM	Dichloromethane
DCMU	3,4 Dichlorophenyl-1-1dimethyl urea
DDD	Dichlorodiphenyldichloroethane
DDE	Dichlorodiphenyldichloroethylene
DDSs	Drug delivery systems
DDT	Dichlorodiphenyltrichloroethane
DNA	Deoxyribonucleic acid
DNRA	Dissimilatory nitrate reduction to ammonia
DO	Dissolved oxygen
DU	Dobson unit
EA	Estrogenic activity
EC	Electrical conductivity
EDCs	Endocrine disrupting compounds
EDOs	Extradiol ring cleavage dioxygenases
EGFR 2/HER-2	Epidermal growth factor receptor 2/herceptin-2
ELISA	Enzyme-linked immunosorbent assay
EPA	Environmental Protection Agency
EPAR	Environmental Protection Agency Report
EPS	Extracellular polymeric substance
ESBL	Extended spectrum beta-lactamase
ESC	Embryonic stem cell

ESR	Electron spin resonance spectroscopy
ETDSP	Electrothermal dynamic stripping process
EU	European Union
FAO	Food and Agriculture Organization
FBR	Fluidized bed reactor
FRET	Fluorescence resonance energy transfer
GC	Gas chromatography
GE	Genetic engineering
GESTEC	Genetically engineered stem cell
GFP	Green fluorescent protein
GFRP	Glass fibre-reinforced plastics
GHGs	Greenhouse gases
GLSP	Good large-scale practices
GM	Genetically modified
GMOs	Genetically modified organisms
GRAS	Generally regarded as safe
Gt	Giga ton
HCG	Human chorionic gonadotropin
HCH	Hexachlorocyclohexane
HEPA	High-efficiency particulate air
HIV	Human immunodeficiency virus
HPLC	High-performance liquid chromatography
HR	Hypersensitive reaction
HRT	Hydraulic retention time
HUP	Hydrogen uranyl phosphate
HWM	Hazardous waste management
ICCVAM	Interagency Coordinating Committee on the Validation of Alternative Methods
ICRISAT	The International Crops Research Institute for the Semi-Arid Tropics
ID50	Infectious dose required to produce infection in 50 percent of the experimental subjects
IFBBR	Inverse fluidized bed biofilm reactor
IL-6	Interleukin-6
IL-8	Interleukin-8
IPM	Integrated pest management
IS	Indian standards
ISL	*In* situ leaching
ISO	International Organization for Standardization
ISR	Induced systemic resistance
ISTD	*In situ* thermal desorption
IUCN	International Union for Conservation of Nature
JCU	Jackson candle unit
LASs	Linear alkylbenzene sulfonates
LD_{50}	Lethal dose which causes the death of 50 % of a group of test animals

LiP	Lignin peroxidase
LPG	Liquefied petroleum gas
LTRs	Long terminal repeats
mAb	Monoclonal antibody
MBR	Membrane bioreactor
MBT	Mechanical biological treatment
MCO	Multicopper oxidase
MCRT	Mean cell residence time
MDR	Multidrug resistance
MEA	Millennium ecosystem assessment
MEOR	Microbially enhanced oil recovery
MF	Microfiltration
MLVSS	Mixed liquor volatile suspended solids
MnP	Manganese-dependent peroxidase
MoEF	Ministry of Environment and Forest
MS	Murashige and Skoog's
MSW	Municipal solid waste
MSWI	Municipal soil waste incineration
NBOD	Nitrogenous biochemical oxygen demand
NF	Nanofiltration
NICICO	National Iranian Copper Industries Company
NMR	Nuclear magnetic resonance
NOx	Oxides of nitrogen
NSCs	Neural stem cells
NSR	Net smelter return
NTU	Nephelometric turbidity unit
OD	Optical density
ODS	Ozone depleting substance
OECD	Organization for Economic Co-operation and Development
OPAA	Organophosphorus acid anhydrolase
OPDA	Organophosphate degrading enzyme
OPH	Organophosphorus hydrolase
ORF	Open reading frame
PAGE	Polyacrylamide gel electrophoresis
PAHs	Polycyclic aromatic hydrocarbons
PBAT	Poly (butylene adipate-co-terephthalate)
PBDEs	Polybrominated diphenyl ethers
PBS	Polybutylene succinate
PCBs	Polychlorinated biphenyls
PCDDs	Polychlorinated dibenzo-p-dioxins
PCDFs	Polychlorinated dibenzofurans
PCE	Perchloroethene
PCP	Pentachlorophenol
PCR	Polymerase chain reaction
PE	Polyethylene

PET Polyethylene terephthalate
PGPR Plant growth-promoting rhizobacteria
PHA Polyhydroxyalkanoate
PHB Polyhydroxybutyrate
PIA Polysaccharide intercellular adhesions
PIPs Plant-incorporated protectants
PLA Polylactic acid
POPs Persistent organic pollutants
PP Polypropylene
PPM Part per million
PQQ Pyrroloquinoline quinone
PSA Prostate-specific antigen
PSII Photosystem II
PV Photovoltaic
PVC Polyvinyl chloride
PWM Plastic waste management
RAFT Reversible addition fragmentation-chain transfer
RBC Rotating biological contractor
RDT Recombinant DNA technology
RIA Radioimmunoassay
RNA Ribonucleic acid
RNAi RNA interference
RPM Revolution per minute
rRNA Ribosomal RNA
SAR Systemic acquired resistance
SAW Surface acoustic wave
SBR Sequence batch reactor
SCP Single cell protein
SDS Sodium dodecyl sulphate
SLF Sanitary land fill
SMF Submerged fermentation
SPR Surface plasmon resonance
SRT Solid retention time
SS Suspended solids
SSF Solid state fermentation
SST Single shell tank
SWM Solid waste management
SWNTs Single-wallet carbon nanotubes
TC/61 Technical Committee Section 61
TCLP Toxicity characteristic leaching procedure
TCM Trichloromethane
TCP Trichlorophenol
T-DNA Transfer DNA
TDS Total dissolved solids
TeCE Tetrachloroethene

TeCM	Tetrachloromethane
Ti plasmid	Tumour inducing plasmid
TNT	Trinitrotoluenes
TodD	Toluene dihydrodiol dehydrogenase
TPD	Tons per day
TSS	Total suspended solids
UASB	Upflow anaerobic sludge blanket
UBOD	Ultimate carbonaceous BOD
UF	Ultrafiltration
UNCED	United Nations Conference on Environment and Development
UNESCO	United Nations Educational, Scientific and Cultural Organization
UNFCCC	United Nations Framework Convention on Climate Change
USDA	US Department of Agriculture
UTs	Union territories
UV	Ultraviolet
VAM	Vesicular arbuscular mycorrhizae
Vir	Virulence region
VOC	Volatile organic compounds
VOLR	Volumetric organic loading rate
WHO	World Health Organization
WRI	World Resource Institute
WTE	Waste to energy
WVO	Waste vegetable oil
WWF	World Wildlife Fund
WWTW	Wastewater treatment works
YMB	Yeast malt broth
YSI	Yellow spring instruments
β-hCG	β-Human chorionic gonadotropin
μg	Microgram
μM	Micromolar

About the Author

Professor Ram Lakhan Singh is presently holding the positions of Dean, Faculty of Science, Professor & Head, Department of Biochemistry, Head, Department of Environmental sciences and Coordinator, Biotechnology Programme at Dr Ram Manohar Lohia Avadh University, Faizabad. He completed his Master's degree in Biochemistry from Lucknow University and joined Indian Institute of Toxicology Research, Lucknow, India in 1980. He worked extensively on the toxicity of synthetic dyes and their metabolites and has been awarded Ph.D. degree in 1987. Professor Singh worked in UP Pollution Control Board as Scientific Officer and studied the effects of effluents discharged from various industries on air and water qualities. He moved to G.B. Pant University of Agriculture & Technology, Pantnagar in 1988 as Assistant Professor of Biochemistry and worked on the toxicity of pulp and paper mill effluents on plant and animal systems.

Professor Singh joined Dr. Ram Manohar Lohia Avadh University, Faizabad as Associate Professor of Biochemistry in 1994 and became full Professor in 2002. He developed the undergraduate and postgraduate courses in Biochemistry, Environmental Sciences and Biotechnology. He led and completed various projects funded by DST, UGC, DOE, UPCAR, UPCST etc. and guided 24 Ph.D. students. His main area of research is Environmental Biochemistry and Toxicology. He published more than 85 research papers in peer reviewed journals and participated in various National and International scientific conferences, meetings, symposia and workshops, chaired scientific and technical sessions and presented more than 70 research papers.

Professor Singh has been awarded IUTOX Senior Fellowship by International Union of Toxicology during XI International Congress of Toxicology at Montreal, Canada in 2007. He has been admitted to the Fellowship of the Society of Toxicology, India in 2011 and Fellowship of Academy of Environmental Biology, India in 2015.

Introduction to Environmental Biotechnology

1

Ram Lakhan Singh

Contents

Abstract

Environment is a complex mixture of many variables including all the physical and biological surroundings and their interactions. Each organism is affected by environmental problems like depletion of ozone layer, global warming, overpopulation, depletion of natural resources, loss of biodiversity, etc. Current environmental problems make us vulnerable to disasters and tragedies, now and in the future. The endangered status of environmental health could be changed only through the understanding of interactions among various living organisms and physical, and chemical phenomena. Environmental biotechnology is concerned with the application of biotechnology as an emerging technology in the context of agriculture, resource conservation, environmental protection, monitoring of contaminated environment, and waste management. It can be considered as a driving force for integrated environmental protection leading to sustainable

R.L. Singh (✉)
Department of Biochemistry, Faculty of Science,
Dr. Ram Manohar Lohia Avadh University, Faizabad 224001, India
e-mail: drrlsingh@rediffmail.com

R.L. Singh (ed.), *Principles and Applications of Environmental Biotechnology for a Sustainable Future*, Applied Environmental Science and Engineering for a Sustainable Future, DOI 10.1007/978-981-10-1866-4_1

development. Sustainable development defines progress in human well-being that can be extended or prolonged over many generations rather than just a few years. It requires a framework for integrating environmental policies and development strategies in a global context. Environmental biotechnology may revamp the possibilities for the prevention of pollution, treatments of solid waste and wastewater, manufacturing with less pollution or less raw materials, ensuring the health of the environment through biomonitoring, and genetic engineering. Since environmental biotechnology has a large potential to contribute to the prevention, detection, and remediation of environmental pollution and degradation of waste, it is a sustainable way to develop clean processes and products, less harmful, with reduced environmental impact than their forerunners. It's role is important with reference to clean technology options in the industrial, agroforestry, food, raw material, and mineral sectors.

Keywords
Environmental biotechnology • Sustainable development • Pollution • Bioremediation • Environmental monitoring

Environment is a complex mixture of many variables including all the physical and biological surroundings and their interactions. It encompasses all living and nonliving things occurring naturally on earth or some region thereof (Fig. 1.1). Our environment is constantly changing. An environmental problem arises whenever there is a change in the quality or quantity of any environmental factor which directly or indirectly affects the health and well-being of human in an adverse manner. Each organism is affected by environmental problems like depletion of ozone layer, global warming, overpopulation, depletion of natural resources, loss of biodiversity, etc. Current environmental problems make us vulnerable to disasters and tragedies, now and in the future. Human population is growing at alarming rate, and this leads to the problems including shortage of our resources, overcrowding, food shortages, social conflict and health, and survival of other species. The rapidly growing human population is accelerating the destruction of tropical forests, grasslands, wetlands, and other biologically rich landscapes for their needs, resulting in to loss of most of the wild species and habitat alteration as well as a reduction of biological varieties and abundance that could severely limit our future options. It was the industrial revolution that gave birth to environmental pollution as we know it today. Pollution is the introduction of contaminants into the natural environment that causes adverse change. The agents which cause environmental pollution are known as pollutant, a physical, chemical, or biological substance unintentionally released into the environment which is directly or indirectly harmful to human and other living organisms or nonliving materials. The major source of environmental pollution is human

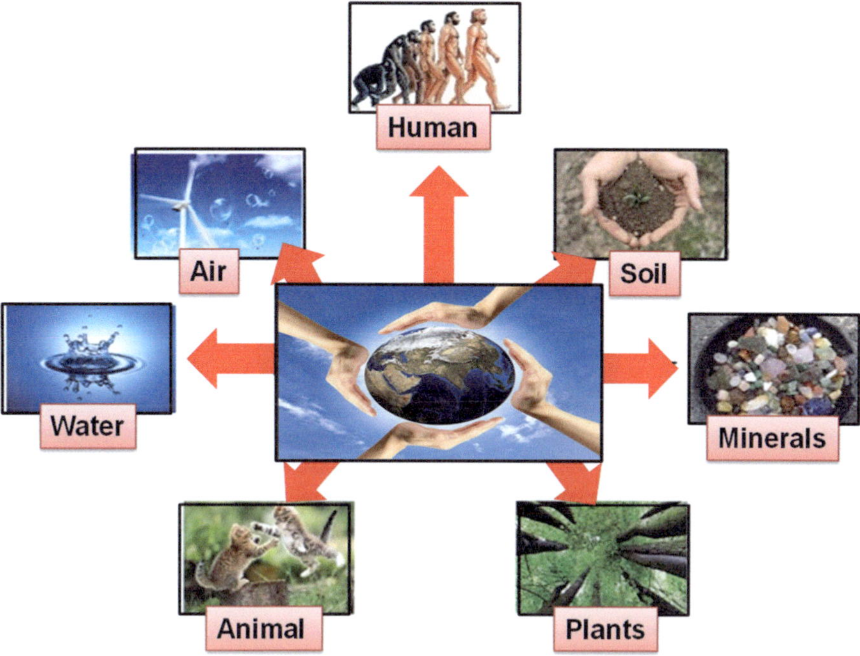

Fig. 1.1 Our environment and its components

activities that directly or indirectly affect the environment adversely. The development of multiple human activities (in industry, transport, agriculture, domestic space), change in the living standard, and higher consumer demand increase the pollution of air (with CO_2, NOx, SO, greenhouse gases, particulate matters), water (with chemical and biological pollutants, nutrients, leachate, oil spills), and soil (due to the disposal of hazardous waste, spreading of pesticides). The burning of fossil fuels releases the CO_2 and other greenhouse gases that causes the environmental pollution leading to global warming which may be responsible for failure of monsoon, acid rain, cyclones, tsunamis, floods, and earthquakes. Chlorinated compound like chlorofluorocarbon used in refrigeration and air conditioning contributes to depletion in ozone layer. The industrial revolution brought an infusion of untreated chemicals and wastes into local streams that served as the water supply source. The polluting industries are pulp and paper industries, fertilizer plants, dyeing industries, tanneries, cement industries, oil refineries, etc. These industries are polluting the environment beyond repair, as well as they are hazardous to human safety and health. Waste generation from domestic and industrial activities is a side effect of consumption and production activities and tends to rise with the level of economic advancement. Pollutants from these wastes affect the natural environment which consequently affects the ecosystem. Some of these pollutants can be readily degraded or removed by means of biotechnological applications, which involve the action of microbes and plants under certain conditions.

1.1 Role of Environmental Biotechnology

Biotechnology is the integrated application of scientific and engineering principles in order to achieve the useful products and services by using the biological systems. It is a key area which has greatly impacted various technologies based on the application of biological processes in manufacturing, agriculture, food processing, medicine, environmental protection, and resource conservation. Actual endangered status of environmental health could be changed by only through the understanding of main interactions between biological, biophysical, and biochemical phenomena and processes directly involved in biotechnological applications.

Environmental biotechnology has been found as a key enabling technology for sustainable environmental protection in the beginning of the twenty-first century. The International Society for Environmental Biotechnology defines environmental biotechnology as the development, use, and regulation of biological systems for remediation of contaminated environments (land, air, water) and for environment friendly processes (green manufacturing technologies and sustainable development). Environmental biotechnology is concerned with the application of biotechnology as an emerging technology in the context of agriculture, resource conservation, environmental protection, monitoring of contaminated environment, and waste management.

As the concentration of toxic chemicals is increasing day by day in the environment, there is a potential risk that these chemicals may cause significant damage to our environment and human health. It represents the destruction of natural life of flora and fauna which is directly related to concentration of pollutants and the time of exposure to it.

Pollution sources and their effects on environment are not always entirely defined by the initial nature of the contamination, since the reaction or breakdown products of a given pollutant can sometimes be more dangerous than the original substance. Environmental persistence of chemicals is a particularly important factor in pollution and is often linked to mobility and bioaccumulation. Highly toxic chemicals which are environmentally unstable and break down rapidly are less harmful than persistent substances, even though these may be intrinsically less toxic. The degree of toxicity can vary depending on the place where organism is found within its food web. Bioaccumulation occurs when some pollutants, although present in very small quantities within the environment, can be taken up by a living organism and become concentrated by it, over time leading to the establishment of a trophic cascade and the biomagnification of specific toxicants. Harmful effects of chemicals and biological agents can include toxicants from pollutants including insecticides, pesticides, and fertilizers, all of which can affect an organism and its community through shifts in species diversity and abundance.

Environmental biotechnology can be considered as a driving force for integrated environmental protection leading to sustainable development (Fig. 1.2).

Environmental biotechnology is allowing major improvements with new solutions in remediation and monitoring of contaminated environment, minimizing future waste release generated by industrial establishments, and creating pollution

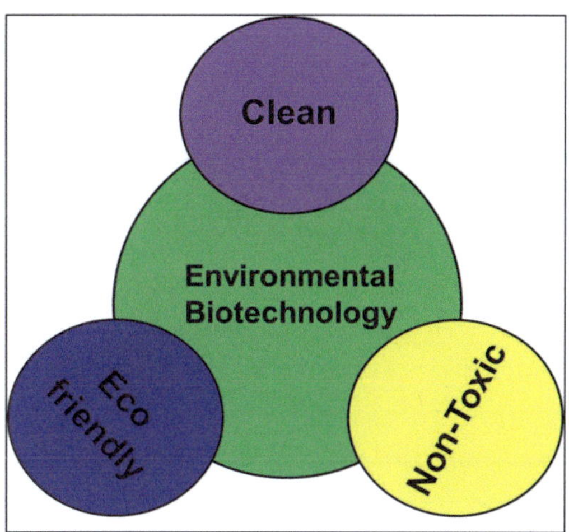

Fig. 1.2 Role of environmental biotechnology

prevention alternatives. There is as an urgent need for changing the current environmental crisis by improving the efficiency of using biotechnology to convert the hazardous wastes and contaminants into useful by-products or degrade them as harmless metabolites by microbial communities and plant species. The use of biological cleaning agents is another area of potential benefit, especially where there is a need to remove oils and fats from process equipment, work surfaces, or drains. Aside from typically reducing energy costs, this may also avoid the need for toxic or dangerous chemical agents. The application of environmental biotechnology across various industrial sectors has invariably led to both economic and environmental benefits including less expensive processing, enhanced product quality, and environmentally sustainable processing relative to conventional operations.

1.2 Application of Environmental Biotechnology for Sustainable Future

Sustainable development means progress in human well-being that can be extended or prolonged over many generations rather than just a few years. The World Commission on Environment and Development defined the sustainable development as "meeting the needs of the present without compromising with the ability of future generations to meet their own needs." Sustainable development requires a framework for integrating environmental policies and development strategies in a global context (Fig. 1.3). The social, environmental, and economic benefits go hand in hand to contribute to development of a sustainable society. Compared to conventional production, sustainable processes and production systems should be more profitable because they would require less wasteful use of materials and energy, resulting in less emission of greenhouse gases and other pollutants and enabling greater and more efficient use of renewable resources, to lesser dependence on nonrenewable resources.

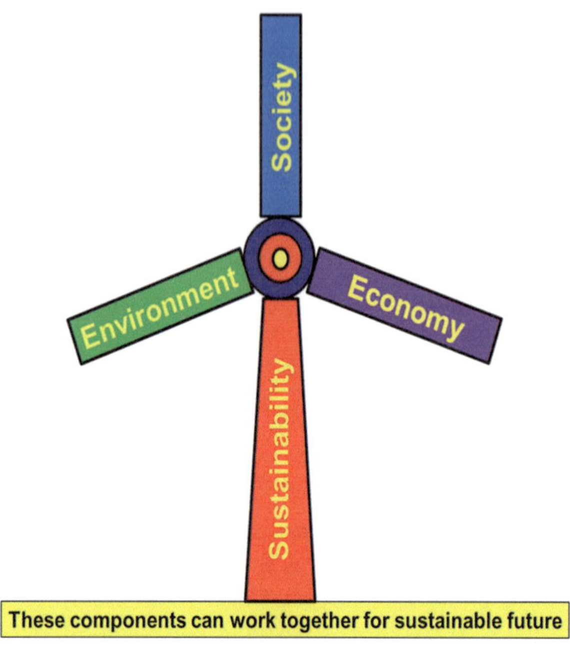

Fig. 1.3 Environmental biotechnology for sustainable future

Environmental biotechnology improved the possibilities for the prevention of pollution, treatments of solid waste and wastewater, manufacturing with less pollution or less raw materials, ensuring the health of the environment through biomonitoring, and genetic engineering for environmental protection and control and holds a promise for a further development of bioremediation. Environmental biotechnology may also be used to prevent hazardous waste production by using biotechnological analogs, develop biodegradable materials for environmental sustainability, produce fuels from biomass and organic wastes, and reduce toxicity by bioimmobilization of hazardous wastes. The efficiency of actual biotechnological application depends on its design, process optimization, and cost minimization.

The following sections explain the use of environmental biotechnology for enhanced sustainability (Fig. 1.4).

1.2.1 Bioremediation

Environmental hazards and risks that occur as a result of accumulated toxic chemicals or other waste and pollutants could be reduced or eliminated through the application of environmental biotechnology in the form of biotreatment or bioremediation as well as through pollution prevention and control practices. This involves wastewater treatment, municipal waste management, hazardous waste treatment, and bioremediation of toxic and recalcitrant compounds. All forms of life can be considered as having a potential function in environmental biotechnology. Bioremediation is

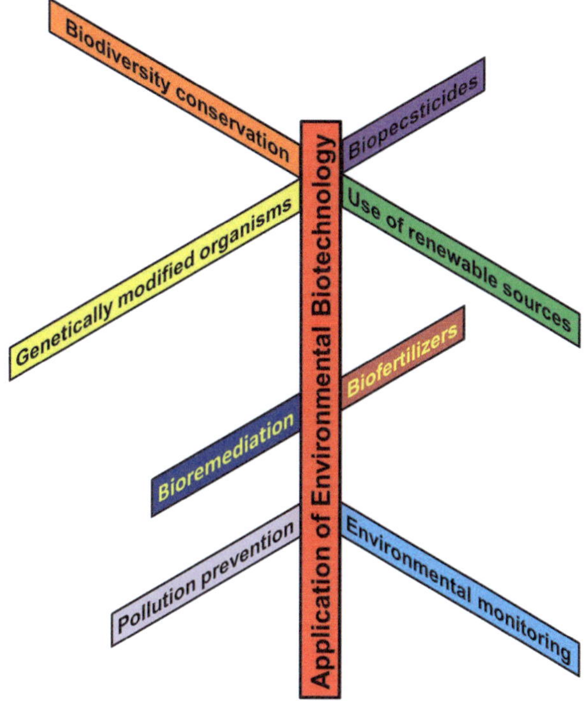

Fig. 1.4 Applications of environmental biotechnology

the use of biological systems (mainly microbes and certain plants) for the reduction of pollution from air or from aquatic or terrestrial systems. It is defined by the US Environmental Protection Agency (USEPA) as "a managed or spontaneous practice, in which microbiological processes are used to degrade or transform contaminants to less toxic or nontoxic forms, thereby remediating or eliminating environmental contamination." Biological agents used for bioremediation of toxic waste include bacteria, fungi, algae, and protozoa. Naturally occurring microorganisms are being used to treat organic and inorganic contaminants in soil, groundwater, and air. Bacteria are the most active organism participating in the biodegradation of organic waste and are used in all areas of environmental biotechnology. Fungi are important degraders of biopolymers and are used in solid waste treatment, especially in composting, or in soil bioremediation. Fungal biomass can also be used as an adsorbent of heavy metals. An alga assimilates light energy and is used in environmental biotechnology for the removal of organic matter and nutrients from waste exposed to light energy. Protozoa are also used in the treatment of industrial hazardous wastes by grazing on bacterial cells, thus maintaining adequate bacterial rate of oxygen consumption and ratio of BOD to COD (biological to chemical oxygen demand ratio) in the treatment systems and helping to reduce cell concentration in the waste effluents. Microbes can break down the pollutants/xenobiotics for their growth and/or energy needs. In some cases, metabolic pathways which organisms normally use for growth and energy supply may also be used to break down the pollutant molecules. These cases are known as co-metabolism, in which microorganism does not benefit directly. The rate of xenobiotic biodegradation in the environment may range from partial to complete degradation from days and weeks to years and decades. A complete biodegradation results in detoxification by mineralizing pollutants to carbon dioxide, water, and harmless inorganic salts. Incomplete biodegradation will yield breakdown products which may or may not be less toxic than the original pollutant. Bioremediation using plants is called phytoremediation. This technique is generally used for removal of metals from contaminated soils and groundwater. Bioremediation processes have been established for both *in situ* (in its original place) and *ex situ* (somewhere else) treatment of contaminated soil and groundwater, and it can offer significant cost and environmental benefits in comparison with alternative technologies.

The use of biological system to remove contaminants from wastewater is largely dependent on wastewater source and characteristics. The effluent components may be of chemical, physical, or biological nature, and they can induce an environmental impact, which includes changes in aquatic habitats and species structure as well as in biodiversity and water quality. Microorganisms in treatment plants remove the pollutants from wastewater before it is discharged into main water stream. The costs of wastewater treatment can be reduced by the conversion of wastes into useful products. For example, most anaerobic wastewater treatment systems produce useful biogas.

1.2.2 Environmental Monitoring

Environmental monitoring deals with the assessment of environmental quality, essentially by measuring a set of selected parameters on a regular basis. Generally, physicochemical and biological methods are used to detect pollution incidents and for quantifying the extent of pollution as well as for the continuous monitoring of pollutants. Environmental monitoring is of great importance for its protection. The harmful effect of toxic chemicals on natural ecosystems has led to an increasing demand for early-warning systems to detect those toxicants at very low concentration levels. Long-established biological methods include counting the number of plant, animal, and microbial species, counting the numbers of individuals in those species, or analyzing the levels of oxygen, methane, or other compounds in water. Recently, biological detection methods for environmental monitoring using biosensors and immunoassays have been developed. Rapid and highly specific detection of numerous pollutants has become possible by using biosensors. Biosensors are most popular approach of environmental biotechnology that are used in biomonitoring, i.e., monitoring of biodegradability, toxicity, mutagenicity, concentration of hazardous substances, and monitoring of concentration and pathogenicity of microorganism in wastes and in the environment. Biosensors are analytical devices built onto a microchip composed of biological and electronic component. The biological component is a biomarker (an enzyme or antibody, DNA) or even a colony of bacteria or a membrane which is in contact with a physical transducer (optical, mass, or electrochemical). Both components relate the concentration of an analyte in the form of a measurable electrical signal. Toxicity can be monitored specifically by whole cell sensors whose bioluminescence may be inhibited by the presence of hazardous substance. The most popular approach uses cells with an introduced luminescent reporter gene to determine changes in the metabolic status of the cells following intoxication. These whole cell toxicity sensors are very sensitive and may be used online to monitor and optimize the biodegradation of hazardous soluble substances. Similar sensors can be used for the measurement of the concentration of specific pollutants. Antibodies coupled with fluorochromes are useful in monitoring the fate of bacteria released into the environment for the treatment of a polluted site. Fluorescent or enzyme-linked immunoassays have been derived and can be used for a variety of contaminants, including pesticides and chlorinated polycyclic hydrocarbons. Moreover, integration of environmental biotechnology with information technology has advanced the capacity to monitor and control processes at molecular levels in order to achieve real-time information and computational analysis in complex environmental systems.

1.2.3 Biopesticides

The excessive use of chemical pesticides (insecticides, fungicides, herbicides) applied in plant protection as an integral part of intensive agriculture resulted in contamination of agroecosystem and agriculture. Wide and persistent usage of these chemical pesticides causes environmental hazards due to its low biodegradability, pollute soil, and water sources, brings on ecological damage, destroys entomological communities, and causes pathologic changes in bird and animal populations, as well as it is dangerous to human health. These are recalcitrant xenobiotics that persist in the environment for many years. A prominent example of a very persistent xenobiotic is the insecticide DDT (1,1,1-trichloro-2,2 bis [P chlorophenyl] ethane), which was used extensively from the 1930s until its ban in 1979. DDT was found to persist with an average half-life of 4.5 years in field soils and a half-life in anoxic soils of about 700 days. Stable metabolites of DDT may be detected in soil, in groundwater, and in the tissue of organisms even today. To reduce the environmental impact of chemical pesticides, more attention should be paid to developing and establishment of eco-friendly alternatives. One of the promising alternatives has been the use of biopesticides.

Biopesticides offer powerful tools to create a new generation of sustainable agriculture products. Biopesticides generally tend to be specific toward their target, do not leave toxic residues, reduce the risk of resistance development in the target species, and produce a lesser overall impact on the environment than traditional problematic chemical pesticides. They offer solutions to concerns such as pest resistance, conventional chemical pesticides, public concern about side effects of pesticides on the surrounding environment, and ultimately on human health.

Biopesticides are derived from natural materials such as animals, plants, bacteria, and certain minerals. They are categorized in microbial pesticide (biopesticides include microorganisms that control pests), biochemical pesticides (naturally occurring substances that control pests), and plant incorporated protectants (pesticidal substances produced by plants containing added genetic material).

Microbial Pesticides The active ingredient of microbial pesticides is a microorganism (bacterium, fungus, virus or protozoa, nematodes, yeast) that either occurs naturally or is genetically engineered. Microbial pesticides can control many different kinds of pests although each separate active microbial ingredient is relatively specific for its target pests. For example, there are fungi that kill specific insects and other fungi that control certain weeds. The most widely used microbial pesticides include subspecies and strains of *Bacillus thuringiensis* or Bt (bioinsecticides), *Trichoderma* (biofungicides), and *Phytopthora* (bioherbicides).

Biochemical Pesticides Biochemical pesticides are naturally occurring substances such as plant extracts, fatty acids, or pheromones that control pests by nontoxic mechanisms. These pesticides include substances that interfere with growth or mating, such as insect sex pheromones which interfere with their mating, plant growth regulators, as well as various scented plant extracts that attract insect pests to traps.

Plant-Incorporated Protectants These are pesticidal substances that are produced naturally on genetic modification of plants, for example, incorporation of Bt gene into the plant genome so that the transgenic plant instead of the Bt bacterium synthesizes Bt pesticidal protein that destroys the targeted pest. The genetically modified plant varieties which are resistant to pest and produce natural biodegradable proteins with no harmful effect on animals and human beings may considerably diminish the use of hazardous pesticides.

1.2.4 Biofertilizers

Biofertilizers are attracting attention as inexpensive and safe alternative to chemical fertilizers that are used to deliver inorganic nutrients (nitrogen, phosphorus, potassium, sulfur, etc.) required for crop growth. Biofertilizers are those substances which contain living microorganisms which colonizes the rhizosphere of the plant and promote growth by increasing the supply or availability of primary nutrients and/or growth stimulus to the target crop. Microorganisms having capability to fix the atmospheric nitrogen and solubilize phosphate and mycorrhizae are the main sources of biofertilizer. Bacteria (*Bacillus*, *Pseudomonas*, nitrogen fixing bacteria), fungi (*Trichoderma*), yeast, and algae are used as biofertilizers. These have shown great potential as a renewable and eco-friendly source of plant nutrient, and its enhancement is expected to contribute significantly in reducing pollution, energy, and resource consumption associated with the use of conventional fertilizers.

1.2.5 Bioplastics

Conventional plastics based on petroleum-based synthetic polymers (polyethylene, polystyrene, polyvinyl chloride, etc.) have become an integral part of our lives. These plastics are common in every aspect of our lives from the household use to pharmaceutical industry. Production of petrochemical plastics consumes large quantities of nonrenewable resources like petroleum, coal, and natural gas as raw materials. These plastics represent a major environmental problem as they are non-biodegradable, resulting in worldwide growth of plastics wastes. Environmental, economic, and safety concerns highlight the need for renewable biologically derived polymers (Bioplastics). Bioplastics are environment friendly, biodegradable, produced from natural renewable resources, able to be recycled, reused and composted (degrade to organic matter after disposal) or burned without producing toxic by-products, etc., making it an excellent alternative to traditional plastic products. Biopolymers limit the emission of greenhouse gases such as carbon dioxide during their production which is one of the prime sources of air pollution and leads to environmental issues such as global warming, climate change, etc. Bioplastics can be made from many different sources and materials including cellulose, corn starch, potato starch, plant oil, sugarcane, hemp, poly lactic acid (PLA), poly-3-hydroxybutyrate (PHB), etc. which avoids the use of nonrenewable resources. All of these renewable resources can be obtained, modified, and processed into bio-based plastics. Microbes can be induced to produce enzymes needed to convert plant and vegetable raw materials into building blocks for bioplastic. Biopolymers with enhanced properties and microbial strains for producing them are being developed.

1.2.6 Genetically Modified Organisms (GMOs)

The genetically modified organisms (GMOs) are most commonly used term for crop plants as well as animals and microorganisms that have been genetically engineered (GE) to produce some desired trait by the application of recombinant DNA technology or genetic engineering techniques. Genetic modification techniques allow novel traits to be introduced into animals, crops, and microorganisms. Genetically modified animals and plants are potentially versatile chemical factories. They offer the genetic improvement of plant varieties and animal populations to increase their yields or efficiency, genetic characterization and conservation of genetic resources, plant or animal disease diagnosis, vaccine development, and improvement of feeds. Genetically modified plant varieties offer many advantages over their conventional counterparts including lower demand for fertilizers and pesticides, better tolerance to adverse environments and pests, resistance to disease, superior yields, improved nutrition and other functional qualities, ability to generate products that a crop does not produce naturally, and reduced production cost. Genetically modified (GM) crops carrying novel traits (pest resistance, herbicide resistance, disease resistance) have been developed and released for commercial agriculture production. These include pest-resistant cotton, maize, and canola, herbicide glyphosate-resistant soybean and cotton, and viral disease-resistant potatoes and papaya. For the development of pest-resistant plant, Bt (*Bacillus thuringiensis*) gene that produces crystal protein toxic to insect pests is transferred to the plants. The resulting plant contains the Bt toxin in its cells. When the plant is eaten by the target insect, the toxin binds to receptors in the insect's gut, causing the gut wall to break down and allowing toxin spores and normal gut bacteria to enter the body. As spores and bacteria proliferate in the body, the insect dies. In addition, various GM crops with traits for phytoremediation and production of pharmaceuticals, such as rice with high level of carotenoid for production of vitamin A (e.g., golden rice) and bananas with vaccines, are under consideration. Thus, GM crops are considered as the solution to yield deficits, use of pesticides, as well as negative impacts of their residues, pests, plant diseases, and other problems.

GM crops are grown in varying amounts in different countries due to controversy posed by GM technology regarding the ecological stability of the GMOs, genetic contamination/interbreeding, ecosystem impacts, etc.

1.3 Concluding Remarks

New environmental challenges continue to evolve, and new technologies for environmental protection and control are obviously always under development. Also, new approaches continue to grow harnessing the potential of microorganism and plants as eco-efficient and robust cleanup agents in a variety of practical situations.

Along with the wide levels of technology with potential to accomplish the objectives of sustainability, environmental biotechnology continues to play an important role to offer renewable raw material and energy, pollution prevention, and bioremediation. Since environmental biotechnology has been proved to have a large potential to contribute to the prevention, detection, and remediation of environmental pollution and degradation of waste, it is a sustainable way to develop clean processes and products, less harmful, with reduced environmental impact than their forerunners, and this role is illustrated with reference to clean technology options in the industrial, agroforestry, food, raw material, and mineral sectors. An evaluation of the consequences, opportunities, and challenges of modern environmental biotechnology is important both for policy makers and the industries for implementation of ways for sustainable development.

The future of environmental biotechnology seems to be very strong and bright. Scientists have come to believe that environmental biotechnology is a boon for human society, and it has the great potential to solve many of the environmental issues.

Global Environmental Problems 2

Ram Lakhan Singh and Pradeep Kumar Singh

Contents

R.L. Singh (✉)
Department of Biochemistry, Faculty of Science, Dr. Ram Manohar Lohia
Avadh University, Faizabad 224001, India
e-mail: pkbt99@gmail.com

P.K. Singh
Department of Biochemistry, Dr. Ram Manohar Lohia
Avadh University, Faizabad 224001, India
e-mail: drrlsingh@rediffmail.com

R.L. Singh (ed.), *Principles and Applications of Environmental Biotechnology for a Sustainable Future*, Applied Environmental Science and Engineering for a Sustainable Future, DOI 10.1007/978-981-10-1866-4_2

Abstract

Everything that surrounds or affects an organism during its lifetime is collectively referred to as its environment. It comprises both living (biotic) and nonliving (abiotic) components. Human civilisation and globalisation are the dominant culprits of constant change in the global environment in present scenario. Various processes that can be said to contribute to the global environmental problems include pollution, global warming, ozone depletion, acid rain, depletion of natural resources, overpopulation, waste disposal, deforestation and loss of biodiversity. Almost all these processes are the result of the use of natural resources in unsustainable manner. These processes have highly negative impact on our environment. One of the major impacts is the release of large quantities of carbon dioxide and other greenhouse gases in atmosphere as the result of burning of fossil fuels by industries and automobiles. The result is the worldwide pollution problem, temperature fluctuation of our planet, ozone hole and possible change in Earth's climate. Loss of forests, damage to water bodies (lakes and ponds) and their ecosystems by acid rain, over-exploitation of natural resources, massive extinction of species due to habitat destruction and other well-known causes worldwide are connected with environmental issues globally. The rapidly growing demographic structure and globalisation are leading to a number of environmental issues because of the uncontrolled urbanisation, industrialisation, deforestation and loss of useful agriculture land. The global environmental health impact remains profoundly perturbing. Unsafe water, poor sanitation and hygiene conditions, air pollution and global climate change accounts for nearly a tenth of deaths and disease burden worldwide. Due to above-mentioned environmental issues, our planet is facing severe environmental crisis. Current environmental problems lead to disasters and tragedies now, will also be the reason of casualties in future and require urgent attention from the responsible authorities/nations to frame appropriate laws to overcome these issues and also by making people aware to use natural resources in sustainable manner.

Keywords
Climate change • Deforestation • Ozone depletion • Acid rain • Renewable and nonrenewable resources

2.1 Introduction

Environment is virtually everything which is around us. The term environment describes sum total of physical and biotic conditions influencing the responses in the organism. The natural environment mainly consists of four interlinking systems, namely, hydrosphere, lithosphere, atmosphere and biosphere. These four systems are in a constantly changing state and such changes are affected by human activities and vice versa.

2.1.1 Hydrosphere

Hydrosphere describes combined mass of water found on, under and over the surface of a planet. Hydrosphere includes all water bodies found on earth such as lakes, ponds, rivers, streams, ocean, etc. It functions in a cyclic nature, which is termed as hydrological cycle or water cycle.

2.1.2 Lithosphere

Lithosphere means the mantle of rocks constituting the earth's crust which is the outermost solid surface of the planet. Earth's lithosphere includes the crusts, mantle and core (outer and inner). Lithosphere mainly contains soil, earth rocks, mountain, etc.

2.1.3 Atmosphere

The cover of the gases that envelope the earth is known as the atmosphere. Atmosphere is a thin layer which contains gases like oxygen, carbon dioxide, nitrogen, argon, hydrogen, helium and ozone which protects the solid earth and human beings from the harmful radiations of the sun. There are five concentric layers within the earth's atmosphere and each layer has its own characteristics. These layers include the troposphere, stratosphere, mesosphere, thermosphere and exosphere which can be differentiated on the basis of variation in air and temperature.

2.1.4 Biosphere

The biosphere can also be termed as the zone of life on Earth. It refers to all organisms on the earth's surface and their interaction with water and air. Biosphere is the global sum of all ecosystems consisting of plants, animals and microorganisms, ranging from the tiniest microscopic organism to the largest whales in the sea. The richness of biosphere depends upon a number of factors like rainfall, temperature, geographical reference, etc.

An environmental problem arises whenever there is a change in the quality or quantity of any environmental factor which directly or indirectly affects the health and well-being of living organisms in an adverse manner. Global environmental issues present a comprehensive and stimulating introduction to the key environmental issues presently threatening our global environment. Some of the environmental problems which are critical at the present time are fairly widely known because of the growing awareness at every level of society, including governments, general public and the scientific community. However, the present information on the structure and function of biosphere is not sufficient to allow an accurate evaluation of the total situation expect to indicate some broad problem areas. At the same time, different regions do face different problems. One key distinction is between the environmental threats faced by developed nations, such as the United States and western European countries, and developing nations, such as India and Mexico. The environmental problems faced by developed nations are largely the result of their economic strength and higher standards of living. In contrast, the environmental crises faced by developing nations are the result of poverty. Human impact on the environment includes impacts on biophysical environment, biodiversity, and other resources. However, as human populations continue to grow, material consumption intensifies and production technology further expands; by consequence the quantity and quality of environmental resources keep steadily decreasing.

The major environmental problems are pollution, global warming, ozone layer depletion, acid rain, natural resource depletion, overpopulation, waste disposal, deforestation and loss of biodiversity.

2.2 Pollution

Pollution means any solid, liquid or gaseous substance present in such concentration as may be, or tend to be, hazardous or detrimental to the environment. Pollutants, the components of pollution, can be either foreign substances or naturally occurring contaminants. Environmental pollution is a worldwide problem and this is scrambled with the unsustainable anthropogenic activities, resulting in substantial human health problems (Fereidoun et al. 2007). Human activities and natural forces are the main reasons of pollution. There are two different types of pollution:

2.2.1 Point Source Pollution

Point source pollution is pollution caused from a stationary location or fixed facility where pollutants are discharged, any single identifiable source of pollution. This type of pollution can be easily tracked back to its source and it may destroy habitats or kill life on large scales, because it is concentrated amount of pollutants. The examples are oil spill from ship, smoke/effluent discharge from a factory, etc.

2.2.2 Non-point Source Pollution

Non-point source pollution comes from diffuse sources (i.e. without a single point of origin or not introduced into a receiving stream from a specific outlet). It is the type of pollution that cannot be easily tracked back to its source. The pollutants are generally carried off the land by storm water. Common non-point sources are agriculture, forestry, urban, mining and construction activities, dams, channels, land disposal and city streets. This type of pollution takes place every day in all communities. Non-point source pollution does not always immediately destroy habitats or kill large amounts of life at one time but the habitats are usually destroyed over longer periods of time. The examples are fertilisers, pesticides, animal waste, etc.

2.2.3 Classification of Pollution

Environmental pollution is classified into various categories (Table 2.1). The major groups are:

Table 2.1 Different categories of pollution

Category	Pollutants	Health effects
Air pollution	Smoke, dust, toxic substances (SOx, NOx), CFCs, ozone, PAN, etc.	Respiratory infection, bronchitis, asthma, heart disease
Water pollution	Industrial and sewage wastewater, wastewater, pesticide, fertiliser, detergents, heavy metals, etc.	Deoxygenation of water bodies, noxious odours, poisoning, disease like cholera
Soil pollution	Chemicals used in agriculture, pesticides, petroleum hydrocarbons, arsenic, lead and other heavy metals, human and animal waste, etc.	Congenital disorders, poisoning
Radioactive pollution	Radiation (X-ray), radionuclides (strontium-90, iodine-129, cesium-137 and different other isotopes)	Damage DNA, eye cataract, cancer
Noise pollution	Machines in factories, transportation systems, motor vehicles, aircrafts, trains, late-night commercial operations, etc.	Headaches, hearing loss, cardiovascular effects, depression, discomfort, impaired development

2.2.3.1 Air Pollution

Air pollution is a result of industrial and certain domestic activities. An ever increasing use of fossil fuels in power plants, industries, transportation, mining, construction of buildings, and stone quarries had led to air pollution. It may be defined as the introduction of particulates and chemicals into the atmosphere in such concentration that may be directly and indirectly injurious to human health or other living organisms, plants or interferes with the normal environmental processes.

2.2.3.1.1 Air Pollutants

Sulphur oxides (SO_2, SO_3); nitrogen oxides (NO_2, NO_3); carbon monoxide (CO) produced by industry and motor vehicles; carbon dioxide (CO_2) from burning of fossils fuels; benzopyrene from tobacco smoke, charcoal and gasoline exhaust; volatile organic compounds like methane, toluene and xylene; chlorofluorocarbon (CFC); ozone (O_3); peroxyacetyl nitrate (PAN); ammonia (NH_3); particulate matter or fine dust, metals, odours; etc. are the most common air pollutants.

2.2.3.1.2 Effects of Air Pollution

Air pollution is a significant risk factor for the environment and a number of health conditions. It is responsible for global warming, ozone depletion and acid rain. It primarily affects the body's respiratory system and the cardiovascular system causing respiratory infections, asthma, bronchitis, heart disease, stroke and lung cancer.

2.2.3.2 Water Pollution

Water pollution is the result of contamination of water bodies (e.g. lakes, rivers, oceans and groundwater). Any physical, chemical or biological change in waterbodies that adversely affects living organisms or makes water unsuitable for desired uses can be considered as water pollution. This form of environmental degradation occurs when pollutants are directly or indirectly discharged into water bodies without adequate treatment to remove harmful compounds. Water pollution is caused by a variety of human activities such as industrial, agricultural and domestic. It affects the health and quality of soils and vegetation (Carter 1985).

2.2.3.2.1 Sources and Pollutants

Agricultural runoff laden with excess fertilisers and pesticides, industrial effluents with toxic chemical substances (e.g. detergents, heavy metals) and sewage water with human and animal wastes and pathogens pollute our water bodies thoroughly. Natural sources of water pollution are soil erosion, leaching of minerals from rocks and decaying of organic matter. Rivers, lakes, seas, oceans, estuaries and ground water sources may be polluted by point or non-point sources.

2.2.3.2.2 Effects of Water Pollution

Water pollution affects the environment and public health. The main problems caused by water pollution are death of aquatic animals due to toxic chemicals and deoxygenation of water, disruption of food chains, destruction of ecosystem and diseases like cholera due to pathogens.

2.2.3.2.3 Minamata Disease

The tragic incident of Minamata, a neurological syndrome in Japan, is well known. A paper factory using mercury compounds carelessly dumped its waste effluents into the sea, which in turn was consumed by the sea fish like swordfish and tuna. The Japanese people who consumed such fish showed symptoms of organic compound methyl mercury poisoning like gingivitis, vomiting, fever, diarrhoea and, in extreme cases paralysis, coma and death. The formation of methyl mercury from inorganic mercury takes place by methylating microorganisms that are present in the sediments of lakes, rivers and other natural water sources. Due to methylation, solubility of mercury in lipid increases. Thus lipid-soluble organic mercury is the culprit of this disease.

2.2.3.3 Soil Pollution

Soil is formed by weathering of rocks. It is the most important abiotic factor that holds most advantageous microbes and supports life of plants that rely on it for their nutrition, water supply and mineral supply. As the population increase, there is need of increase in food supply so both agricultural and industrial activities throughout the world would also increase. These activities lead to degradation of soil, water and air. Soil degradation causes decline in soil productivity through adverse change in nutrient status, structural stability, concentration of solutes, etc. The effect of soil degradation is not restricted to soil alone but have a number of off-site implications, for example, soil erosion is often associated with increased flooding and siltation of rivers and lakes.

2.2.3.3.1 Sources and Pollutants

Soil quality is greatly affected by agrochemicals. Both nitrogenous and phosphate fertilisers have impact on soil activity. Excessive use of nitrogen fertilisers has lead to the soil acidity. Mostly phosphorus use in agriculture is in the form of single super phosphate and ammonium phosphate fertilisers. It is important that mechanisms involved in the acidifying effects of different fertilisers may be studied before any decision can be made on the choice of the fertilisers in relation to their effect on soil acidification. Four major nutrients that are used in fertilisers are nitrogen, potassium, sulphur and phosphorous which are involved in soil acidification as these are in the form of complex salts like ammonium sulphate, ammonium chloride, ammonium nitrate, potassium nitrate, calcium nitrates, sodium nitrates, etc. Pesticides are the chemicals that are used in agriculture, farming, and indoor gardening for destroying the insects and pests. Herbicides, insecticides and fungicides are the three main pesticide groups that are used in agriculture to kill the harmful pests for the crop plants. Pesticides have direct effect on surrounding area, direct contamination of users and residues on food when they are sprayed on crops. Synthetic products such as chlorinated hydrocarbons including DDT and dieldrin are highly toxic to wildlife and often to human beings, so their use is totally banned in most of the developed countries. It is estimated that around 10,000 deaths are attributed to pesticides every year worldwide. These pesticides are retained in soil; concentrate in vegetables, crops, cereals and fruits; and ultimately affect the human and environmental health.

2.2.3.3.2 Effects of Soil Pollution

Soil pollution has direct effect on all life forms of aquatic and terrestrial habitat. When soil pollutants runs off into rivers and lakes, it may kill the fishes, plants and other aquatic organisms. Soil pollutants impair soil stability due to which it does not support growth of crops in field, and toxic soil particle may have harmful effect on human health.

2.3 Global Warming

Before the Industrial Revolution, human activities used to release very few gases, in minute quantities, into the atmosphere and all climatic changes happened naturally. After the Industrial Revolution, the natural composition of gases in the atmosphere is getting affected due to human activities mostly by fossil fuel combustion, changing agricultural practices and deforestation which began to alter the climate and environment significantly. Global warming is defined as an increase in the average temperature of the earth's atmosphere, especially a sustained increase to cause changes in global climate. The term global warming is synonymous with enhanced greenhouse effect, implying an increase in the amount of green house gases (CO_2, CFCs, CH_4, etc.) in the earth's atmosphere, leading to entrapment of more and more solar radiations, and thus increasing the overall temperature of the earth. Over the last 100 years, it is observed that the earth is getting warmer and warmer, unlike previous 8000 years when temperatures have been relatively constant. The present temperature is 0.3–0.6 °C warmer than it was 100 years ago.

2.3.1 Greenhouse Gases and Greenhouse Effect

Some greenhouse gases occur naturally in the atmosphere, while others result from human activities. Naturally occurring greenhouse gases include water vapour, carbon dioxide, methane, nitrous oxide and ozone (Fig. 2.1). The key greenhouse gas responsible for causing global warming is carbon dioxide. Chlorofluorocarbons (CFCs) are also significant contributors to global warming even though they exist in very small quantities. Carbon dioxide is one of the most prevalent greenhouse gas released to the atmosphere by anthropogenic activities such as combustion of solid waste, fossil fuels (oil, natural gas and coal) and wood.

Greenhouse effect is the phenomenon whereby the earth's atmosphere traps solar radiation mainly in the ultraviolet (UV), visible (vis) and infrared (IR) regions of the electromagnetic spectrum. This is mediated by greenhouse gases such as carbon dioxide, water vapour and methane present in the atmosphere that allow incoming

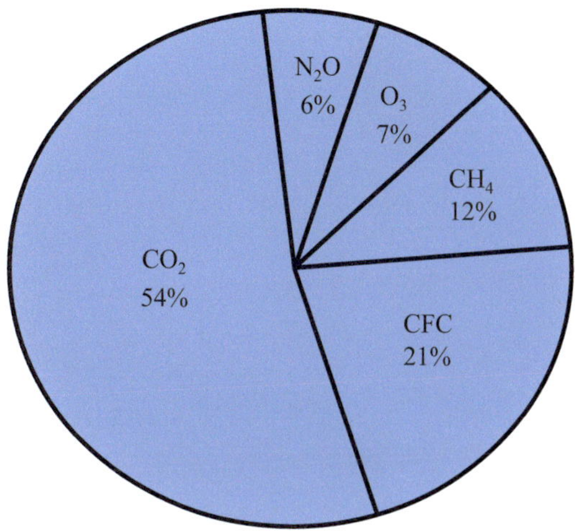

Fig. 2.1 Relative contributions of greenhouse gases (GHGs) in global warming

sunlight to pass through but absorb the heat radiated back from the earth's surface. When solar radiations reach the earth, part of it is reflected back into space and part of it is absorbed by the earth's surface. The part which is absorbed heats up the earth which in turn then radiates some of its energy out into space in the IR region of spectrum. A steady state is reached where the earth absorbs and radiates energy at the same rate, resulting in a fairly constant average temperature. However, the IR radiation emitted by the earth can be absorbed by greenhouse gases in the troposphere and become trapped. The radiation is then re-emitted in all directions; some back towards the earth, which is known as 'greenhouse effect' (Fig. 2.2). This leads to an increase in temperature and global warming, making the average surface temperature of the earth about 286 K or 13 °C. If there were no greenhouse effect at all, the temperature of earth surface would have been about 256 K or −17 °C and life as we know could not exist without water, because at this temperature water would have been in solid state. Thus the greenhouse gases (GHGs) provide a blanketing effect in the lower strata of the earth's atmosphere, and this blanketing effect is being enhanced because of the human activities like burning of fossil fuels, etc.

2.3.2 Impacts of Global Warming

In general, the faster the climate changes, the greater will be the risk of damage. The major impacts of global warming are:

Rise in global temperature: Climate models predict that the global surface temperature is likely to rise by about 6 °C during the twenty-first century. There is strong evidence now that most of the observed warming over last 50 years is caused by human activities.

Rise in sea level: It is one of the most certain impacts of global warming. The mean sea level around the world is predicted to rise by 9–88 cm by the year 2100 which

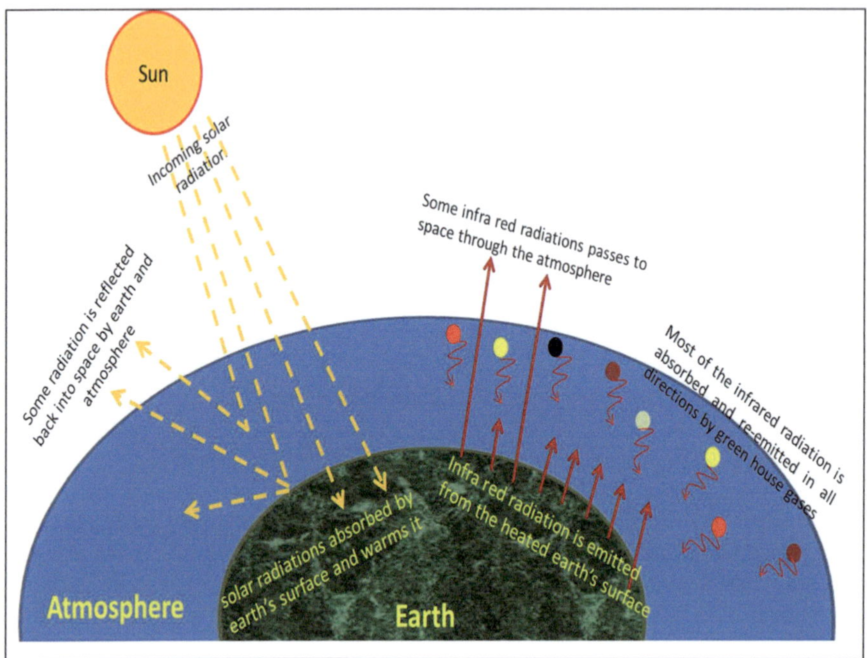

Fig. 2.2 Green house effect

is many times greater than the average rate over the last 3,000 years. Scientists estimated that sea level will continue to rise as a result of man-made greenhouse gas pollution and could reach an additional 3.5 in. to 3 ft (9–88 cm) by the end of this century, with even further rises in subsequent centuries as sea level gradually adjusts to the warmer climate (Church et al. 2001).

Retreat of glaciers: In almost every mountainous region across the world, glaciers are retreating in response to the warming climate. The ice shelf acts as a dam for glaciers on land; its breakup is causing a worrisome speed up of glacier flow into the ocean, which could raise global sea level.

Decreasing crop yield: Water resources are affected as precipitation and evaporation patterns change around the world. This will affect agricultural output, threat to food security from decreasing crop yields. Food security is likely to be threatened and some regions are likely to experience food shortages and hunger.

Disease outbreaks: Higher temperatures accelerate the maturation of disease-causing agents and the organisms that transmit them, especially mosquitoes and rodents leading to disease and loss of human life. For example, the incubation period required for a mosquito to be able to spread dengue fever virus after it has been infected ranges from 12 days at 30 °C to 7 days at 32–35 °C and this will lead into a potential threefold increase in the transmission rate of disease (Patz et al. 1996).

Other likely effects of the global warming include probable expansion of subtropical deserts; retreat permafrost and sea ice; more frequent extreme weather events including heat waves, droughts, wildfires, heavy rainfall, flooding and heavy snowfall; ocean acidification; and species extinctions due to shifting temperature regimes.

2.3.3 Kyoto Protocol

There is a scientific consensus that human activities are causing global warming that could result in significant impacts such as sea level rise, changes in weather patterns and adverse health effects. The climate policy negotiations resulted in to the Kyoto Protocol, the first legally binding international agreement on climate protection. The Kyoto Protocol was adopted in Kyoto, Japan, on December 11, 1997, and came into force on February 16, 2005. There are currently 192 parties to the Protocol. The Kyoto Protocol implemented the objective of the United Nations Framework Convention on Climate Change (UNFCCC) to fight global warming, committing the industrialised nations to specify, legally binding reductions in emissions of six greenhouse gases in the atmosphere to 'a level that would prevent dangerous anthropogenic interference with the climate system'. The six major greenhouse gases covered by the protocol are carbon dioxide (CO_2), nitrous oxide (N_2O), methane (CH_4), hydrofluorocarbons (HFCs), perfluorocarbons (PFCs) and sulphur hexafluoride (SF_6).

2.3.4 Carbon Credit

The concept of carbon credits came into existence as a result of increasing awareness of the need for controlling greenhouse gas emission. The mechanism of carbon credit was formalised in the Kyoto Protocol. Carbon credit is a general term for any tradable certificate or permit, representing the right to emit one tonne of carbon dioxide or the mass of another greenhouse gas with a CO_2 equivalent (tCO_2e), equal to one tonne of carbon dioxide. One carbon credit is equal to one tonne of carbon dioxide or carbon dioxide-equivalent gases.

Carbon credits and carbon markets are a component of national and international attempts to diminish the growth in concentrations of greenhouse gases. Carbon trading is an application of an emission trading approach. The goal is to allow market mechanisms to drive industrial and commercial processes in the direction of low emissions or less carbon-intensive approaches than those used when there is no cost of emitting CO_2 and other greenhouse gases.

2.4 Ozone Layer Depletion

Earth's atmosphere is divided into three regions, namely, troposphere, stratosphere and mesosphere. The stratosphere extends from 10 to 50 km from the earth's surface. The ozone (O_3) layer that is present in stratosphere in atmosphere protects earth from harmful ultraviolet radiations. The depletion of this layer is another major global atmospheric issue that is attributed to anthropogenic activities. Solar radiation emits electromagnetic waves with a wide range of energies and wavelength. Ultraviolet (UV) wavelength is slightly shorter than the wavelength of violet light, which are shortest wavelengths that are visible to the human eyes. UV-B radiation is of wavelength 280–320 nm, whereas UV-A radiation is of 320–400 nm, and since energy is inversely proportional to wavelength, UV-B is more energetic and therefore more dangerous than UV-A radiation.

On penetrating the atmosphere and being absorbed by biological (living) tissues, UV radiation damages protein and DNA molecules. Suppose that if full quantity of ultraviolet radiations falling on the stratosphere reaches earth's surface, it will destroy all living organisms on earth. Because most UV radiation (>99 %) is absorbed by ozone in the stratosphere, stratospheric ozone is commonly considered as the ozone shield. However, small portion of UV-B radiation that escapes from stratosphere and reach us is responsible for sunburns and most cases of skin cancers.

2.4.1 Formation of Ozone Shield

Ozone shield is a protective covering present in our atmosphere and its formation takes place when ozone is formed in the stratosphere by the reaction between UV radiation and oxygen (O_2) molecules. The high-energy UV radiation first break some molecular oxygen (O_2) into free oxygen (O) atom (nascent) and these atom further combine with molecular oxygen to form ozone (O_3) via fallowing reaction:

$$O_2 \xrightarrow{\text{UV-B}} O + O\left(\text{atomatic oxygen}\right)$$
$$O + O_2 \rightarrow O_3\left(\text{ozone}\right)$$

Not all of the molecular oxygen is converted to ozone, because free oxygen atoms may also combine with ozone molecules to form two oxygen molecules in the following reaction:

$$O + O_3 \rightarrow O_2 + O_2\left(\text{molecular oxygen}\right)$$

Finally, when ozone absorbs UV-B, it is converted back to free oxygen and molecular oxygen:

$$O_3 \xrightarrow{\text{UV-B}} O + O_2$$

Thus, the amount of ozone in the stratosphere is dynamic in nature.

2.4.2 Depletion of Ozone Shield

Chlorofluorocarbons (CFCs) are a type of halogenated hydrocarbons. These are nonreactive, non-inflammable, nontoxic organic molecules in which both chlorine and fluorine atoms have replaced some hydrogen atoms at room temp. CFCs are gases under normal pressure but they liquefy under modest pressure. Its major sources are aerosol propellants, cleaning solvents, refrigerants and plastic blowing agents.

Some CFCs and their sources are:

1. Dichloromethane, dichlorodifluoromethane – Aerosol propellants
2. Trichlorofluoroethane – Cleaning solvents
3. Dichlorodifluoromethane – Refrigerants freon
4. Trichlorofluoromethane – Plastic blowing agents

All these applications release CFCs into the atmosphere, where these are mixed with normal atmospheric gases and reach to stratosphere and cause damage to ozone. Rowland and Molina (Got Nobel Prize in 1995) proposed that CFCs would be stable in the troposphere, and in stratosphere, they would be subjected to intense UV radiations which would break them apart, releasing free chlorine atoms via the fallowing reaction:

$$CFCl_3 \xrightarrow{UV} Cl + CFCl_2$$

Ultimately, all the chlorine of a CFC molecule would be released as a result of further photochemical break down. The free chlorine atom reacts with stratospheric ozone to form chlorine monoxide (ClO) and molecular oxygen:

$$Cl + O_3 \rightarrow ClO + O_2$$

Furthermore, two molecules of chlorine monoxide may react to release more chlorine and oxygen molecules:

$$ClO + ClO \rightarrow 2Cl + O_2$$

Upper last two reactions are called ‘chlorine catalytic cycle’, because chlorine acts as a catalyst which is continuously regenerated and reacts with ozone. The chlorine atom is highly stable and it can last from 40 to 100 years in stratosphere, and single atom of chlorine have potential to break down 100,000 molecule of ozone.

Now it is very clear that any substance carrying reactive halogens to the stratosphere has potential to destroy the ozone. These substances include halons, methyl chloroform, carbon tetrachloride and methyl bromide. Chemically similar to chlorine, bromine also attacks on ozone and forms a monoxide (BrO) in a catalytic reaction (bromine is 40 times as potent as chlorine in ozone destruction).

2.4.3 The 'Ozone Hole'

In 1985 a British scientist working in Antarctica reported a gaping 'hole' (actually, thinning of one area) in the stratospheric ozone layer over the South Pole where ozone levels were 50 % lower than normal. The CFCs, NOx, methyl bromide and chlorofluorocarbons are the agents which are busy in destroying the ozone layer over the South Pole. The ozone hole came as a surprise, and if it had occurred anywhere but over the South Pole, the UV damage would have been extensive. Ozone concentration is measured in Dobson unit (DU); the concentration of ozone has been gradually decreasing since 1970, from 306 DU to 245 DU (1971) and 240 DU (1993). The ozone layer is being depleted globally except over tropical areas, and the rate of depletion is higher in the highest latitude regions. In Japan also, a statistically significant trend has been verified in Sapporo and in the South Pole; the largest ozone hole in history was observed in 2000.

2.4.4 Effect of Ozone Depletion

It is well understood that ozone present in stratosphere protects humans from the harmful UV radiation coming from the sun. Ozone depletion has not only harmful effects on human being but also on all life forms including plants. Past (and current) emissions of ozone-depleting substances (ODS) result in increase in ultraviolet radiation reaching the Earth's surface which can cause several health effects:

- In humans different types of cancers are the result of harmful UV radiation, for example, conjunctiva tumours, basal cell carcinoma, melanoma and non-melanoma skin cancers.
- UV radiation leads to skin sunburns.
- More exposure time to UV radiation can caused leukaemia and breast cancer.
- O_3 depletion have great impact on plants life; intense UV radiation causes high rate of transpiration leading to decrees in moisture of soil and also decrease in the crop yield.
- Reduce effectiveness of the immune system.
- Damage to oceanic ecosystems and reduced fish yield (by killing microbial organisms in the ocean).
- Impact on nutrition (e.g. reduced plant yield).

2.4.5 Montreal Protocol

In 1987 United Nations under its environmental programme convened a meeting in Montreal, Canada, to address ozone depletion. Member nations reached an agreement known as the Montreal Protocol to decrease production of ozone-depleting substances back to 50 % by the year 2000 and 184 countries have signed the original agreement in Montreal, Canada.

2.5 Acid Rain

Most people of the world became aware of the problems associated with acid precipitation (the deposition of wet acidic solution or dry acidic particle from the air) within the last few decades. The term 'acid rain' was coined by Robert Angus Smith in his studies of air chemistry in Manchester, England, in the 1850s. Acid rain is actually the mixture of mainly H_2SO_4 and HNO_3 in which H_2SO_4 is the major constituents about 60–70 % and HNO_3 contribute about 30–40 %. This is an issue that has largely dropped from sight, after a flurry of concern in the 1970s and early 1980s over sharp declines in the populations of some fish and frogs and extensive signs of plant stress and dieback in many forests. But acid rain is still there, despite substantial efforts to reduce acidifying emissions of sulphur dioxide and NOx.

2.5.1 Causative Agent of Acid Rain

Acid rain is caused by air pollution and mainly attributed to the oxides of sulphur and oxides of nitrogen when these oxides dissolve or react with H_2O in upper atmosphere to form H_2SO_4 (sulphuric acid) and HNO_3 (nitric acid). This rainwater contains lower pH and falls on the earth surface as an acid rain.

The reactions by which sulphuric acid and nitric acid are formed are as follows:

$$2SO_2 + O_2 \rightarrow 2SO_3$$
$$SO_3 + H_2O \rightarrow H_2SO_4 \text{ (Sulfuric acid)}$$
$$2NO + O_2 \rightarrow 2NO_2$$
$$4NO_2 + 2H_2O + O_2 \rightarrow 4HNO_3 \text{ (Nitric acid)}$$

By the 1940s, it was known that pollutants, including atmospheric acids, could be transported to long distances by wind current.

2.5.2 Effects of Acid Rain

We describe acidity in terms of pH. The pH scale ranges from 0 to 14 and its midpoint is 7. Normal, unpolluted rain generally has a pH of about 5.6 due to carbonic acid created by CO_2 in air. There are several objects that are main target for acid rain as described below:

Effects on Aquatic Life: Principally H_2SO_4 and HNO_3 generated by industrial and automobiles emissions in Europe are carried by wind to Scandinavia where they are deposited in rain, snow and dry precipitation. The thin acidic soils and oligotrophic lakes and streams in the mountains of southern Norway and Sweden have been severely affected by this acid deposition. Some 18,000 lakes in Sweden are now so acidic that they no longer support fish's life. These fishless lakes are now known as 'fish graveyards'.

Reproduction is the most important stage in fish life cycles. Eggs of many species are killed when the pH of aquatic water drops.

- Acid rain causes decrease in fish food by destroying aquatic plants, insects and invertebrates on which fish depends for food.
- Acidity destroys fish's gills and prevent O_2 uptake, causes bone decalcification and disrupts muscle contraction and other serious hazards.
- It leads to the leaching of toxic metals, such as mercury and aluminium out of soil and rocks which are toxic to different aquatic life.

Effects on Forest Ecosystem: Forest of West Germany, Czechoslovakia, Poland, Austria and Switzerland are severely damaged due to acid rain. Acid rain causes reduced photosynthesis, decrease in uptake of nutrients from soil and also retarded growth of different plants of forest ecosystems. It causes leaching of some beneficial nutrients like calcium, phosphorus, etc. from soil and thus decreases the soil fertility that ultimately affects the growth of plants. It also causes chlorotic spots (coloured patches) on the leaves by which rate of photosynthesis decreases which lead to decreased net productivity of forest ecosystem. Toxic metals such as aluminium may be solubilised by acidic ground water. Fungi living in soil that form mutualistic association (mycorrhiza) with tree roots may be damaged by acid rain. Other air pollutants such as sulphur dioxide, ozone or toxic organic compounds may also damage trees.

Effects on Buildings and Monuments: Acid rain causes extensive damage to the oldest and glorious building throughout the world. Smoke and soot coat these buildings, paintings and textiles. Structural component of these building like limestone, slate, mortar, etc. are slowly damaged by atmospheric acid at an alarming rate.

$$CaCO_3 + \underset{(\text{Acid rain})}{H_2SO_4} \rightarrow CaSO_4 + H_2O + CO_2$$

The Taj Mahal in Agra, the Parthenon in Athens, the Colosseum in Rome, the mediaeval cathedrals in Europe and the Washington Monument in Washington DC are slowly being dissolved by acid fumes in air.

2.6 Natural Resource Depletion

Natural resources are vital to all forms of wildlife and the ecosystems in which they live. In human history, societies were always dependent on the use of natural resources. Natural resources, including materials, water, air, energy, fertile land, coal, oil, natural gas, minerals, etc. occur naturally in environment and are essential or useful to humans. Every manmade product at its fundamental level is composed of natural resources. People use natural resources as raw materials to manufacture or create a range of modern conveniences. Water and food provide humans with sustenance and energy; fossil fuels generate heat as well as energy for transportation and industrial production. Many of the same natural resources used by people are important to plants and wildlife for survival as well. A nation's access to natural resources often determines its wealth and status in the world economic system.

2.6.1 Classification

On the basis of their renewability, natural resources can be categorised as either renewable or nonrenewable.

2.6.1.1 Renewable Resources

Natural resources that can be replenished naturally are termed renewable resources. Renewable resources are sunlight, air, water, soil, forests, etc. These resources are continuously available and their quantity is not noticeably affected by human consumption. Thus many renewable resources do not have such a rapid recovery rate; these are susceptible to depletion by unsustainable use. Resources from human use perspective are classified as renewable only so long as the rate of replenishment/recovery exceeds that of the rate of consumption.

2.6.1.2 Nonrenewable Resources

Nonrenewable resources are exhaustible and are extracted faster than the rate at which they are formed. These resources are either formed slowly or not naturally formed in the environment. These resources include fossil fuels (coal, oil and natural gas), diamonds and other precious gems, minerals, metals and ores which exist in finite quantities and can only be replaced by processes that take millions of years.

2.6.2 Impact of Natural Resource Depletion

Globalisation may lead to depletion of natural resources that are used in the production of super quality goods. Different forms of societies consume dramatically different levels of natural resources. Resource depletion is the consumption of a resource faster than it can be replenished. Unfortunately, our society is consuming natural resources at a very fast speed for their conveniences, for example, electricity, transportation and industrial production, as well as materials for basic survival in an unchecked manner. Due to this, these resources may not be available in the future. Rapid population growth, a higher standard of living and technology all contribute to increased use of natural resources. Extracting, processing and using natural resources can cause environmental problems such as the disruption or destruction of ecosystems; climate change; shrinkage of fresh water reserves and forests; decrease in biodiversity; destruction of fertile land; and soil, water and air pollution. In order to continue to thrive on this planet, our lifestyles need to become more sustainable, so that we are able to protect our natural resource base and the fragile ecosystems on our planet.

2.7 Overpopulation

A population is a total of all the organisms within a species, which live in a particular geographical area whose individuals have a higher tendency of interbreeding among themselves. The growth of population depends upon fertility, mortality and migration. The three fundamental demographic factors of births, deaths and migration produce changes in population size, composition and distribution. Overpopulation is a function of the number of individuals compared to relevant resources they need to survive on. It can result from an increase in births, a decline in mortality rates, an increase in immigration or an unsustainable biome and depletion of resources. Human overpopulation occurs if the number of people in a group exceeds the carrying capacity of a region occupied by that group.

The world faces a serious overpopulation problem and the effects exist throughout the world, especially in Africa, Asia and Latin America (Table 2.2). The world's population has grown exponentially, from 1 billion people in 1800, to 2.2 billion in 1950, to 6 billion in 1999 and is estimated to reach 8–10 billion by the year 2150. The current rate of population growth is 1.2 % (Population Reference Bureau 2011),

Table 2.2 Top 20 countries by population as projected for 2015

Rank	Country	Population	% of world population
1.	China	1,369,102,000	18.9 %
2.	India	1,269,445,000	17.5 %
3.	United States	320,634,000	4.40 %
4.	Indonesia	255,461,700	3.53 %
5.	Brazil	204,119,000	2.82 %
6.	Pakistan	189,767,000	2.62 %
7.	Nigeria	183,523,000	2.54 %
8.	Bangladesh	158,272,000	2.19 %
9.	Russia	146,267,288	2.02 %
10.	Japan	126,910,000	1.75 %
11.	Mexico	121,005,815	1.67 %
12.	Philippines	101,246,800	1.40 %
13.	Vietnam	90,730,000	1.25 %
14.	Ethiopia	90,076,012	1.24 %
15.	Egypt	88,300,000	1.22 %
16.	Germany	80,925,000	1.12 %
17.	Iran	78,246,000	1.08 %
18.	Turkey	77,695,904	1.07 %
19.	Congo	71,246,000	0.98 %
20.	France	66,109,000	0.91 %

Population figures are based on National Census Authority and projection for 2015 by the Population Division of the United Nations Department of Economic and Social Affairs which means it requires a doubling time of 58 years. Thus, the current world population of 7 billion is expected to double to 14 billion in less than 60 years.

Six of Earth's seven continents are permanently inhabited on a large scale. Asia is the densest populous continent, with its 4.3 billion peoples accounting for 60 % of the world population. The world's two most populated Asian countries alone, China and India, together constitute about 37 % of the world's population. India is the second most populous country in the world after China. Africa is the second most populated continent, with around 1 billion people, or 15 % of the world's population. Europe's 733 million people make up 12 % of the world's population as of 2012, while the Latin American and Caribbean regions are home to around 600 million (9 %). The United States and Canada have a cumulative population of around 352 million (5 %), and Oceania, the least dense populated region, has about 35 million peoples (0.5 %).

2.7.1 Causes and Effects of Overpopulation

The main causes of overpopulation are decline in death rate, immigration, better medical facilities and lack of family planning.

Rapid increase in human population raised the demand for development and increased the consumption of various natural resources resulting in environmental degradation. Globally overpopulation is responsible for many serious environmental problems. These include depletion of natural resources, degradation of environment, land/soil degradation, habitat destruction, loss of biodiversity, global warming and climate change, water scarcity and water pollution, air pollution, food shortage and health problems.

2.8 Waste Disposal

All types of human activities produce different residues called 'wastes', and these differ in terms of quantity and quality and in their properties from one country to another. Waste is an object that can have negative impacts on the environment. According to United Nations Statistics Division, wastes are materials that are not prime products (i.e. products produced for the market) for which the initial user has no further use in terms of his/her own purposes of production, transformation or consumption and which he/she wants to dispose. Wastes may be generated during the extraction of raw materials, the processing of raw materials into intermediate and final products, the consumption of final products and other human activities, for example, municipal solid waste (household trash/refuse), wastewater (such as sewage, industrial effluents and surface runoff), radioactive waste, etc.

The total amount of waste generated has gradually increased with the growing population as well as economic, industrial and urban development. Rapid growth of industries resulted in generation of increasing volume of hazardous wastes. Dumping of hazardous wastes by the industries may cause severe environmental problems. In most developed and developing countries, it remains a major challenge to collect, recycle, treat and dispose of increasing quantities of solid waste and wastewater.

2.8.1 Classification of Wastes

Wastes are classified into two types according to their nature: solid and liquid wastes.

2.8.1.1 Solid Wastes

Solid wastes are the solid or semisolid substances left behind from human activities and are disposed of at source as residues of no value, because they are not considered to be worth keeping. Solid wastes can be classified into different types depending on their source:

(a) Municipal waste (household waste)
(b) Industrial waste (hazardous waste)
(c) Biomedical waste (infectious waste)

2.8.1.1.1 Municipal Waste

Municipal solid waste consists of household waste, construction and demolition debris, sanitation residue and waste from gardens, markets, hotels and recreational places and from streets. It also covers waste of small factories and camps. This garbage is generated mainly from residential and commercial complexes, for example, waste tyres, scrap metal, latex paints, furniture and toys, garbage, appliances and vehicles, oil and antifreeze, empty aerosol cans, paint cans and compressed gas cylinders, construction and demolition debris, asbestos, etc. These wastes also contain some hazardous substances such as drug residues, expired medicines, chemicals, paints, household insecticides and their empty containers, used dry batteries and electrical and electronic equipments. With rising urbanisation and change in lifestyle and food habits, the amount of municipal solid waste has been increasing rapidly and its composition changing. Over the last few years, the consumer market has grown rapidly leading to products being packed in cans, aluminium foils, plastics and other such nonbiodegradable items that cause incalculable harm to the environment.

2.8.1.1.2 Industrial Waste

Industrial wastes are considered hazardous as they may contain toxic substances. The term ‘hazardous waste’ is used as an indication of all residues representing hazard in connection with human health and environment upon use, storage, treatment or disposal. Industrial solid wastes are classified into organic wastes and inorganic wastes based on their components and into solid, semisolid, liquid and gaseous wastes based on their species. Because many industrial solid wastes hold hazardous properties, they usually receive special attention. Industrial solid waste resulting from medium and large-scale industrial activities contains hazardous ingredients, such as chemicals and heavy metals. The major generators of hazardous wastes are metal, chemical, paper, dye, refining, pesticide and rubber goods industries. Hazardous wastes could be highly toxic to humans, animals and plants. Hazardous wastes can be identified by the characteristics that they exhibit, for example, ignitability, corrosivity, reactivity or toxicity. It is presumed that about 10–15 % of wastes produced by industry are hazardous and the generation of hazardous wastes is increasing at the rate of 2–5 % per year.

Heavy metals are elements that are present in both natural and contaminated environments. In natural environments, they occur at low concentrations, whereas in contaminated environments they occur at high concentrations. These elements including mercury, lead, chromium, cadmium, arsenic, zinc, nickel and copper are considered as hazardous and show toxic effect on human health (Table 2.3). Heavy metals may be released into the environment from refining industries, metal smelting, scrap metal, rubber and plastic industries and from burning of waste containing these elements.

2.8.1.1.3 Biomedical Wastes

These types of wastes are generated in hospitals during the diagnosis, treatment or immunisation of human beings or animals or in research activities in these fields or in the production or testing of biologicals. It may include wastes like sharps, disposables, anatomical waste, cultures and discarded drugs or other pharmaceutical products, chemical wastes, etc. These are in the form of disposable syringes, swabs, bandages, body fluids, human excreta, etc. These wastes are highly infectious and can be a serious threat to human health if not managed in a scientific and careful manner.

2.8.1.2 Liquid Wastes

Liquid wastes are the remains of daily activities in liquid form, such as industrial effluents and agricultural drainage. Sediments known as sludge resulting from sanitary wastewater and industrial effluent treatment processes constitute the solid strand of these liquid wastes.

2.8.1.2.1 Drainage/Sewage Wastewater

This results from use of water for domestic, commercial and other municipal purposes. Sewage is the waste produced by toilets, bathing, laundry or culinary operations or thc floor drains associated with these sources and includes household cleaners, medications and other constituents in sewage restricted to amounts normally used for domestic purposes.

2.8.1.2.2 Industrial Wastewater/Effluents

These are liquids resulting from water consumed or used in major industrial activities. Industrial wastewater is the water or liquid resulting from industry, manufacture, trade, automotive repair, vehicle wash, business or medical activity. These

Table 2.3 Sources and health effects from exposure to heavy metals

Heavy metal	Sources of environmental exposure	Health effects
Mercury	Industrial wastewater, electronics, plastic waste, pesticides, pharmaceutical and dental waste, fossil fuel combustion, mining, smelting and solid waste combustion	Gastrointestinal disorders, respiratory tract irritation, allergic reactions, skin rashes, renal failure and neurotoxicity
Chromium	Leather, textile and steel manufacturing, ore refining, cement-producing plants, automobile brake lining, electro painting and chemical manufacturing	Respiratory irritation, kidney and liver damage, gastrointestinal effect, skin rashes, upset stomachs and ulcers
Lead	Industrial, vehicular emissions, paints, burning of plastics, papers, batteries	Impairment of neurological development, miscarriages and subtle abortions, suppression of the haematological system, kidney damage, behavioural disruptions
Cadmium	Manufacturing process, electronics, plastics, batteries and contaminated water	Irritation of the lungs and gastrointestinal tract, kidney damage, abnormalities of the skeletal system, psychological disorders, reproductive failure, cancer of the lungs and prostate
Arsenic	Industrial, contaminated water, mineral ore, electronic wastes, pesticides, wood preservative, burning of fossil fuels	Gastrointestinal and hepatic disorders, renal failure, lung irritation, cardiovascular injury, dermal lesions and neurotoxicity
Zinc	Industrial processes like galvanisation, automobile industry, watercolours or paints, rubber industry, military smoke bombs	Respiratory disorder, metal fever, stomach cramps, skin irritations, vomiting, nausea, anaemia and arteriosclerosis
Nickel	Fossil fuel combustion, nickel metal refining, sewage sludge incineration, manufacturing facilities	Allergic reactions, dermatitis respiratory failure, birth defects, asthma, bronchitis
Copper	Combustion of fossil fuels, mining, metal production, wood production and phosphate fertiliser production	Irritation of the nose, mouth and eyes; headaches; stomachaches; dizziness; vomiting and diarrhoea; liver and kidney damage

liquids may contain toxic or hazardous constituents including organic pollutants and chemical substances posing, in most cases, environmental and health risks.

2.8.1.2.3 Agricultural Wastewater

These are liquids resulting from water used in irrigation. Agriculture wastewater contains:

Sediment Runoff

Excess sediment causes high levels of turbidity in water bodies, which can inhibit growth of aquatic plants, clog fish gills and smother animal larvae.

Nutrient Runoff

Nitrogen and phosphorus are key pollutants found in runoff and they are applied to farmland in several ways, such as in the form of commercial fertiliser, animal manure or municipal or industrial wastewater or sludge. These chemicals may also enter runoff from crop residues, irrigation water, wildlife and atmospheric deposition.

2.8.2 Impacts of Waste Disposal

- With increase in the global population and the increased demand for food and other essentials, there has been a rise in the amount of solid wastes, liquid discharges (wastewater) and gaseous effluents. Consequently they lead to transfers of pollutants to natural environment which in turn are responsible for threatening the biological equilibrium of ecosystems.
- Household waste is directly thrown into municipal waste collection centres from where it is collected by the local municipalities to be further settled into the landfills and dumps. Improper disposal management can lead to adverse effect on human health and causes spread of infectious diseases and problems to the surrounding environment.
- Landfills release hazardous gas including methane, toluene, methylene chloride, etc. Methane is a powerful greenhouse gas, and in addition to its effect in the ozone layer, it is highly combustible gas that may be responsible for various explosion hazards in and around landfills.
- Industrial and agricultural waste can also impose serious health risks. Co-disposal of industrial hazardous waste with municipal waste can expose people to chemical and radioactive hazards. Wastes dumped near a water source also cause contamination of the water body or the groundwater resource. Direct dumping of untreated wastes in rivers, seas and lakes results in the accumulation of toxic substances which cause a large number of harmful effects on public health and on the biodiversity of aquatic environments.
- Untreated sewage may contain contaminated water, pathogens (including bacteria, viruses and protozoa), helminths (intestinal worms), nutrients (nitrogen and phosphorus), solids (including organic matter), oils and greases, heavy metals (including mercury, cadmium, lead, chromium, copper), and many toxic chemicals including PAHs, pesticides, phenols and chlorinated organics cause threat to both human health and the environment.

2.9 Deforestation

Deforestation is the conversion of forest to an alternative land use such as agriculture or urban development. Deforestation is primarily a concern for the developing countries of the tropical region due to shrinking areas of the tropical forests (Barraclough and Ghimire 2000) that lead to loss of biodiversity and increase in the greenhouse effect (Angelsen 1999). Deforestation includes not only the conversion to non-forest but also process of degradation by which reduction in forest quality, the structure and density of the trees, the biomass of plants and animals, the species diversity and the genetic diversity takes place. The year 2011 was 'The International Year of Forests' and this event has accelerated great attention towards the forests worldwide. We know that forests are a precious renewable resource and have much contribution in economic development by providing goods and services to people at large and also the forest-based industries. It was estimated that 30 % of the earth's land area or about 3.9 billion hectares is covered by forests at present and the original forest cover was approximately 6 billion hectares. The Russian Federation, Brazil, Canada, China and United States of America were the most forest-rich countries having almost 53 % of the total forest area of the world. Another 64 countries with a combined population of two billions were reported to have less than 10 % forest cover. Also, there are countries with almost no forest-covered land.

Asian countries especially India and China, due to their large-scale tree plantation programme in the last few years, reported a net gain in forest area. India has very diverse forest vegetation ranging from the moist evergreen forests in the North-East, along the West Coast and the Andaman and Nicobar Islands to the temperate and alpine vegetation in the Himalayan regions. However, these forests were degraded due to overgrazing, over-exploitation, forest fire, encroachments and indiscriminate sitting of development projects in the forest areas. Forest resources can provide long-term national economic benefits. For example, at least 145 countries on the globe are currently involved in wood production. In present scenario, sufficient evidence is available to prove that the whole world is facing an environmental crisis due to heavy deforestation.

2.9.1 Cause of Deforestation

Competition between humans and other species for the present ecological resources on land and in coastal regions leads to the conversion of forest land to other uses such as agriculture, urban development, industry, infrastructure, etc. So it is very important to discriminate between the agents of deforestation and its causes in order to understand the major culprits of deforestation. The agents of deforestation are those slash and burn farmers, firewood collectors, commercial farmers, loggers, infrastructure developers and others who are cutting down the forests for their benefit. There are many forces that cause deforestation as mentioned below.

2.9.1.1 Agricultural Activities

Agricultural activities are a must for human survival and also one of the major agents of deforestation. Due to increasing demand for food and food products, large numbers of trees are cut down to grow crops and for cattle grazing. Shifting agriculture, also called slash and burn agriculture, is the removal of forests for raising or growing the crops until the soil nutrients are not exhausted and/or the site is overtaken by weeds and then moving on to clear more forest. About 60 % of the deforestation of tropical moist forests is for agriculture with logging and other reasons like roads, urbanisation and fuel wood accounting for the rest. However, as the land degrades, people migrate and utilise new forests that lead to increased deforestation rate.

2.9.1.2 Urbanisation/Industrialisation

Urbanisation requires land to establish the new cities and towns necessary to support growing population which is normally done by way of deforestation. Overpopulation has direct effects on forest; as with the expansion of cities, more land is needed to establish new colonies and settlements. Infrastructure developments for various industrial activities require the construction of roads, railways, bridges and airports that result into deforestation.

2.9.1.3 Mining

Oil and coal mining also require forest land. Mining is very intensive and very destructive processes (Sands 2005). The area of land involved is quite small and it is not seen as a major cause of primary deforestation. Mining is a beneficial activity promoting developmental processes which may finally attract population growth and result into deforestation. The deforestation rate due to massive mining activities in Guyana from 2000 to 2008 increased 2.77 times according to an assessment by the World Wildlife Fund Guianas (Staff 2010). Similarly, in the Philippines, mining, along with logging, has been the main reason behind the country's loss of forest recourses. Massive mining of coal, iron ore and bauxite in Jharkhand (India) led to large-scale deforestation and created a huge water scarcity.

2.9.1.4 Fires

Fire is a major tool of forest degradation. A large number of trees are lost each year due to forest fires in different regions of the world. This happens due to extreme warm summers and milder winters. The fire engulfed a large forest area in India and Canada in the summers of 2016. Fires, whether caused by man or naturally, results in huge loss of forest resources. It is considered that fire is a good servant but a poor master. If fire used carefully, it can be a valuable tool in agricultural and forest management, but if abused, it can be a destroyer of forest cover (Repetto 1988; Rowe et al. 1992). Based on the data available from 118 countries representing 65 % of the global forest area, an average of 19.8 million hectares or 1 % of all forests are reported to be significantly affected each year by forest fires.

Besides above-mentioned causes for deforestation, there are also several other culprits such as logging, overpopulation, poverty, air pollution, wars, tourism, etc.

2.9.2 Effect of Deforestation

- Deforestation affects wind flows, water vapour flows and absorption of solar energy, thus significantly influencing local and global climate.
- Deforestation alters normal weather patterns that impose hotter and drier weather resulting in desertification and drought, crop failures, melting of the polar ice and coastal flooding.
- Deforestation also disturbs the global water cycle. With removal of part of the forest, the area cannot hold as much water producing a drier climate. Water resources affected by deforestation include drinking water, fisheries and aquatic habitats.
- Forests especially of tropical regions are rich in biodiversity, and consequently deforestation, fragmentation and degradation damage the biodiversity as a whole and habitat for migratory species including the endangered ones.

2.10 Loss of Biodiversity

Biodiversity is the variety of life on earth and includes variation at all levels of biological organisation from genes to species to ecosystems. Genetic, organism and ecological diversity are all elements of biodiversity with each including a number of components. In other words biodiversity is 'the variability among living organisms from all sources including terrestrial, aquatic, marine and other ecosystems. It includes diversity within species, between species and of ecosystems'.

It is believed that the 2.1 million species are presently known (Table 2.4) which represents only a fraction of the total number that exists. It is estimated that between 3 million and 100 million different species alive today (Heywood 1995), interestingly 30 million are tropical insects.

About 70 % of all known species are invertebrates (animals without backbone such as insect, sponges, worms, etc.). Out of total globe species, only 10–15 % live in North America and Europe. India has only 2.4 % of the total land area of the world; the known biodiversity of India contributes 8 % of the known global biological diversity. It is 1 of the 12 large biodiversity regions in the world. In terms of the number of mammalian species, India ranks tenth in the world, and in terms of the endemic species of higher vertebrates, it ranks eleventh. It stands seventh in the world for the number of species contributed to agriculture and animal husbandry.

Table 2.4 Number of living species by taxonomic group

Group	Identified species	Estimated total
Bacteria and viruses	5,800	10,000
Fungi	80,000	1,500,000
Protozoa and algae	100,000	250,000
Nonvascular plants	150,000	200,000
Vascular plants	250,000	300,000
Invertebrates	1,500,000	7–50 million
Fishes	20,000	23,000
Amphibians and reptiles	12,000	13,000
Birds	9,100	9,200
Mammals	250,000	300,000
Total	**2,125,300**	**9–52 million**

Source: UNEP data, 1993–1994

2.10.1 Causes of Loss of Biodiversity

2.10.1.1 Habitat Loss

Humans have increased the species extinction rate by as much as 1000 times over background rates typical over the planet's history. Ten to thirty percent of mammal, bird and amphibian species are currently threatened with extinction. The biggest cause for the increase in extinction is habitat loss. The destruction of forest, wetland and other diverse ecosystem cause elimination of millions of species throughout the globe. Humans have had an effect on every habitat on Earth, particularly due to the conversion of land for agriculture.

2.10.1.2 Invasive Species Introductions

When exotic (invasive) species are introduced into natural habitat, they are the most dangerous agent of habitat destruction in the world. Species such as Nile Perch, Carp have been released internationally because they look attractive but at the same time they pose threat to the natural habitants of that particular habitats.

2.10.1.3 Pollution

Toxic pollutants of nitrogen, phosphorus, sulphur, pesticide and toxic metals has emerged as one of the most important drivers of ecosystem change in terrestrial, freshwater and coastal ecosystems and have disastrous effect on population of organisms.

2.10.1.4 Hunting

Overharvesting of biological species is responsible for extinction of many species. For example, in 1830 in North America, the American pigeon (*Ectopistes migratorius*) was the most abundant bird with population of three to five billion, but by 1914, the entire population got extinct due to extensive hunting.

2.10.1.5 Diseases

Pathogens may also consider as predators. Bacteria, viruses, fungi, protozoan and many other harmful microorganisms are the major culprits for global biodiversity loss.

References

Angelsen A (1999) Agricultural expansion and deforestation: modeling the impact of population, market forces and property rights. J Dev Econ 58:185–218

Barraclough SL, Ghimire KB (2000) Agricultural expansion and tropical deforestation-poverty, international trade and land use. Earthscan Publications, London

Carter FW (1985) Pollution problems in post-war Czechoslovakia. Trans Inst Br Geogr 10(1):17–44

Church JA, Gregory JM, Huybrechts P et al (2001) Changes in sea level. In: Climate change 2001: the scientific basis. Contribution of working group I to the third assessment report of the IPCC, UNO

Fereidoun H, Nourddin MS, Rreza NA, Mohsen A et al (2007) The effect of long-term exposure to particulate pollution on the lung function of Teheranian and Zanjanian students. Pak J Physiol 3(2):1–5,

Heywood VH (1995) Global biodiversity assessment. Cambridge University Press, Cambridge

Patz JA, Epstein PR, Burke TA, Balbus JM (1996) Global climate change and emerging infectious diseases. J Am Med Assoc 275:217–223

Population Reference Bureau (2011) http://www.prb.org/pdf11/2011population-data-sheet_eng.pdf

Repetto R (1988) The forest for the trees? Government policies and the misuse of forest resources. World Resource Institute, Washington, DC

Rowe R, Sharma NP, Bowder J (1992) Deforestation: problems, causes and concern. In: Sharma NPP (ed) Managing the world's forest: looking for balance between conservation and development. Kendall/Hunt Publishing Company, Iowa, pp 33–46

Sands R (2005) Forestry in a global context. CABI Publishing, Wallingford

Staff S (2010) Mining deforestation nearly tripled between 2000–08. Archives Wednesday, October 13 2010

Suggested Further Readings

Allaby M (2002) Basics of environmental science, 2nd edn. Taylor & Francis, London

Anand SV (2013) Global environmental issues. Open Access Sci Rep 2:632. doi:10.4172/scientificreports, 632

Cassar M (2005) Climate change and the historic environment. Centre for sustainable heritage, University College London, London

Cunningham WP, Saigo BW (1999) Environmental sciences, 5th edn. WCB/McGraw-Hill, New York

Danny Harvey LD (2000) Global warming: the hard science. Prentice-Hall, Harlow, p 336

Devinny JS, Deshusses MA, Webster TS (1999) Biofiltration for air pollution control. CRC-Lewis Publishers, Boca Raton, pp 28–34

Harris F (2012) Global environmental issues, 2nd edn. Wiley, New York

Houghton JT, Ding Y, Griggs DJ et al (2001) IPCC climate change 2001: the scientific basis. Contribution of working group I to the third assessment report of the intergovernmental panel on climate change. Cambridge University Press, Cambridge/New York, p 881

Sheram K (1993) The environmental data book. The World Bank, Washington, DC

Vlek C, Steg L (2007) Human behavior and environmental sustainability: problems, driving forces and research topics. J Soc Issues 63(1):1–19

Microbes and Environment

3

Ankit Gupta, Rasna Gupta, and Ram Lakhan Singh

Contents

A. Gupta (✉)
National Institute of Allergy and Infectious Disease, National Institute of Health, 12735 Twinbrook Parkway, Rockville, MD, USA
e-mail: guptaankitbiochem07@gmail.com

R. Gupta
Department of Biochemistry, Dr. Ram Manohar Lohia Avadh University, Faizabad 224001, India
e-mail: rasna.gupta.biochem@gmail.com

R.L. Singh
Department of Biochemistry, Faculty of Science, Dr. Ram Manohar Lohia Avadh University, Faizabad 224001, India
e-mail: drrlsingh@rediffmail.com

R.L. Singh (ed.), *Principles and Applications of Environmental Biotechnology for a Sustainable Future*, Applied Environmental Science and Engineering for a Sustainable Future, DOI 10.1007/978-981-10-1866-4_3

Abstract

Microbes are omnipresent in the biosphere, and their presence invariably affects the environment in which they grow. The effects of microbes on their environment can be beneficial or harmful or inapparent with regard to human measure or observation. The most significant effect of the microbes on earth is their ability to recycle the primary elements that make up all living systems, especially carbon, oxygen, and nitrogen (N). Primary production involves photosynthetic organisms which take up CO_2 from the atmosphere and convert it to organic (cellular) material. The process is also called CO_2 fixation, and it accounts for a very large portion of organic carbon available for synthesis of cell material. Decomposition or biodegradation results in the breakdown of complex organic materials to other forms of carbon that can be used by other organisms. There is no naturally occurring organic compound that cannot be degraded by some microbe, although some synthetic compounds such as Teflon, plastics, insecticides, and pesticides are broken down very slowly or not at all. Through the microbial metabolic processes of fermentation and respiration, organic molecules are eventually broken down to CO_2 which is returned to the atmosphere for continuous process of primary production. Biological nitrogen fixation is a process found only in some bacteria which remove N_2 from the atmosphere and converts it to ammonia (NH_3), for use by the plants and animals. Nitrogen fixation also results in replenishment of soil nitrogen removed by agricultural processes. Thus along with all these benefits, microbes greatly contribute in maintaining sustainability of environment. This chapter mainly focuses on beneficial and harmful impacts of microbes on environment and their role to maintain quality, health, and sustainability of environment.

Keywords

Microbe • Environment • Species interaction • Nutrient cycle • Bioremediation • Pathogen • Disease

3.1 Introduction

The Earth is known as a "closed system" where materials cycle between lithosphere (rocks), atmosphere (air), hydrosphere (water), and biosphere (organism) (Fig. 3.1). Together, they make up all the components of our planet, both living and nonliving. Earth produces everything it needs to ensure the survival and growth of its residents. Environment is defined as the circumstances or conditions that surround an organism or group of organisms. Environment is the complex of social or cultural conditions that affects an individual or community. Since humans inhabit the natural world as well as the built or technological, social, and cultural world, all constitute an important part of our environment.

Environmental studies need to understand the life processes at the microscopic level and ecologist levels from species to ecosystem. Species refer to organisms of the same kind that are genetically similar enough to breed in nature and produce live, fertile offspring. Population consists of all members of the same species living in a given area in the same time. All the populations of organism living and interacting in a particular area make up a biological community. An ecological system or ecosystem is composed of a biological community and its physical environment. The environment includes abiotic factor such as climate, water, minerals, and sunlight as well as biotic factors such as organisms, their products, and effect in a given area.

Photosynthesis is the basis of energy economy of all, but a few specific ecosystem and ecosystem dynamics are based in how organisms share food resources. In

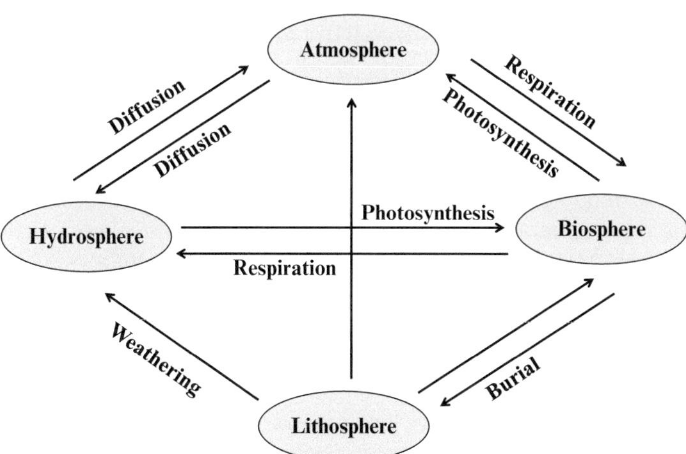

Fig. 3.1 Exchange of materials between different spheres of earth

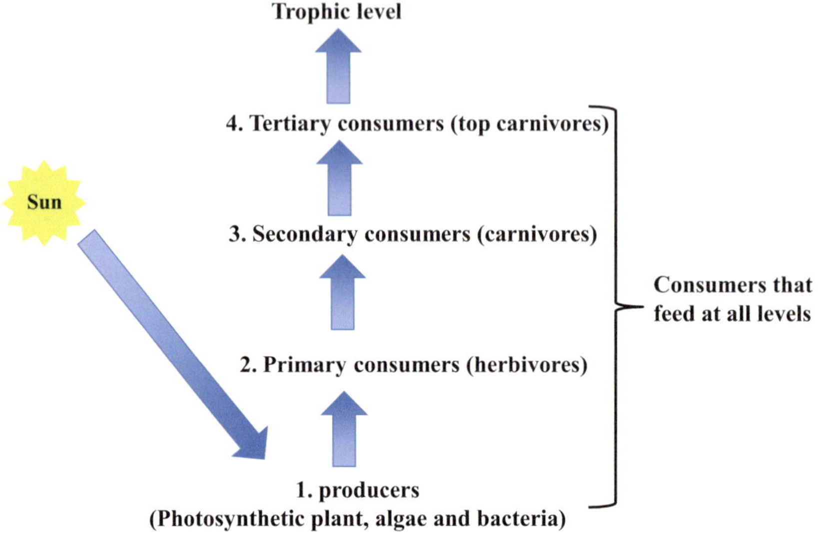

Fig. 3.2 Trophic cascade

fact one of the major properties of an ecosystem is its productivity, the amount of biomass produced in a given area during a given period of time. The rate of production of food creates a linked series of feeding known as food chain, whereas when individual food chains become interconnected, they form food web. An organism's feeding status in an ecosystem can be expressed as trophic level shown in Fig. 3.2.

3.2 Environmental Communities and Their Factors

The groups of similar species create population which results in a community. Microbial community means all microbial populations in a habitat. The activities of complex communities of microbes affect biogeochemical transformations in natural, managed, and engineered ecosystem. Microbial communities are very important for the rigorous progress in the field of agriculture which increases the rate of crop production. Microbial community may be terrestrial or aquatic.

3.2.1 Terrestrial Communities

A community of microbes and their environment that occurs on the landmasses of continents and islands form a terrestrial microenvironment. Terrestrial microenvironment is distinguished from the aquatic microenvironment ecosystems by the lower availability of water and the consequent importance of water as a limiting factor. Rain forests are the most diverse and productive terrestrial microenvironment, but their soil is nutrient deficient due to extensive leaching by rainwater.

3.2.1.1 Soil

Soil formation is a slow process that involves physicochemical weathering and biological processes over millions of years. Microbes play an important role in soil aggregate formation and soil stability that confer fertility and productivity to soil. The soil microbes participate in these processes through many ways, e.g., filamentous microbes assemble clay particles using extensive network of hyphae resulting into soil aggregates. Additionally, some microbes secrete exopolysaccharides or cause compaction of clay particles that promote soil aggregation. The surface soil is always rich in indigenous population of bacteria (including actinomycetes), fungi, algae, and protozoans. Additionally, human and animal activities also introduce specific microbes in the soil by several ways. Human activity directly adds bacteria as biodegradative agents or applying sewage sludge to agricultural fields. Animals introduce microbes through bird dropping or excretion.

3.2.1.2 Air

The atmosphere is an inhospitable climate for microbes because of stress due to dehydration. This results in a limited time frame for microbes to be active; however, some microbes get resistance to these stresses through specific mechanisms promoting loss of their biological activity. Spore-forming bacteria, molds, fungi, and cyst-forming protozoans all have specific mechanisms through which they are protected from these harsh gaseous environments. Therefore, viability is highly dependent on the environment, time they spend in the environment, and type of microbes. However, many other factors also influence the viability of microbes such as humidity, temperature, oxygen content, specific ions, UV radiation, various pollutants, and other air-associated factors (AOFs).

3.2.1.3 Relative Humidity

The relative humidity or relative water content of the air is critical for survival of airborne microbes. Most of the gram-negative bacteria associated with aerosols are able to survive for longer period at low relative humidity, whereas in contrast gram-positive bacteria remain viable longer in association with high relative humidity. The ability of microbes to survive in aerosol is related to the organism's surface biochemistry. One possible explanation of this fact could be a structural change in lipid bilayers of the cell membrane in response to very low humidity. During loss of water, the cell membrane bilayer changes from the typical crystalline structure to a gel phase and affects the surface protein configuration resulting into inactivation of the cell. The viruses with enveloped nucleocapsids (e.g., influenza virus) have longer airborne survival in low relative humidity below 50 %, whereas viruses without nucleocapsids (e.g., enteric viruses) are able to survive in high relative humidity above 50 %.

3.2.1.4 Temperature

Temperature is a critical factor influencing the activity of microbes. In general, high temperature leads to inactivation due to desiccation and protein denaturation, whereas lower temperature promotes longer survival rates. At very low temperature, some microbes lose viability because of ice crystal formation on their surface due to freezing.

3.2.1.5 Radiation

Mostly radiation at low wavelengths, e.g., UV radiation and ionic radiation (X-rays), is harmful for microbes causing DNA damage. These radiations target DNA by producing single or double strand breaks and changing the structure of nucleic acid bases. UV radiation causes damage by forming intra-strand thymidine dimers causing inhibition of biological activity such as replication of genome, transcription, and translation. Several mechanisms including association of microbes with large airborne particles, pigments or carotenoids, high relative humidity, cloud cover, etc. protect microbes from these harmful radiations. However, many microbes (e.g., *Deinococcus radiodurans*) have evolved mechanisms to repair DNA damage caused by UV radiation.

3.2.1.6 Oxygen, OAF, and Ions

Oxygen, open-air factors (OAF), and ions combine to inactivate many species of airborne microbes. Some reactive forms of oxygen including superoxide radicals, hydrogen peroxide, and hydroxide radicals are produced due to lighting, UV radiation, or pollution and cause DNA damage by producing mutations. Similarly, OAFs (mixture of ozone and hydrocarbons) also cause inactivation of microbes by damaging nucleic acids and enzymes. In addition to these factors, positive ions cause only physical decay, e.g., inactivation of cell surface proteins, whereas negative ions confer both physical and biological damages such as DNA damage.

3.2.2 Aquatic Communities

Aquatic microenvironments occupy more than 70 % of the earth's surface including mostly ocean but also others such as estuaries, harbors, river, lakes, wetlands, streams, springs, aquifers, etc. The microbiota, living in aquatic environment, are the primary producers (responsible for approximately half of all primary production on earth) and primary consumers as well. A large variety of microbial communities live in aquatic environments such as the planktonic, sediment, microbial mat, biofilm communities, etc. Planktons refer to photoautotrophic microbial community including both eukaryotes (algae) and prokaryotes (cyanobacteria) and heterotrophic community including bacteria (bacterioplankton) and protozoans (zooplankton). Phytoplanktons are the primary producers in the food web using their ability to fix CO_2 into organic matter through photosynthesis. Aquatic microenvironment is further classified into three microenvironments, occupied by microbes living in freshwater, brackish water, and marine water.

3.2.2.1 Freshwater

The study of freshwater microenvironment is known as micro-limnology. There are two types of freshwater environment: standing water or lentic habitats (e.g., lakes, ponds, bogs) and running water or lotic habitats including springs, rivers, and streams. Lentic habitats are dominated by phytoplankton, forming distinct community gradients based upon the wavelength and the amount of light that penetrates to a depth, e.g., *Chlorobium*. *Chlorobium* can utilize longer wavelength than other phototrophs and survive with little or no oxygen by consuming H_2S instead of H_2O for photosynthesis. In freshwater environment, two types of lakes are present: eutrophic and oligotrophic lakes. Oligotrophic lakes have higher rate (20–120 mg carbon/m^3/day) than eutrophic lakes (1–30 mg carbon/m^3/day) because eutrophic lakes have much higher levels of organic matter causing turbidity and interfering with light penetration. However, in terms of secondary productivity, eutrophic lakes have much higher rates (190–220 mg carbon/m^3/day) as compared to oligotrophic lakes (1–80 mg carbon/m^3/day).

3.2.2.2 Brackish Water

Brackish water environment is more saline than freshwater but less saline than marine water environment. An estuary, a part of river that meets with sea, is the best example of brackish water environment. Estuaries are highly variable environments because salinity changes drastically over a relatively short distance. Despite this, estuaries are highly productive environments, e.g., mangrove swamps in the Everglades of Florida, USA. Estuaries are generally turbid due to the large amount of organic matter brought by rivers and the mixing action of tides; therefore, light penetration is poor. Primary producers vary from 10^0 to 10^7organisms/ml and in relation to depth and proximity to littoral zones. Despite the low primary productivity, substrate availability is not limited, and heterotrophic activity is high ranging from 150 to 230 mg carbon/m^3/day.

3.2.2.3 Marine Water

Marine water environments are highly diverse and contain 33–37 % salinity. The ocean is divided into two zones on the basis of light availability: photic zone, where light can penetrate, and aphotic zone with lower light. Marine microenvironment is further divided into four habitats: neuston, pelagic, epibiotic, and endobiotic. Habitat at the surface of sea (air-water interface) is termed as neuston. On the basis of the precise depth, pelagic habitat is subdivided into epipelagic and benthopelagic zones. Epipelagic zone is found in upper 100 m of the water column, and a large proportion of organisms living in it are photosynthetic, whereas benthopelagic zone is sea-sediment interface. The third major habitat is the epibiotic habitats referring to surfaces on which attachment of communities occur, while the fourth is the endobiotic habitat with organisms (e.g., *Epulopiscium*) found within the tissues of other larger organisms such as fish.

3.2.3 Extremophilic Communities

The organisms living in physically or geochemically extreme conditions that are detrimental to most life on earth are termed as extremophiles. Most of the extremophiles are microbes and belong to the domain Archaea. Here are some extreme environmental conditions where extremophiles survive.

3.2.3.1 High Temperature

Environments with high temperature (>70 °C) including terrestrial and submarine springs with a temperature of 100 °C, hydrothermal vents with a temperature more than 300 °C are inhospitable for most forms of life except some bacteria and archaebacteria, e.g., *Thermus*, *Methanobacterium*, *Sulfolobus*, *Pyrodictium*, and *Pyrococcus*. *Pyrodictium* and *Pyrococcus* are capable of surviving at temperature >100 °C. Another example of such renowned extremophiles is *Thermus aquaticus* having thermotolerant DNA polymerase, which is widely used in the polymerase chain reaction (PCR). These thermophiles have developed such characteristic mechanisms facilitating proteins in folded state even at high temperatures due to increased salt bridges (cations that bridge charges between amino acid residues).

3.2.3.2 High Solute

Some organisms require salt concentrations substantially higher than that found in seawater for their growth, and they are known as halotolerant. *Halobacterium* and *Halanaerobium* are two examples of halotolerant bacteria; however, some algae and fungi also exhibit halotolerance feature. The main mechanism of salt tolerance operates by internal sequestration of high balancing solute (K^+ in bacteria and glycerol in halotolerant eukaryotes) equal to external salt concentration. A second mechanism of salt tolerance involves proteins with acidic and low proportion of nonpolar amino acids. These proteins require high salt concentrations to balance their charge for their optimal activity. Therefore, some obligate halophiles are unable to survive in the environment lacking high salt concentration due to these macromolecular modifications.

3.3 Microbes

There is a wide range of microbes present in our biosphere depending on their physical and other characteristics. Microbes fall into two groups, prokaryotes and eukaryotes, depending upon whether they have nucleus or not. Prokaryotes lack this membrane around their genetic material, and this group includes viruses, bacteria, and related archaea. The other category of microbes includes algae, fungi, protists, and other microscopic animals, having cell nucleus.

3.3.1 Bacteria and Archaea

Bacteria and archaea are the smallest free-living, unicellular organisms present on the earth. Their cell sizes typically range from 0.5 to 1.0 μm in diameter. Both exist in various cell shapes, e.g., cocci, rods, or spirals, and some soil bacteria form branching filaments, e.g., actinomycetes. Their DNA is found free in the cell

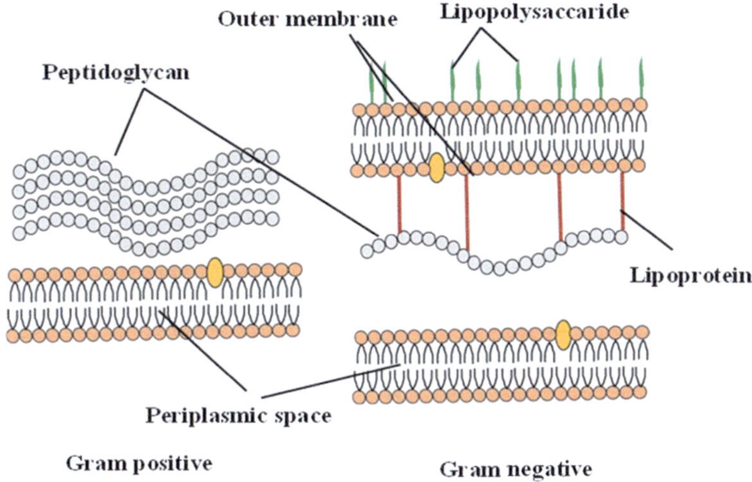

Fig. 3.3 Difference in gram-positive and gram-negative bacteria

cytoplasm and lack a true nuclear envelope, and the genome is mainly composed of single double-stranded DNA molecule with smaller DNA elements known as plasmids. The size of bacterial genome typically ranges from four to six million nucleotides in length and enable to code 3,000–4,000 genes. A bacterial cell envelope is composed of two layers, the inner layer is cell membrane made of phospholipids and the outer layer is cell wall made of proteins, carbohydrates, and lipids, but its composition varies based on the type of organism. Most of the microbes move through flagella (whiplike extensions from the cell) and file filaments, e.g., pili. The pili enable them to attach with each other or to soil particles. Additionally these pili are also involved in transfer of genetic material between bacterial cells, known as conjugation. These microbes usually reproduce asexually, e.g., binary fission, resulting in the formation of two genetically identical bacterial cells. On the basis of gram staining, bacteria are of two types: gram positive and gram negative; both vary in cell structure and physiology (Fig. 3.3). Bacteria and archaea both require carbon as building blocks of their cellular materials and energy to drive the reactions involved in cell biosynthesis and metabolism. Most of the bacteria utilize oxygen, whereas some bacteria and archaea grow anaerobically by using alternative electron acceptors, e.g., nitrate and sulfate.

Basically microbes are classified into autotrophs and heterotrophs. Autotrophs utilize sunlight or inorganic compounds such as Fe^{2+}, nitrate, or nitrite as energy source to fix atmospheric carbon dioxide to produce carbohydrates, fats, and proteins. However, heterotrophic bacteria use organic compounds as a source of carbon and energy. Archaea were originally known to be found in extreme environments and termed as “extremophiles,” but now they are widely distributed and are found in many environments including soil.

It is hard to distinguish both archaea and bacteria on the basis of their morphology. But most recently their classification using molecular phylogenetic tools based on a comparison of 16S ribosomal rRNA sequences has revealed three separate domains of life: eukaryotes, bacteria, and archaea. Archaea are closely related to eukaryotes (all multicellular organisms) than the bacteria (Woese et al. 1990).

3.3.2 Fungi

Fungi belong to eukaryotes and, therefore, are more closely related to plants and animals than bacteria or archaea. Fungal cell consists of membrane-bound nucleus with chromosomes containing genetic material, e.g., DNA, membrane-bound organelles, e.g., mitochondria, and a cell wall composed of glucans and chitin. Fungi are basically heterotroph organisms meaning thereby that they derive their food from nonliving organic sources, e.g., saprophytic fungi, which feed on dead or decaying organic materials. Few fungi also exist as unicellular organisms, e.g., yeast, which grow through cylindrical threadlike structures (2–10 cm in diameter) known as hyphae. These hyphae may be either septate, e.g., compartmentalized through cross walls, or nonseptate. The hypha is a main part of fungus and constitutes a mycelium. Finely branched mycelium occupies a large surface area in the soil and produces a range of enzymes acting on soil organic matter to produce nutrients and energy required for fungal growth. Fungi can reproduce by both sexually, e.g., through spores, and asexually, e.g., budding or binary fission. Fungi are highly diverse and play a wide range of role in their surrounding environment such as decomposers, mutualists, endophytes of plants, pathogens, and predators. Fungal hyphae are the basic components of soil food webs since they constitute a food source for soil biota, whereas fungal sporocarps provide food for larger animals.

3.3.3 Protists

Protists are unicellular eukaryotes having characteristic organelles but lacking cell wall. Protists may be free living, parasitic, or opportunistic based on environmental changes. The size of protists varies from 2 μm to several centimeters. These are classified into four groups: flagellates (Mastigophora), amoebae (Sarcodina), sporozoans (Sporozoa), and ciliates (Ciliophora). Flagellates are characterized by the presence of flagella as a locomotive tool, multiplication by binary fission, and having both heterotrophic and autotrophic feeding mechanisms. Flagellates are further classified into two major divisions: photosynthesizing (Phytomastigophora) and non-photosynthesizing (Zoomastigophora). Autotrophic flagellates possess chloroplast and synthesize their own food or nutrients by photosynthesis such as *Euglena*, whereas heterotrophic flagellates are parasitic and take their nutrients from their host, for example, *Leishmania* and *Giardia*. Amoebae have pseudopodia as their locomotive tool. Pseudopodia help in ingestion of food materials and provide an extended basis of further classification. The Rhizopoda move by a fluid endoplasm, whereas the Actinopoda have a spikelike pseudopodium for their movement and feeding. These organisms are vacuolated and covered in a shell-like outer layer known as "test." These shells may be composed of proteinaceous, siliceous, or calcareous substances and have either a single or multiple chambers. Some testate amoebae are also found in soil, building their shells by excreting substances capable of aggregating soil particles. Some important free-living soil amoebae are grouped in the family *Vahikampfia*. *Entamoeba* is well-known parasitic amoeboid causing dysentery in humans. The ciliates have hairlike structures in an ordered array surrounding the cell known as cilia that divides by transverse fission. Cilia help in their locomotion and feeding. Ciliates are generally free living such as *Paramecium*, but some species are adapted to parasitic life cycles. Sporozoans are mostly parasitic and form spores. Few members are adapted in a symbiotic relationship and have no locomotive organ; therefore, they rely on vectors or direct contact with susceptible host to continue their growth and replication. For example, some species have evolved to enable digestion in the gut of domestic livestock. However, parasitic members are totally dependent on host for their nutrition, for example, *Toxoplasma*, *Isospora*, and *Plasmodium*. Most of parasitic protists are of obvious public health concern causing deadly diseases such as malaria, sleeping sickness, Chagas disease, leishmaniasis, giardiasis, cryptosporidiosis, etc. These parasites have adapted themselves for surviving and reproducing in their hosts by evading host immune responses. Many flagellated protists are capable of forming cysts that are known to survive conventional methods of disinfection and can be transmitted to their host via a water route. *Entamoeba histolytica* is a parasitic amoeba and causes diarrhea and dysentery. *Naegleria* is a free-living amoeba in freshwater, causing infections in nasal passage of humans and capable of invading brain tissues. *Toxoplasma* is an invasive protist that causes blindness and serious illness or death in unborn fetuses. *Cryptosporidium* is responsible for a number of epidemics including the largest US waterborne outbreak in Milwaukee. *Plasmodium* is a main causative agent of mosquito-borne disease, malaria. Domestic animals are also at risk of serious illness and death from parasitic protozoans, for example, *Histomonas*, *Trichomonas*, etc.

3.3.4 Viruses

Viruses are small, obligate, intracellular pathogens that require a host cell for their growth and replication. They can survive outside the host cell but not multiply without a host. Viruses are generally species specific and can infect all types of life forms: bacteria, plants, and animals. Public focus is most often on (1) plant viruses affecting important crops such as tobacco, potatoes, and tomatoes or (2) animal viruses causing deadly diseases: herpes, smallpox, rabies, mumps, measles, meningitis, hepatitis, encephalitis, influenza, diarrhea, yellow and dengue fever, etc. The basic virus structure includes a capsid protein coat and an internal nucleic acid (RNA or DNA), but some viruses may also have protein and lipid envelopes, glycoprotein spikes, or more complex tail and sheath structures. A number of protein capsomers held together by non-covalent bonds form a capsid coast surrounding the nucleic acid molecule. The size of capsids ranges from 18 μm, as in small parvovirus of animals, to several hundred nanometers, as in some filamentous plant viruses.

This outer coat protects and shields the viral nucleic acid and harbors specific receptor sites for attachment on hosts. The viral capsids have two types of symmetrical organization: helical and icosahedral. The helical viruses look like a spiral or a helix with a cylindrical shape, whereas icosahedral viruses adopt a 20-triangular-sided spherical shape when viewed with an electron microscope. Viruses have either RNA or DNA in the double-stranded or single-stranded form. Viral nucleic acid length varies from 1.7 to over 200 kb, and it encodes 4–200 genes.

3.4 Microbial Diversity

Early studies on diverse soil bacteria and fungi are mainly focused on what could be easily cultured from soils, but the fact is that less than 10 % of the soil bacteria could be cultured, suggesting the requirement of other approaches. Norman Pace and colleagues in 1980 found that microorganisms could be identified in naturally occurring microbial populations without culturing them (Hugenholtz et al. 1998), but this process requires the PCR amplification of the rRNA genes using rRNA specific primers and RNA extracted directly from the cells present in soil. These specific primers may differentiate among various microbial communities at level of different domains such as Bacteria, Eukarya, and Archaea or phylum (e.g., *Actinobacteria* or *Bacteroidetes*) (Fig. 3.4). However, a range of approaches could be adopted in order to separate and sequence the rRNA genes. The advanced high-throughput DNA sequencing now allows the identification of each individual in thousands of samples within a short duration (Caporaso et al. 2012). Once we compare these rRNA gene sequences from cultivated species using various online databases, e.g., GenBank, allowing identification of evolutionary (phylogenetic) relationships

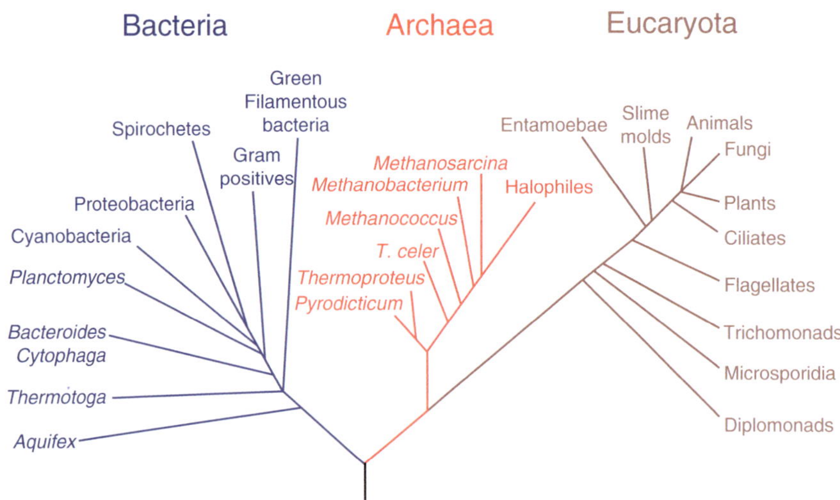

Fig. 3.4 Phylogenetic classification of living world based on 16S and 18S rRNA gene sequences

among various unknown and known organisms, it displays an estimate of the genetic diversity of organisms in a particular community. It is easy to speculate about the organism's characteristics and its closest cultivated relative on the basis of sequencing details. Phylogenetic information sometimes also provides details about the physiology, e.g., all cyanobacteria constitute a monophyletic group in a similar way as sulfate-reducing bacteria, halophiles, and methanogenic archaea do.

3.4.1 Bacterial Phyla

The composition of in situ environmental bacterial communities has been investigated using various molecular tools. The relative abundance of the major phyla varies among diverse soils and environmental conditions such as some members of the phyla, i.e., *Proteobacteria*, *Acidobacteria*, and *Actinobacteria*, are abundant and widely distributed; however, members belonging to *Verrucomicrobia*, *Bacteroidetes*, *Firmicutes*, *Chloroflexi*, *Planctomycetes*, and *Gemmatimonadetes* are comparatively less prevalent.

The *Proteobacteria* is a diverse group of organisms among various subphyla out of which, α-, β-, γ-, and δ-*Proteobacteria* are most commonly found in soil. The members belonging to α-, β-, and γ-subphyla are copiotrophs (an organism able to grow in nutrient-rich environments particularly carbon in contrast to oligotrophs, those found in environments with much lower carbon concentration). The *Proteobacteria* are more prevalent in those area rich in resource availability, e.g., rhizosphere soils.

The α-*Proteobacteria* consist of various metabolically diverse heterotrophic and autotrophic bacteria, e.g., heterotrophic bacteria such as *Sphingomonas*, that are able to degrade various toxic compounds including pentachlorophenol and polyaromatic hydrocarbons and also involved in weathering of minerals. Some heterotrophs that are also able to fix atmospheric nitrogen by forming symbiotic relationships with legumes belong to family Rhizobiaceae such as *Rhizobium*, *Mesorhizobium*, and *Bradyrhizobium*. The autotrophic α-*Proteobacteria* also include soil methane oxidizers, e.g., *Methylobacter* and *Methylophilus*, nitrite oxidizers, e.g., *Nitrospira* and *Nitrobacter*, and phototrophs, e.g., *Rhodospirillum* and *Rhodobacter*.

The β-*Proteobacteria* is also found into three groups: heterotrophs, autotrophs, and methanotrophs. Some of the known heterotrophs found in soil belong to the genera *Burkholderia*, *Alcaligenes*, and *Acidovorax*. Out of these, *Burkholderia* spp. is metabolically diverse and uses simple amino acids, sugars, and recalcitrant aromatic and phenolic compounds as a source of carbon and therefore plays a major role in carbon turnover. *Burkholderia* spp. is also known to promote plant growth by fixing atmospheric nitrogen. Some examples of heterotrophic β-*Proteobacteria* are *Collimonas*, able to degrade live hyphae by producing chitinase, and autotrophic β-*Proteobacteria* are *Nitrosospira* (ammonia oxidizer), *Thiobacillus* (iron oxidizer), *Methylomonas* (methane producer), *Rhodocyclus* (phototroph), etc.

The γ-*Proteobacteria* are also categorized into heterotrophs, lithotrophs, and phototrophs. *Pseudomonas* and *Xanthomonas* are well-known heterotrophs.

Pseudomonas spp. has a remarkable nutritional versatility since most of them are able to grow on more than 50 or 100 different substrates including sugars, amino acids, fatty acids, alcohols, and hydrocarbons. The *γ-Proteobacteria* also include various photolithotrophs, e.g., *Thiocapsa* and *Chromatium*, that utilize sulfide or elemental S as an electron donor and CO_2 as a source of carbon under anaerobic conditions in light.

The *δ-Proteobacteria* consist mainly of SO_4^{-2}- and Fe-reducing bacteria. *Desulfovibrio*, a sulfate-reducing bacteria, grow anaerobically by utilizing lactate or ethanol as carbon sources, found in oxygen-depleted soil. This group also includes a parasitic bacteria named as *Bdellovibrio*.

Acidobacteria are widely distributed and found in abundance in the soil with low pH. Since they are poorly present in soil culture collections, a little knowledge is available about their metabolic capabilities. The complete genome sequencing of cultured soil *Acidobacteria*, e.g., *Acidobacterium capsulatum*, implies that they may be oligotrophs and able to metabolize a range of simple and complex carbon sources. These bacteria are also found in low nutrient conditions, tolerant to changes in soil moisture. They also play a role in nitrate and nitrite reduction but not in denitrification or nitrogen fixation.

Verrucomicrobia, also found commonly in soil, are oligotrophic in nature; however, the ecology of Verrucomicrobia is poorly understood. The majority of bacteria of this group are *Chthoniobacter flavus* and *Opitutus terrae*. Genome sequencing of a free-living heterotroph bacteria found in aerobic soil, e.g., *Chthoniobacter flavus*, suggests that it is able to metabolize plant polysaccharides but not amino acids except pyruvate. However, genome sequencing of a verrucomicrobium from rice paddy soil, e.g., *Opitutus terrae*, implies that it is an anaerobic bacterium, capable of producing propionate through fermentation of polysaccharides from plants.

Gram-positive bacteria are abundant in soil culture collections and categorized into two groups: *Actinobacteria* and *Firmicutes*. The *Actinobacteria* are commonly found in soil and further classified into three subphyla: *Actinobacteridae*, *Acidimicrobidae*, and *Rubrobacteridae*. The abundance of *Actinobacteridae* in soil relatively increases with addition of labile carbon sources. The *Actinobacteridae* includes metabolically diverse aerobic heterotrophs, e.g., *Arthrobacter*, *Rhodococcus*, *Streptomyces*, and *Mycobacterium*. *Streptomyces* is a well-known bacterium producing antimicrobial compounds. The *Rubrobacteridae* includes two genera not represent in soil culture collections, e.g., *Rubrobacter* and *Solirubrobacter*. *Acidimicrobium ferrooxidans* is an acid-tolerant ferrous iron oxidizer bacterium and has been detected in soil culture collections among few members of *Acidimicrobidae*.

The *Firmicutes* include bacteria that are able to form endospores such as *Bacillus* and *Clostridium*, and because of endospore production, they are able to survive longer in the soil during dry periods. *Bacillus* is capable of degrading various carbon sources, including polysaccharides from plants, whereas *Clostridium* may ferment sugars, starch, pectin, and cellulose. The addition of recalcitrant C compounds in soil favors the growth of Clostridiales.

Planctomycetes multiply through budding and lack peptidoglycan in their cell walls. These bacteria are also involved in ammonium oxidation (Anammox) in soil under anaerobic conditions. *Planctomycetes*, Verrucomicrobia, and Chlamydia are important from evolutionary point of view since they share numerous features, e.g., presence of membrane-coat-like proteins and condensed DNA, rarely found in bacteria but more common in Archaea and Eukaryotes. *Sphingobacteria* are known to be involved in aerobic degradation of plant materials present in soil and complex organic molecules, e.g., starch, proteins, cellulose, and chitin. Members belonging to genus *Chitinophaga* are filamentous and chitinolytic and exhibit gliding movement.

3.4.2 Archaeal Phyla

Archaea are known to be widely distributed in the soil on the basis of 16S rRNA gene sequences (Bates et al. 2011), and additionally these gene sequences suggest that members belonging to the phylum Crenarchaeota are found in abundance in the marine environment. All cultured Crenarchaeota are thermophilic or hyperthermophilic organisms: having the ability to survive at a temperature up to 113 °C. Crenarchaeota is sulfur-dependent extremophile; one of the best-characterized members is *Sulfolobus solfataricus*. This organism was originally isolated from geothermally heated sulfuric springs in Italy and grows at 80 °C and pH of 2–4. These organism stains are gram negative and are morphologically diverse having rod-, cocci-, filamentous-, and oddly shaped cells. The *Crenarchaea* play a role in ammonia oxidation in the soil. Mesophilic ammonia-oxidizing archaea (AOA) is abundant in a diverse range of marine environments, including the deep ocean, as revealed by the quantification of the archaeal *amoA* gene encoding the alpha-subunit of the ammonia monooxygenase. *Nitrososphaera viennensis* is a most recent ammonia-oxidizer *Crenarchaea* that was extracted from garden soil, and subsequent phylogenetic analysis confirmed its taxonomic affiliation.

Methanogens are strict soil anaerobes and grow in association with bacteria that participate in the anaerobic food chain and convert complex organic molecules to methane (CH_4) and CO_2. Methanogens generate methane through various pathways. They display various activities such as reduction of carbon dioxide and methanol, cleavage of acetate, and methane production from methylated compounds. Members belonging to genera *Methanosarcina*, *Methanosaeta*, and *Methanocella* are widely distributed in the environment, and both *Methanosarcina* and *Methanosaeta* produce methane through acetate reduction.

3.4.3 Fungal Phyla

Seven fungal phyla are currently recognized: *Chytridiomycota*, *Blastocladiomycota*, *Neocallimastigomycota*, *Glomeromycota*, *Ascomycota*, *Basidiomycota*, and parasitic endobionts, e.g., *Microsporidia*. Among these fungal phyla, the first three have flagellated cells at least during one stage of their life cycle and thereby turned into terrestrial organisms in contrast to higher fungi that lack motile cells. Among these phyla, *Chytridiomycota* is a basal group and may degrade chitin and keratin. *Batrachochytrium dendrobatidis* is a known pathogenic unicellular chytrid of many amphibian species resulting into decline of worldwide amphibian population. *Blastocladiomycota* differ from the *Chytridiomycota* in reproduction since they exhibit different form of meiosis.

Glomeromycota exhibit some features identical to "lower" fungi, e.g., they have multinucleate aseptate mycelia and most of them have no known sexual stages. They reproduce through large thick-walled asexual spores, commonly found in soils, and may germinate in the presence of a plant root. This phylum includes arbuscular mycorrhizal (AM) fungi that form obligate biotrophic symbioses with mosses, approximately 80 % of all land plants, and cyanobacterium *Nostoc* forming cyano-lichens. These higher fungal phyla have a characteristic feature having two compatible nuclei in a hyphal cell also known as dikaryon.

Ascomycota is one of the largest fungal phyla with 64,000 or more number of species. The members of this phylum have a characteristic spore-bearing saclike structure also known as asci, produced in large numbers during sexual reproduction. Ascomycetes mostly reproduce asexually and are rarely found to reproduce through sexual mating. Ascomycetes have a typical haploid mycelium with septate hyphae and cell wall made up of chitin and β-glucans. Few macroscopic ascomycetes exhibit well-known reproductive structures such as morels, truffles, etc. However, many ascomycetes are microscopic unicellular organisms, e.g., yeast or *Saccharomyces*, and filamentous fungi, e.g., *Aspergillus*. Most of the Ascomycetes are saprobes and have various enzymes to degrade complex substrates such as cellulose, keratin, and collagen. Therefore, ascomycetes are known to play a critical role in decomposing and nutrient recycling. Approximately 18,000 species of ascomycetes form lichens through symbiotic relationship with green algae and cyanobacteria. However, other *Ascomycota* constitute ectomycorrhizal and/or ectendomycorrhizal associations through symbiosis with woody plants. Some ascomycetes are known plant parasites and predators. The family *Orbiliaceae* includes carnivorous fungi having specialized hyphae to trap prey including a range of soil mesofauna, protists, nematodes, and arthropods.

Members belonging to *Basidiomycota* are also known as club fungi and produce spores during sexual reproduction on club-like stalks known as basidia. These microscopic basidia are basically clustered on specialized structures also known as sporocarps. Numerous haploid spores are produced after meiosis and released in the environment resulting into a new haploid mycelium after germination.

3.5 Microbial Application to Environment

3.5.1 Contribution of Microbes to Nutrient Cycling

3.5.1.1 Microbes in Carbon Cycle

Microbes play a critical role in carbon cycle on the global scale that is a key constituent of all living organisms (Fig. 3.5). Microorganisms avail carbon for living

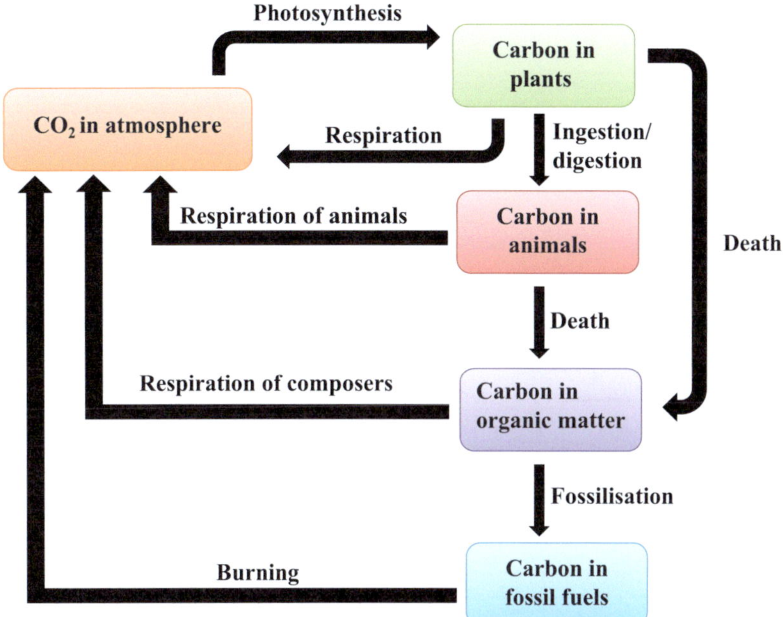

Fig. 3.5 Role of microbes in carbon cycle

organisms and for themselves as well through extracting it from nonliving sources. In aquatic habitats, microbes convert carbon anaerobically, present at oxygen-free zones such as deep mud of lakes and ponds. Carbon dioxide (CO_2) is the most common form of carbon that enters into a carbon cycle. CO_2 is a water-soluble gas present in the atmosphere. Plants and photosynthetic alga use CO_2 during photosynthesis to synthesize carbohydrates. Additionally chemoautotrophs such as archaea and bacteria also utilize CO_2 to synthesize sugars. This carbon, present in the form of sugar, is further processed through a chain of reactions during respiration known as tricarboxylic acid cycle resulting into energy. Microbes may also use carbon under anaerobic conditions to produce energy through a process called as fermentation.

Plants are the primary producers in a terrestrial ecosystem; however, free-living planktons, cyanobacteria, and symbionts such as lichens also contribute in fixing carbon in some ecosystems. Nonliving organic material is recycled by heterotrophic bacteria and fungi, whereas saprobes utilize organic material and produce CO_2 during respiration, thereby contributing to carbon cycle. However, higher animals, e.g., herbivores and carnivores, also digest organic materials to obtain energy using gut microbiota residing in their intestinal tracts; the process is known as decomposition, resulting into inorganic products such as CO_2, ammonia, and water.

Actinobacteria and *Proteobacteria* are capable of degrading soluble organic compounds, e.g., organic acids, amino acids, and sugars (Eilers et al. 2010). Similarly, bacteroidetes are also involved in degradation of more recalcitrant carbon compounds, e.g., cellulose, lignin, and chitin, and utilize higher level of available

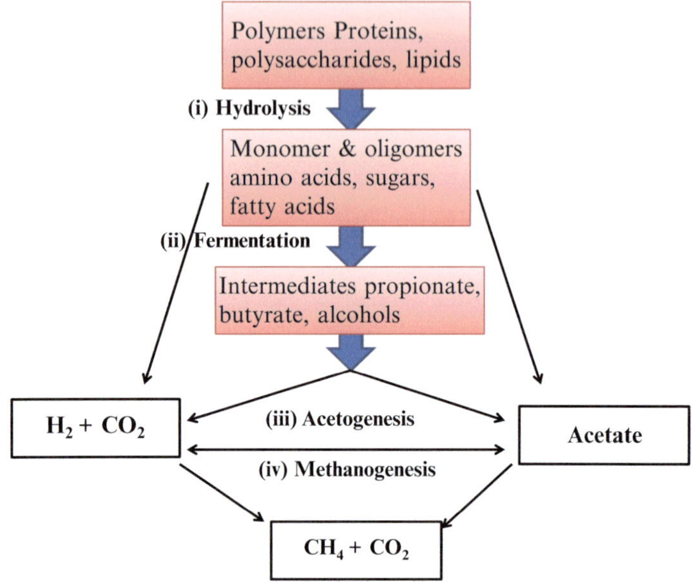

Fig. 3.6 Production of methane by microbes

nitrogen that help in production of extracellular and transport enzymes (Treseder et al. 2011). On the contrary, bacteria living in low-nitrogen environments are more able to metabolize nitrogen-rich organic compounds, e.g., amino acids. The abundance of *α-Proteobacteria* and *Bacteroidetes* favors carbon mineralization, whereas *Acidobacteria* oppose it (Fierer et al. 2007).

3.5.1.2 Microbes in Methane Production

Some microbes execute anaerobic or fermentative degradation of organic compounds into organic acids and some gases, e.g., hydrogen and CO_2. Methanogens are able to use that hydrogen to reduce CO_2 into methane under strict anaerobic conditions (Fig. 3.6). In order to complete cycle, methane-oxidizing bacteria, e.g., methanotrophs, transform methane to CO_2, water, and energy under aerobic conditions.

Other microbes such as green and purple sulfur bacteria participate in carbon cycle by degrading hydrogen sulfide (H_2S) into compounds having carbon during energy production (see in reaction). Some bacteria, e.g., *Thiobacillus ferrooxidans*, derive energy from oxidation of ferrous iron to ferric iron and thereby contribute to carbon cycle. Few microbes such as *Bacteroides succinogenes*, *Clostridium butyricum*, and *Syntrophomonas* spp. make a collaborative effort (also known as interspecies hydrogen transfer) for anaerobic degradation of carbon to produce CO_2 and methane in bulk. The following reaction shows anaerobic photoautotrophism in purple sulfur bacteria:

$$2CO_2 + H_2S + 2H_2O \rightarrow 2(CH_2O) + H_2SO_4$$

3.5.1.3 Microbes in Nitrogen Cycle

Nitrogen is an essential element present in protein and nucleic acid structure. Microorganisms play a critical role in nitrogen cycle through various processes such as nitrogen fixation, nitrate reduction, nitrification, denitrification, etc. (Fig. 3.7). The microbial processes limit the productivity of an ecosystem because nitrogen availability is a limiting factor for plant biomass production. Ammonification involves decomposition of organic nitrogen into ammonia. Both bacteria and archaea are capable of fixing atmospheric nitrogen through reduction into ammonium (Fig. 3.7). Nitrogenase is an oxygen-sensitive enzyme that catalyzes nitrogen fixation under low oxygen environment. N-fixation requires energy in form of ATP (16 mol) per mole of fixed nitrogen.

$$N_2 + 8H^+ + 8e^- + 16\ ATP = 2NH_3 + H_2 + 16\ ADP + 16Pi$$

The free-living microbes such as *Azotobacter*, *Burkholderia*, and *Clostridium* have an ability to fix nitrogen, and few of them form a symbiotic relationship with the rhizosphere of plants such as *Rhizobium*, *Mesorhizobium*, and *Frankia. Sophora* and *Clianthus* are native legumes and form a symbiotic relationship with *Mesorhizobium* or *Rhizobium leguminosarum*. The symbiotic rhizobia are able to fix nitrogen by two or three orders of magnitude higher than free-living soil bacteria.

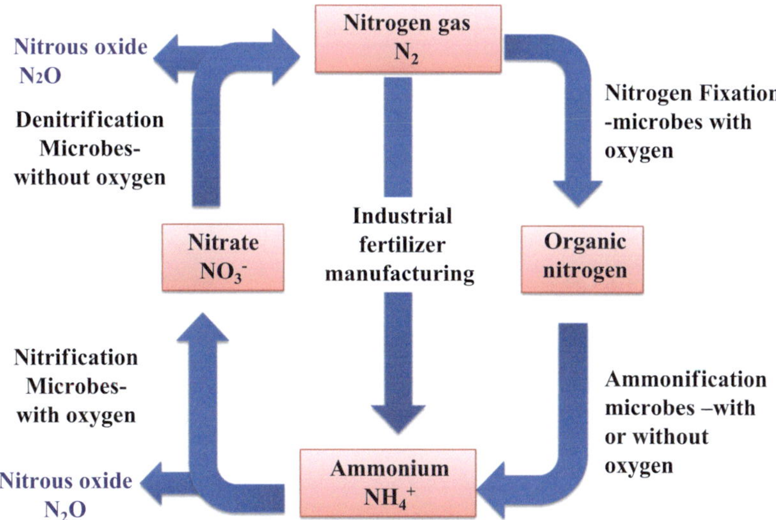

Fig. 3.7 Role of microbes in nitrogen cycle

Nitrification involves two steps: first, ammonia is oxidized to nitrite and then to nitrate. The oxidation of ammonia to nitrite is carried out by few soil bacteria, e.g., *Nitrosospira*, *Nitrosomonas*, *Crenarchaeum*, or *Nitrososphaera*, and thereafter nitrite is oxidized to nitrate by some bacteria, e.g., *Nitrobacter* and *Nitrospira* (Fig. 3.7). Nitrification also changes the ionic state of soil from positive to negative through oxidation of ammonia to nitrite and release of energy, which is used by nitrifying microbes to assimilate CO_2.

Denitrification involves sequential reduction of nitrate (NO_3^-), nitrite (NO_2^-), and nitric oxide (NO) to the greenhouse gas nitrous oxide (N_2O) or benign nitrogen gas (N_2). Since this process requires limiting oxygen, therefore, it occurs mostly in waterlogged areas that provide anaerobic environment. Nitrogen cycle involves denitrification process through which fixed nitrogen returns back to the atmosphere from soil and water in order to complete the nitrogen cycle. Denitrification involves a range of soil microbiota belonging to *Proteobacteria*, *Actinobacteria*, and *Firmicutes* and other soil eukaryotes. Most of the bacteria lack single or multiple enzymes involved in denitrification and known to be incomplete denitrifier, for example, most of the fungi and bacteria lack nitrous oxide reductase and thereby produce N_2O as a final product. Therefore, incomplete denitrification results into emission of greenhouse gases.

3.5.1.4 Microbes in Sulfur Cycle

Sulfur is an important component of a couple of vitamins and essential metabolites, and it is found in two amino acids, cysteine and methionine. In spite of its paucity in cells, it is an absolutely essential element for living systems. Like nitrogen and carbon, the microbes can transform sulfur from its most oxidized form (sulfate or SO_4) to its most reduced state (sulfide or H_2S) (Fig. 3.8). The sulfur cycle, in

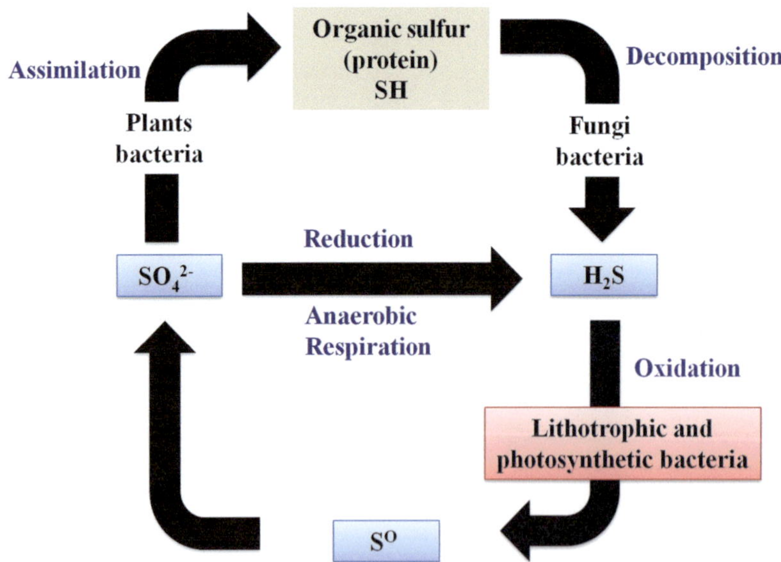

Fig. 3.8 Role of microbes in sulfur cycle

particular, involves some unique groups of prokaryotes. Two unrelated groups of prokaryotes oxidize H_2S to S and S to SO_4. The first is the anoxygenic photosynthetic purple and green sulfur bacteria that oxidize H_2S as a source of electrons for cyclic photophosphorylation. The second is the "colorless sulfur bacteria" (now a misnomer because the group contains many archaea) which oxidize H_2S and S as sources of energy. In either case, the organisms can usually mediate the complete oxidation of H_2S to SO_4.

$$H_2S \rightarrow S \rightarrow SO_4 \text{litho or phototrophic sulfur oxidation}$$

Sulfur-oxidizing prokaryotes are frequently thermophiles found in hot (volcanic) springs and near deep-sea thermal vents that are rich in H_2S. They may be acidophiles as well, because they acidify their own environment by the production of sulfuric acid. Since SO_4 and S may be used as electron acceptors for respiration, sulfate-reducing bacteria produce H_2S during a process of anaerobic respiration analogous to denitrification. The use of SO_4 as an electron acceptor is an obligatory process that takes place only in anaerobic environments. The process results in the distinctive odor of H_2S in anaerobic bogs, soils, and sediments where it occurs. Sulfur is assimilated by bacteria and plants as SO_4 for use and reduction to sulfide. Animals and bacteria can remove the sulfide group from proteins as a source of S during decomposition. These processes complete the sulfur cycle.

3.5.1.5 Microbes in Phosphorus Cycle

Phosphorus is a critical element of various building blocks such as nucleic acids, e.g., DNA and RNA, ADP, ATP, and phospholipids. Phosphorus is a rare element in the environment because of its tendency to precipitate in the presence of divalent and trivalent cations at neutral and alkaline pH.

Microorganisms (bacteria and fungi) mineralize organic phosphate in the form of phosphate esters into inorganic phosphate through a process driven by phosphatase enzymes (Fig. 3.9). Additionally, they also convert insoluble phosphorus into soluble form by a reaction with resulting byproducts such as organic acids. Mycorrhizal fungi help plants to overcome phosphorus limitation through its mobilization from insoluble mineral form by producing oxalate, e.g., various ectomycorrhizal basidiomycetous fungi express phosphate transporters in their extraradical hyphae during phosphorus deficiency in surrounding environments.

3.5.2 Contribution of Microbes in Recycling Wastes and Detoxification

Bacteria and fungi are able to biodegrade or detoxify substances through various ways; thereby, microbial processes are extensively used for bioremediation.

3.5.2.1 Biodegradation

Bioremediation/biotransformation is a waste management tool that involves naturally occurring organisms to remove or neutralize hazardous waste into less toxic or

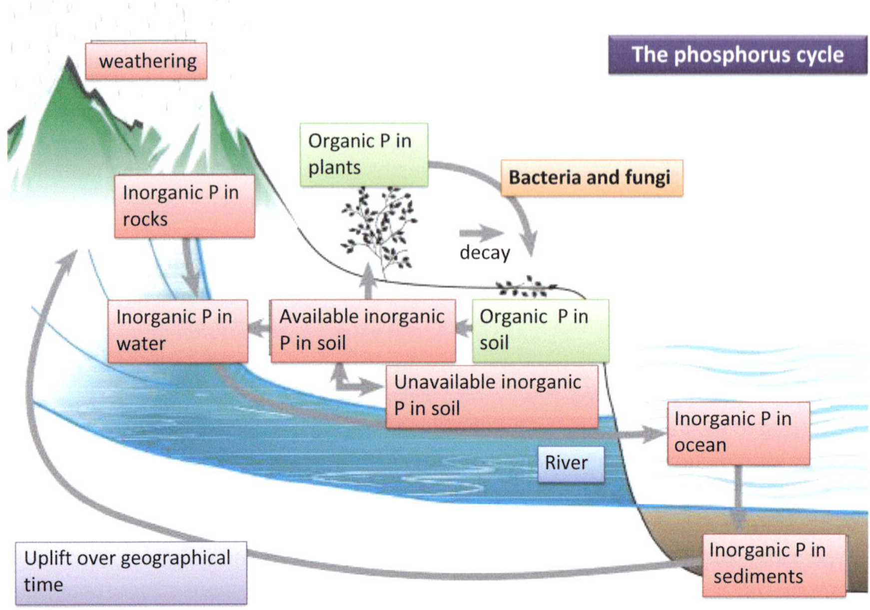

Fig. 3.9 Role of microbes in phosphorus cycle

nontoxic substances. The most commonly used microorganisms are *Flavobacterium*, *Arthrobacterium*, and *Azotobacter*. Bioremediation focuses on different sources and hence is called with different names:

$$\text{Plant} \rightarrow \text{Phytoremediation}$$
$$\text{Fungi} \rightarrow \text{Mycoremediation}$$

Biotechnological treatment of waste management involves use of microorganisms to detoxify air, water, and soil pollutants and carried out at lower temperature and pressure; therefore, it requires less energy than the conventional physicochemical treatment method. Depending upon the types of contaminants' site of monitoring and favorable environmental conditions, bioremediation may be carried out either *in situ* or *ex situ* (Table 3.1).

Biostimulation and bioaugmentation processes promote the rate of degradation of organic and inorganic pollutants (Fig. 3.10). These treatment technologies have been found eco-friendly and cost-effective means of pollution control leading to increased public acceptance and compliance with environmental legislation.

Heterotrophic microbes such as *Pseudomonas*, *Sphingomonas*, and *Mycobacterium* are known to be involved in oil degradation. *Pseudomonas* is one of well-studied bacteria capable of degrading alkanes, monoaromatics, naphthalene, and phenanthrene under aerobic conditions. The hydrocarbon-degrading bacteria are dominant in soil contaminated with oil; however, higher concentration of hydrocarbons may deplete available nitrogen and phosphorus in that area since these elements are assimilated during biodegradation.

Table 3.1 Bioremediation and its types

Type	Example	Benefits	Limitations	Factors involved
In situ	Bioventing	Cost efficient, noninvasive, relatively passive, natural attenuation, treats soil and water	Environmental constraints, extended treatment time difficulties	Biodegradation abilities of indigenous microorganism, presence of metals and other inorganics, biodegradability of pollutants, chemical solubility, geological factor and pollutant distribution
	Biosparging			
	Biostimulation			
	Bioaugmentation			
	Rhizofiltration			
Ex situ	Landforming	Cost-effective, less time	Space requirement, extended treatment time, abiotic loss, mass transfer problem, bioavailability limitation	
	Composting			
	Biopiles			
Bioreactor	Slurry reactor	Rapid degradation, optimized environmental parameters, enhanced mass transfer, effective use of intoxicant and surfactants	Soil require excavation, relatively high cost	As above, toxicity of amendment, toxic concentration of contaminants
	Aqueous reactor			

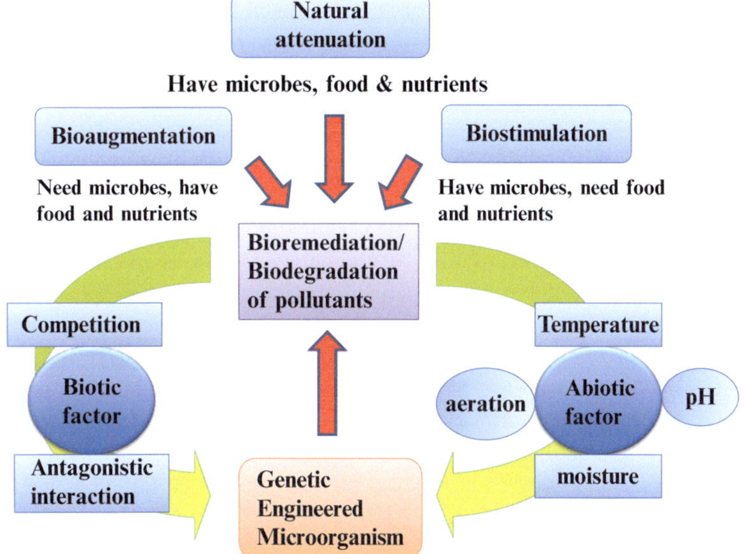

Fig. 3.10 Overview of bioremediation

Microbes (bacteria and fungi) are able to degrade a range of biodegradable pesticides such as atrazine, which is degraded by a bacterium, e.g., *Arthrobacter nicotinovorans*, and related derivatives such as simazine, propazine, and cyanazine (Aislabie et al. 2005). Non-biodegradable pesticides, e.g., DDT (dichlorodiphenyltrichloroethane), are not readily degraded and still persist in the soil. Some fungi having ability to degrade lignin, such as *Phanerochaete chrysosporium*, are able to degrade various contaminants such as pentachlorophenol and dioxin, and the best example are *Zygomycetes* that degraded various contaminants during wood-treating operation in Whakatane (Thwaites et al. 2006).

Biodegradation of a contaminant depends upon its chemical structure and physical state since various contaminants, e.g., oil are readily degradable, but synthetic contaminants, e.g., DDT and aldrin, are nondegradable and persist in the environment. The ability for degradation also depends upon rare and novel structures and water solubility since less soluble compounds are difficult to degrade. Additionally, poorly water-soluble or hydrophobic contaminants also readily bind to clay particles and, therefore, are easily available to microbes present in soil. These soil microbes utilize these contaminants as energy source, present at higher concentration, and these could be toxic for them, resulting into slow biodegradation. Biodegradation also involves a contact between contaminants and microbes. Some microbes, e.g., chemotactic bacteria and fungi, have an ability to sense and move toward them.

3.5.2.2 Metal Detoxification

The microbial ability to withstand metal toxicity and their physiological adaptation to metal stress has an important significance. Indeed, the expression and activity of proteins involved in metal uptake are crucial for metal resistance, and different bacteria adapt distinct complements of these systems. Bacteria have evolved some regulatory mechanisms to control membrane transporter activity that take up metals, and some of these activities are determined by regulators that bind to metal ions with femtomolar activities. In response to metal exposure, some microorganisms upregulate the expression of extracellular polymers or siderophores containing functional groups that are capable of coordinating metal ions and may be subject to reduced uptake or increased efflux by membrane transporters upon binding to toxic metals. Many microbes have ability to precipitate metals as metal oxides, metal sulfides, metal protein aggregates, or metal crystals forming particulates in close association with cytoplasmic membranes. In addition, some microbes can use cytoplasmic proteins, e.g., bacterioferritin and metallothioneins, to bind, sequester, or store metals (Carrondo 2003). Some microbes use metals in specific redox and covalent reactions that convert toxic metal species into less toxic forms either oxidative or reductive metabolism (Fig. 3.11).

Few microbes have evolved detoxification mechanisms during their exposure to heavy metals, e.g., copper, mercury, lead, zinc, cadmium, etc. One of the known examples is cadmium accumulation in agricultural soils in New Zealand due to extensive use of superphosphate fertilizer (Loganathan et al. 2003). Due to metal toxicity, microbes have evolved few defense mechanisms such as metal

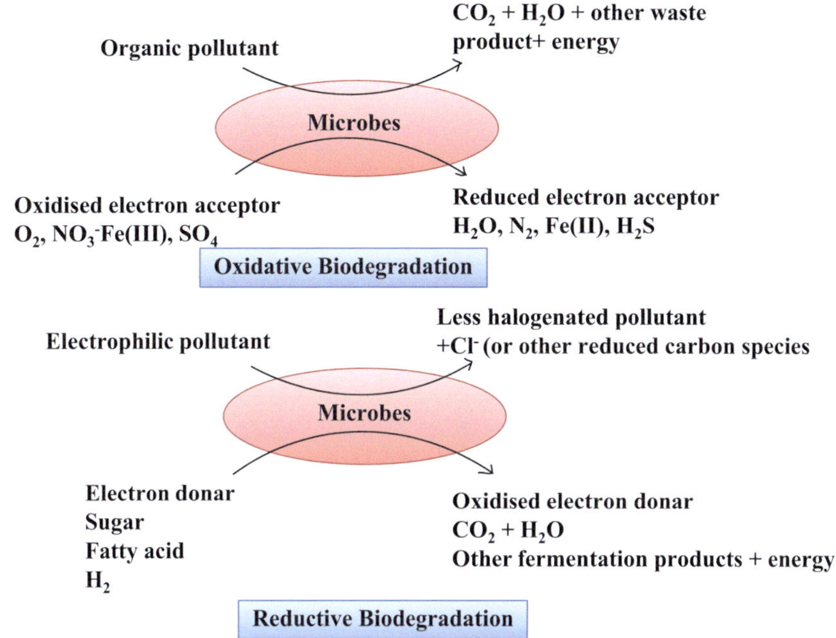

Fig. 3.11 Oxidative and reductive biodegradation of environmental pollutants

sequestration, detoxification, and efflux of ions. Bacteria sequester heavy metals through their binding with cell membrane, cell wall, and extracellular polysaccharides (Harrison et al. 2007). Microbes may also detoxify toxic metals through reduction using various cellular enzymes, e.g., mercury oxidase reduces Hg^{+2} to Hg, which has a low evaporation point and, therefore, diffuses from cell (Nies 1999). Few gram-negative bacteria, e.g., *Alcaligenes eutrophus*, have evolved a mechanism to fight with metal toxicity by expelling them from cytoplasm to external environment through cation/proton antiporter, present at cell membrane (Silver and Phung 1996). Nowadays, the microbial ability to transform heavy metals is being extensively used as a tool for bioremediation.

3.6 Microbial Enzymes in Bioremediation

Microbes are ultimate garbage disposal of nature that clean up or transform contaminants into non-hazardous or less hazardous substances. Various microorganisms such as bacteria and fungi detoxify hazardous substances by secreting various enzymes (Table 3.2), also involved in various industrial applications.

Table 3.2 Industrial applications of microbial enzymes

S. No.	Enzyme	Subclass	Substrate	Reaction(s)	Applications
1.	Oxygenase	Monooxygenase	Alkanes, steroids, fatty acids, and aromatic compounds	Use substrates as a reducing agent by incorporating single oxygen atom, i.e., desulfurization, dehalogenation, denitrification, ammonification, hydroxylation, etc.	Bioremediation, protein engineering, synthetic chemistry, etc.
		Dioxygenase	Aromatic compounds	Incorporating two oxygen atoms to the substrate and resulting into aliphatic products	Bioremediation, pharmaceutical industry, synthetic chemistry, etc.
2.	Laccase	N/A	Ortho- and paradiphenols, aminophenols, polyphenols, polyamines, lignins, and aryldiamines	Oxidation, decarboxylation, and demethylation of substrates	Bioremediation, food and paper industry, textile industry, cosmetics, synthetic chemistry, etc.
3.	Peroxidase	Lignin peroxidase	Halogenated phenolic, polycyclic, aromatic, and other aromatic compounds	Substrate oxidation using H_2O_2 as a co-substrate and mediator like veratryl alcohol	Bioremediation, food and paper industry, textile industry, pharmaceutical industry, etc.

		Manganese peroxidase	Lignin and other phenolic compounds	Oxidizes Mn^{2+} into Mn^{3+} which results into an Mn^{3+} chelateoxalate that in turn oxidizes the phenolic substrates	Bioremediation, food and paper industry, textile industry, pharmaceutical industry, etc.
		Versatile peroxidase	Methoxybenzenes and phenolic aromatic compounds	Catalyzes the electron transfer from an oxidizable substrate with reduction of complex I and II intermediates	Bioremediation and industrial biocatalyst
4.	Hydrolase	Lipase	Organic pollutants, e.g., oil spill	Hydrolyzes triglycerides to glycerols and free-fatty acids	Detergent production, baking and paper industry, personal products, etc.
		Cellulase	Cellulosic substances	Hydrolyzes substrate to simple carbohydrates	Bioremediation, paper and textile industry, detergent production
		Protease	Proteins	Hydrolyzes peptide bonds	Leather, laundry, biocatalyst, bioremediation, etc.

3.6.1 Oxidoreductases

Microbial oxidoreductases detoxify toxic xenobiotics such as phenolic compounds, produced from the decomposition of lignin through polymerization, copolymerization with other substances, or binding to humic substances. Most of the metal-reducing bacteria reduce the radioactive metals into insoluble forms that appear as a precipitant with the help of an intermediate electron donor (Leung 2004). The paper and pulp industry produces chlorinated phenolic compounds upon partial degradation of lignin during pulp bleaching process. These recalcitrant wastes are removed by the action of various fungal extracellular oxidoreductase enzymes such as laccase, manganese peroxidase, and lignin peroxidase that are released from fungal mycelium into their neighborhood environment. The plants belonging to families such as Fabaceae, Gramineae, and Solanaceae release oxidoreductases, which are recruited in the oxidative degradation of certain soil constituents.

3.6.2 Oxygenases

Oxygenases recruit oxidation of reduced substrates through oxygen transfer from molecular oxygen (O_2) using various co-substrates (FAD, NADH, NADPH). Oxygenases fall into two major categories on the basis of number of oxygen atoms used during oxygenation: monooxygenases and dioxygenases. Oxygenases mediate dehalogenation of halogenated pollutants, methanes, ethanes, and ethylenes through association with multifunctional enzymes (Fetzner and Lingens 1994).

Monooxygenases play an important role in bioremediation process as a biocatalyst due to their high selectivity and stereoselectivity on the wide range of substrates. Most of the monooxygenases are known to have cofactors but few act without them. Monooxygenases incorporate single atom of oxygen molecule into the substrate and are further classified into two subgroups based on the presence of cofactor. P_{450} monooxygenases are heme-containing oxygenases, while flavin-dependent monooxygenases consist of flavin as a prosthetic group and require NADP or NADPH as a coenzyme. Monooxygenases catalyze desulfurization, dehalogenation, denitrification, ammonification, hydroxylation, and biotransformation of various aromatic and aliphatic compounds. Methane monooxygenase is the best-characterized monooxygenase involved in the degradation of various hydrocarbons. Monooxygenases exhibit differential activity in the presence or absence of oxygen. Monooxygenases catalyze oxidative dehalogenation under oxygen-rich conditions, whereas under low oxygen conditions, they catalyze reductive chlorination.

Microbial dioxygenases primarily oxidize aromatic compounds and are involved in bioremediation process. Aromatic hydrocarbon dioxygenases belong to Rieske non-heme iron oxygenase family and are involved in oxygenation of various substrates. An example is naphthalene dioxygenase having Rieske (2Fe-2S) cluster and mononuclear iron molecule in each alpha-subunit (Dua et al. 2002). One of the best nature's strategies for bioremediation is the catechol dioxygenase found in soil bacteria which is involved in transformation and degradation of aromatic molecules into aliphatic products.

3.6.3 Laccases

Laccases are the members of multicopper oxidase family, produced by certain plants, fungi, insects, and bacteria and catalyze oxidation of a wide range of phenolic and aromatic substrates. Most of the microbes produce intra- and extracellular laccases, catalyzing the oxidation of aminophenols, polyphenols, ortho- or paradiphenols, lignins, aryl diamines, etc. These enzymes act not only by oxidizing phenolic and methoxy-phenoic acids but also through their decarboxylation and demethylation. Laccases are also involved in depolymerization of lignin resulting into various phenols that are utilized by microorganisms.

3.6.4 Peroxidases

Microbial peroxidases catalyze oxidation of lignin and other phenolic compounds in the presence of hydrogen peroxide (H_2O_2) and a mediator. Among all microbial peroxidases, lignin peroxidase (LiP), manganese-dependent peroxidase (MnP), and versatile peroxidase (VP) have shown potent activity to degrade toxic substances.

Lignin peroxidases are heme-containing proteins secreted by white rot fungi during secondary metabolism and play an important role in degradation of lignin from plant cell wall. Manganese-dependent peroxidase is an extracellular heme-containing enzyme secreted by basidiomycete fungi. Mn^{2+} stimulates MnP production and itself oxidizes to Mn^{3+} by MnP. This results into Mn^{3+} chelateoxalate, which in turn oxidizes various phenolic substances. Versatile peroxidases directly oxidize Mn^{+2}, methoxy benzenes, phenolic aromatic substrates similar to MnP and LiP. VP exhibits broad substrate specificity and is able to oxidize substrates even in the absence of manganese as compared to other peroxidases. Hence, bioremediation and biotechnological applications for industrial processing need efficient VP production.

3.6.5 Hydrolases

Microbial hydrolases play an important role in bioremediation process and act by disrupting chemical bonds in toxic compounds and thereby reduce their toxicity up to some extent. These enzymes are readily available and do not need any cofactor for stereoselectivity. Some extracellular hydrolases such as amylases, proteases, lipases, DNases, and xylanases exhibit potential role in various sectors, e.g., food industry, biomedical sciences, and chemical industries. Other hydrolases, e.g., hemicellulases, cellulases, and glycosidases, have shown more importance because they are involved in biomass degradation.

3.6.6 Lipases

Lipases are capable of degrading lipids (e.g., triglycerides) derived from a range of microbes, plants, and animals, into glycerol and free-fatty acids. Recent reports suggest a close association of lipase with organic pollutants present in the soil, and its activity results into reduced hydrocarbon content in the contaminated soil. Microbial lipases are extensively used in industries since these enzymes catalyze various chemical reactions such as hydrolysis, esterification, alcoholysis, aminolysis, etc. Lipase activity is an important indicator or parameter for testing hydrocarbon content present in the soil. Lipases are widely used in pharmaceutical, food, chemical, cosmetic, and paper industries, but the cost of their production limits their potent application in the industries.

3.6.7 Cellulases

Cellulases convert cellulosic waste materials to glucose and have been implicated in intense research for bioremediation processes. Some bacteria and fungi express extracellular cellulases, hemicellulases, and pectinases at very low levels, e.g., *Bacillus* strains produce alkaline cellulases and *Trichoderma* and *Humicola* fungi produce neutral and acidic cellulases. Cellulases are widely used in paper and pulp industry for ink removal during paper recycling, in ethanol production from cellulosic biomass, and in brewing industry to enhance juice release from fruit pulp.

3.6.8 Proteases

Proteases hydrolyze the proteinaceous substances in the atmosphere resulting from animal death, shedding, and molting of appendages, as a by-product of poultry, fishery, and leather industries. Proteases are divided into two groups: exopeptidases and endopeptidases. Exopeptidases are further classified into aminopeptidases and carboxypeptidases depending on their site of cleavage either at N- or C-terminus of a peptide chain. Endopeptidases are also grouped based on the position of active site such as serine endopeptidases, cysteine endopeptidases, aspartic endopeptidases, and metallopeptidases.

Microbial proteases have been employed in cheese and detergent manufacturing industries since many years. Some proteases have been used in production of non-calorific artificial sweetener, e.g., dipeptide aspartame. Alkaline proteases are extensively used in leather industry for removal of hairs and parts on animal skin. Some proteases are also used in combination with broad-spectrum antibiotics in the treatment of wounds, cuts, and burns.

3.7 Contribution of Microbes in Food Web Maintenance

Food chains show the relationships between producers, consumers, and decomposers, showing who eats whom with arrows. The arrows show the movement of energy through the food chain. For example, in the food chain shown below, the small fish (silverside) gets its energy by eating the plankton and the large fish (bluefish) gets its energy by eating the small fish. Finally, the bacterium eats the fish after it dies, getting its energy from the large fish. The bacterium also returns nutrients back to the environment for use by the phytoplankton.

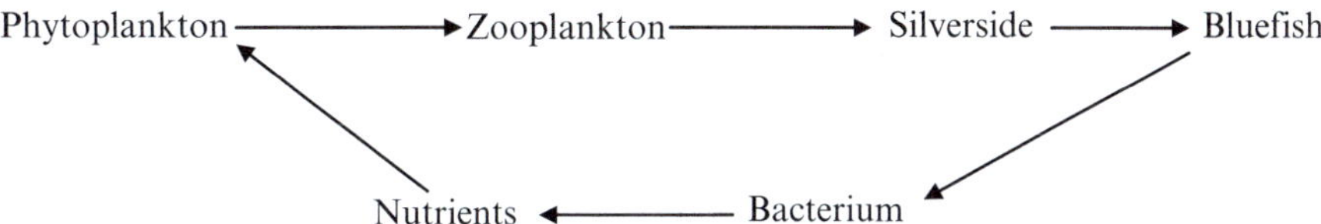

Thus the food chain becomes a complete circle. Animals may eat more than one type of food. They may eat many different types of plants or many different animals. This makes everything more complicated, and the food chain becomes a food web. A food web is made up of interconnected food chains. Most communities include various populations of producer organisms that are eaten by a number of consumer populations (Fig. 3.12). The green crab, for example, is a consumer as well as a

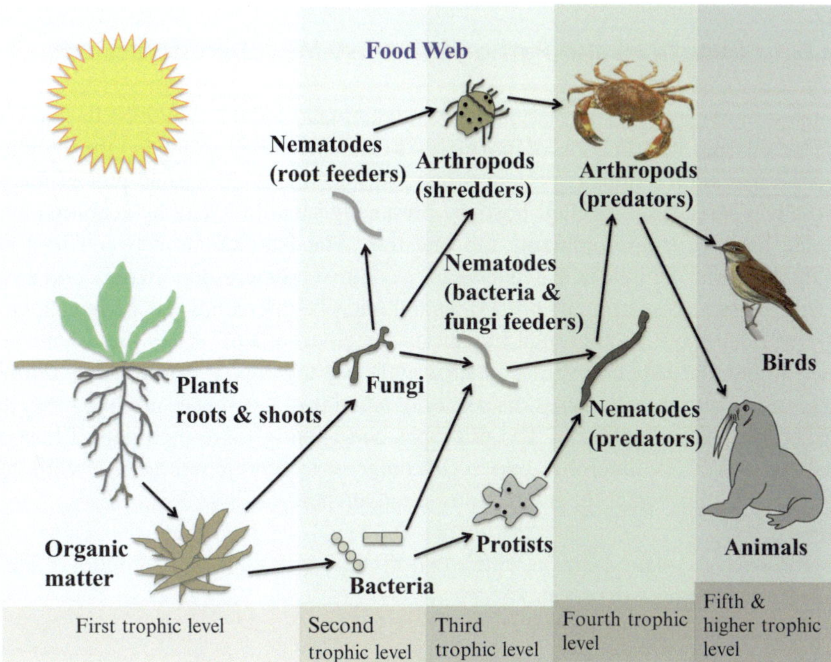

Fig. 3.12 Food web: an interaction between different species in ecosystem

decomposer. The crab will eat dead things or living things if it can catch them. A secondary consumer may also eat a number of primary consumers or producers. This nonlinear set of interactions which shows the complex flow of energy in nature is more easily visualized. In a food web, nutrients are recycled by decomposers in the end. Animals like shrimp and crabs can break the materials down to detritus. Then bacteria reduce the detritus to nutrients. Decomposers work at every level, setting free nutrients that form an essential part of food chain. Large number of primary producers such as bacteria and algae can maintain the base of pyramid to balance the biomass in trophic levels. In a food chain, energy is lost in each step of the chain in two forms, first by the organism producing heat and doing work and second by the food that is not completely digested or absorbed. Therefore, the food web depends on the constant supply of energy from producer and nutrients that are recycled by the decomposition of organism.

As food is passed along the food chain, only about 10 % of the energy is transferred to the next level. For example, 10 % of the energy phytoplankton received from the sun can be used by zooplankton at the next level. From one level to the next, about 90 % of the energy used by the previous level is lost. This means that there has to be a lot more organisms at the lower levels than at the upper levels. The number of organisms at each level makes a pyramid shape and is called a food pyramid. To better understand this energy loss, it is helpful to look at a food pyramid. Thus food web may create the capacity of coexistence which was responsible for species evolution and maintenance of microbial diversity.

3.8 Beneficial and Pathogenic Host-Microbe Interactions

Host-microbe symbiosis exists in almost all animals, and the symbiotic bacteria can be profitable, harmful, or of no effect to the host. For example, the harmless *Escherichia coli* strains commonly found in intestine are a normal part of the gut flora and can advantage their hosts by producing vitamin K and by keeping pathogenic bacteria from colonizing the intestine. The interactions between host and microbe form complicated networks. By contrast, some others like *E. coli* strain O26 can cause disease in its host. Host and microbe interaction can be beneficial or harmful but have an important impact on the environment. When the organism is able to produce disease even in an apparently healthy host, it is referred as primary pathogen, but when it causes disease only when host's defenses are impaired, it is called secondary pathogen. The microbes consistently associated with a host are called flora. These microbes have a full range of symbiotic interactions with their hosts. Some host-microbe interactions are given below (Table 3.3).

Symbiosis A relationship in which two dissimilar organisms (symbionts) live in close association with one another.

Table 3.3 Types of host-microbe interactions

Types of interaction	Response	
	Species A	Species B
Neutralism	0	0
Mutualism	+	+
Commensalism	+	0
Competition	–	–
Amensalism	–	0
Predation	+	–
Parasitism	+	–
Parasitoidism	+	–

0 (No effect), – (harmful effect), + (beneficial effect)

Table 3.4 List of some useful plant-associated bacteria

Taxon	Functions
Agrobacterium radiobacter K84 and *K1026*	Biological control
Agrobacterium spp. *M4*	Source for a drug for cholesterol degradation
Agrobacterium radiobacter J14	Biodegradation of atrazine (herbicide)
Agrobacterium tumefaciens	Plasmid vector for plant transformation
Erwinia amylovora	Source of harpin, an elicitor of disease resistance in plants
Xanthomonas campestris pv. *campestris*	Xanthan gum used in food, cosmetics, etc.
Other plant-associated bacteria	Restriction endonucleases and other enzymes

Commensalism A relationship between two species in which one is benefited and the other is not affected, neither negatively nor positively.

Mutualism Mutually beneficial relationship between two species.

Parasitism A relationship between two species in which one benefits (parasite) from the other (host); it usually involves some detriment to the host.

3.9 Beneficial Plant-Microbial Interactions

Beneficial plant-microbial interactions are categorized into three parts. First, microbes either through direct interaction with the plants or indirectly through influencing biotic or abiotic parameters of soil supply minerals to support plant growth. Second, some microbes inhibit the growth and activity of plant pathogens to promote plant growth. Third, few microbes produce phytohormones that stimulate the growth of plants (Table 3.4). Additionally, some saprophytic microbes establish neutral interactions with plants without directly benefiting or harming them. These microbes enrich soil nutrient levels by decomposing organic components and thereby influence their productivity and growth. The plant rhizosphere is the major soil ecological environment for plant-microbe interactions. In rhizosphere different microbes colonize around growing roots, which may either result in symbiotic, neutralistic, or parasitic interactions depending upon nutritional status of soil, soil environment, plant defense mechanism, and the type of microbial proliferation in the rhizosphere zone. The microbial community living in the rhizosphere zone benefits plant by promoting their growth and are also known as plant growth-promoting rhizobacteria (PGPR). These PGPRs include various bacteria, e.g., *Azospirillum*, *Bacillus*, *Pseudomonas*, *Rhizobium*, *Serratia*, *Stenotrophomonas*, and *Streptomyces* and fungi, e.g., *Ampelomyces*, *Coniothyrium*, *Trichoderma*, etc. These PGPRs support plant growth by increasing soil fertility, secreting phytohormones, and protecting them from various diseases by producing antibiotics and inducing plant defense system. *Pseudomonas* and *Bacillus* are well-studied PGPRs and dominating bacteria, present in rhizosphere. PGPR bacteria have following roles to play:

1. PGPR bacteria suppress the growth of pathogenic microbes by lowering iron availability through secretion of low molecular weight siderophores.
2. PGPR can reduce the activity of pathogenic microorganisms by activating the plant to induced systemic resistance (ISR) or systemic acquired resistance (SAR). These plant resistance systems are induced by signaling molecules, e.g., jasmonic acid, ethylene, and salicylic acid.
3. PGPR bacteria enhance the production of phytohormones (e.g., auxin, cytokinin, gibberellins), besides having nitrogen-fixing ability. These phytohormones play a critical role in root initiation, cell division, and cell growth. Auxin is most prominently secreted by *Azospirillum* spp.
4. Several commercial PGPRs support plant growth by several means such as bioprotectants, biostimulants, and biofertilizers.
5. PGPR bacteria, e.g., *Azospirillum*, also provide nutrition to plants by liberating phosphorous from organic compounds and thereby support plant growth.

Root-colonizing microbes are guided by chemical plant signal overlap. For example, plant flavonoids act as chemoattractants for nitrogen-fixing bacteria, mobile zoospores, and symbiotic fungi. During interaction of microbes with plant epidermis, plants secrete signal molecules in the form of flavonoids and flavones in the rhizosphere that drive the differentiation between pathogenic, associative, symbiotic, or neutralistic adaptation of microbes with the plants.

In legume *Rhizobium* symbiosis, the rod-shaped soil bacterium *Rhizobium* induces nitrogen-fixing nodules on the roots of leguminous plants that convert approximately 80% of chemically inert nitrogen present in the atmosphere into ammonia through reduction process using bacterial enzyme nitrogenase in nitrogen-deficient condition (Zahran 1999). During this symbiotic relationship, plant root releases elicitors of *nod gene* expression, bacteria releases Nod factor, and plant root demonstrates ion flux, expresses nodulin proteins, and undergoes nodule morphogenesis. The plant supports metabolism of bacterial endosymbionts by providing a micro-aerobic environment for effective functioning of the oxygen-sensitive nitrogenase, encoded by bacterial *nif genes* and carbohydrates. In return, bacteria fix atmospheric nitrogen for plants to meet their biological needs. The other diazotrophs such as *Azotobacter*, *Azospirillum*, as well as rhizosphere fungi and bacteria especially *Pseudomonas* and *Bacillus* also interact with *Rhizobium* affecting nodulation and nitrogen fixation and help in creating a beneficiary region where interacting microbes get benefit from additional nutrient resources. Therefore, a mutualistic relationship exists between *Azotobacter* and *Azospirillum* where both interact with *Rhizobium* to improve plant growth, and these beneficiary effects are mainly attributed to improvements in root development, increase in water and mineral uptake by roots, the displacement of fungi and pathogenic bacteria, and, to lesser extent, biological nitrogen fixation (Heath and Tiffin 2009). Nodule formation involves expression of rhizobia specific genes: bacterial genes (*nod* genes) and plant genes (*nodulin genes*). The component, enzymes, and their function are leghemoglobin (protection against oxygen), nitrogenase (N_2 fixation), glutamine synthetase (N-detoxification), and uricase (N-detoxification).

A mycorrhiza is a symbiotic association of a fungus and roots with vesicular plant. In a mycorrhizal association, the fungus colonizes the host plant roots either intracellularly as in arbuscular mycorrhizal fungi (AMF or AM) or extracellularly as in ectomycorrhizal fungi (Sikes 2010). In ectomycorrhiza, a fungus does not enter into plant cell, whereas it colonizes the outer cell layers and forms a hartig net. Hartig net is a soil network that connects several organisms and protects against pathogenic fungi and soil bacteria. In this association, fungi form a net around the roots (hairs) to extend their access to soil nutrients. Ectomycorrhiza promotes growth of tree seedlings and germination of seeds. This mutualistic association provides the fungus with relatively constant and direct access to carbohydrates such as glucose and sucrose. The carbohydrates are translocated from their source (usually leaves) to root tissue and on to the plant's fungal partners. In return, the plant gains the benefits of the mycelium's higher absorptive capacity for water and mineral nutrients due to the comparatively large surface area of the mycelium/root ratio, thus improving the plant's mineral absorption capabilities. Additionally, mycorrhizal plants are often more resistant to disease caused by soilborne pathogens and metal toxicity. The mycorrhizal fungi, especially the vesicular arbuscular mycorrhizae (VAM) belonging to the *Zygomycetes* class, play an important role in phosphorous mobilization in soils having a relatively low level of available phosphorous for the better growth of cereals as well as legumes. The fundamental characteristics of fungal species that form VAM are:

- They all belong to *Glomales* (*Zygomycetes*).
- Initiation of interaction through germinating spores on plant plasma membrane.
- Hyphae form appressorium (attachment site).
- Formation of an extracellular hyphal system in the apoplast.
- Formation of haustorium: penetration into plant cell (intracellular arbuscules).
- Enlargement of interaction surface.
- Lifetime of arbuscle: a few days.

Extracellular hyphae of the fungal species collect nutrients and transfer them to the fungus. The association of mycorrhizal fungi with legumes has a great impact on root and shoot development and phosphorous uptake resulting in the enhancement of nodulation and nitrogen fixation. Benefited fungi activate the defense genes that encode defensin proteins and may produce the reactive oxygen species through NADH oxidase to protect crops against pathogenic microbes. The yield of crop plants may increase four times higher with mycorrhizal fungi.

3.10 Pathogenic Microbes

Infectious diseases are caused by pathogenic microbes that attack and obtain their nutrition from the host they infect. A pathogen is a microorganism that has the potential to cause disease. An infection is the invasion and multiplication of pathogenic microbes in an individual or population. An infection does not always result in a disease. When an infection causes damage to the individual's vital functions in system, it leads to a disease. Ability to cause disease is pathogenicity, whereas the degree of pathogenicity is known as virulence. To cause disease, a pathogen must gain an access to the host, adhere to host tissues, penetrate or evade host defense system, and damage the host either directly or indirectly by accumulation of microbial wastes.

Numerous fungi, bacteria, viruses, and nematodes are pathogenic in nature and caused many plant and animal diseases (Tables 3.4 and 3.5). Disease triangle is one of the first concepts published by Stevens in 1960 and recognized the interaction among the host, pathogen, and environment (Fig. 3.13).

This triangular relationship is unique to phytopathology in comparison to veterinary and medical sciences because terrestrial plants possess little thermal storage capacity, and their immobility prevents escape from inhospitable environment. The sophisticated immune system found in mammals is absent in plant, and this places an emphasis on the earth's genetic constitution. Finally the predominance in the phytopathology of fungi, which are also highly dependent on environment, may have contributed to the development of this paradigm. Any disease caused by a pathogen is a chain of events involved in the development of pathogen and the effects of disease on the host. All infectious disease-causing agents go through a disease cycle. A generalized disease cycle is illustrated (Fig. 3.14).

3.11 Mechanism of Pathogenesis

Mechanisms of pathogenesis determine the relationship between virulence and components of parasite fitness, such as transmission to new hosts and survival within the host. By making explicit how the biochemical mechanisms of pathogenesis set the relations between parasite fitness and virulence, we expand the

Table 3.5 List of human pathogenic microbes

Microbes	Name of disease	Parasite name
Bacteria	Anthrax	*Bacillus anthracis*
	Abscess	*Bacteroides* spp.
	Whooping cough	*Bordetella pertussis*
	Lyme disease	*Borrelia burgdorferi*
	Campylobacter enteritis	*Campylobacter* spp.
	Trachoma, conjunctivitis, respiratory infection	*Chlamydia spp.*
	Botulism, tetanus, gangrene	*Clostridium spp.*
	Diphtheria	*Corynebacterium diphtheriae*
	Gastroenteritis, urinary infection, meningitis	*Escherichia coli*
	Vaginitis, vulvitis	*Gardnerella* spp.
	Meningitis, pneumonia	*Haemophilus influenzae*
	Peptic ulcer	*Helicobacter pylori*
	Pneumonia	*Klebsiella pneumoniae*
	Pontiac fever	*Legionella* spp.
	Conjunctivitis	*Moraxella lacunata*
	Tuberculosis, leprosy	*Mycobacterium* spp.
	Fatal pneumonia	*Mycoplasma pneumoniae*
	Gonorrhea	*Neisseria gonorrhoeae*
	Pasteurellosis	*Pasteurella* spp.
	Dermatitis, enteritis	*Pseudomonas aeruginosa*
	Rocky mountain spotted fever	*Rickettsia* spp.
	Food poisoning or typhoid fever	*Salmonella* spp.
	Dysentery	*Shigella* spp.
	Wound infection and food poisoning	*Staphylococcus aureus*
	Rheumatic fever	*Streptococcus pyogenes*
	Syphilis	*Treponema pallidum*
	Cholera	*Vibrio cholerae*
	Plague	*Yersinia pestis*
Fungi	*Aspergillosis*	*Aspergillus niger*
	Blastomycosis	*Blastomyces dermatitidis*
	Candidiasis	*Candida albicans*
	Coccidioidomycosis	*Coccidioides* spp.
	C. neoformans infection	*Cryptococcus neoformans*
	C. gattii infection	*Cryptococcus gattii*
	Fungal eye infection	*Cryptococcus neoformans*
	Histoplasmosis	*Histoplasma capsulatum*
	Mucomycosis	*Mucoromycotina* spp.
	Pneumocystis pneumonia	*Pneumocystis jirovecii*
	Ringworm	*Trichophyton*, *Microsporum*, and *Epidermophyton*
	Sporotrichosis	*Sporothrix schenckii*

(continued)

Table 3.5 (continued)

Microbes	Name of disease	Parasite name
Virus	Influenza (Flu)	Influenza (flu) virus
	Small pox	Variola virus
	Chicken pox	Varicella zoster virus
	Mumps	Paramyxovirus
	Measles	Rubeola virus
	German measles	Rubella virus
	Yellow virus	*Flavivirus*
	Severe acute respiratory syndrome	*Coronavirus*
	Swine flu	Triple-reassorted flu virus A (H_1N_1)
	Genital herpes	Herpes simplex virus
	Genital warts	Human papilloma virus
Protozoa	Anemia	*Nector americans*
	Hay fever	*Enterobius vermicularis*
	Trichinosis	*Trichinella spiralis*
	Cysts	*Trichuris trichiura*
	Onchocerciasis	*Wuchereria bancrofti*
	Blindness	*Onchocerca volvulus*
	Ulcer	*Dracunculus medinensis*
	Ascariasis (intestinal infection)	*Ascaris lumbricoides*

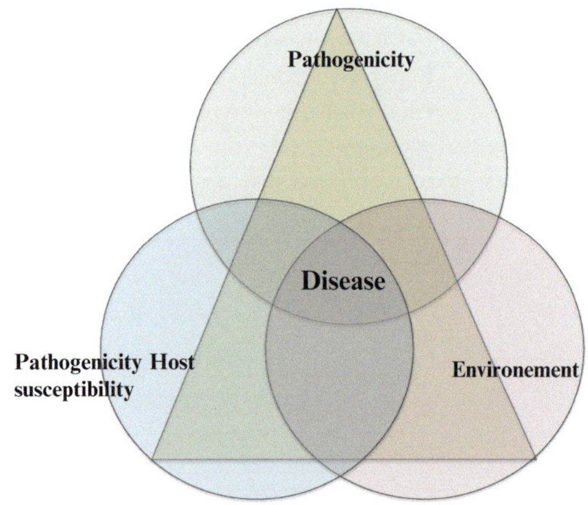

Fig. 3.13 Disease triangle, illustrating the factors required for disease development

conceptual framework of parasite virulence to encompass many cases that are not addressed by the prior theories of virulence. Human pathogens may enter into the host by different routes such as the mucous membranes, skin, and parental route and cause many diseases (Table 3.4). Most microbes must enter through their preferred portal of entry in order to cause disease, whereas some can cause disease from many routes of entry. The likelihood of disease increases as the number of invading pathogens increases. Infectious dose (ID_{50}) and lethal dose (LD_{50}) are used to determine

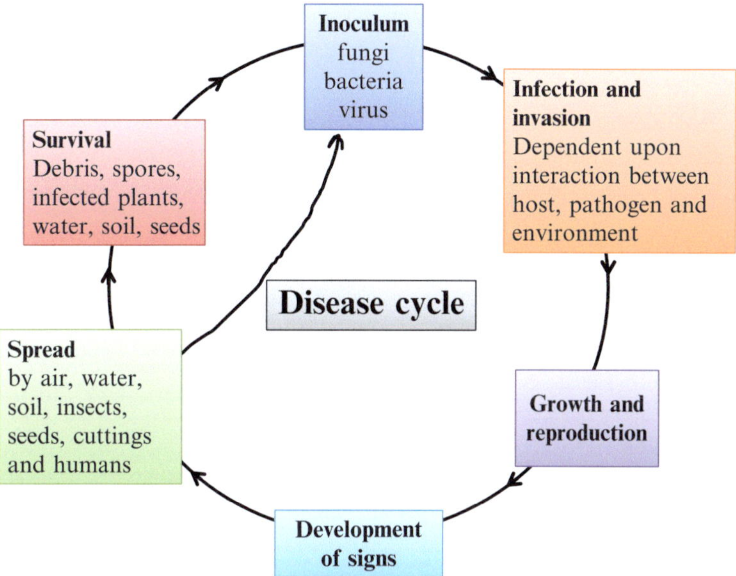

Fig. 3.14 Disease cycle

the number of microbes. ID_{50} is the number of microbes required to produce infection in 50 % of the population, whereas LD_{50} is the amount of toxin or pathogen necessary to kill 50 % of the population necessary in a particular time frame.

The role of microbes in plant diseases has been recorded as far back 700 BC (Table 3.5). Whereas some bacteria cause hormone-based distortion of leaves and shoots known as fasciations or crown gall, a proliferation of plant cells leading to swelling at the intersection of stem and soil and on roots happen as well. However, symptoms may vary with photoperiod, variety of plants, temperature, humidity, and infective dose. Some plant pathogenic microbes cause severe economically damaging disease such as spots, mosaic patterns, or pustules on leaves and fruits, smelly tuber rots, etc. (Table 3.5). Most of the plant pathogens induce a hypersensitive reaction (HR) in nonhosts or indicator plants, and this HR acts as a plant defense mechanism elicited by the presence of a pathogen in nonhost tissue. Although most plant disease is caused by fungi (85 %), only a small fraction of fungi in the environment cause disease. Plant pathogenic fungi have played a powerful role in human and natural history. Fungal pathogens produce toxins causing allergies, e.g., mushrooms produce hallucinogenic mycotoxins (black molds). Protozoa can grow inside host cells causing lysis. They may use host cells, food source, and microbial wastes (Table 3.6).

In sum, it may be said that microbes play a significant role to maintain our environmental sustainability by maintaining biogeochemical and nutrient cycles. Microbes protect our environment from hazardous compounds by using a technique known as bioremediation and keeping our environment healthy. This chapter also provides evidences to explore PGPRs in sustainable agriculture to improve

Table 3.6 List of plant diseases caused by microbes

Microbes	Name of disease	Name of pathogenic microbe
Bacteria	Black rot	*Xanthomonas campestris* pv. *campestris*
	Bacterial canker	*Clavibacter michiganensis* pv. *michiganensis*
	Bacterial soft rot	*Pseudomonas spp.*, *Erwinia* spp.
	Bacterial leaf spot	*Xanthomonas campestris* – various strains
	Bacterial wilt	*Ralstonia solanacearum*
	Bacterial blight	*Pseudomonas syringae* – various strains
	Bacterial blight	*Pseudomonas syringae* pv. *pisi*
	Bacterial speck	*Pseudomonas syringae* pv. tomato
	Bacterial brown spot	*Pseudomonas syringae* pv. *syringae*
	Crown gall disease	*Agrobacterium* spp.
	Bacterial speck of tomato plant	*Pseudomonas syringae*
	Scabs	*Burkholderia* spp.
Fungi	Late blight	*Phytophthora infestans*
	Early blight	*Alternaria solani*
	Gray mold	*Botrytis cinerea*
	Downy mildew	*Plasmopara viticola*
	Powdery mildew	*Uncinula necator*
	Stem rust	*Puccinia graminis*
	Glume blotch	*Staganospora nodorum*
Protists	Root knot	*Meloidogyne* spp
	Root lesions	*Pratylenchus coffeae* and *Helicotylenchus multicinctus*
	Cluster of sprouts on tubers	*Ditylenchus dipsaci*
	Reduced plant growth	*Globodera rostochiensis*
	Devitalized buds	*Aphelenchoides fragariae*
	Leaf discoloration	*Aphelenchoides besseyi*
	Root surface necrosis	*Tylenchulus semipenetrans*
	Curly tip	*Xiphinema* spp.
	Root rot	*Ditylenchus destructor*
	Discoloration of foliage	*Pratylenchus coffeae*
Virus	Fruit distortion on eggplant fruit	*Tomato bushy stunt virus*
	Bark scaling	*Citrus psorosis virus*
	Yellow vein banding	*Grapevine fanleaf virus*
	Yellow mosaic symptoms on lettuce	*Lettuce mosaic virus*
	Sugarcane leaf mosaic	*Sugarcane mosaic virus*
	Black necrotic ring spots in cabbage	*Cauliflower mosaic virus*
	Maize dwarf syndrome	*Maize dwarf mosaic virus*
	Peanut dwarfing or stunting	*Peanut stunt virus*
	Barley yellow dwarf	*Barley yellow dwarf virus*
	Yellow streaking of leaves	*Tobacco mosaic virus*
	Alfalfa mosaic	*Alfalfa mosaic virus*
	Curly top	*Beet curly top virus*

productivity and other environmental prospects. Therefore, current agricultural practices need to be improved through use of biopesticides and biofertilizers in order to minimize environmental and health problems.

References

Aislabie J, Bej AK, Ryburn J, Lloyd N, Wilkins A (2005) Characterization of *Arthrobacter nicotinovorans* HIM, an atrazine-degrading bacterium, from agricultural soil New Zealand. FEMS Microbiol Ecol 5:279–286

Bates ST, Berg-Lyons DB, Caporosa JG, Walters WA, Knight R, Fierer N (2011) Examining the global distribution of dominant archaeal populations in soils. ISME J 5:908–917

Caporaso JG, Lauber CL, Walters WA, Berg-Lyons D, Huntley J, Fierer N et al (2012) Ultra-high throughput microbial community analysis on the Illumina HiSeq and MiSeq platforms. The ISME J 2016:1621–1624.

Carrondo MA (2003) Ferritins, iron uptake and storage from the bacterioferritin viewpoint. EMBO J 22:1959–1968

Dua M et al (2002) Biotechnology and bioremediation: success and limitations. Appl Microbiol Biotechnol 59:143–152

Eilers KG, Lauber CL, Knight R, Fierer N (2010) Shifts in bacterial community structure associated with inputs of low molecular weight carbon compounds to soil. Soil Biol Biochem 42:896–903

Fetzner S, Lingens F (1994) Bacterial dehalogenases: biochemistry, genetics and biotechnological applications. Microbiol Rev 58:641–685

Fierer N, Bradford MA, Jackson RB (2007) Towards an ecological classification of soil bacteria. Ecology 88:1354–1364

Harrison JJ, Ceri H, Turner R (2007) Multiresistance and tolerance in microbial biofilms. Nat Rev Microbiol 5:928–938

Heath KD, Tiffin P (2009) Stabilizing mechanisms in legume-rhizobium mutualism. Evolution 63:652–662

Hugenholtz P, Goebel BM, Pace NR (1998) Impact of culture-independent studies on the emerging phylogenetic view of bacterial diversity. J Bacteriol 80:4765–4774

Leung M (2004) Bioremediation: techniques for cleaning up a mess. J Biotechnol 2:18–22

Loganathan P, Hedley MJ, Grace ND, Lee J, Cronin SJ, Bolan NS et al (2003) Fertiliser contaminants in New Zealand grazed pasture with special reference to cadmium and fluorine: a review. Aust J Soil Res 41:501–532

Nies DH (1999) Microbial heavy metal resistance. Appl Microbiol Biotechnol 51:730–750

Sikes BA (2010) When do arbuscular mycorrhizal fungi protect plant roots from pathogens? Plant Signal Behav 5:763–765

Silver S, Phung LT (1996) Bacterial heavy metal resistance: new surprises. Annu Rev Microbiol 50:753–789

Stevens RB (1960) In: Horsfall JG, Dimond AE (eds) Plant pathology, an advanced treatise, vol 3. Academic, NY, pp 357–429

Thwaites JM, Farrell RL, Duncan SD, Lamar RT, White RB (2006) Fungalbased remediation: treatment of PCP contaminated soil in New Zealand. In: Singh SN (ed) Environmental bioremediation technologies. Springer, New York, pp 465–479

Treseder KK, Kivlin SN, Hawkes CV (2011) Evolutionary trade-offs among decomposers determine responses to nitrogen enrichment. Ecol Lett 14:933–938

Woese C, Kandler O, Wheelis M (1990) Towards a natural system of organisms: proposal for the domains Archaea, Bacteria, and Eucarya. Proc Natl Acad Sci U S A 87:4576–4579

Zahran HH (1999) Rhizobium-legume symbiosis and nitrogen fixation under severe conditions and in an arid climate. Microbiol Mol Biol Rev 63:968–989

Suggested Further Readings

Belén Rodelas González M, Gonzalez-López J (2013) Beneficial plant-microbial interactions: ecology and applications. CRC Press, 468pp, ISBN 9781466587175

Bertrand JC, Caumette P, Lebaron P, Matheron R, et al (2015) Environmental microbiology: fundamentals and applications. Springer, 933pp, ISBN 978-94-017-9118-2

Grant WD, Long PE (1981) Environmental microbiology, Springer, 215pp, ISBN 978-1-4615-8537-4

Gupta VK, Schmoll M, Maki M, Tuohy M, Mazutti MA (2013) Applications of microbial engineering. CRC Press, 504pp, ISBN 9781466585775

Helga ST, Sergiu F (eds) (2012) Adaption of microbial life to environmental extremes. Springer, 284pp, ISBN 978-3-211-99691-1

Kubicek CP, Druzhinina IS (eds) (2007) Environmental and microbial relationships. Springer, 350pp, ISBN 978-3-540-71839-0

Satyanarayana, Tulasi, Johri, Bhavdish Narain, Anil Prakash (eds) (2012) Microorganisms in environmental management. Springer, 819pp, ISBN 978-94-007-2229-3

Satyanarayana, Tulasi, Littlechild, Jennifer, Kawarabayasi, Yutaka (eds) (2013) Thermophilic microbes in environmental and industrial biotechnology. Springer 954pp, ISBN 978-94-007-5899-5

Singh SN, Tripathi RD (2004) Environmental bioremediation technologies. Springer, pp:ISBN 978-3-540-34793-4

Stolp H (1988) Microbial ecology: organisms, habitats, activities. Cambridge University Press, 308pp, ISBN 9780521276368

Bioremediation of Plant Refuges and Xenobiotics

4

Soni Tiwari, Ashutosh Tripathi, and Rajeeva Gaur

Contents

S. Tiwari • A. Tripathi • R. Gaur (✉)
Department of Microbiology (Centre of Excellence), Dr. Ram Manohar Lohia Avadh University, Faizabad 224001, UP, India
e-mail: rajeevagaur@rediffmail.com

R.L. Singh (ed.), *Principles and Applications of Environmental Biotechnology for a Sustainable Future*, Applied Environmental Science and Engineering for a Sustainable Future, DOI 10.1007/978-981-10-1866-4_4

Abstract

Industrialization, urbanization, and agricultural practices have created noxiousity of xenobiotics compounds in the atmosphere, seriously affecting the health of all living systems. The hazards created by such compounds are alarming and must be controlled/treated through bioremediation, a safe, economical, and rapid method for the treatment of almost all types of xenobiotic compounds. Microbial systems mainly bacteria, fungi, yeast, actinomycetes, and algae have diversified enzyme system for metabolizing such compounds into nontoxic forms and mineralizing up to the level of plant nutrients. The knowledge of such xenobiotics compounds, their existence and persistence in natural ecosystem, risk created by such compounds, biomagnifications, and biodegradation/bioremediation are essential for its effective control from different ecosystems. A very comprehensive classification of compounds and their degrading/metabolizing microbial enzymes along with the list of microorganisms has been discussed which is very essential for the environmental scientists, microbiologist, biochemist as well as agricultural scientist for their awareness and adopting remedial measures. Several new xenobiotic compounds are being synthesized in the natural ecosystem through polymerization and other organic reactions which must be identified and treated with specific suitable microbial consortia of different tolerance capabilities to temperature, pH, O_2, etc. The current scenario of bioremediation of xenobiotics compounds is greatly facilitated by the consortia of competent strain of various groups of microorganisms having the ability to coexist for longer periods without affecting their growth and metabolism, resulting in better remediation from natural ecosystem. The industrial effluents and solid waste treatment as well as field application of such microorganisms are showing effective results; therefore, large-scale cultivation and long-term preservation of these microorganisms either alone or in consortium are another area of research for safe and effective applications.

Keywords

Xenobiotic compounds • Textile dye • Pesticides • Aromatic compounds

4.1 Introduction

Fast industrialization, mainly in emergent countries like India, Brazil, and Bangladesh, has led to severe ecological pollution and has become a danger to the local ecosystem. Bioremediation is an efficient and cost-effective approach to eliminate organic pollutants from water as well as soil by selective microorganisms. Biological degradation has several advantages over the physicochemical treatments, as it is economical, is eco-friendly, and has less sludge-producing properties and maximum degradation of dye molecules into CO_2 and H_2O. Presently, a widespread study paying attention on microbial cultures is as economical as promising for the elimination of dyes from contaminated water. Biodegradation can be used successfully to decrease toxicity of contaminants, mobility, or volume to levels that are secure to being health and ecosystem. Advantages of using microbes (bacteria, fungi, and algae) for bioremediation include their natural habitat, ease of manipulation, high adaptive ability, and cost-effective treatment of industrial wastewater at industrial scale (Kot-Wasik et al. 2004; Jadhav and Govindwar 2006; Tiwari et al. 2014; Tiwari and Gaur 2014).

Xenobiotic compounds are purely man-made synthetic compounds which are generally not found in nature [DDT (dichlorodiphenyl tetrachloroethane), BHC (benzene hexachloride), organophosphates, phenols, aromatic hydrocarbons, metals], except a few, i.e., lignin as well as other phenolic compounds present in plants. Most xenobiotic compounds are recalcitrant and some of them are biomagnified to noxious levels. Biomagnification is the occurrence of progressive raise in the concentration of xenobiotic compounds as it passes through the food chain, for example, DDT, which is absorbed by plants, microorganisms, and other aquatic living systems (phytoplanktons, zooplanktons, and other living systems). The DDT not only passes on in the food chain but goes on accumulating mostly in body fats. Moreover, certain seabirds accumulated DDT in their livers to a toxic level, damaging their fertility and causing other diseases (Singh et al. 1990, 1991). In addition, regular exposure to low levels of such chemicals would also lead to their accumulation in human and other natural entities. Most of the xenobiotic compounds like pesticides, however, are recalcitrant, the compounds that resist biodegradation and persist in atmosphere, and some of them are biomagnified to lethal level. Certain chemicals are very persistent in environment, and they remain in the nature, for example, aldrine (5 year), chlordane (12 year), DDT (10 year), hexachlorocyclohexane (HCH, 11 year), and zineb (75 year); as these compounds are toxic for different living beings, therefore, integrated pest management is being recommended for sustainable environment and are most essential to detoxify the pesticides from food commodities. Several industrial effluents like distillery, textile, paper, and pulp containing color compounds like melanoidin, lignin contents, and dyes having xenobiotic characteristics should only be removed by microbial consortium. The selective microbial preparations including bacteria, yeast, and fungi are being used for efficient removal of such xenobiotic compounds from soil and water (Tiwari et al. 2014).

4.1.1 Hazards from Xenobiotic

- Various xenobiotics, i.e., halogenated and aromatic hydrocarbons, are lethal to prokaryotes, lower eukaryotes, and humans. Their small concentrations may cause skin and reproductive diseases.
- Various halogenated hydrocarbons have carcinogenic properties.
- Several xenobiotics are nondegradable and persist for longer period because they hold its molecular integrity, and hence physical, chemical, and functional characteristics in environment may remain transported and spread for significant period of time.

- Various xenobiotics, like DDT and PCBs (polychlorinated biphenyls), are recalcitrant and lipophilic and showed biomagnifications. Biomagnifications occur because such compounds are frequently taken up from the atmosphere and collected in the lipid body, e.g., plankton accumulates DDT from water (100-fold).

4.1.2 Characteristic of Xenobitic Compounds and Their Degradation Patterns

These compounds contain different numbers of halogen (e.g., Cl, Br, F, I) atoms in the place of H atoms ($CHCl_3$, ferons, CCl_3F, CCl_2F_2, $CClF_3$, CF_4, insecticides like DDT, BHC, lindane).

4.1.2.1 Polychlorinated Biphenyls (PCB)

These compounds have two covalently linked benzene rings having halogens substituting for H. PCB are used as plasticizers, as insulator coolants in transformers, and as heat exchange fluids. They are both biologically and chemically inert to various forms, which rise with the figure of chlorine atoms present in the molecule.

4.1.2.2 Synthetic Polymers

These compounds are produced as plastics, e.g., polyethylene, polystyrene, polyvinyl chloride, etc., and nylons which are used in clothes, packaging materials, etc. They are recalcitrant mainly due to their molecular size and water insolubility.

4.1.2.3 Alkylbenzyl Sulfonates

These are surface-active detergents superior to soaps. The sulfonate group present at one end resists microbial degradation while the other end (non-polar alkyl end) becomes recalcitrant if it is branched (resistance increases with the degree of branching). At present, alkylbenzyl sulfonates having nonbranched alkyl ends are used. These are biodegraded by oxidation from their alkyl ends.

4.1.2.4 Oil Mixtures

Oil is a natural product and degraded at different rates. Biodegradation is able to handle small oil seepages. But when huge spills occur, the problem of pollution becomes acute, though it is recalcitrant commonly because of its water insolubility and due to their toxicity of some of its constituent.

4.1.2.5 Other Xenobiotics

Other pesticides are based on aliphatic, cycliening structures having substitution of nitro, sulfonate, methoxy, amino, and carbamoyl groups and also contain halogens which include polychlorinated biphenyls (PCBs), synthetic polymers, alkylbenzyl sulfonates, oil mixture, and pesticides like compounds. These functional groups made them recalcitrant. In ancient times, the large number of new, man-made synthetic chemicals came into application to control the spread of disease. Classifications of pesticides on the basis of chemical nature are:

(i) *Organochlorines*: These are chlorinated hydrocarbons. Hence, they are called organochlorines. Organic compounds like organochloride, organochlorine, chlorocarbon, chlorinated hydrocarbon, and chlorinated solvent contain at least one chlorine atom. Their broad structural and chemical properties lead to an extensive range of applications. Various derivatives are contentious because these compounds affected surroundings. Most of organochlorines degrade slowly and can persist in the ecosystem for a long time. Some of the examples of organochlorine pesticides include alpha-BHC, beta-BHC, gamma-BHC (lindane), delta-BHC, heptachlor, endosulfan, methoxychlor, Aroclor, and dichlorodiphenyltrichloroethane (DDT).

(ii) *Organophosphates*: Organophosphates consist of phosphorus bonded to carbon atoms of organic radicals. Various organophosphates react with acetylcholinesterase and other cholinesterases which results in disruption of nerve impulses and killing the insect. Organophosphates have a preservative noxious effect to flora and fauna, so multiple exposures to the chemicals amplify the toxicity. Some of the organophosphorus compounds include acephate, azinphos-methyl, fenthion, chlorethoxyfos, bensulide, diazinon, dicrotophos, dimethoate, disulfoton, ethoprop, fenamiphos, fenitrothion, chlorpyrifos, methamidophos, fosthiazate, parathion, monocrotophos, and malathion.

(iii) *Carbamates*: Carbamates are phenols, esters (N-methyl and N,N-dimethyl carbamic), and heterocyclic enols. Carbamates are closely related to the organophosphorus compounds in mode of action and resistance development. Symptoms of the carbamate poisoning are similar to that of organophosphorus poisoning. Bendiocarb, carbaryl, carbofuran, fenobucarb, fenoxycarb, isoprocarb, dioxacarb, pirimicarb, and methomyl are some of the examples of carbamate insecticides.

(iv) *Pyrethroids*: The chemical compound, pyrethroid pesticides, has been developed to imitate the insecticidal activity of pyrethrum. Pyrethroids are degradable, act as sodium channel modulators, and are less toxic than others. These compounds are used against domestic pests. Cypermethrin, fenvalerate, cyfluthrin, etofenprox, permethrin, phenothrin, deltamethrin, prallethrin, resmethrin, and tetramethrin are some of the pyrethroids.

(v) *Neonicotinoids*: Neonicotinoids are synthetic analogues of insecticide nicotine (with a lesser severe mammalian toxicity and greater field persistence). Neonicotinoids are nicotinic acetylcholine receptor agonists. These are extensive range insecticides and have rapid action. Neonicotinoids are used as sprays, drenches, and seed and soil treatments. Insects exhibit leg tremors, rapid wing movement, stylet removal (aphids), confused movement, paralysis, and death due to exposure of neonicotinoids. Acetamiprid (Fig. 4.1), clothianidin, imidacloprid (Fig. 4.1), nitenpyram, nithiazine, thiacloprid, and thiamethoxam are some of neonicotinoids.

(vi) *Trizines*: These pesticides have some limitations as they fail to control different kinds of disease caused by different causal organisms; besides that sometimes the pathogenic organism undergoes mutations and hence develops resistance to different pesticides. Shortly their drawback was realized because these

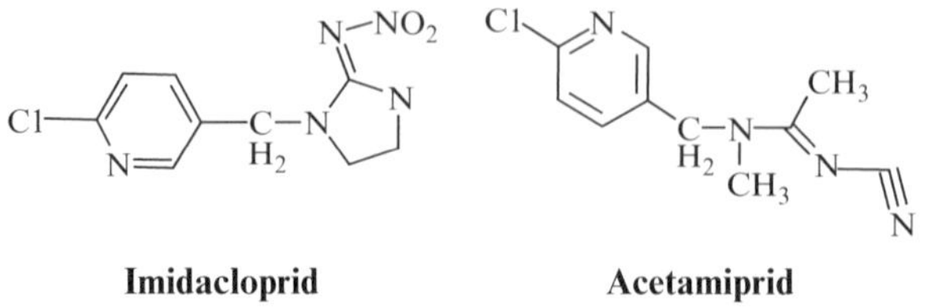

Imidacloprid **Acetamiprid**

Fig. 4.1 Structure of imidacloprid and acetamiprid

chemicals were extremely toxic and persistent in nature, and it not only kills the damaging pest but several friendly pests which are helpful in many activities of food chain, biodiversity, and pollen distribution.

Xenobiotics include a broad range of compounds, and their degradation occurs through different catabolic pathways. Degradation of alkenes and aromatic hydrocarbons usually occurs in the following manner: (1) firstly, hydroxyl group of xenobiotic compound is reactivated by oxygenase; (2) then hydroxyl group is oxidized to a carboxyl group; (3) in the case of cyclic compounds, the ring structure is opened up; and (4) in the last step, linear molecule is split into acetyl-CoA by β-oxidation. Likewise, an alicyclic hydrocarbon (cyclohexane) is oxidized in the following manner: (1) first, an OH group in alicyclic hydrocarbon is activated by oxygenase; (2) next, another oxygenase forms a lactone ester; and (3) then ring structure is hydrolyzed into a linear structure. In both oxidation processes, oxygen binds at single position in the molecule by monooxygenases. In contrast, oxidation of benzene ring may involve a dioxygenase which adds oxygen at two positions in the molecule in a single step. Monooxygenase reacts with short-chain alkanes, while dioxygenase reacts with cyclic alkanes. These enzymes are not specific, and each enzyme oxidizes a limited range of compounds. Therefore, xenobiotics are degraded by a wide range of microorganisms, each of which degrades a little variety of compounds.

Pesticides are classified into insecticides, herbicides, fungicides, rodenticides, and others. The pesticides are classified depending upon target species and mode of action. According to the application, the main classes of pesticides include herbicides (that kill weeds), fungicides (that kill fungi), and insecticides (that kill insects). Depending upon the chemical nature, the major classes of pesticides include organochlorine, organophosphorus, thiocarbamates, pyrethroids, and neonicotinoids. Generally, the oxidation of xenobiotics occurred in the presence of rubredoxin or cytochrome P-450. In addition, the halogen or other functional groups are either modified/removed as one of the first reactions, or sometimes it is attained later in the process.

4.1.3 Hydrocarbon Degradation

1. Halomethanes are converted into methanol by enzyme methane monooxygenase which uses them as substrate. This enzyme presents in methylotrophs. On the

other hand, glutathione-dependent hydrolases catalyze oxidative dechlorination of halomethanes into methanol. This reaction is anaerobic and uses O_2 derived from H_2O. Methanol is oxidized to CO_2 and H_2O with formaldehyde and formic acid.

2. Cyanide (HCN) is toxic to biological system, and still microorganism capable of degrading cyanide cannot resist a high concentration of HCN. Some of the cyanides, e.g., HCN and CH_3CN, are volatile. It is degraded by fungal hydratase into $HCONH_2$ and by *Pseudomonas fluorescence* into CO_2 and NH_2.
3. Aliphatic hydrocarbon may be saturated or unsaturated. N-alkanes of 10–24 carbons are lethal biodegradable. Similarly, aliphatics are easier to degrade than unsaturated ones, and branched chain show decreased biodegradation. Biodegradation of n-alkanes is catalyzed by oxygenases to produce carboxylic acids, which is then oxidized by β-oxidation. Oxidation may involve the methane group at one end of n-alkane molecule, or it may occur as beta methylene group. Sometimes, both terminal methyl groups are oxidized to yield a dicarboxylic acid.
4. Alicyclic hydrocarbons are present naturally in waxes from plants, crude oil, and microbial lipids and are represented by xenobiotics used as pesticides and also in petroleum products.
5. Aromatic hydrocarbons are rather stable and oxidized by dioxygenases to catechol which is further metabolized by two different pathways: (i) in the case of ortho-linkage cleavage pathway, a 1,2-dioxygenase cleaves the ring between the two adjacent hydroxyl groups, and sequential catabolism of the product cis, cis-muconate yields succinate and acetyl-CoA. (ii) Alternatively, the enzyme 2,3-dioxygenase cleaves the ring between the C atoms having an OH group and adjacent carbon lacking an OH group (meta-cleavage). The product at the end of the reaction is acetaldehyde and pyruvate. Aromatic hydrocarbon degradation occurs with the help of ortho- and meta-cleavage. Benzene is degraded by meta-cleavage pathway.
6. Polycyclic hydrocarbon contains two or more rings. Generally one of the terminal rings is attacked by dioxygenase. Degradation of complex molecule containing aliphatic, aromatic, alicyclic, or heterocyclic components is difficult to simplify, but the following features are observed:
   (i) Amide, ester, or other bonds are first attacked and further degradation of the product generated takes place.
   (ii) If these bonds are absent or inaccessible, aliphatic chains are degraded.
   (iii) If aliphatic chains are branched the aromatic component of the complex molecule may be attacked.
   (iv) The position and mode of reaction depend on the molecular structure, the microorganism involved, and the environmental conditions; in general recalcitrance of various benzene derivatives increases with substituent groups (at meta-position): $COOH = OH < -NH_2 < -O\text{-}CH_3 < -SO_3 < -NO_2$. Further the larger the number of substituent groups on the benzene ring, the higher the rate of recalcitrance. The position of substituent group also affects recalcitrance as meta > ortho > para.

4.2 Pesticide Pollution: A Rising Threat

Pesticides are applied for controlling crop pest to reduce losses of agricultural yield and manage insect vectors to stop the eruption of human and animal epidemics. Foodstuff storage has resulted in increased use of the pesticide and herbicides in agriculture. India is the principal user of pesticide among the South Asian countries, and the rate of pesticide consumption for crop protection accounts for 3 % of the world consumption. High pesticide concentration in soil adversely affects human health and causes serious environmental pollution. In India, the most commonly used pesticides include organophosphates, organochlorines, neonicotinoids, etc. The variety of pesticides are increasingly produced and released to the environment at a growing rate, which may pollute the soil, water, and air in the world if it is not used properly (Yao et al. 2006; Rani and Dhania 2014). Nowadays, lots of proofs are reported that some of pesticides are hazardous to humans and other living things and have undesirable side effects to the ecosystem. About one million deaths and chronic illnesses are reported per year due to pesticide poisoning worldwide. Ideal pesticide must be deadly to the targeted insects, but regrettably, this is not so; hence, the divergence of use and misuse of pesticides has to be faced.

The overview of the present situation demands the promising tools for remediation of these hazardous compounds from the environment. The governments and higher authorities of different countries are taking concrete steps toward these issues nowadays. Several pesticides are banned permanently; the organochlorines and organophosphorus compounds are the core among them. The restrictions over the use organochlorines and organophosphorus have made the neonicotinoids favorites. Although the neonicotinic compounds are less toxic to mammals, they pose severe toxicity toward nontarget insects, especially economic insects like honeybees and silkworms. Extensive use of pesticides results in interruption of ecosystem life. It is estimated that only 5 % of the applied pesticide reaches to the target species, and the remaining pesticide reaches to the soil and groundwater, and if immobile it gets accumulated in soil surface and exerts toxic effects on plants, animals, and humans.

Persistent pesticides cause a risk to the well-being of the surroundings and to being healthy. The organochloride insecticides accumulate in human adipose tissue. Some insecticides, like chlordane, can be absorbed dermally. The triazine pesticides are human carcinogens and are persistent in water and mobile in environment (EPA 1979). Organophosphate insecticides can affect nervous system. Nitrophenolic and nitrocresolic herbicides (dinoseb) can be absorbed by the skin and cause skin cancer. The promising risk to human health makes the degradation of pesticide-contaminated places essential. Most of the pesticides are banned under pesticide pollution and toxicity prevention act which resulted in the use of alternatives like neonicotinoids which are less toxic to the mammals. But, extreme use of these compounds resulted in to a new problem concerning harmful effect on nontarget economical insects like honeybees and silkworms.

4.2.1 Detoxification of Pesticide

There are several ways for detoxification of pesticide which are specially applied to different pesticides:

(i) Physical
(ii) Chemical
(iii) Biological

The physicochemical methods comprise of washing; cooking; peeling; sun drying; brushing; solvent washing/removal; washing with water, soaps, adjuvant, or surfactant; burial, and disposal. The chemical alterations include oxidation, reduction, nucleophilic displacement, hydrolysis, and high energy decomposition.

Low-temperature thermal desorption is an *ex situ* remediation method, commonly used to remediate pesticide-polluted places. This method is capable of eliminating semi-volatile and volatile organic compounds (pesticides) from soils and removing pesticides from sludge, sediments, and filter cakes. The organics in the polluted gas stream are completely destroyed by either passing through an afterburner or condenser or adsorption by carbon adsorption beds. The condenser transforms the gas into a liquid stage for further treatment although the carbon adsorption beds immobilize, but do not degrade the contaminants. Low-temperature thermal desorption needs well-specialized facilities and carries a relatively high cost.

Incineration is also a significant technology that has generally been used to remediate pesticide infected places. The initial stage of incineration heats the polluted media at temperatures between 1000 and 1800 °F that result in oxidation and the volatilization of the organics. In the second stage, complete destruction of organics occurs at 1600 and 2200 °F temperatures. After that, resultant ash can be disposed of in a landfill, if it meets well-being regulations. Incineration has the ability to complete remediation of the contaminant.

Phytoremediation is a cost-effective and artistically pleasing technique for remediation of polluted sites (Ziarati and Alaedini 2014). Herbicides are designed to kill plants; hence, the use of phytoremediation to remediate herbicides can be a difficult chore. Numerous studies have reported the effectiveness of remediation of remaining pollutants with different plant species.

Biological methods are an innovative attractive tool that is frequently applied for the cleanup of unhygienic sites due to its eco-friendly cost (Murali and Mehar 2014). The microbial degradation of pollutants is enhanced when supplementing these microorganisms with nutrients, carbon sources, or electron donors. This can be done by native microorganisms or augmented culture of microorganisms that have specific characteristics to degrade the contaminant at a higher rate. Complete oxidation of pollutant waste occurs through bioremediation into H_2O and CO_2 without the buildup of any intermediate. Bioremediation methods can be classified into *ex situ* and *in situ*. *Ex situ* bioremediation tools comprise bioreactors, biofilters, land farming, and few composting methods. *In situ* bioremediation tools involve bioventing, biosparging, biostimulation, liquid delivery systems, and several composting methods. The most favorable part of this technology is its effectiveness and low cost. The bioremediation can be applied to the solid sludge, soil, and sediment as well as groundwater pollution. The major sources involved in the pesticide biodegradation are bacteria and fungi.

The absolute biodegradation of the pesticide through oxidation results in H_2O and CO_2, and this process gives the energy to the microbes for their metabolism. Degradation of chemical compounds by the microbes is mediated through the enzymes either intracellularly or extracellularly. Persistence of the particular pesticide in the soil is due to the absence of the microbial systems that bears the pesticide-degrading enzymes in that particular soil. In such cases, where innate microbial population of the soil cannot manage pesticides, the external addition of pesticide-degrading microflora is recommended. Degradation of the pesticides by the microbes not only depends upon the enzyme systems but also on the different environmental conditions such as temperature, pH, water potentials, and available nutrients (Rani and Dhania 2014). Some of the pesticides are readily degraded by the microbes; however, some are recalcitrant in nature. Many of the pesticide-degrading genes are reported to be harbored by the plasmid DNA. The plasmid-encoding degrader genes are known as the catabolite plasmids, and several bacterial species, viz., *Pseudomonas*, *Flavobacterium*, *Alcaligenes*, *Acinetobacter*, *Klebsiella*, *Moraxella*, *Rhodococcus*, and *Arthrobacter*, containing catabolite plasmids are reported. Recently, the bacterium *Raoultella* sp. X1 is reported for cometabolic degradation of organophosphorus compound dimethoate (Liang et al. 2009). Similarly, Li et al. (2010) also stated that biodegradation of dimethoate with detailed biochemical pathway is also demonstrated using *Paracoccus* sp. Lgjj-3 (Fig. 4.2).

Bioremediation of toxic chemicals through fungus from the innate source is also an efficient method when bacterial species fails to achieve the goal. *Phanerochaete* and other fungi produce extracellular enzyme like peroxidase that degraded the wide range of chemical compounds (Fragoeiro and Magan 2008). *Phanerochaete chrysosporium* is one of the well-known fungi that degrade different chemicals like polychlorinated biphenyl (PCB), pesticides, and polycyclic aromatic hydrocarbons (PAH) (Coelho-Moreira et al. 2013).

Chlorinated endosulfan pesticide is generally utilized in India for the protection of cotton, tea, sugarcane, and vegetables. It was noted that fungal isolate *Aspergillus niger* could tolerate 400 mg $ml^{-1}$of endosulfan during degradation process. The white rot fungus *Phanerochaete chrysosporium* is well known for xenobiotic metabolism, and it was reported to have the ability to degrade isoproturon, a herbicide, via solid-state fermentation. Besides these, various fungal species including *Trametes* sp. and *Polyporus* sp. were able to remediate a variety of chemicals (pesticides). Pesticide degradation was also reported using *Aspergillus flavus*, *A. fumigates*, *A. sydowii*, *A. terreus*, *Fusarium oxysporum*, and *Penicillium chrysogenum*.

White rot fungi, which can remediate lignin present in wood, have also shown efficiency to remediate wide varieties of different chemicals, even those which did not have any structural relationship with lignin. Various species of *Phanerochaete chrysosporium* could degrade several environmentally persistent organic pollutants like DDT, lindane, benzopyrene, azo dyes, and dioxin (Coelho-Moreira et al. 2013). Many researchers reported that *Cyathus bulleri* and *Phanerochaete sordida* showed

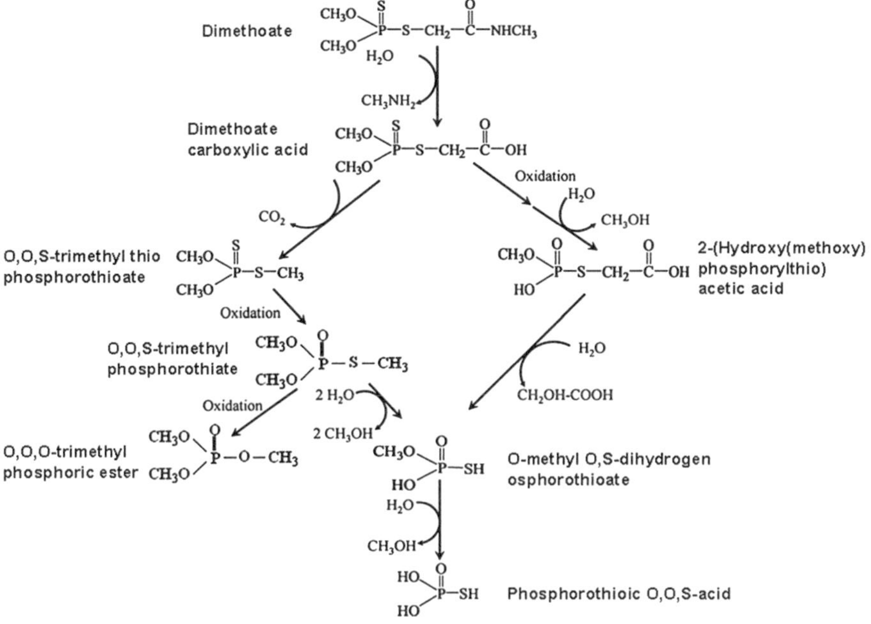

Fig. 4.2 Degradation of dimethoate by *Paracoccus* sp.

effectiveness for degradation of lindane. Bioremediation of various organophosphorus compounds via different fungi has also been documented.

Apart from bacteria and fungi, algae are also used in pesticide cleansing though to slighter degree. Several green and blue-green algae have the ability to degrade organophosphorus insecticide chlorpyrifos, monocrotophos, and quinalphos, isolated from soil or water. *Pseudomonas sp.* and *Aspergillus niger* could hydroxylate 2,4-D present in a factory waste. Degradation of three benzonitrile herbicides, bromoxynil (3,5-dibromohydroxybenzonitrile), ioxynil (3,5 diiodo-4 hydroxybenzonitrile), and dichlobenil (2,6-dichlorobenzonitrile), and their mixture was efficiently degraded in batch culture by *Agrobacterium radiobacter*.

4.2.2 Present Scenario of Microbial Enzyme Status for Pesticide Degradation

The biological methods included are through biocatalysis, enzymatic oxidation, and microbial degradation either by using them as a carbon source in place of other substrate as a carbon or by cometabolism (Rao et al. 2010). Various mechanism involved in chemical bond cleavage of xenobiotics compounds. Xenobiotic compounds may be recalcitrant due to the following reasons:

1. When halogens replace the hydrogen atom in the molecule, the carbon–halogen linkage is extremely stable and its degradation needs extensive energy.

2. Replacement of hydrogen atom by other functional group like sulfonate, nitro, amino, methoxy, and carbonyl.
3. Aromatic compounds, cycloalkanes, and heterocyclic compounds are highly recalcitrant than aliphatic compounds.
4. Branched aliphatic compounds resist biodegradation. Biodegradation involves oxidation by cytochrome P-450; the most common monooxygenase reaction are those employing heme protein P-450. Like mitochondrial cytochrome oxidase, cytochrome P-450 can react with O_2 and bind CO. (CO complex of its reduced form absorbs 450 nm, thus the name P-450) Cytochrome P-450 catalyzes hydroxylation (OH) in which organic substrate, RH, is hydroxylated to R-OH, incorporating one oxygen atom of O_2.

Detoxification of xenobiotics may involve one of the following mechanisms:

Hydrolysis: breaks the ester bond (C-OC) in malathion; activation of herbicide (e.g., dichlorofop methyl). *Hydroxylation*: OH group replace H group in 2,4-D.

Dehalogenation: It involves reduction (removal of H group), and hydrolysis (removal of OH group) of DDT into DDE.

Deamination: It cleaves ethers, e.g., conversion of nitrile to amide in 2,4-D reduces phytotoxicity.

Conjugation: e.g., pyrene and glucose to glucoside conjugate microorganisms involved in biodegradation.

4.2.2.1 Organophosphorus Hydrolase (OPH)

The organophosphorus hydrolase purified from *B. diminuta* and *Flavobacterium* sp. is identical in amino acid sequences. *A. radiobacter* produces another type of organophosphorus hydrolase called OP-degrading enzyme (OPDA). Organophosphorus hydrolases are amidohydrolase-type enzyme and contain (α–β) eight barrel structural folds. The active site of organophosphorus hydrolase contains two transition metal ions, like Zn, Fe, Mn, or Co.

4.2.2.2 Organophosphorus Acid Anhydrolase (OPAA)

Organophosphorus acid anhydrolase is another organophosphorus-degrading enzyme [encoded by *opaA* (OPAA)], generally produced by *Alteromonas undina* and *Alteromonas haloplanktis*. Organophosphorus acid anhydrolase is related to the dipeptidase class.

4.2.2.3 Aryl Acylamidase

Aryl acylamidase was purified from *Bacillus sphaericus* with the 75 kDa molecular weight. The enzyme was found to have the herbicide and fungicide degrading capacity.

4.2.3 Chlorinated Pesticide

Chlorinated pesticides represent the largest group of compounds in the list of priority pollutants due to its toxicity, persistence, and ubiquitous transmission in the ecosystem which affects quality of life (EPA 1979). Chlorinated pesticides are accumulated in the ecosystem in huge amount due to their widespread use as herbicides, insecticides, fungicides, solvents, hydraulic and heat-transfer fluids, plasticizers, cleaning agents, fumigants, aerosol propellants, gasoline additives, and degreasers (Table 4.1). Several naturally occurring chlorinated compounds also present at lower concentrations in the biosphere. For example, widespread genera of wood-rotting fungi release chlorinated anisyl metabolites in the environments. From higher plants and ferns, mostly 130 chlorinated pesticides have been isolated.

The persistence of chlorinated pesticides in the biosphere is largely determined by their susceptibility to biotransformation by microorganisms. Most of the pesticides that are not remediated by microorganisms show potential to persist in the surroundings for longer periods of time. Therefore, recognition and use of novel microorganisms that use chlorinated compounds for growth have become a significant area of research nowadays.

4.2.3.1 Degradation of Chlorinated Pesticides by Microorganisms

The biodegradation of hazardous materials is also based on the principles that include circulation, transformation, and assimilation of energy and matter that maintain all ecosystems. Microorganisms degrade complex organic compounds into CO_2 or other simple organic compounds through their central metabolic pathway. During degradation (oxidation), energy and reducing equivalents are produced that involved intermediate production (assimilation) which is utilized by microorganisms for their growth. Degradation of natural origin compounds is generally easily achieved by microorganisms that can be easily isolated from their natural ecosystem. However, xenobiotic compounds that are different from naturally occurring compounds are more difficult to degrade. However, in the recent years, a group of microorganisms has been identified that use chlorinated pesticides for their survival.

4.2.3.1.1 Energy Metabolism

Various bacterial strains have been reported for coupling reductive dechlorination and utilized chlorinated compounds for energy metabolism. *Desulfomonile tiedjei* uses H_2 and formate as an electron donor and 3-chlorobenzoate as a terminal electron acceptor in a respiratory process (Cobb and Bouwer 1991). This strain can synthesize ATP by coupling hydrogen oxidation to reduction of the C–Cl bond of 3-chlorobenzoate. Holliger et al. (1992) have showed that dechlorination of tetrachloroethene (TeCE) is growth associated and demonstrated that isolate PER-K23 has ability to convert perchloroethene (PCE) to cis-1,2-dichloroethene (1,2-DCE) by trichloroethene (TCE) and also synthesize energy through electron transport phosphorylation (Cupples et al. 2003). Consumption of chlorinated pesticides is not based on energy metabolism, while microorganisms must get energy from degradation of toxic materials.

Table 4.1 Commercial applications of chlorinated hydrocarbons

Chlorinated hydrocarbon	Major use
Chloromethanes	
Monochloromethane	Production of silicones, tetraethyllead, methylcellulose, other methylation
Dichloromethane	Degreasing agent, paint remover, pressure mediator in aerosol, extract technology
Trichloromethane	Production of monochloro-difluoromethane (for the production of tetrafluoroethene which is used for manufacture of Hostaflon and Teflon), extractant of pharmaceutical product
Chloroethanes	
Monochloroethanes	Production of tetraethyllead, production of ethyl cellulose; ethylating agent for fine chemical production, solvent for extracting processes
1,1-Dichloroethane	Feedstock for the production of 1,1,1-trichloroethane
1, 2-Dichloroethane	Production of vinyl chloride production of chlorinated solvents such as 1,1,1-trichloroethane and tri- and tetrachloroethane, synthesis of diethylenediamines
1,1,1-Trichloroethane	Dry cleaning, vapor degreasing, solvent for adhesives and metal cutting fluids; textile processing
1,1,2-Trichloroethane	Intermediate for production of 1,1,1-trichloroethane and 1,1-dichloroethane
Chloroethenes	
Monochloroethene	Production of polyvinyl chloride (PVC), production of vinyl chloride chlorinated solvents, primarily 1,1,1-trichloroethane
Trichloroethene	Solvent for vapor degreasing in the metal industry and for dry cleaning, extraction solvent, solvents in formulations for rubbers, elastomers, and industrial paints
Tetrachloroethene	Solvent for dry cleaning, metal degreasing, textile finishing, dyeing, extraction processes, intermediate for the production of trichloroacetic acid and some fluorocarbons
2-Chloro-1,2-butadiene (chloroprene)	Starting monomer for polychloroprene rubber
Chlorinated paraffins	Plasticizers in PVC; flameproofing agents in rubber textiles, plastics, H2O repellent, and not preventive agents; elastic sealing compounds paints and varnishes; metalworking agents (cutting oils); leather finishing
Chlorinated aromatic hydrocarbon	
Monochlorobenzene	Production of nitrophenol, nitroanisole, chloroaniline, phenylenediamine for the manufacture of dyes, crop protection products, pharmaceuticals, and rubber chemicals

(continued)

Table 4.1 (continued)

Chlorinated hydrocarbon	Major use
1,2-Dichlorobenzene	Production of 1,2-dichloro-4-nitrobenzene for the production of dyes and pesticides; production of disinfectants, room deodorants
1,4-Dichlorobenzene	Production of disinfectants, room deodorants, moth control agent; production of insecticides; production of 2,5-dichloronitrobenzene for the manufacture of dyes; production of polyphenylenesulfide-based plastics
Chlorinated toluenes	Hydrolysis of cresol, solvent for dyes; precursors for dyes, pharmaceuticals, pesticides, preservatives, and disinfectants
Chlorophenols	Preparation of agricultural chemical herbicides, etc.
Chloro-phenoxy alkanoic acids	Herbicides
Side-chain chlorinated aromatic hydrocarbon	
Chloromethylbenzene (benzyl chloride)	Production of plasticizer, benzyl alcohol, phenylacetic acid, quaternary ammonium salts, benzyl esters, triphenylmethane dyes, dibenzyl disulfide, benzylphenol, benzylamines
Dichloromethylbenzene (benzal chloride)	Production of benzaldehyde
Trichloromethylbenzene (benzotrichloride)	Production of benzyl chloride; production of pesticides; UV stabilizers and dyes
Pesticides, herbicides, and fungicides	For seed treatment, for treatment of diseases of plants, animals, and humans

4.2.3.1.2 Cometabolism

It is a process of degradation of aromatic hydrocarbons in the presence of other organic materials which serve as primary energy source. The cometabolism process prevails in ecosystem between several groups of chemoauto- and chemoheterotrophs, e.g., *Nitrosomonas europaea* can cometabolize dichloromethane (DCM), trichloromethane (TCM), 1,1,2-trichloroethane, 1,1,1-trichloroethane, and 1,2,3-trichloropropane, where ammonia serves as the primary substrate. Several organic compounds like toluene, methane, and ammonia are oxidized by several bacteria in natural ecosystem by cometabolism with TCE, DCE, and vinyl chloride. Mostly *Pseudomonas cepacia* G4 uses toluene and degrades TCE cometabolically. *E. coli* degrades fumarate along with *Pseudomonas* through cometabolic processes through electron transport chain.

The cometabolism process has been certified by the generation of broad-specificity enzymes. In cometabolism process, both the primary substrate and the chlorinated compound compete for the same enzyme. In cometabolism process, the degradation rate of the chlorinated compound is dependent on the electron flow from the primary substrate.

4.2.3.1.3 Aerobic Degradation

Aerobic degradation of chlorinated compounds by microorganisms requires oxygen molecules as the electron acceptor. Several chlorobenzenes (containing one, two, three, or four chlorine atoms) can only be degraded under aerobic conditions. Chlorophenol [4-chlorophenol (4-CP)] has been completely degraded by several aerobic bacteria like *Pseudomonas*, *Alcaligenes*, *Rhodococcus*, and *Azotobacter* and fungi like *Phanerochaete* spp., *P. chrysosporium* respectively (Chen et al. 2014; Padhye and Chakrabarti 2015).

Microbes play a vital role in the bioconversion and total breakdown of pesticides. Among the microbial population, bacteria and fungi are major degraders of pesticides. Yeasts, microalgae, and protozoa are less frequently encountered in the remediation process. Microbes responsible for the remediation of various pesticides have been described in Table 4.2. Among bacteria, *Pseudomonas* is considered to be the most efficient group in bioremediation (Table 4.3).

A number of microorganisms are recognized to degrade DDT anaerobically. The primary metabolic mechanism was studied for the reductive dechlorination of DDT, with the formation of 1,1-dichloro-2,2-bis(4-chlorophenyl) ethane (DDD) or dichlorodiphenyldichloroethane. Nadeau et al. (1994) has reported the aerobic bioremediation of DDT to 4-chlorobenzoic acid by *Alcaligenes eutrophus* A5. Numerous actinomycetes like *Nocardia* spp., *Streptomyces aureofaciens*, *Streptomyces cinnamoneus*, and *Streptomyces viridochromogenes* degraded DDT to DDD. These organisms, however, required another carbon source to facilitate degradation. Soil fungi not only produced DDD and small amounts of dicofol (4,4′-dichloro-α-(trichloromethyl) benzhydrol), but some variants could produce DDA (bis(4-chlorophenyl)acetic acid) or DDE (1,1-dichloro-2,2-bis(4-chlorophenyl)ethene) exclusively. DDD was obtained under both aerobic and anaerobic conditions when DDT was incubated with *Enterobacter aerogenes*. *Escherichia coli* dechlorinated 50% of DDT to DDE when grown in various broths or skimmed milk. In aerobic conditions the major product of DDT metabolism, in *Bacillus cereus*, *B. coagulans*, and *B. subtilis*, was DDD, while DDMU (1-chloro-2,2-bis(4-chlorophenyl)ethylene), DDMS (1-chloro-2,2-bis(4-chlorophenyl)ethane), DDNU (2,2-bis(4-chlorophenyl)ethane), DDOH (2,2-bis(4-chlorophenyl)ethanol), DDA, and DBP (4,4′-dichlorobenzo phenone) were in trace amounts and were found under anaerobic conditions.

4.2.3.1.4 Anaerobic Degradation

Anaerobic degradation of pesticides requires a molecule other than O_2 as electron acceptor. These molecules are CO_2, NO^{-3}, SO_4^{2-}, Fe^{3+}, H^+, S, fumarate, trimethylamineoxide, and any organic compound. The word "dehalorespiration" is an alternative term for anaerobic bacteria which is a reductive dehalogenation process of chlorinated aliphatic and aromatic compounds for ATP production through an electron transport system. Reductive dechlorination/dehydrogenolysis, a common pathway, generally occurs under methanogenic conditions for production of chloroaliphatic compound containing one or more carbon atoms. Some chlorinated

Table 4.2 Microorganisms for pesticide remediation

Pesticide	Microorganism
2,4-D	*Alcaligenes eutrophus*, *Alcaligenes xylosoxidans*, *Flavobacterium* spp., *Pseudomonas putida*, *Pseudomonas cepacia*, *Comamonas* spp.
2,4,5-T	*Pseudomonas cepacia*
DPA	*Flavobacterium* spp.
Mecoprop	*Sphingomonas herbicidivorans* MH
Mecocarp	*Alcaligenes denitrificans*
Organochlorines	
DDT	*Aerobacter aerogenes*, *Alcaligenes eutrophus* A5, *Agrobacterium tumefaciens*, *Arthrobacter* spp., *Bacillus cereus*, *Bacillus coagulans*, *Bacillus megaterium*, *Bacillus subtilis*, *Clostridium pasteurianum*, *Clostridium michiganense*, *Enterobacter aerogenes*, *Erwinia amylovora*, *Escherichia coli*, *Hydrogenomonas* spp., *Klebsiella pneumoniae*, *Kurthia zopfii*, *Micrococcus* spp., *Nocardia* spp., *Pseudomonas aeruginosa* DT-Ct1, *Pseudomonas aeruginosa* DT-Ct2, *Pseudomonas fluorescens* DT-2, *Serratia marcescens*, *Serratia marcescens* DT-1P, *Streptomyces annomoneus*, *Streptomyces aureofaciens*, *Streptomyces viridochromogens*, *Xanthomonas* spp., *Phanerochaete chrysosporium* (fungus), *Trichoderma viride* (fungus)
Γ-HCH	*Bacillus cereus*, *Bacillus megaaterium*, *Citrobacter freundii*, *Clostridium rectum*, *Escherichia coli*, *Pseudomonas fluorescens*
	Pseudomonas putida, *Pseudomonas paucimobilis*, *Pseudomonas* spp., *Anabaena* spp., (*Cyanobacteria*), *Nostocellipssosun* (*Cyanobacterium*), *Phanerochaete chrysosporium* (fungus), *Trametes versicolor* (fungus), *Phanerochaete sordida* (fungus)
	Cyathus bulleri (fungus)
Organophosphates	
Parathion	*Flavobacterium* spp., *Pseudomonas aeruginosa*, *Pseudomonas diminuta*, *Pseudomonas melophthara*, *Pseudomonas stutzeri*
Phorate	*Rhizobium japonium*, *Rhizobium meliloti*, *Streptomyces lividans*
	Bacillus megaterium
Carbamates	
Carbaryl	*Pseudomonas cepacia*, *Pseudomonas melophthora*, *Pseudomonas aeruginosa*, *Gliocladium roseum* (fungi), *Aspergillus flavus* (fungi), *Aspergillus terreus* (fungi) *Culcitalna* spp. (fungi), *Halosphaeria* spp. (fungi), *Fusarium solani* (fungi), *Rhizopus* spp. (fungi), *Penicillium* spp. (fungi)
Carbofuran	*Achromobacter* spp. WMIII, *Arthrobacter* spp., *Flavobacterium* spp., *Pseudomonas cepacia*, *Pseudomonas stutzeri*, *Bacillus pumilus*
s-Triazines	*Pseudomonas* spp., *Klebsiella pneumoniae*, *Rhodococcus corallinus*, *Rhizobium* spp. PATR, *Phanerochaete chrysosporium* (fungus)

aromatic compounds (2,4,5-trichlorophenoxyacetate, 3-chlorobenzoate, 2,4-dichlorophenol, 4-chlorophenol, 2,3,6-trichlorobenzoate, and dichlorobenzoates) also efficiently degraded under methanogenic conditions. *Dehalococcoides* spp. have ability to completely detoxify vinyl chloride.

Table 4.3 Bioconversion of xenobiotics by pseudomonads

Mode of action	Species
Hydrolysis of carbaryl, dichlorphos, diazinon, parathion	*P. melophthora*
Hydrolysis of parathion	*P. stutzeri*
Dehalogenation of halide acetate	*P.* species
Total dehalogenation of DDT, aromatic ring cleavage	*P. aeruginosa*
Total degradation of 3-chlorobenzoate	*P. putida*
Oxidative dehalogenation of lindane	*P. putida*
Reduction of nitro group in 4,6-dinitro-o-cresol	*P.* species
Total degradation of 2,4,5-T	*P. cepacia*
Degradation of toluene, xylene, styrene, α-methylstyrene	*P. putida*
	P. aeruginosa

4.2.3.1.5 Enzymes for Dechlorination

Microorganisms have developed a diverse potential to produce an array of enzymes that degraded both aliphatic and chloroaromatic compounds and dehalogenation. Enzymes are classified on the basis of reactions catalyzed given below:

(i) *Oxidative Dehalogenation*

Certain enzymes like mono- or dioxygenase lead to oxidative dechlorination of aliphatic chlorinated compounds through cometabolic reactions. The chlorinated hydrocarbon competes with the growth material of the microorganism for the catalytic site of the oxygenase. The primary step in the aerobic transformation of chlorinated alkenes is generally epoxidation of the C = C bond. The best example of this kind of dehalogenation is methane monooxygenases (MMO), which catalyze the conversion of haloalkenes (such as trichloroethane to its epoxide), and after that it undergoes isomerization or hydrolysis (Fig. 4.3).

Microbial oxidation of trichloroethane was catalyzed by toluene 2,3-dioxygenase, toluene-2-monooxygenase, toluene-4-monooxygenase, phenol hydroxylase, 2,4-dichlorophenol hydroxylase, and propane monooxygenase. Similarly two-component dioxygenases such as 4-chlorophenyl acetate, 3,4-dioxygenase, and 2-halobenzoate 1,2-dioxygenase specially catalyze chloroaromatic compounds. Degradation of 1,2,4,5-tetrachlorobenzene by *Pseudomonas* strain PS14 results in the elimination of HCl during rearomatization of the dihydrodiol, yielding 3,4,6-trichlorocatechol through initial 5,6-dioxygenating attack.

Dioxygenolytic dechlorination of 2,2-dichlorobiphenyl, 2,3-dichlorobiphenyl, and 2,5,2-trichlorobiphenyl at the *ortho*-position is catalyzed by biphenyl 2,3-dioxygenase. All these dioxygenases catalyze the formation of *cis*-diols, which suddenly rearomatize with Cl elimination, yielding a catechol product.

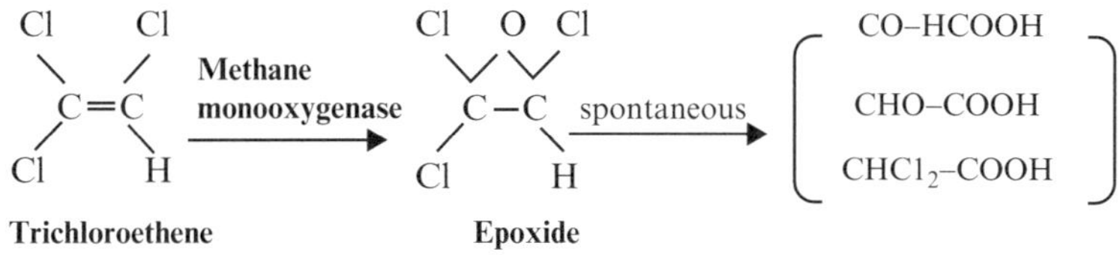

Trichloroethene → (Methane monooxygenase) → Epoxide → (spontaneous) → CO–HCOOH, CHO–COOH, $CHCl_2$–COOH

Fig. 4.3 Oxidative dehalogenation reaction

(ii) *Dehydrohalogenation*

In dehydrohalogenation, HCl is released from the haloorganic compounds, leading to the production of a double bond. The removal of HCl from γ-HCH and γ-pentachlorocyclohexene (intermediate metabolite) is catalyzed by a dehydrochlorinase (LinA). The enzyme catalyzes the release of three chloride ions per molecule of γ-HCH. Dehydrochlorinase (LinA) catalyzes the stereoselective dehydrochlorination of HCH with a *trans* and diaxial pair of H^+ and Cl^-. Glutathione-dependent DDT dehydrochlorinase from the housefly and the 3-chloro-D-alanine dehydrochlorinase have also been reported as dehydrochlorinase enzyme. Ortho-cleavage of chlorocatechols into chlorinated *cis*-muconates by dehydrohalogenases is also reported, which are cycloisomerized to diene lactones.

(iii) *Substitutive dehalogenation*

Substitutive dehalogenation of chlorinated compounds includes the following processes:

1. Hydrolytic processes
2. Thiolytic processes
3. Intramolecular substitution reactions

Hydrolytic Processes

The halidohydrolases are the key enzyme for hydrolytic dehalogenation for heterocyclic, aromatic, and aliphatic compounds. Hydrolytic dechlorination of 1,2-dichloroethane (haloalkanes) by haloalkane dehalogenase from the nitrogen-fixing hydrogen bacterium *Xanthobacter autotrophicus* GJ10 has been reported. Hydrolytic process involves halide elimination by nucleophilic displacement with H_2O. Asp-24 is the nucleophilic residue attacking the substrate. During this process, an ester (intermediate) is formed; after that alcohol molecule is released in the presence of H_2O molecule. The hydrolytic dechlorination mechanism of 1,2-dichloroethane is given as follows (Fig. 4.4).

Another example for hydrolytic dehalogenation reaction is the conversion of 4-chlorobenzoate to 4-hydroxybenzoate which involves three enzymes, i.e., 4-chlorobenzoate coenzyme A (CoA) ligase, 4-chlorobenzoyl-CoA dehalogenase, and 4-hydroxybenzoyl-CoA thioesterase. In the initial step, 4-chlorobenzoate ligase catalyzes the adenylation of the carboxyl group followed by production of the AMP,

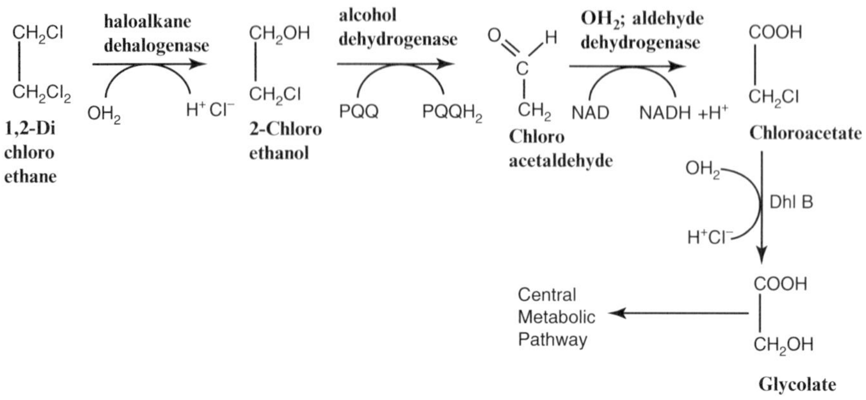

Fig. 4.4 Hydrolytic dehalogenation reaction (*PQQ* pyrroloquinoline quinone)

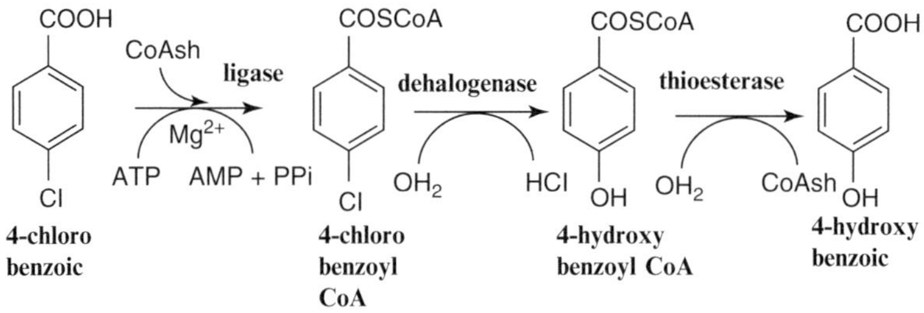

Fig. 4.5 Hydrolytic dehalogenation reactions

a thiol group from CoA, and leading to the production of 4-chlorobenzoyl-CoA (thioester). The production of the CoA ester activates the functional group in the *para* position for nucleophilic attack; after that chlorine is replaced by a hydroxyl group from H_2O by dehalogenase enzyme (Fig. 4.5). In the last step 4-hydroxy benzoic formed from the 4-hydroxybenzoyl-CoA in the presence of the last enzyme thioesterase, and CoA is released by H_2O molecule.

Thiolytic Processes

The thiolytic substitutive dehalogenation process of dichloromethane by facultative methylotrophic bacteria is catalyzed by glutathione-*S*-transferase enzymes. Dichloromethane is converted into formaldehyde and chloride with *S*-chloromethylglutathione (intermediate) in the presence of formaldehyde dehalogenase enzyme. After that, formaldehyde is converted into CO_2 molecule in the presence of formate dehydrogenase enzyme (Fig. 4.6).

Intramolecular Substitution Reactions

Intramolecular substitution reactions are catalyzed by halohydrin–hydrogen halide lyases/halohydrin epoxidases. Halohydrin dehalogenase converted C-2 and C-3 chloroalcohols into its epoxides and was active with chloroacetone and

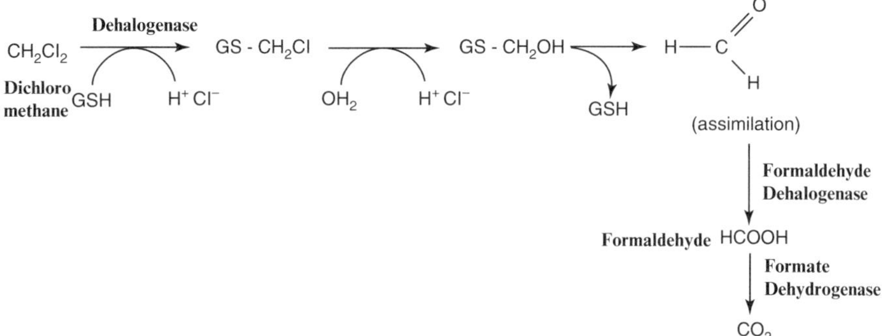

Fig. 4.6 Thiolytic substitutive dehalogenation reaction

1,3-dichloroacetone. This reaction does not require any cofactors or O_2 for the dehalogenation. Thus, the reaction mechanism was proceeding via intramolecular substitution.

4.2.3.2 Phytoenzymes for Bioremediation

Phytoremediation is the *in situ* use of plants, their enzymes, their roots, and their associated microorganisms to remediate toxic pollutants present in different ecological systems. With respect to their direct roles in remediation, plants may employ different mechanisms to efficiently remove both organic and inorganic pollutants from biosphere:

(a) Rhizofiltration
(b) Absorption
(c) Concentration and precipitation of heavy metals by roots
(d) Phytoextraction, i.e., extraction and accumulation of pollutants in plant tissues (roots and leaves)
(e) Phytodegradation, i.e., degradation of complex organic molecules in CO_2 and H_2O which are utilized by plant
(f) Rhizodegradation or plant-assisted bioremediation, i.e., stimulation of microbial degradation by the production of root enzymes and exudates in the rhizosphere
(g) Phytostabilization, i.e., adsorption and precipitation of metals and reduction of their mobility

The synergic interaction between plants and microorganisms specifically occurs in the soil ecosystem influenced by plant roots or rhizosphere. Although plants may be poor in catabolic pathways for the complete degradation of pollutants when compared with microorganisms, research efforts have been devoted to engineer plants with genes that can confer them additional and enhanced degradation abilities. The efficiency of phytoremediation can be enhanced by overexpressing the genes involved in metabolism, uptake, or transport of specific pollutants in plants.

Furthermore, desired genes may be expressed in roots to augment the rhizodegradation of recalcitrant compounds (Abhilash et al. 2009).

Another promising approach to enhancing phytoremediation ability is the cultivation of transgenic plants which produce enzymes for the rhizoremediation of xenobiotics (Abhilash et al. 2009). In these plants xenobiotic-degrading genes have been inserted in their root, and therefore plants secrete degrading enzymes into the rhizosphere. Further rhizosphere effects may contribute to the enhancement of pollutant degradation. Microbial density, diversity, and/or metabolic activity may increase because of the release of plant root exudates, mucigel, and root lysates. The physical and chemical properties of the polluted soil can be augmented by plants as well as by the contact between the root-associated microorganisms and the soil contaminants. Recently, application of plant growth-promoting rhizobacteria (PGPR), i.e., bacteria capable of promoting plant growth by colonizing the plant root, has paid much attention for their use in bioremediation of contaminated soils (Zhuang et al. 2007). Various bacteria associated with plants like wheat, alfalfa, tall fescue, *Brassica juncea*, Indian mustard, canola, and others have been successfully used in the bioremediation of crude oil, PAHs, total petroleum hydrocarbons, TCE, PCBs and lead, zinc, nickel, and cadmium (Zhuang et al. 2007). Therefore, phytoremediation in conjunction with rhizospheric microbes may provide sustainable, eco-friendly, and proficient rhizoremediation for polluted biosphere.

4.2.3.3 Catabolic Enzymes of Degradation Pathways

Catabolic enzymes have been grouped into peripheral and ring-cleavage or lower pathway enzymes. The peripheral enzymes convert a xenobiotic compound into metabolites, which are degradable. These enzymes use products as substrates, and thus these have to be modified to suit the chemo- and region specificities of a variety of xenobiotics. The products of these enzyme-catalyzed reactions are called central metabolites, such as catechol, gentisate, protocatechuate, or their derivatives. Some such important enzymes are:

(A) *Peripheral Enzymes*

(i) *Aromatic Ring-Oxygenases*

These enzymes attach molecular dioxygen into the aromatic ring and require cofactors like NADH and NADPH during reaction. These dioxygenases play an important role in the bacterial catabolism of naturally occurring xenobiotic compounds. Dioxygenases enhance the reactivity of xenobiotic compound through incorporation of two OH groups into the aromatic ring which make them susceptible to enzymatic ring fission reactions. On the basis of structure and component of xenobiotics, these enzymes can be divided into the following subgroups.

(ii) *Multicomponent Dioxygenase*

Multicomponent dioxygenase contains three subunits, a terminal oxygenase also known as iron–sulfur protein or hydroxylase protein, a ferredoxin, and an NADPH-ferredoxin reductase. Multicomponent dioxygenase forms electron transport chains with flavins and iron–sulfur clusters as redox components (Fig. 4.7). The initial component of the chain is a flavoprotein that oxidizes reduced pyridine nucleotides and passes the electrons to the terminal dioxygenases via a ferredoxin electron carrier. The ferredoxin and dioxygenase contained [2Fe-2S] redox centers also known as Rieske type iron–sulfur center, which is either associated with the oxygenase. In xylene degradation by *Pseudomonas putida*, mt-2 requires ferredoxin [2Fe-2S] that reactivates oxygen-reactivated catechol 2,3-dioxygenase (Hugo et al. 2000). A subgroup of ferredoxins [2Fe-2S] also reactivates intrinsically labile extradiol ring cleavage of various aromatic hydrocarbons.

There are also two-component dioxygenases, such as toluate dioxygenase and benzoate dioxygenase from *P. putida* and *Acinetobacter calcoaceticus*, in which the electron transfer function is carried out by a

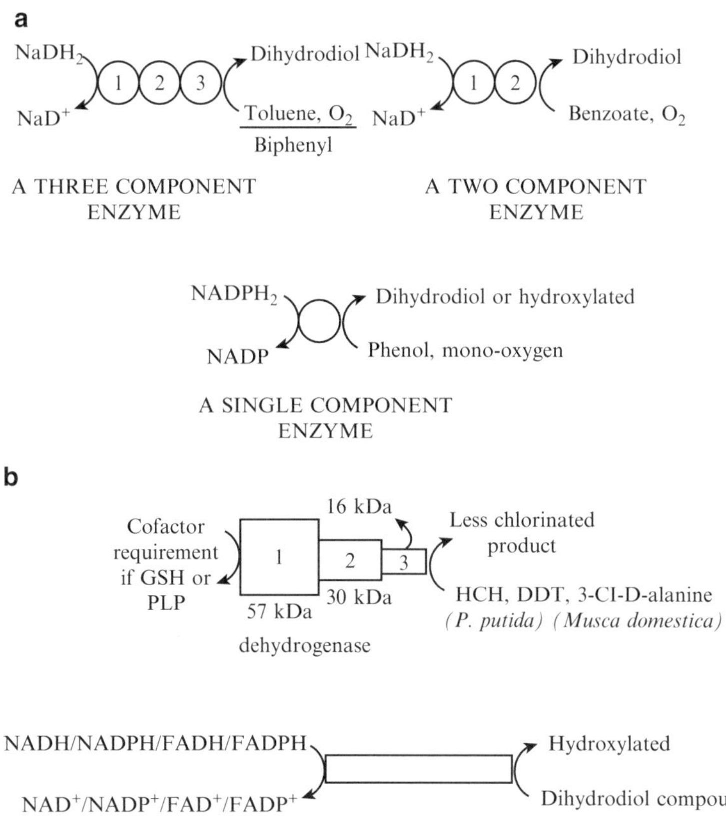

Fig. 4.7 Catabolic enzymes designated as peripheral or upper pathway enzymes with their basic structure. (**a**) Structure of various dioxygenases and mono-oxygenases. (**b**) other hydroxylating systems, such as dechlorinase and dehydrogenase

single protein possessing a ferredoxin-like structure. A novel two-component 2-halobenzoate 1, 2 dioxygenase from *Pseudomonas cepacia* 2CBS has activity toward *ortho*-substituents of chlorobenzoates. Another two-component dioxygenase, 4-chlorophenylacetate 3,4 dioxygenase from *Pseudomonas* spp., strain CBS3, shows dehalogenation activity. These enzymes are members of the short-chain alcohol dehydrogenase family.

(iii) *Multicomponent Monooxygenase*

Phenol hydroxylase is a multicomponent monooxygenase that converts phenol to catechol. Some dioxygenases such as toluene dioxygenase also function as monooxygenases. These multicomponent monooxygenases are structurally related to methane monooxygenases. The existence of six polypeptides has been found to be involved in the activity of hydroxylase in the initial conversion of phenol into catechol. The multicomponent nature of phenol hydroxylase has been intriguing, because, in general, monohydroxylated ring structures such as phenol are oxygenated by single-component flavoprotein monooxygenases.

(iv) *Single-Component Monooxygenase*

Various single-component hydroxylases and monooxygenases have been reported and were found to share conserved domains. Salicylate hydroxylase NahG was shown to be 25 % homologous in amino acid sequence to *p*-hydroxybenzoate hydroxylase from *P. fluorescens*. Similarities have also been reported in the salicylate hydroxylase and phenol hydroxylase.

(B) *Dehalogenase*

These key enzymes catalyze dehalogenation of aromatic hydrocarbons by cleaving the carbon–halogen bond. The haloacid dehalogenases differ with respect to their substrate specificities, electrophoretic mobilities, and inhibition by sulfhydryl-blocking agents. Based on substrate range, reaction type, and gene sequences, the dehalogenating enzymes are classified in different groups, including hydrolytic dehalogenases, glutathione transferases, monooxygenases, and hydratases. A hydration type of dehalogenation reaction has been proposed for aromatic compounds and aliphatic acrylic acids. The enzyme is able to dehalogenate substrates bearing fluorine, chlorine, bromine, and iodine in the 4-position, although the rate of dehalogenation of 4-fluorobenzoyl-CoA is quite slow, while three polypeptides with sizes of 57, 30, and 16 kDa were investigated to consist of the 4-chlorobenzoate dehalogenase activity in *Pseudomonas* spp., strain CBS3. This activity was proposed to be the sum of the individual activities of a 4-chlorobenzoate:CoA ligase, a chlorobenzoate:CoA dehalogenase existing as a heterodimer of 57- and 30-kDa components, respectively, and a 16-kDa 4-hydroxybenzoate:CoA thioesterase. Some oxygenases, such as 2-chlorobenzoate and 4-chlorophenoxyacetate dioxygenases and pentachlorophenol monooxygenase, have also been implicated in the dehalogenation of their substrates. The dehalogenation reaction is believed to be a nucleophilic aromatic substitution in which chloride substituent is replaced by a hydroxyl group derived from water.

(C) *Dehydrogenase*

Dehydrogenases are members of short-chain alcohol family, which have their N-terminal similar to known adenosine triphosphate-binding motifs of NAD +-binding domains. On the basis of the known three-dimensional structures of five proteins out of the 15 or 20 family members, the dehydrogenases containing such motifs, an anionic side chain close to the C-terminal end of ß-α-ß fold of dehydrogenases. This anionic side chain functions as a hydrogen bond acceptor for 2′-OH group of adenosine moiety of NAD^+ but acts unfavorably with the 2′-phosphate group of $NADP^+$. The substrates of BphB, toluene dihydrodiol dehydrogenase (TodD), benzene dihydrodiol dehydrogenase (BnzE), and benzoate dihydrodiol dehydrogenase (EntA) and dihydrodiol dehydrogenases of biphenyls, toluene, benzene, and benzoate degradation, respectively, differ only in their substituents at one of their carbon atoms next to the dihydrodiol carbons. Based on sequence comparison, polychlorinated biphenyl (PCB) degrading dihydrodiol dehydrogenase share about 61 % amino acid with toluene dihydrodiol dehydrogenase (Tod) and benzene dihydrodiol dehydrogenase (*Bnz*E) while only 28 % with benzoate dihydrodiol dehydrogenase (EntA).

4.2.3.4 Aromatic Ring-Cleavage Dioxygenase

These enzymes incorporate two atoms of dioxygen into aromatic substrates, and aromatic ring is cleaved. This reaction does not require any external reductant, such as NAD or NADPH or others. Based on the cleavage of the aromatic ring, they are classified into two types as follows:

(A) *Extradiol Enzymes*

The extradiol ring-cleavage dioxygenases (EDOs) seem to form a superfamily of enzymes that catalyze *meta*-cleavage of catechols. This enzyme consists of four identical subunits of 32 kDa and contains one catalytically essential Fe (II) ion per subunit. The substrate range of this enzyme is relatively broad: this enzyme oxidizes 3-methyl, 3-ethyl, 4-methyl, and 4-chlorocatechol. 3-Chloro and 4-ethycatechol, in contrast, are not efficiently oxidized by this enzyme. Other dioxygenases of this superfamily include catechol 2,3-dioxygenase, 1,2-dihydroxynaphthalene dioxygenase, and 2,3-dihydroxybi-phenyl dioxygenase (BphCs)

(i) *Protocatechuate 4, 5-Dioxygenase*

Protocatechuate 4,5-dioxygenase catalyzes extradiol cleavage of protocatechuate. The enzyme consists of an equal number of two different subunits, α and β, 18 and 34 kDa, respectively, and its quaternary structure may be $(\alpha ß)_2$ Fe^{+2}. The amino acid sequences of the subunits of protocatechuate 4,5-dioxygenase differ from C2, 3O. Investigation of the Fe^{2+} environment of this enzyme from *C. testosterone* using electron spin resonance (EPR) spectroscopy revealed that electron delocalization in the ternary complex, enzyme-Fe (II)-O-O, of a hypothetical reaction sequence is assumed to polarize dioxygen, thus preparing the distal oxygen atom for nucleophilic attack on the aromatic ring of the substrates. The iron peroxy-substrate intermediate, enzyme-Fe (II)-O-O-S, thus produced initiates a sequence of reaction, resulting in the ring fission of the substrate.

(ii) *Intradiol Enzymes*
This group of enzymes consists of catechol 1,2 dioxygenases and protocatechuate 3,4-dioxygenases. Both these enzymes contain Fe^{2+} as cofactors and contain a nonheme, noniron sulfur Fe^{3+} as a prosthetic group. Usually, C1, 2O from many bacteria consist of nonidentical α- and β-subunits (αß-Fe^{3+}), whereas in some bacterial strains, C1, 2O consist of a single polypeptide chain (αα-Fe^{3+}). Chlorocatechol 1, 2-dioxygenase (Clc-C12O) is another class of intradiol enzyme, characterized by broad substrate specificity. It degrades both catechol as well as chlorocatechol, while $C_{12}O$ is not able to catalyze chlorocatechols.

4.3 Aromatic Compounds in the Atmosphere

A wide range of aromatic hydrocarbon compounds coexist in the environment. Plants have such compounds in the form of lignin, alkaloids, turpentine, etc. There are three chief categories: polycyclic aromatic hydrocarbons (PAHs), heterocyclics, and substituted aromatics. The PAHs have two or more complex aromatic rings either in linear or angular to cluster form. The chemical nature is different on the basis of the number of rings and molecular weight. Higher molecular weight reduces the volatility and water solubility of aromatic compounds. PAHs are difficult to degrade in soil and generally come through petroleum, biosystems, and burning process. The biosystem contributes aromatic amino acid lignin compounds and other plant resins, tannins, etc. (Table 4.4).

PAHs may be collected in high concentrations in terrestrial environments near coal gasification sites and tar oil distillation plants. Major sources of PAHs are incomplete burning of organic materials, gas production, wood treatment services, and waste burning. PAHs are formed during thermal geologic reactions linked with fossil fuel and mineral production and during burning of vegetation in forest and bush fires. Anthropogenic sources, mainly fuel combustion, automobiles, spillage of petroleum products, and waste incinerators, are major sources of PAHs into the environment. Tobacco cigarette smoking is a major source of PAH exposure to smokers and secondary smokers. Petroleum refining and transport activities are chief contributors to localized loading of PAHs into the environment.

Dibenzothiophene and carbazole are sulfur- and nitrogen-containing heterocyclic compounds found in creosote, crude oils, and shale oils and often coexist in the environment with PAHs and other aromatic compounds. Dibenzofuran and its substituted analogues are found in numerous woody plants as stress chemicals, so-called phytoalexins. These compounds are reported as carcinogenic and toxic to all the living system and bioaccumulate in the food chain. Persistent organic pollutants (POPs) are referred to as persistent, bioaccumulative, and toxic chemicals (PBTs) such as aldrin, brominated flame retardants, chlordane, DDT, dieldrin, endrin, and mirex and organometallic compounds such as tributyltin, PAHs, heptachlor, hexachlorobenzene, polychlorinated biphenyls (PCBs), polychlorinated dibenzo-*p*-dioxins (PCDDs), polychlorinated dibenzofurans (PCDFs), and fly ash (Yang et al. 2007).

Table 4.4 Bacterial degradation of aromatic compounds

Bacterial species	Aromatics
Achromobacter sp. NCW	Carbazole
Alcaligenes denitrificans	Fluoranthene
Arthrobacter sp. F101	Fluorine, alkylated
Arthrobacter sp. P1-1	Dibenzothiophene, carbazole, phenanthrene
Arthrobacter sulphureus RKJ4	Phenanthrene
Acidovorax delafieldii P4-1	Phenanthrene
Bacillus cereus P21	Pyrene
Brevibacterium sp. HL4	Phenanthrene
Burkholderia sp., *Burkholderia* sp.	Phenanthrene
Burkholderia cepacia	NAP, phenanthrene, pyrene
Burkholderia cocovenenans	Phenanthrene, benzoate, biphenyl
Burkholderia xenovorans	
Chryseobacterium sp. NCY	Carbazole
Cycloclasticus sp. P1	Pyrene
Janibacter sp. YY-1	Dibenzofuran, fluorene, alkylated dibenzothiophene, phenanthrene, anthracene, dibenzo-*p*-dioxin
Marinobacter NCE312	Naphthalene
Mycobacterium sp.	Pyrene, benzo[*a*]pyrene
Mycobacterium sp. JS14	Fluoranthene
Mycobacterium sp. 6PY1, KR2, AP1	Pyrene
Mycobacterium sp. RJGII-135	Pyrene, benz[*a*]anthracene, benzo[*a*]pyrene
Mycobacterium sp. PYR-1, LB501T	Fluoranthene, pyrene, phenanthrene, anthracene
Mycobacterium sp. CH1, BG1, BB1, KR20	Phenanthrene, fluorene, fluoranthene, pyrene
Mycobacterium flavescens	Pyrene, fluoranthene
Mycobacterium vanbaalenii PYR-1	Phenanthrene, pyrene, dimethylbenz[*a*]anthracene
Mycobacterium sp. KMS	Pyrene
Nocardioides aromaticivorans IC177	Carbazole
Pasteurella sp. IFA	Fluoranthene
Polaromonas naphthalenivorans CJ2	Naphthalene
Pseudomonas sp. C18, PP2, DLC-P11	Naphthalene, phenanthrene
Pseudomonas sp. BT1d	3-Hydroxy-2-formylbenzothiophene
Pseudomonas sp. B4	Biphenyl, chlorobiphenyl
Pseudomonas sp. HH69	Dibenzofuran
Pseudomonas sp. CA10	Carbazole, chlorinated dibenzo-*p*-dioxin
Pseudomonas sp. NCIB 9816-4	Fluorene, dibenzofuran, alkylated dibenzothiophene
Pseudomonas sp. F274	Fluorine
Pseudomonas paucimobilis	Phenanthrene
Pseudomonas vesicularis OUS82	Fluorine
Pseudomonas putida P16, BS3701, BS3750, BS590-P, BS202-P1	Naphthalene, phenanthrene

(continued)

Table 4.4 (continued)

Bacterial species	Aromatics
Pseudomonas putida CSV86	Methyl naphthalene
Pseudomonas fluorescens BS3760	Phenanthrene, chrysene, BaA
Pseudomonas stutzeri P15	Pyrene
Pseudomonas saccharophila	Pyrene, phenanthrene
Pseudomonas aeruginosa	
Ralstonia sp. SBUG 290, U2	Dibenzofuran, naphthalene
Rhodanobacter sp. BPC-1	Benzo[*a*]pyrene
Rhodococcus sp.	Pyrene, fluoranthene
Rhodococcus sp. WU-K2R	Naphthothiophene, benzothiophene
Rhodococcus erythropolis I-19	Alkylated dibenzothiophene
Rhodococcus erythropolis D-1	Alkylated dibenzothiophene
Staphylococcus sp. PN/Y	Phenanthrene
Stenotrophomonas maltophilia VUN 10,010	Pyrene, fluoranthene, benzo[*a*]pyrene
Stenotrophomonas maltophilia VUN 10,003	Pyrene, fluoranthene, benz[*a*]anthracene, pyrene, dibenz[*a*,*h*]anthracene, coronene
*Sphingomonas yanoikuyae*R1	Pyrene
Sphingomonas yanoikuyae JAR02	Benzo[*a*]pyrene
Sphingomonas sp. P2, LB126	Fluorene, phenanthrene, fluoranthene, anthracene
Sphingomonas sp.	DBF, alkylated dibenzothiophene, CBZ
Sphingomonas paucimobilis EPA505	Fluoranthene, naphthalene, anthracene, phenanthrene
Sphingomonas wittichii RW1	Chlorinated dibenzo-*p*-dioxin
Terrabacter sp. DBF63	Dibenzofuran, chlorinated dibenzothophene, chlorinated dibenzo-*p*-dioxin, fluorine
Xanthomonas sp.	Pyrene, benzo[*a*]pyrene, carbazole

4.3.1 Aerobic Bacterial Degradative Pathways of Polycyclic Aromatic Hydrocarbons

4.3.1.1 Naphthalene

Several aerobic bacteria, viz., *Alcaligenes*, *Burkholderia*, *Mycobacterium*, *Polaromonas*, *Pseudomonas*, *Ralstonia*, *Rhodococcus*, *Sphingomonas*, and *Streptomyces*, degrade naphthalene via PAH catabolic pathways. Several dioxygenases attack on the aromatic ring mainly during naphthalene degradation and form *cis*-(1R, 2S)-dihydroxy-1,2-dihydronaphthalene (*cis*-naphthalene dihydrodiol). After that, dehydrogenated of *cis*-naphthalene dihydrodiol takes place and forms 1,2-dihydroxynaphthalene by a *cis*-dihydrodiol dehydrogenase. In next step, oxidation of 1,2-dihydroxynaphthalene occurs via nonenzymatic process, and it is converted into 1,2-naphthaquinone. On the other hand, 1,2-dihydroxynaphthalene is metabolized to salicylate. Lastly, salicylate is decarboxylated to catechol, which is further metabolized by ring fission in *meta*- and *ortho*-pathways to succinate and acetyl coenzyme A or acetaldehyde and pyruvate (Fig. 4.8) (Baboshin et al. 2008).

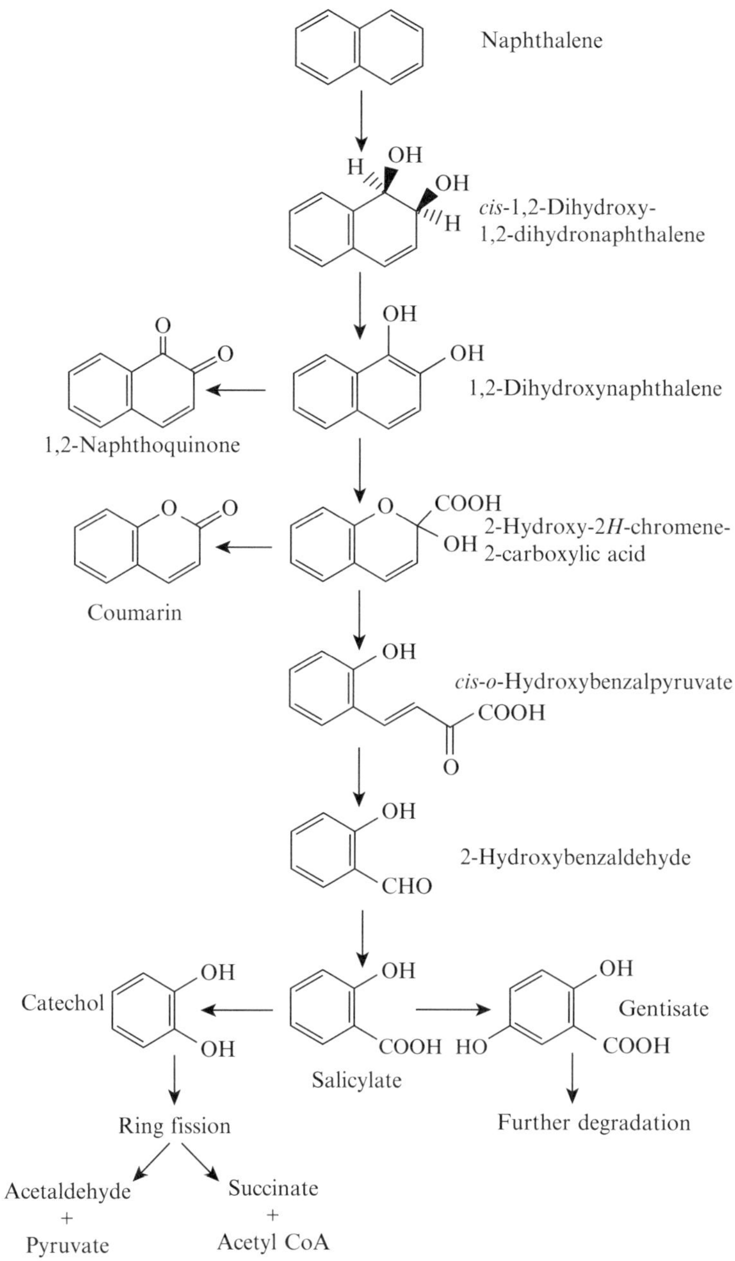

Naphthalene

H OH OH H

cis-1,2-Dihydroxy-1,2-dihydronaphthalene

O O OH OH

1,2-Naphthoquinone

1,2-Dihydroxynaphthalene

O O O COOH OH

Coumarin

2-Hydroxy-2*H*-chromene-2-carboxylic acid

OH COOH O

cis-*o*-Hydroxybenzalpyruvate

OH CHO

2-Hydroxybenzaldehyde

OH OH OH COOH HO OH COOH

Catechol

Salicylate

Gentisate

Ring fission

Further degradation

Acetaldehyde + Pyruvate

Succinate + Acetyl CoA

Fig. 4.8 Degradation of naphthalene

4.3.1.2 Fluorene

Another important aromatic oxygenic compound is fluorene found in fossil fuels and coal derivatives degraded by several bacteria like *Arthrobacter*, *Brevibacterium*, *Burkholderia*, *Mycobacterium*, *Pseudomonas*, and *Sphingomonas* (Baboshin et al. 2008). In initial path fluorene is converted into fluorene-1,2-diol by dioxygenation and is further transformed to 3-chromanone via 2-hydroxy-4-(2-oxo-indan-1-ylidene)-2-butenoic acid, 1-formyl-2-indanone, 2-indanone-1-carboxylic acid, and 2-indanone. During the second path, 3,4-dioxygenation takes place, and fluorene is converted into salicylate as end product through the formation of intermediate like 2-hydroxy-4-(1-oxo-indan-2-ylidene)-2-butenoic acid, 2-formyl-1-indanone, 1-indanone-2-carboxylic acid, 1-indanone, 2-chromanone, and 3-(2-hydroxyphenyl)-propionic acid. In the last step, a subsequent angular carbon dioxygenation occurs, leading to the formation of phthalate that is further converted into protocatechuate (Fig. 4.9).

4.3.1.3 Phenanthrene

A three-aromatic ring system is found in phenanthrene which is a xenobiotic compound degraded by several bacterial strains like *Acidovorax*, *Arthrobacter*, *Brevibacterium*, *Burkholderia*, *Comamonas*, *Mycobacterium*, *Pseudomonas*, and *Sphingomonas*. Phenanthrene contains bay- and K-regions able to form an epoxide, which is assumed to be a vital carcinogen. The microbial degradation of

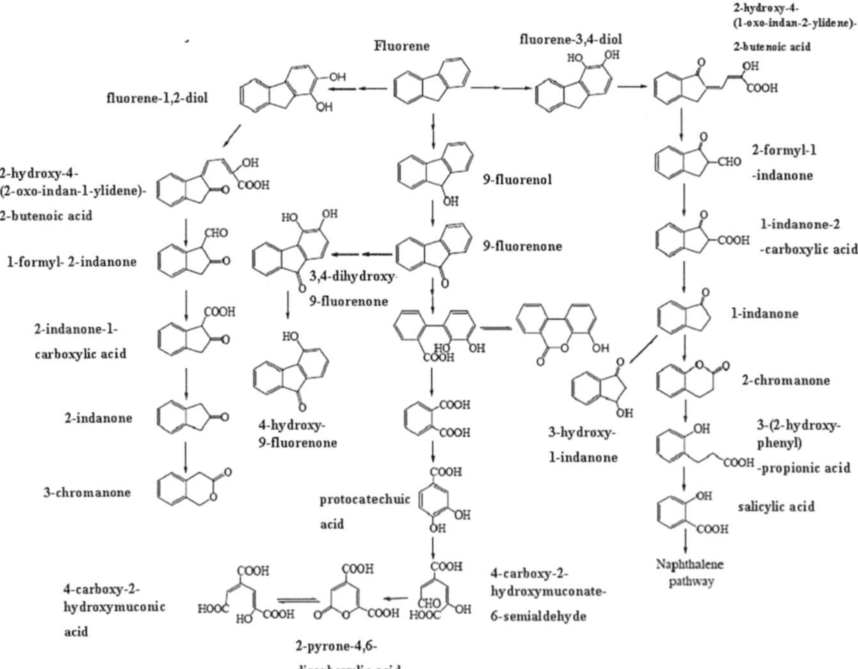

Fig. 4.9 Degradation pathways of fluorene

phenanthrene is facilitated by 3,4-dioxygenation to yield *cis*-3,4-dihydroxy-3,4-dihydrophenanthrene, which undergoes enzymatic dehydrogenation to phthalic acid and 3,4-dihydroxyphenanthrene and finally into protocatechuic acid through *ortho*-cleavage (Seo et al. 2006a, b) (Fig. 4.10).

They also reported that 2-(2-carboxy-vinyl)-naphthalene-1-carboxylic acid is formed by *ortho*-cleavage and is further degraded to naphthalene-1,2-diol via naphthalene-1,2-dicarboxylic acid and 1-hydroxy-2-naphthoic acid. It is possible that 1,2-dioxygenation of phenanthrene forms *cis*-1,2-dihydroxy-1,2-dihydrophenanthrene, which undergoes enzymatic dehydrogenation to 1,2-dihydroxyphenanthrene and finally degraded into salicylic acid through *meta*-cleavage (Fig. 4.10). Mallick et al. (2007) have reported that, a novel *meta*-cleavage of 2-hydroxy-1-naphthoic acid to form *trans*-2,3-dioxo-5-(2′-hydroxyphenyl)-pent-4-enoic acid in *Staphylococcus* sp. PN/Y.

4.3.1.4 Fluoranthene

A four-aromatic ring-containing compound is fluoranthene, a xenobiotic group of PAHs in the environment mainly degraded by *Mycobacterium*, *Burkholderia*, *Pasteurella*, *Rhodococcus*, *Sphingomonas*, and *Stenotrophomonas*. Fluoranthene is

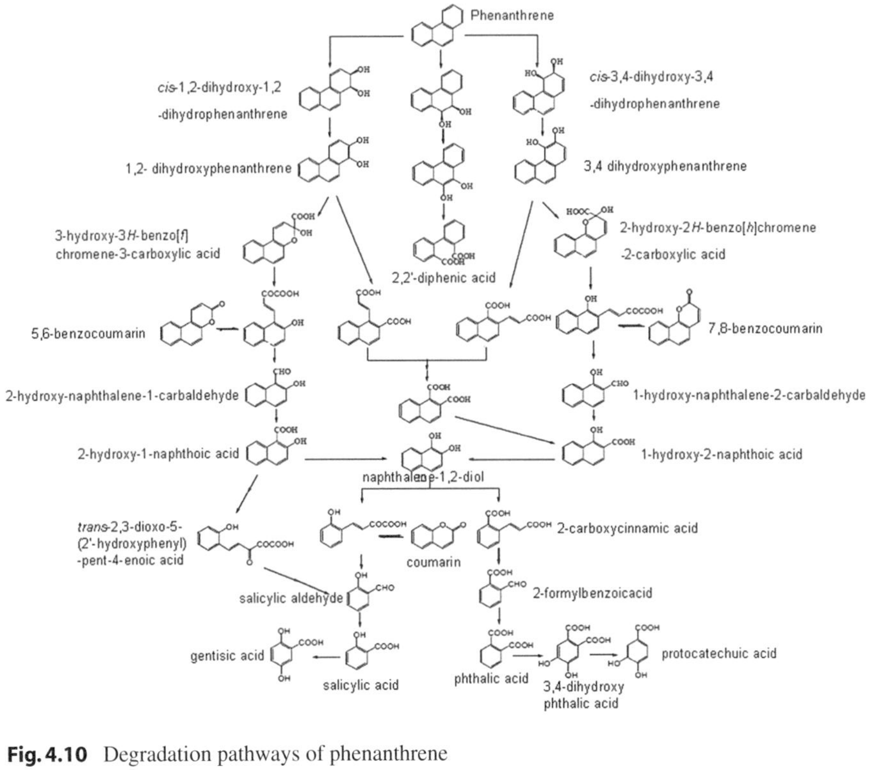

Fig. 4.10 Degradation pathways of phenanthrene

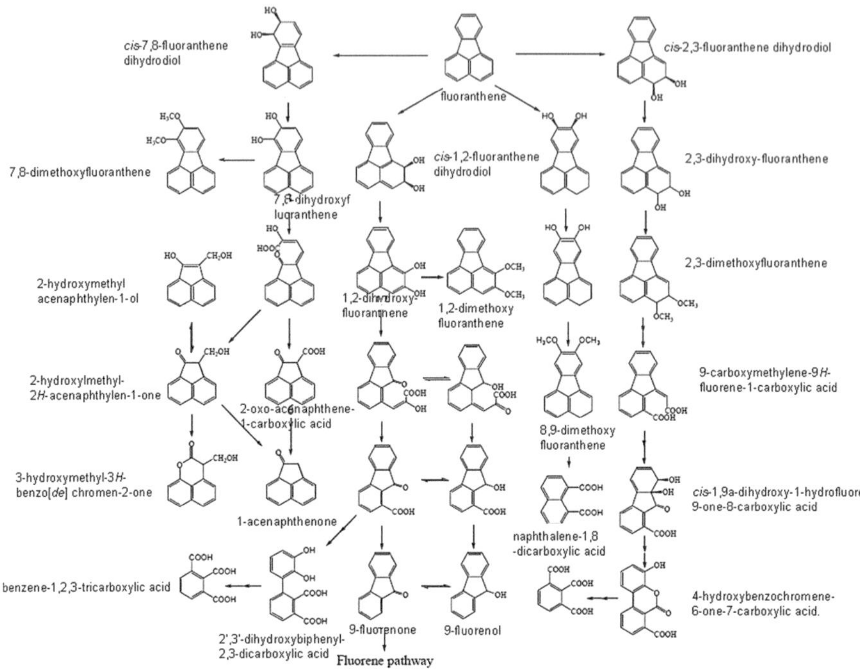

Fig. 4.11 Degradation pathways of fluoranthene

degraded into 1,2- or 7,8-dioxygenation by various microbial system (Fig. 4.11). The compounds having *cis*-1,2-fluoranthene dihydrodiol or *cis*-7,8-fluoranthene dihydrodiol are dehydrogenated to 1,2-dihydroxyfluoranthene or 7,8-dihydroxyfluoranthene, respectively. After that, 7,8-dihydroxy fluoranthene is converted into 1-acenaphthenone and 3-hydroxymethyl-3*H*-benzo[*de*]-chromen-2-one via *meta*-cleavage.

Lee et al. (2007) reported that *Mycobacterium* sp. JS14 showed dioxygenation of fluoranthene at four possible positions: 1,2-position, 2,3-position, 7,8-position, and 8,9-position. 1,2-Dihydroxy-fluoranthene is also further converted into 9-fluorenone by *meta*-cleavage. In the last step, 9-fluorenone is converted into 9-fluoreno.

In several cases, the 2,3-dioxygenation of fluoranthene produced several compounds after degradation, i.e., *cis*-2,3-fluoranthene dihydrodiol, 9-carboxymethylen e-9*H*-fluorene-1-carboxylic acid, *cis*-1,9a-dihydroxy-1-hydrofluorene-9-one-8-carboxylic acid, and 4-hydroxybenzochromene-6-one-7-carboxylic acid (Rehmann et al. 2001). The *cis*-2,3-fluoranthene dihydrodiol is degraded into benzene-1,2,3-tricarboxylic acid by *ortho*-cleavage (Lee et al. 2007).

4.3.1.5 Pyrene

A four-benzene ring-containing compound, pyrene, a by-product of gasification processes and other incomplete combustion of petroleum product, is contaminating the environment. In the first step, pyrene is converted into pyrene cis-1,2-dihyrodiol, pyrene *cis*-4,5-dihydrodiol, and pyrene-*trans*-4,5-dihydrodiol by dioxygenase and

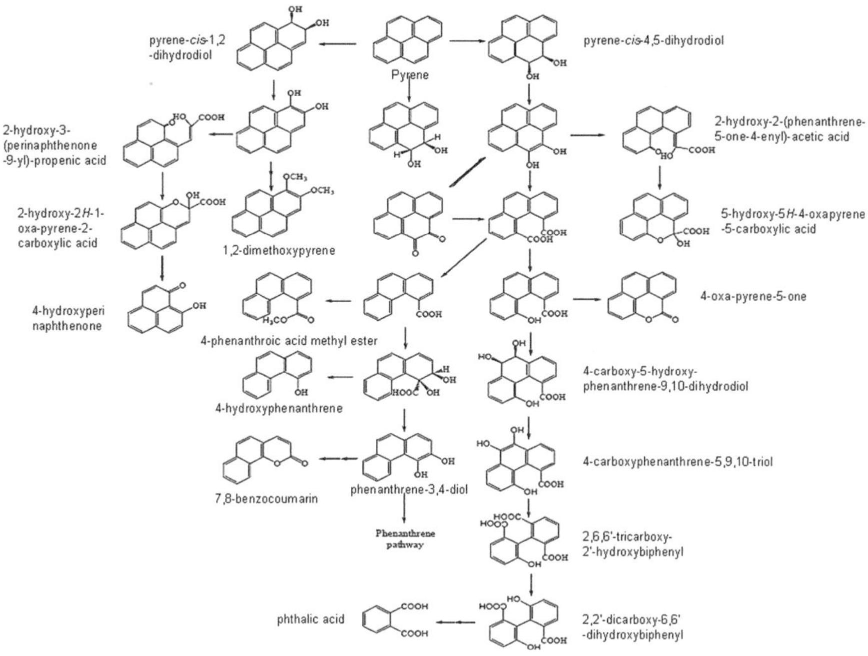

Fig. 4.12 Degradation pathways of pyrene

monooxygenase through multiple oxidative attacks on pyrene (Heitkamp et al. 1988) (Fig. 4.12). Pyrene cis-1,2-dihyrodiol is converted into 4-hydroxyperinaphthenone and 1,2-dimethoxypyrene (Kim and Freeman 2005). Pyrene-4,5-diol is degraded into phenanthrene-4,5-dicarboxylic acid and 5-hydroxy-5*H*-4-oxa-pyrene-5-carboxylic acid by *ortho*-cleavage and *meta*-cleavage (Liang et al. 2006).

4.3.1.6 Benzo[a]pyrene

A five-aromatic benzene ring compound is benzo[*a*]pyrene which is carcinogenic in nature with low water solubility (0.0023 mg/L) and a high octanol/water partition coefficient (LogKow, 6.06). Gibson et al. (1975) reported that at initial stage of degradation of benzo[*a*]pyrene, *cis*-4,5-benzo[*a*]pyrene-dihydrodiol, *cis*-7,8-benzo[*a*]pyrene-dihydrodiol, *cis*-9,10-benzo[*a*]pyrene-dihydrodiol, and *cis*-11,12-benzo[*a*]pyrene-dihydrodiol were formed (Fig. 4.13). They found that benzo[*a*] pyrene *cis*-4,5-dihydrodiol is converted into 4,5-chrysene-dicarboxylic acid which further degraded into chrysene-4-carboxylic acid or 5-carboxylic acid through *ortho*-cleavage (Fig. 4.13). Benzo[*a*]pyrene *cis*-7,8-dihydrodiol is degraded into 7,8-dihydropyrene-8-carboxylic acid via *meta*-cleavage. In the third step, benzo[*a*] pyrene *cis*-9,10-dihydrodiol is converted into 7,8-dihydropyrene-7-carboxylic acid via *meta*-cleavage. In the fourth step, benzo[*a*]pyrene is converted into benzo[*a*] pyrene *cis*-11,12-dihydrodiol and benzo[*a*]pyrene-*trans*-11,12-dihydrodiol by

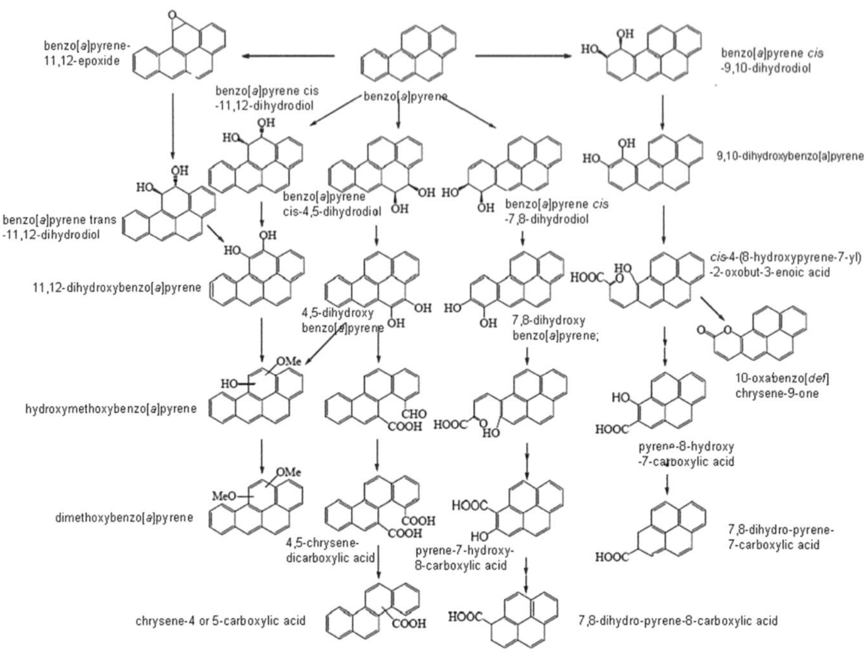

Fig. 4.13 Degradation pathways of benzo[*a*]pyrene

dioxygenase and monooxygenase which further degraded into dimethoxybenzo[*a*]pyrene and benzo[*a*]pyrene 11,12-epoxide.

4.3.2 Aerobic Bacterial Degradative Pathways of Heterocyclic Compounds

The heterocyclic compounds are more recalcitrant in nature. Dibenzofurans, dibenzodioxins, and dibenzothiophenes are highly hazardous pollutants in biosphere. They are generally degraded by *Pseudomonas* spp. and *Sphingomonas* spp. with the help of monooxygenase, dioxygenase, phenoxydase, peroxidases, and laccases. Several enzymes catalyzed angular and lateral dioxygenation reactions of dibenzofuran and carbazole. Dioxygenases of *Pseudomonas* sp. CA10 and *Sphingomonas wittichii* RW1 catalyze angular insertion of oxygen, while dioxygenase of *Pseudomonas* sp. NCIB 9816-4 catalyzes only lateral dioxygenation (Guo et al. 2008).

The broad category of dioxygenase is PAH dioxygenase which catalyzes various reactions including reduction and mono- and dioxygenation. Numerous degradative pathways can produce various metabolites in the presence of different enzymes. For example, *Terrabacter* sp. DBF63 can degrade fluorene and dibenzofuran by the same dioxygenase into phthalate and salicylate, respectively.

Degradation of dibenzo-*p*-dioxins is catalyzed by dioxygenases, an angular dioxygenation to produce trihydroxy diphenyl ethers (Nam et al. 2006). Chlorophenols are highly toxic heterocyclic compounds and affect various living organisms. Chlorophenols degrade into glutathione or other sulfur-containing primary metabolites during dioxin metabolism by different microorganisms.

Petroleum products also produce several high molecular products like dibenzothiophene which catabolized by different enzymes in two pathways (Fig. 4.14). In the first pathway, sulfur oxidation occurs, also known as 4S pathway, in which fast desulfurization is attained by desulfinase. Flavin-containing monooxygenases, noted as DszA and DszB (or SoxA and SoxB), are generally distributed in bacteria and catalyze a successive addition of single oxygen atom. These monooxygenases require FAD as a cofactor, accompanied by a specific flavin reductase. FAD-containing monooxygenases are very common in all biota and catalyze various detoxification steps. In second pathway, dibenzothiophene degraded by lateral dioxygenation and successive reactions by different microbial enzymes (Kodama pathway, Fig. 4.14).

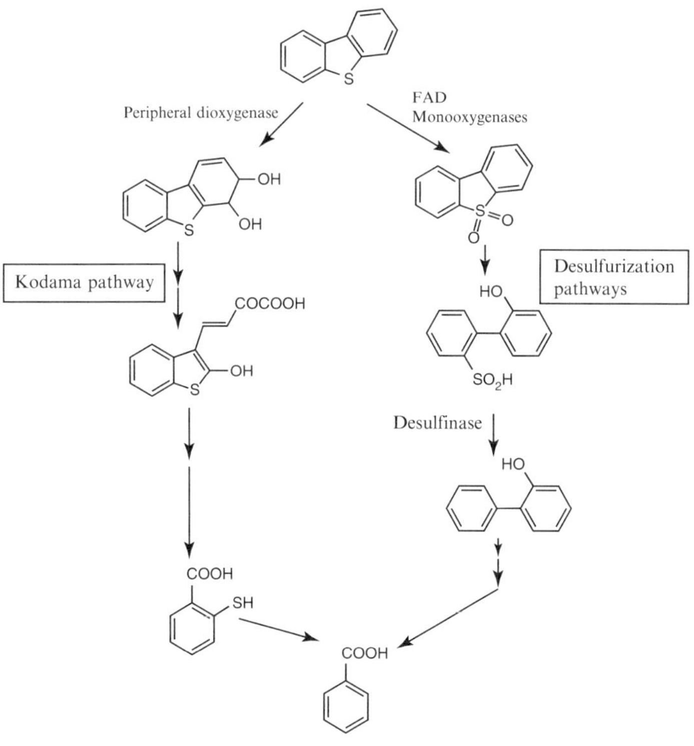

Fig. 4.14 Bacterial degradation of dibenzothiophene

4.3.3 Bacterial Degradative Pathways of Alky PAHs

The PAHs having alkyl and nitro group are common substituted PAHs which produce numerous problems in their degradation and have substantial toxicity. Their existence may inhibit proper orientation and accessibility of the PAHs into dioxygenases. In general, methyl-/ethyl-naphthalenes and phenanthrenes are prevalent contaminants in the ecosystem, and, though, limited numbers of studies have been done in relation to bacterial degradation. Therefore, a microbial consortium with several enzymes is used for catabolism of alkyl PAHs in which oxidation of methyl group to alcohol, aldehyde, or carboxylic acid, decarboxylation, demethylation, and dioxygenation take place. Aerobic bacteria are limited to halogenated pollutants such as PCBs, PBDEs (polybrominated diphenyl ethers), PCDDs/Fs, and halogenated solvents. However, PAHs and their alkyl derivatives can be remediated by various anaerobes through novel catabolic pathways (Fig. 4.15). The presence of

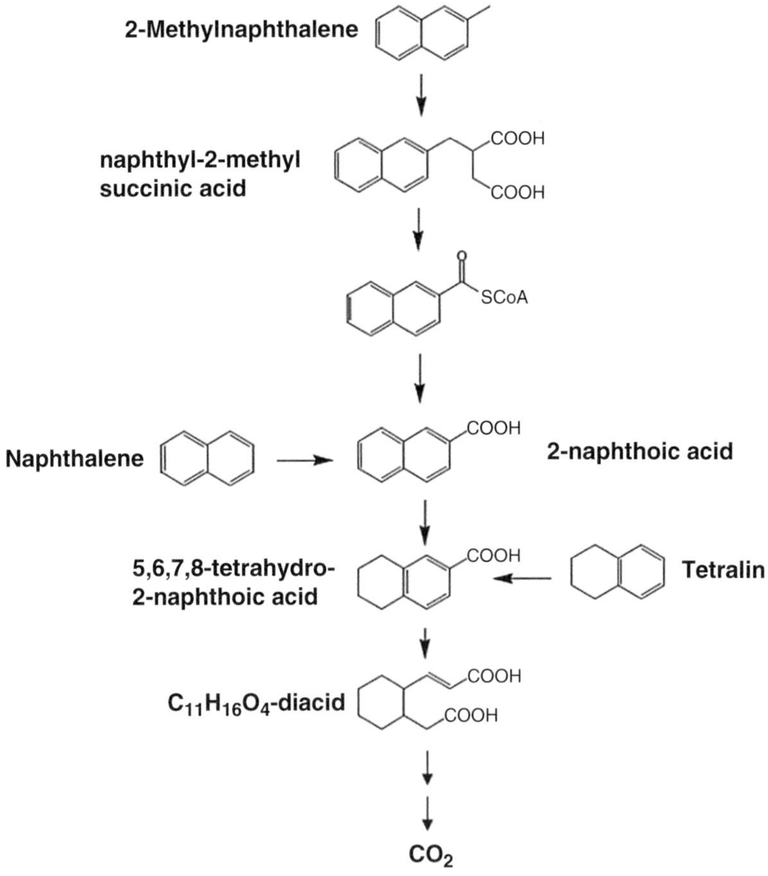

Fig. 4.15 Degradation of naphthalene, methylnaphthalene, and Tetralin by bacteria

ring-cleavage products (e.g., derivatives of cyclohexane) suggests that oxygen molecule or oxygen equivalents may be required in these consecutive catabolic steps (Fig. 4.15). Metabolism of alkyl benzenes, alkanes, and other hydrocarbons are mostly facilitated by anaerobic bacteria.

It is noteworthy that a large amount of nitro-PAHs (e.g., 1-nitropyrene) are produced from automobile exhaustion. It is well known that nitropyrene can be transformed into mutagenic metabolites, including amino-pyrene, nitroso-pyrene, and hydroxyamin-pyrene.

4.4 Phenol

Common name of phenol is hydroxybenzene, an aromatic compound having one hydroxyl group attached to the benzene ring, and its molecular formula is C6H5OH with 94.14 g/mol molecular weight. Phenol is also known as carbolic acid, phenic acid, phenylic acid, phenyl hydroxide, or oxybenzene. It is a white crystalline solid which is soluble in most organic solvents, volatile at room temperature, and quite flammable. Melting and boiling point of phenol is 43 and 181.8 °C. It has a very strong odor (acrid odor) with an odor threshold of 0.04 PPM and a sharp burning taste. Phenol is a weak acid, and in its ionized form, it is very sensitive to electrophilic substitution reactions and oxidations. Phenol is produced both naturally and synthetically by chemical processes. Naturally, phenol has been extracted from coal tar distillation. Synthetically, cumene oxidation accounts for 95 % of phenol production worldwide.

Phenol is used in the production of slimicides, disinfectants, antiseptics, and medicine such as ear and nose drops, mouthwashes, and sore throat lozenges. It is also used in some cream and shaving soap for its germicidal and local anesthetic properties, as an internal antiseptic and gastric anesthetic in veterinary medicine, as a peptizing agent in glue, as an extracting solvent in refinery and lubricant production, as a blocking agent for blocked isocyanate monomers, as a reagent in chemical analysis, and as a primary petrochemical intermediate. Phenol is used for the production of phenolic resins like phenol–formaldehyde resins (Bakelite) which are low-cost thermosetting resins applied as plywood adhesive; it is also used in construction, automotive, and appliance industries. It is broadly used in the health-care, food, and pharmaceutical sectors to check futile microorganisms from causing disease. Phenol acts as the precursor for the production of various epoxy resins. Phenol reacts with acetone and is converted into bisphenol A, a monomer for epoxy resins. It is also used for production of cyclohexanone and cyclohexanone–cyclohexanol mixtures by hydrogenation. Cyclohexanone is later converted into its oxime and further to caprolactam, the monomer for nylon 6. The mixture cyclohexanone–cyclohexanol is oxidized by nitric acid to adipic acid, one of the monomers for the production of nylon-6,6. Phenol is also used to produce polyphenoxy and polysulfone polymers, corrosion-resistant polyester, and polyester polyols. Phenol is also a building block for the synthesis of pharmaceuticals, such as aspirin. Phenol is used along with chloroform (a commonly used mixture in molecular biology for DNA and RNA purification from proteins) and also used for cell disruption and lysis purpose.

Most of phenolic compounds are used as intermediate-level disinfectants to treat noncritical medical devices which pose the lowest risk of transmission of infection and generally contact only whole skin. They hold more activity in the presence of organic material than iodine or chlorine-containing disinfectants. These are usually used in commercial animal production units including hatchery and equipment sanitation and footbaths. Phenolic disinfectants (including cresols and pine oil) are usually safe, but prolonged exposure to the skin may cause irritation.

4.4.1 Toxicity of Phenol

Generally phenol is toxic at low concentrations for microbial cells. Acute exposure of phenol causes central nervous system disorders. Acute exposure of phenol by the oral route leads to damage in the blood, liver, and kidney and cardiac toxicity including weak pulse, cardiac depression, and reduced blood pressure. Ingestion of 1 g phenol is reported to be lethal for humans. Different researchers reported that phenol causes hypothermia, myocardial depression, burning effect on skin, irritation of the eyes, gastrointestinal disorder, diarrhea, and excretion of dark urine. Furthermore, the use of water containing high phenolic compounds may lead to cancer as it is a susceptible carcinogen. Due to these adverse health effects of phenolics, as per the rules of the World Health Organization, the maximum permissible level for phenol in environment is 0.1 mg/l.

Phenol and its substituent compounds are the characteristics pollutants in the wastewater generated from crude oil refineries, ceramic plants, steel plants, coal conversion processes, manufacturing units of phenolic resins, pesticides and explosives, etc. (Table 4.5).

Table 4.5 Phenol concentrations in industrial effluents

Industry	Phenol concentration (mg/L)
Coking operations	28–3900
Coal processing	9–6800
Petrochemicals	2.8–1220
Pulp and paper	0.1–1600
Gas production	4000
Refineries	6–500
Pharmaceuticals	1000
Benzene manufacturing	50

4.4.2 Treatment Methods for the Elimination of Phenolic Wastes

In view of the wide occurrence of phenols in different wastewaters and their toxicity to human and animal life at low concentrations, it is very essential to remove them prior to release of wastewater into water bodies. Hence, it is essential to use suitable approaches for wastewater treatment in order to compensate these rising environmental troubles. Several treatment tools have been employed in this regard. The applied treatment, which could be a single treatment or a combination of these treatments, must assure the elimination of phenol to permissible limits. The option of treatment depends upon the concentration, volume of the effluent, and cost of the treatment.

4.4.2.1 Physical and Chemical Methods for Elimination of Phenol

Different treatment methods, such as adsorption, wet oxidation, chemical oxidation, etc. have been used for elimination of phenols from aqueous solutions. Several treatment methods that are available for treating the phenolic waste include granular activated carbon processes, reverse osmosis, anaerobic processes, electro-Fenton method, combined applications of flotation and coagulation processes, stripping, and oxidation.

The physicochemical treatment technologies mentioned above were found to have inherent drawback. These processes are high energy consuming, noneconomic, and discharge effluents and wastewaters which need further treatment and hence are alarming for the environment. Phenol is removed by the ion-exchange resins only in the alkaline medium, while the maximum phenol removal was obtained by the non-functionalized resin in acidic medium. Furthermore revival of these ion-exchange resins was a tasking process. Further ion exchange is also highly sensitive to pH of the solution (Liotta et al. 2009).

In adsorption, phenol in the wastewater is selectively transferred into the solid phase (adsorbent) instead of eliminated. It once again produces a large amount of solid waste, which further needs a safe removal. As mentioned previously, the use of activated carbon is not cost effective as high cost factors are related with the revival of activated carbon particles from the treated wastewater. There are many drawbacks linked with the chemical oxidation process like the high cost of the chemicals, release of different unsafe by-products, and production of risky component like secondary effluent problem along with the production of dangerous gases. In the case of chemical oxidation of phenol, different oxidizing agents such as hydrogen peroxide, Fenton's reagents, etc. are used. Hydrogen peroxide when used alone has low reactivity and causes incomplete oxidation of many organic contaminants.

Therefore, the technology that emphasizes detoxification and degradation of phenol without the abovementioned problem has become the focus of the research. Biological treatment with pure and mixed microbial isolates is considered to be an attractive and proficient choice for the treatment of polluted wastewaters containing recalcitrant substances such as phenolics since it produces no toxic end products and it is cost effective.

4.4.2.2 Microbial Degradation of Phenol

Biodegradation is the breakdown of toxic organic contaminants to nontoxic and simpler elements by microbial activity. These contaminants can be used as the microbial food source/substrate. Biodegradation of any organic compound can be through a pathway that finally results in the oxidation of the parent compound that often results in the production of energy. Microorganisms have the ability to degrade all naturally occurring compounds; this is known as the principle of microbial infallibility. Although, the biodegradation is restricted in the number of toxic materials it can hold, but where applicable, it is cost effective. Biodegradation produces no toxic end products, is cost effective, and most significantly retains phenol concentration below the toxic limit.

The focus on the microbial degradation of phenols in recent years has resulted in the isolation, culture, adaptation, and enrichment of a number of microorganisms that can grow on the compound as a sole carbon and energy source. Phenol is an antimicrobial agent; many of the microbes are susceptible to this compound. However, there are some microbes which are resistant to phenol and have the ability to degrade phenol (Sankar 2015). The wide variety of microorganisms that can aerobically degrade phenol include pure bacterial cultures such as *Acinetobacter* sp., *Alcaligenes eutrophus*, *Arthrobacte*r, *Bacillus stearothermophilus*, *Nocardioides*, *Pseudomonas aeruginosa*, *Pseudomonas cepacia G4* also known as *Burkholderia cepacia G4*, *Pseudomonas fluorescens*, *Pseudomonas pictorum*, *Pseudomonas putida*, *Pseudomonas resinovorans*, and *Ralstonia eutropha*, while *Desulfobacterium phenolicum* sp. can anaerobically degrade phenol (Yang and Lee 2007).

Moreover, some yeast such as *Candida tropicalis*, *Fusarium flocciferium*, and *Trichosporon cutaneum* are also capable of degrading phenol. Among all the microorganisms listed above, genus *Pseudomonas* comprises an important group of bacteria with environmental application in bioremediation and biological control. In *Pseudomonas*, many of its induced enzymes are nonspecific, and its metabolic pathway contains a high degree of convergence. The convergence of catabolic pathways allows for the efficient utilization of a wide range of growth substrates while the nonspecificity of the induced enzymes allows for the simultaneous utilization of several similar substrates without an excess of redundant genetic coding for enzyme induction.

Microorganisms are able to adapt to the presence of toxic organic compounds by using a whole cascade of adaptive mechanisms. Among the adaptive mechanisms, changes in the fatty acid composition of membrane lipids are the most important reactions of bacteria to membrane-active substances. One adaptive mechanism enabling several *Pseudomonas* strains to grow in the presence of membrane-disrupting compounds is the isomerization of *cis*-unsaturated fatty acids to *trans*-unsaturated fatty acids. This mechanism could also be found in *Pseudomonas* sp. strain ADP. The extent of the isomerization, usually expressed as the *trans/cis* ratio of unsaturated fatty acids, apparently correlates with the toxicity of organic compounds. Additionally, a mutual dependence was found between the activation of this system and the induction and activation of other stress response mechanisms.

Therefore, an increase in the *trans/cis* ratio can be used as an indicator of environmental stress.

4.4.2.3 Mechanisms of Biodegradation of Phenol

(A) *Aerobic Biodegradation*

In microbial degradation of phenol under aerobic conditions, the degradation is started via oxygenation in which the aromatic ring is firstly monohydroxylated by a monooxygenase phenol hydroxylase at ortho-position to the preexisting hydroxyl group to form catechol. This is the chief intermediate resulting from metabolism of phenol by different microbial strains. Depending on the type of strain, the catechol then undergoes a ring cleavage that can occur either at the ortho-position, thus initiating the ortho-pathway that leads to the formation of succinyl CoA and acetyl-CoA (Fig. 4.16) or at the meta-position, thus initiating the meta-pathway that leads to the formation of pyruvate and acetaldehyde (Fig. 4.17).

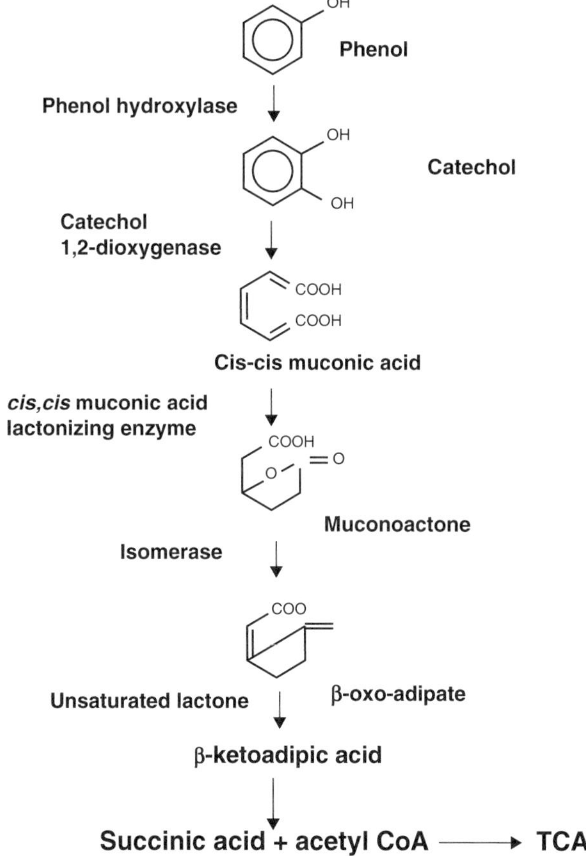

Fig. 4.16 Ortho-cleavage pathway of phenol

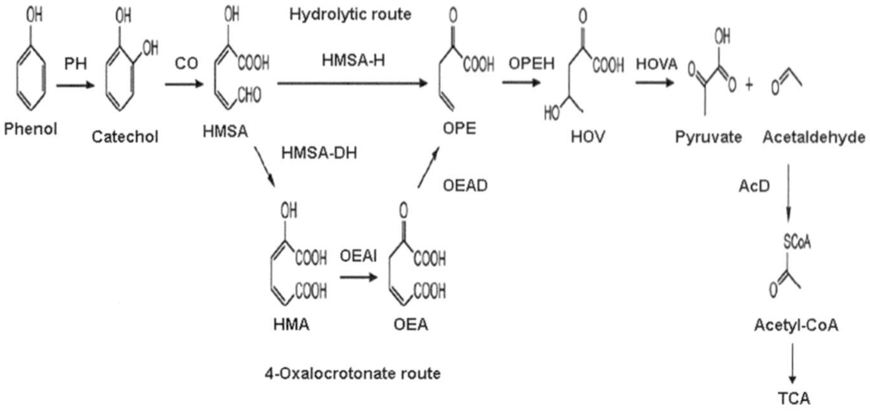

Fig. 4.17 Metacleavage pathway of phenol. *HMSA* 2-hydroxymuconic semialdehyde, *HMA* 2-hydroxymuconic acid, *OEA* 2-oxo-4-enoadipate, *OPE 2-oxopenta-4-enoate*, *HOV* 4-hydroxy-2-oxo-valerate, *PH* phenol hydroxylase, *CO* catechol-2, 3-dioxygenase, *HMSA-DH* 2-hydroxymuconic semialdehyde dehydrogenase, *HMSA-H* 2-hydroxymuconic semialdehyde hydrolase, *OEAI* 2-oxo-4-enoadipate isomerase, *OEAD* 2-oxo-4-enoadipate decarboxylase, *OPEH* 2-oxopent-4-enoate hydratase, *HOVA* 4-hydroxy-2-oxo-valerate aldolase, *AcD* Acetaldehyde decarboxylase, *TCA* tricarboxylic acid cycle

(B) *Anaerobic Biodegradation*

The microbial degradation of phenol under anaerobic conditions is initiated via carboxylation of phenol and has been studied in the denitrifying bacterium *Thauera aromatica*. The phenol carboxylation proceeds in two steps. The first step involves the phosphorylation of the phenol by the addition of a phosphate group from an unknown phosphoryl donor catalyzed by a phosphorylated enzyme called phenyl phosphate synthase (kinase) to form phenyl phosphate as the first intermediate. The second step involves the carboxylation of phenyl phosphate catalyzed by a Mn^{2+} phenyl phosphate carboxylase to form 4-hydroxybenzoate. The production of both the phosphorylating and carboxylating enzymes is strictly regulated. It is very sensitive to oxygen and radical trapping agents; it is not dependent on biotin or thiamine diphosphate and differs from most known carboxylases by using carbon dioxide as substrate and a metal as co-catalyst.

(C) *Enzymes Responsible for the Biodegradation of phenol*

Many microorganisms are capable of degrading phenol by enzymatic activity. These enzymes may include oxygenases, hydroxylases, peroxidases, tyrosinases, and oxidases.

(i) *Oxygenases*

Oxygenases change the hydrophobic organic compound to more water-soluble compounds, and thus it can be broken down by other microorganisms. Two major classes of oxygenases are monooxygenase and dioxygenase. These enzymes are involved in the oxidative metabolism of a wide range of chemicals of pharmaceutical, agricultural, and environmental impact. Some of the most widely recognized substrates for this class of enzymes are the aliphatic and aromatic hydrocarbons of both endobiotic and xenobiotic sources.

(ii) *Monooxygenases*

This class of enzymes inserts one oxygen atom into the substrate, and other oxygen atoms become reduced to water, i.e., two substrates are required. As monooxygenases oxidize two substrates, they are also called mixed-function oxidizes. Also as one of the major substrates becomes hydroxylated, they are also called hydroxylases.

The general stoichiometry is as follows:

$$R-H+NAD(P)H+O_2 \qquad R-OH+NAD(P)+H_2O$$

(iii) *Dioxygenases*

Dioxygenases incorporate both atoms of the oxygen molecule into the substrates. Dioxygenases are very important in initiating the biodegradation of a variety of chlorinated and nitro-aromatic compounds as well as non-substituted PAHs. Many of these compounds are first degraded to catechol or protocatechuate by oxygenases (dioxygenases and monooxygenases). The intermediates are metabolized by ring-cleavage type of dioxygenases to either beta-ketoadipate or 2-keto-4-hydroxyvalerate. These intermediates then enter the TCA cycle.

(iv) *Hydroxylase*

Hydroxylase catalyzes the degradation of phenol via two different pathways initiated either by ortho-or meta-cleavage. Phenol-degrading aerobic bacteria are able to convert phenol into nontoxic intermediates of the tricarboxylic acid cycle via an ortho-or meta-pathway. The monooxygenation of the aromatic ring comprises the first step in the biodegradation of many phenolic compounds. This process is carried out by flavoprotein monooxygenases, which use electrons of NAD(P)H to activate and cleave a molecule of oxygen through the production of an intermediate flavin hydroperoxide and enable the integration of an oxygen atom into the substrate. These reactions can be catalyzed by a single polypeptide chain or by multicomponent enzymes. Monooxygenases consist of a small reductase component that uses NAD(P)H to reduce a flavin that diffuses to a large oxygenase component that catalyzes the hydroxylation of aromatic substrate.

4.5 Dyes

Dyes are colored substances that have an affinity to the substrate to which it is being applied. The dye is commonly applied in an aqueous solution and may require a mordant to improve the fastness of the dye on the fiber. Thousands of synthetic dyes have since been prepared and rapidly replaced the natural dyes. They cost less, they offered a vast range of new colors, and they imparted better properties to the dyed resources. Certain colors need multiple absorption bands; green requires absorption of red and blue violet. This is difficult to attain, and the number of green dyes is relatively less. Black requires a combination of several broad overlapping bands of similar extinction coefficient. The brown, olive green, and dull colors gives different patterns of visible spectrum with different extinction coefficient. The dyes are molecules with delocalized electron systems with conjugated double bonds that contain two groups: the chromophore and the auxochrome. The chromophore is a group of atoms, which controls the color of the dye, and it is usually an electron-withdrawing group. The most important chromophores are C = C, C = N, C = O, N = N, NO_2, and NO groups. The auxochrome is an electron-donating substituent that can intensify the color of the chromophore by altering the overall energy of the electron system and provides solubility and adherence of the dye to the fiber. The most important auxochromes are NH_2, NR_2, NHR, COOH, SO_3H, OH, and OCH_3 groups. Most of the dyes are classified by two ways based on: Chemical constitution and dyeing properties (Fig. 4.18 and Table 4.6).

The routine use of dyes in day-to-day life is rising because of rapid industrialization, most widely in textile, rubber, enamel, plastic, cosmetic, and many other industries. Such industrialization resulted into the server environmental pollution, and water is the prime factor affected, among these. This phenomenon is common

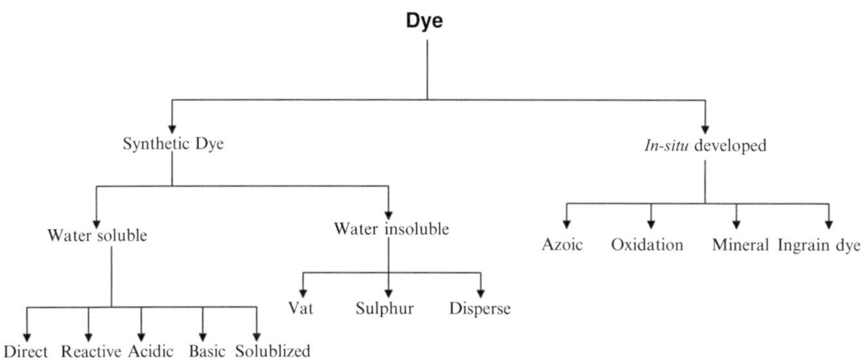

Fig. 4.18 Classification of dyes

Table 4.6 Different classes of dyes on the basis of dispersion

Sr. no	Dye	Class
1	Crystal violet	Triphenylmethane
2	Malachite green	Triphenylmethane
3	Red HE3B	Reactive
4	Green HE4BD	Reactive
5	Direct red 5B	Direct
6	Direct blue GLL	Direct
7	Scarlet RR	Disperse
8	Brown 3RL	Disperse

where the polluting industries like textile dying, leather tanning, paper and pulp mills, and sugarcane industries thrive as a cluster. The effluent release from these industries leads to severe pollution of surface water, groundwater, soil, and other natural resources. The wastewater from the textile and paper pulp industries is a main source of water pollution as they not only contain unsafe chemicals but also the color, which severely affect the aquatic life by reducing the light penetration.

4.5.1 Different Dyes

4.5.1.1 Triphenylmethane Dyes

Triphenylmethane derivatives and their structurally related compounds like fluoresceins and xanthenes belong to synthetic colorants and are used widely in the textile industries for dying cotton, wool, silk, nylon, etc. These are broadly used as industrial dyes for foods, drugs, cosmetics, printing inks, or laboratory indicators as well as nuclear, cytoplasmic, and connective tissue stains. They are considered as the xenobiotic compounds and are very recalcitrant to biodegradation. Triphenylmethane dyes are derived from triphenylmethane (Fig. 4.19).

Triphenylmethane is the hydrocarbon with the formula $(C_6H_5)_3CH$. This colorless solid is soluble in organic solvents. Triphenylmethane has the basic skeleton of many synthetic dyes called triarylmethane dyes; many of them are pH indicators and some display fluorescence. Triphenylmethane dyes are water-soluble organic compounds that contain a colored cation. The intense color of this ion is due to the absolute conjugated system of alternate double and single bonds. The different triphenylmethane dyes used for present study include malachite green, cotton blue, patent blue violet, new fuchsine, light green FCF, crystal violet, bromophenol blue, fuchsinic acid, coomassie brilliant blue, phenolphthalein, bromocresol purple, bromocresol green, victoria blue, etc.

4.5.1.2 Malachite Green

Malachite green is used as a fungicide and ectoparasiticide in aquaculture and fisheries to control fungal and protozoal infections. Malachite green is an organic dye, popular for dyeing materials such as leather, silk, wool, jute, ceramics, cotton, acrylic fibers, and paper. Malachite green and its reduced form, leucomalachite green, may persist in edible fish tissues for extended periods of time (Mitrowska and Posyniak 2004). Short-term exposure (10–20 days) of fishes to malachite green in

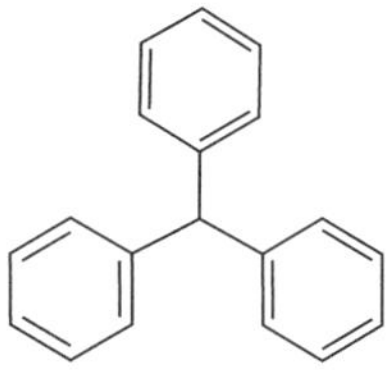

Fig. 4.19 Structure of triphenylmethane

subacute (0.10 mg/l) and sublethal (0.05 mg/l) concentrations leads to decrease in serum calcium and protein level.

4.5.1.3 Cotton Blue

Cotton blue is also known as methyl blue, helvetia blue, and acid blue 93. It is used as a stain in histology and a fungal stain. It is soluble in water and slightly soluble in ethanol. Cotton blue is especially used in textile industries for dyeing cotton and is recalcitrant for biodegradation.

4.5.1.4 Patent Blue

Patent blue is frequently used for lymph node mapping. Allergic and hypersensitivity reactions for patent blue violet used for lymphangiopathy and during sentinel lymph node removal were already reported.

4.5.1.5 New Fuchsine

New fuchsine is intimately related to the dyes that make up the mixture called basic fuchsine, and it can be substituted for them, chiefly. Light green FCF can be used for tinned green peas and other vegetables, jellies, sauces, fish, desserts, and dry bakery mixes at level of up to 100 mg kg^{-1}. This dye has been found to have tumorigenic effects in experimental animals, as well as mutagenic effects in both experimental animals and humans. It furthermore risks irritation of eyes, skin, digestive tract, and respiratory tract in its undiluted form.

4.5.1.6 Methyl Violet

Methyl violet is used in preparation of purple dye textiles and gives deep violet colors in paints and printing ink. It is also used as a pH indicator in chemistry. Methyl violet has the ability of binding DNA. Thus in biomedical sciences, it is used for cell viability assays. Its binding to DNA can cause disruption in DNA replication process leading to mutations and cancers.

4.5.1.7 Bromophenol Blue

Bromophenol blue is used as a pH indicator, as a color marker to monitor the agarose gel electrophoresis or polyacrylamide gel electrophoresis, to stain proteins in wet-mount slides, and also as a dye in textiles. Fuchsinic acid, a magenta dye, is used in dying textiles, as a stain to visualize bacteria and sometimes as a disinfectant. Coomassie brilliant blue, initially, was developed as a wool dye. It is used as laboratory stains to visualize proteins in SDS and native polyacrylamide gel electrophoresis (PAGE). It is an important component of the Bradford method for determining protein concentration in a solution.

4.5.1.8 Phenolphthalein

Phenolphthalein has been used as a laxative, over a century, but now is removed from the market because of its carcinogenicity. Phenolphthalein is used in presumptive blood test, commonly known as the Kastle–Meyer test. Phenolphthalein is used in toys, as disappearing inks or disappearing dye on the Hollywood hair (Barbie hair). It is also used as a pH indicator and to test for signs of carbonation reactions in concrete.

4.5.1.9 Bromocresol Purple

Bromocresol purple is most commonly used as a pH indicator and in medical laboratories to assess albumin. Bromocresol green is used as a pH indicator and as a tracking dye for DNA agarose gel electrophoresis. Victoria blue is used in carbon paper, paper, ink, printing inks, and textile dyeing and used for ink applications generally used in alcohol, glycol, or glycol ether systems. Bromothymol blue is a textile dye derivative and is found to be very harmful in case of ingestion; dangerous in case of skin contact (irritant), of eye contact (irritant), and of inhalation; and slightly hazardous in case of skin contact (permeator). TiO_2 mediated photocatalytic degradation of which is studied.

4.5.1.10 Disperse Dyes

Disperse dyes are incompletely soluble in water. These are generally applied in the form of a dispersion of finely divided dye in a soap solution in the presence of various solubilizing agents like phenol, cresol, and benzoic acid. The assimilation into the fiber is carried out at high temperatures (130 °C) and pressures. Several new methods developed are thermofixation, solvent dyeing, and transfer printing. Disperse dyes are used to dye acetate rayon, Dacron, nylon, cellulose triacetate, polyester, and acrylic fibers. The dyes are finely ground in the presence of a dispersing agent and then sold as a paste or spray dried and sold as a powder. The dyeing rate can be considerably influenced by the variety of dispersing agent used during the grinding. The first disperse dyes were reported for cellulose acetate rayon by British Celanese Corporation in 1920. These dyes were dispersed by numerous techniques like precipitation, milling with surface-active agent, and milling with special kind of sand/glass. The disperse dyes should possess a slight solubility in water for proper dyeing. The dispersing agents include alkyl sulfates, alkylaryl sulfonates, fatty alcohol or amine, and ethylene oxide condensation products; naphthalene, sulfonic acid and formaldehyde condensation products, lignin sulfonate, etc. have been used with disperse dyes.

Disperse dyes are commonly used in the textile industry to color synthetic fabrics such as nylon, orlon, polyester, and cellulose acetate, and these are the only dyes that can be used for dyeing polyester fibers. The decolorization/degradation of disperse dyes has been reported earlier by using physicochemical methods. While, few reports are existing on the degradation of disperse dyes by using biological systems. Earlier researchers have reported the radiation-induced decolorization/degradation of disperse dyes so that radiation technology could be an optional method to elucidate the problem of textile industry, which uses disperse dyes for dyeing. The photochemically active silicadodecatungstic acid ($H_4SiW_{12}O_{40}$) and isopropanol (electron donor) redox system has been considered earlier to study the decolorization of disperse dyes (Fig. 4.20).

4.5.1.11 Reactive Dyes

Reactive dye is a colored compound which has a suitable group capable of forming a covalent bond between a carbon atom of a hydroxy, an amino, and a mercapto group, respectively, of the substrate. Covalent bond between dye and substrate

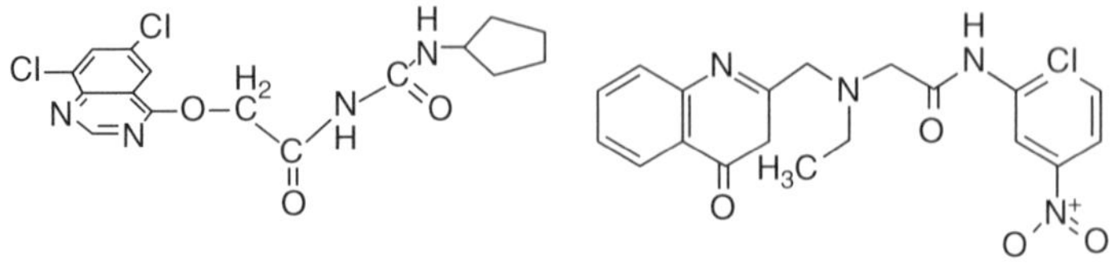

Fig. 4.20 Structures of Brown 3 REL and Scarlet RR

would result in improved wash fastness compared with that of common dye-substrate systems where weaker forces were effective is a previous one. Attempts were made by different dye firms from about 1906 beyond to attain this aim, but it was not until 1956 that the first successful reactive dyes, the Procions, were introduced by ICI for the dyeing and printing of cellulose fibers, following the work of Rattee and Stephen from 1954 onwards.

4.5.2 Toxicity of Dyes

The toxicity of many of these chemicals is known, but limited data are available on biological effects and on the toxicity of fabrics containing these chemicals. An *in vitro* test capable of detecting the combined effects of chemicals on textile products could give useful information for the development of less toxic textile products. Different textile processes can produce different levels of toxicity in products. Reactive dyes have good technical characteristics, but they have been found to cause adverse effects on workers in textile factories and on the environment. Wastewaters and land in an industrial area in India were studied to assess the possible genotoxic health risk and environmental genotoxicity due to textile industry effluents (Mathur et al. 2005). The toxicity was not caused only by textile dyes but by a large number of different textile chemicals. Allergic dermatoses and respiratory diseases are known to be caused by reactive dyes. Contact dermatitis and asthma were also studied by the action of dyes, which affect the change in the immunoglobulin levels. Previous studies have also suggested increased risks of colon and rectum cancers; however, these cancers relate mostly to dyes for synthetic fibers. Besides the toxic effects to the humans, these dyes and dyestuff also affect other forms of life badly especially aquatic life. Acute and short-term toxicity studies of textile dyestuff and wastewaters made on a freshwater fish *Gambusia affinis* revealed the significant reduction in mortality and erythrocyte count of the fish. Similarly strong genotoxic effect of textile effluent on the root cells of *Allium cepa* is also previously demonstrated. A phytotoxicity study of dyes and effluents using plant seedlings is the main and primary toxicity study to assess the toxic nature of dye molecules.

4.5.3 Dye Degradation Techniques

Textile wastewater and dyestuff were recalcitrant in nature. Several physicochemical and biological decolorization methods can be applied for color removal from textile effluent.

4.5.3.1 Physicochemical Techniques

Several physicochemical techniques like membrane filtration, coagulation/flocculation, precipitation, flotation, adsorption, ion exchange, ion pair extraction, ultrasonic mineralization, electrolysis, advanced oxidation (chlorination, bleaching, ozonation, Fenton oxidation, and photocatalytic oxidation), and chemical reduction are applied for dye decolorization/degradation. Among the physicochemical methods, adsorption method is one of the most effective and traditional techniques, which is used for water and wastewater treatment. Adsorption is a rapid phenomenon of passive sequestration and separation of adsorbate from aqueous/gaseous phase into solid phase. For example, chitin is amino nitrogen-containing basic adsorbents, which significantly adsorped acidic dyes.

Membrane filtration methods have the ability to clarify, concentrate, and, most importantly, separate dye continuously from effluent. Ultrafiltration (UF) and nanofiltration (NF) are more effective membrane filtration (pressure-driven) techniques for wastewater treatment. Irradiation method applied for water and wastewater treatments produces high-quality water. While the application of these technologies for remediation of toxic compounds at a large scale is limited due to cost economics and maintenance, decolorization of dye by ion-exchange methods has not been widely used, mainly due to the option that ion exchangers cannot accommodate a wide range of dyes. Through ion-exchange method, both cationic and anionic dyes can be removed from effluents. Oxidation is the most commonly used chemical decolorization process; different oxidizing agents such as chlorines, ozone, Fenton's reagents, UV/peroxide, UV/ozone, or other oxidizing techniques are used for color removal (Abdellatif 2012).

Chlorine is a good dye-oxidizing agent and applied at low capital and operating costs. Hydrolyzing metal salts of iron and aluminum are widely used as primary coagulants to promote the formation of aggregates in effluent and reduce the concentration of colorants and other dissolved organic compounds. Short retention time and low capital cost make chemical coagulation a widely used technique. The high cost of chemicals for precipitation as well as for pH adjustments, problems associated with dewatering and disposing of generated sludge, and high concentration of residual cation levels which remains in the supernatant are some of the limitations of this method.

4.5.3.2 Biological Techniques

Biological decolorization methods use several classes of microorganisms including bacteria, algae, and fungi to degrade the dyes and industrial wastewater. Many workers have reported that several bacteria, fungi, algae, actinomycetes, mixed cultures, or their enzymes participate in dye decolorization (Thummar and Ramani 2014).

4.5.3.2.1 Fungal Decolorization

Several fungi have been reported for the biodegradation of different industrial dyes and dyestuffs and convert them into less toxic compound, with their extracellular enzymes. Many studies have also demonstrated that several fungal isolates are capable of degrading various synthetic dyes such as azo, triphenylmethane, polymeric, phthalocyanine, and heterocyclic dyes. Many researchers used the lignolytic and nonlignolytic fungi for the decolorization of dye containing wastewater. The lignolytic white rot fungi are most efficient microorganisms for dye decolorization. The lignolytic fungi, including *Phanerochaete chrysosporium*, *Trichophyton rubrum* LSK-27, *Ganoderma* sp. WR-1, *Trametes versicolor*, *Funalia trogii*, *Irpex lacteus*, etc., were widely used for the decolorization of textile dyes. White rot fungi produce enzyme like laccase, Mn peroxidase, and lignin peroxidase (LiP), which are responsible for lignin degradation. Dye decolorization by the fungus is mediated by biosorption as well as biodegradation mechanism. The yeast *S. cerevisiae* and waste yeast biomass is also reported for dye decolorization as well as decolorization of textile effluent (Phugare et al. 2010). Various *Penicillium* species were also reported for dye decolorization; *Penicillium ochrochloron* decolorizes cotton blue (50 mg/l) within 2.5 h; newly isolated fungal strain *Penicillium* sp. QQ could decolorize azo dyes significantly (Gou et al. 2009). Fungi represent the promising group of microbes for bioremediation. The white rot fungi do not use lignin as a carbon source for their growth but showed nonspecific mechanisms for the lignin degradation. Table 4.7 summarizes variety of chemicals metabolized by fungi.

Ligninolytic fungi were the possible alternative studied for dye degradation, and white rot fungi were first reported to degrade dyes though few nonligninolytic fungi like various *Aspergillus* sp., and some species of penicillia have been reported to decolorize different dyes and textile effluents. Since then white rot fungi is the most extensively studied fungal group for degradation of dyes. Several enzymes like laccases, polyphenol oxidases, lignin peroxidases, reductases, and methyl transferases from *P. chrysosporium* have ability to degrade sulfonated azo dyes. *Pycnoporus sanguineus* was shown to decolorize azo dyes, orange G, and amaranth partially and triphenylmethane dyes, bromophenol blue, and malachite green completely. Bioremediation of amaranth dye by *Ganoderma* sp. has been reported by Revenkar and Lele (2007). The yeast *Saccharomyces cerevisiae* was shown to decolorize malachite green and methyl red. Laccase enzyme from *Pyricularia oryzae* has ability to oxidize phenolic azo dyes, while lignin peroxidase from *P. chrysosporium* has the potential to decolorize azotriphenylmethane dyes into nontoxic compounds (Kubilay 2009). Several other enzymes like lignin peroxidase and manganese peroxidase from *P. chrysosporium* play a significant role in decolorization of olive mill wastewater. Similarly, *Aspergillus ochraceus* NCIM-1146 mycelium was reported for textile dye (reactive blue-25) decolorization. Table 4.8 represents few of the dye degrading molds and yeasts.

4.5.3.2.2 Bacterial Decolorization

Bacterial decolorization of the dyes is studied extensively, and several bacteria were reported for dye decolorization either alone or in combination. Decolorization of

Table 4.7 Chemicals metabolized by fungi

Chemical	Fungal species
Olive mill wastewater	*Funalia troga*, *Coriolus versicolor*, *C. versicolor* ATCC200801, *F. trogii* ATCC 200800
Phenolic azo dye	*Pyricularia oryzae*
Vinasse	White rot fungi
Insecticide N,N-diethyl-m-toluamide	*Cunninghamella elegans* ATCC 9245, *Mucor ramannianus* R-56, *Aspergillus niger* VKMF-1119, *Phanerochaete chrysosporium* BKM-F-1767
Azatadine	*C. elegans*
Malachite green	*C. elegans*
Jervine	*C. elegans*
Naringenin	*C. elegans*
Diphenhydramine (benadryl)	*C. elegans*
Amoxapine	*C. elegans*
Ethaboxam	*C. elegans*
Vinclozolin	*C. elegans*
Pentachlorophenol (PCP) and creosol	*Phanerochaete chrysosporium*, *P. sordida*, and *Tramates hirsute*
Dodecane and hexadecane	*Candida maltose*
Bisphenol A	*Pleurotus ostreatus*
Bisphenol A, nonylphenol	Lignin degrading
Bisphenol A, 2-4 dichlorophenol, diethyl phthalate	*Trametes* sp.
Diethyl ether and MTBE	*Graphium* sp.
Hydrocarbon	*Pseudallescheria boydii*
Polycyclic aromatic hydrocarbons	Ligninolytic and nonligninolytic fungi
Pyrene and benzo[a]pyrene	Nonbasidiomycete soil fungi
PAH	White rot fungi
Polyethelene	*Penicillium simplicissimum*
Aliphatic hydrocarbons	*Trichoderma asperellum* strain TUB F-1067 (SA4), *Trichoderma asperellum* strain Tr48 (SA5), *Trichoderma asperellum* strain TUB F-756 (SA6), *Penicillium* species (P1), *Aspergillus* species (P9)
Polyvinylamine sulfonate anthrapyridone	*P. chrysosporium*

synthetic dyes from industrial effluents by bacterial system is more eco-friendly than others. Bacterial treatment is relatively cost effective, and the end products after mineralization are not hazardous. Anaerobic degradation/decolorization of azo dyes involves fission of the azo bond and yields aromatic amines as the last product. In the initial step of dye decolorization, bacteria cleave the azo bond by reduction and convert it into colorless aromatic amines. Till date several bacterial species are

Table 4.8 Dye degrading yeasts and molds

Organism	Dyes degraded
Molds	
Penicillium ochrochloron MTCC 517	Cotton blue, malachite green, and textile industrial effluent
P. commune, *P. freii*, *P. allii*	Direct violet
Aspergillus sp.	Reactive blue
A. ochraceus NCIM-1146	Reactive blue-25
Phanerochaete chrysosporium	Congo red
Tinea versicolor	Acid violet 7, acid green 27, indigo carmine
Ganoderma sp.	Amaranth
P. chrysosporium BKM-F-1767 (ATCC 24725)	Poly R-478
Trichophyton rubrum LSK-27	Remazol blue RR and supranol, turquoise GGL
Yeasts	
Saccharomyces cerevisiae MTCC 463	Malachite green
	Methyl red
Galactomyces geotrichum	Methyl red, scarlet RR, brown 3 REL
S. cerevisiae, *Candida tropicalis*, *C. lipolytica*	Remazol blue
Kluyveromyces marxianus IMB3	Remazol black B
C. zeylanoides	Azo dyes

reported for dye decolorization as *Pseudomonas* sp. SUK, *Exiguobacterium* sp. RD3, *Comamonas* sp. VUS, *Bacillus* sp. UVS, *Pseudomonas aeruginosa* strain BCH, and *Staphylococcus hominis* RMLRT03 (Singh et al. 2014, 2015).

Along with the pure bacterial culture, the bacterial consortium is also reported for the efficient decolorization of dye as well as industrial wastewater. The bacterial consortium DAS consisting of *Pseudomonas* sp. LBC2, *Pseudomonas* sp. LBC1, and *Pseudomonas* sp. SUK was reported earlier for removal of color, metals, and toxicity form textile effluent. Similarly, the bacterial consortium PMB11 was reported for bioremediation of reactive blue 59 and red HE3B (Patil et al. 2010). Decolorization of dyes by bacteria is mainly mediated through the bacterial enzyme systems, and lignin peroxidase, azo-reductase, laccase, DCIP reductase, and tyrosinase are among the few dye decolorizing enzymes of the bacteria.

Complete degradation of a diazo dye direct blue-6 by *Pseudomonas desmolyticum* NCIM 2112 within 72 h of incubation was reported previously. The bacterial isolate *Rhizobium radiobacter* MTCC 8161 showed 80–95 % decolorization efficiency for various azo, triphenylmethane, disperse, and reactive textile dyes (Telke et al. 2010). The decolorization of reactive blue 59 by three isolated species from dye contaminated site, namely, *Proteus* sp. SUK7, *Bacillus odysseyi* SUK3, and *Morganella morganii* SUK5, and their consortium has been reported previously (Patil et al. 2008). Similarly the isolated *Exiguobacterium* species was found to remediate navy blue HE2R and reactive yellow 84 A. *Shewanella decolorationis*

S12 and *Acinetobacter calcoaceticus* NCIM 2890 (Ghodake et al. 2009) were found to decolorize brilliant blue KGR and direct brown MR, respectively. Among bacteria, many species of *Pseudomonas* are extensively studied for their capacity to remediate dyes. Few of the examples include *Pseudomonas* sp. SUK1 (reactive red RBL, reactive red 2), *P. luteola*, *P. mendocina*, *P. desmolyticum* NCIM 2112, and *P. putida*.

4.5.3.2.3 Phytoremediation

Phytoremediation is a process of degradation of xenobiotic compounds by the use of various plants which degrade, extract, or immobilize contaminants from soil and water ecosystem. Phytoremediation of dyes is the topic of interest nowadays. Recently the use of various plants for dye removal has been reported: *Phragmites australis* for decolorization of azo dyes from textile wastewater, *Blumea malcolmii* for decolorization of direct red 5B, and *Typhonium flagelliforme* for decolorization of *brilliant blue* R (Kagalkar et al. 2010). *Brassica juncea* is reported to have potential to remediate textile effluent-contaminated sites; similarly, hairy roots of *Tagetes patula* L were used previously for Reactive Red 198 decolorization (Patil et al. 2009).

Not only fungi and bacteria but plants also possess the mechanism of detoxifying contaminants. The opportunity to clean the polluted ecosystem with plants is paying much more attention against the traditional cost-competitive cleanup technologies like excavation, thermal treatment, and chemical soil washings. Phytoremediation is an emerging technology that employs the use of plants for clearing the contaminated areas (Odjegba and Fasidi 2007). During phytoremediation, plants metabolize various molecules in their tissues from the pollutant environment and are found to be very effective for the removal of pollutants from the ecosystem. The advantages of phytoremediation include its solar-driven nature, good image, high public acceptance, *in situ* application, and possibility of combining it with other methods, and the most important above all is that it maintains and stimulates the soil life and promotes ecological rehabilitation of contaminated land.

In tropical and subtropical regions of the world, phytoremediation has been done by floating aquatic plants for nutrient removal. Some floating aquatic plants like *Eichhornia crassipes*, *Lemna minor*, and *Spirodela polyrhiza* have been reported for domestic wastewater treatment. Phytoremediation by aquatic plants should depend on several factors such as its efficiency for nutrient or pollutant removal from a given type of wastewater, its productivity under the particular climatic conditions, its capacity to overgrow other macrophytes in the same ecosystem, the cost of harvesting, and the possible use of harvested biomass. Plant cultivation and harvesting are cost-effective processes compared with traditional engineering approaches involving intense soil management. Applied research has reported that genetically modified plant possesses potential to remove, degrade, and metabolize/immobilize a wide range of dyes. Phytoremediation may be successful by an influence of the vegetation on the physical (water balance, transport processes), the chemical (enzymes, redox potential, pH, complexing agents), and the biological (roots, microbes, mycorrhiza) factors in soil. Based on these processes, several phytoremediation techniques such as phytoextraction, rhizofiltration, phytostabilization, rhizodegradation, phytodegradation, phytovolatilization, hydraulic control, vegetation cover, and buffer stripes have been developed. Phytoremediation is extensively used for cleaning up of metal contaminated sites, removal of organic pollutants, and concentration and uptake of nutrients from domestic and municipal waste.

Phytoremediation of textile dyes, namely, malachite green, red HE8B, methyl orange, reactive red 2, and Direct red 5B, by tissue culture shrub plants of *Blumea malcolmii* Hook was reported previously. The degradation of Reactive Red 198 by hairy roots of *Tagetes patula* was demonstrated (Patil et al. 2009). Degradation of brilliant blue R by *Typhonium flagelliforme* was reported by Kagalkar et al. (2010).

4.5.3.3 Enzymes for Dye Degradation

Microorganisms are omnipresent, well diverse, and carry out a broad range of metabolic activities due to their catalytic activities (enzymes synthesized and secreted by them). The enzymatic treatments are simpler, require less energy, are cheaper, are environmentally friendly, and are efficient when compared with whole cultures and physicochemical treatment. There is limited production of hazardous by-products and, besides no threat of ecological contamination, thus has less impact on the biological system. So they present an attractive option for biobleaching. The efficiency of enzymatic reactions in textile processing has been recognized for many years and increasingly gained importance as biocatalysts in textile wet processing. The identification and characterization of the degradation pathways functioning in the microorganisms has gained more importance as a starting point for biotechnological and environmental applications.

Several oxidoreductases and lignin-modifying enzymes help in dye degradation. Oxidative enzymes like monooxygenases, dioxygenases, and reductases (cytochrome c reductase, NADH-DCIP reductase, MG reductase, flavin reductase, and triphenylmethane reductase) have played a significant role in biotransformation of xenobiotic compounds. The oxido-reductive enzymes from *Exiguobacterium* sp. RD3 represent coordinate action in bioremediation of reactive yellow 84A dye. Some peroxidases and laccases have been used for bioremediation of dyes present in ecosystem. Dye decolorization by peroxidases is cost-competitive processes due to use of hydrogen peroxide as a co-substrate.

The lignin-modifying enzymes are oxidative type of enzyme by their enzymatic mechanisms. These include peroxidases, like lignin peroxidase, manganese peroxidase, and versatile peroxidase, and phenol oxidases. Many wood-rotting fungi (*Phanerochaete chrysosporium*, *Ceriporiopsis subvermispora*, *Trametes versicolor*, *Phlebia radiata*, *Pleurotus ostreatus*, *Pleurotus eryngii*), basidiomycetous fungi such as *Agaricus bisporus* (common button mushroom), and many *Coprinus* and *Agrocybe* species were reported to produce lignin-modifying enzymes. The brown rot fungi are not able to produce lignin-modifying enzymes. Some species of filamentous bacteria such as *Streptomyces viridosporus* T7A, *Streptomyces lavendulae* REN-7, and *Clostridium stercorarium* produce lignin-modifying enzymes.

4.5.3.3.1 Lignin Peroxidase (EC 1.11.1.14)

Lignin peroxidase showed oxidative cleavage of C-C and ether (C-O-C) bonds in a number of lignin compounds (of the diarylpropane and arylpropane-aryl ether type). Lignin peroxidases oxidize the benzyl alcohols to aldehydes, by an aromatic cation radical. Molecular oxygen may be involved in the reaction of some isoenzymes under aerobic conditions. Previously, many researchers have reported the role of peroxidase of fungi, bacteria, and plants in dye degradation.

4.5.3.3.2 Laccase

Laccases are copper-containing oxidases, broadly distributed in many plants, fungi, and microorganisms belonging to the multicopper oxidase (MCO) superfamily, yet their biological importance is unclear. These enzymes catalyze the oxidation of phenolic compounds and reduction of O_2 to H_2O. Attention in laccases has been stimulated by their prospective use in bioremediation of environmental pollutants, wine stabilization, paper processing, enzymatic alteration of chemical intermediates, and the generation of valuable chemicals from lignin.

4.5.3.3.3 Tyrosinase

Tyrosinase (monophenol, dihydroxy-L–phenylalanine: oxygen oxidoreductase) is a monooxygenase containing Cu^{2+} ions at active site. This enzyme is responsible for producing melanins in the orthohydroxylation reaction of monophenols and the oxidation of o-diphenols and p-phenols to o-quinone, showing catecholase and cresolase activities. It is present in the broad group of organisms including fungi, actinomycetes, bacteria, gymnosperms, angiosperms, insects, Chordata, mammals, and human, in which they participate in several biological functions and have considerable heterogeneity.

4.5.3.3.4 NADH-DCIP Reductases and Aminopyrine N-Demethylase

Xenobiotics are metabolized by mixed-function oxidase system. These enzymes like cytochrome P-450, cytochrome b5 reductase, aminopyrine N-demethylase, NADH-DCIP reductase, acetanilide hydroxylase, and glutathione-S-transferase participate in remediation of xenobiotic compounds. Aminopyrine N-demethylase and NADH-DCIP reductase are components of mixed-function oxidase system, and they participate in the degradation of xenobiotic compounds like cotton blue, malachite green, methyl red, reactive red 141, and red BLI. Enzymatic degradation of xenobiotic compounds has not yet been possible at industrial level because of the vast amount of contaminated water.

References

Abhilash PC, Jamil S, Singh N (2009) Transgenic plants for enhanced biodegradation and phytoremediation of organic xenobiotics. Biotechnol Adv 27:474–488

Baboshin M, Akimov V, Baskunov B, Born TL, Khan SU, Golovleva L (2008) Conversion of polycyclic aromatic hydrocarbons by *Sphingomonas* sp. VKM B-2434. Biodegradation 19:567–576

Chen A, Zeng G, Chen G, Hu X, Zhang C, Yan, M, Shang, C and Lunhui Z (2014) Hydrogen sulfide alleviates 2,4-dichlorophenol toxicity and promotes its degradation in *Phanerochaete chrysosporium*. Chemosphere 109:208–212

Cobb GD, Bouwer EJ (1991) Effects of electron acceptors on halogenated organic compound biotransformations in a biofilm column. Environ Sci Technol 25:1068–1074

Coelho-Moreira J, Bracht A, de Souza A, Oliveira R, de Sá-Nakanishi A, de Souza C, and Peralta R (2013) Degradation of Diuron by *Phanerochaete chrysosporium*: role of ligninolytic enzymes and cytochrome P450. BioMed Res Int 2013:9

Cupples AM, Spormann AM, McCarty PL (2003) Growth of a Dehalococcoides like microorganism on VC and cis-dichloroethene as electron acceptors as determined by competitive PCR. Appl Environ Microbiol 69:953–959

Environmental Protection Agency (1979) Phenol ambient water quality criteria. Office of the Planning and Standards, EPA, Washington, DC

Fragoeiro S, Magan N (2008) Impact of Trametes versicolor and Phanerochaete crysosporium on differential breakdown of pesticide mixtures in soil microcosms at two water potentials and associated respiration and enzyme activity. Int Biodeterior Biodegrad 62:376–383

Ghodake GS, Telke AA, Jadhav JP, Govindwar SP (2009) Potential of Brassica juncea in order to treat textile effluent contaminated sites. Int J Phytorem 11:297–312

Gibson DT, Mahadevan V, Jerina DM, Yogi H, Yeh HJ (1975) Oxidation of the carcinogens benzo[*a*]pyrene and benzo[*a*]anthracene to dihydrodiols by a bacterium. Science 189:295–297

Gou M, Qu Y, Zhou J, Ma F, Tan L (2009) Azo dye decolorization by a new fungal isolate, *Penicillium* sp. QQ and fungal-bacterial cocultures. J Hazard Mater 170:314–319

Guo W, Li D, Tao Y, Gao P, Hu J (2008) Isolation and description of a stable carbazoledegrading microbial consortium consisting of *Chryseobacterium* sp. NCY and *Achromobacter* sp. NCW. Curr Microbiol 57:251–257

Heitkamp MA, Freeman JP, Miller DW, Cerniglia CE (1988) Pyrene degradation by a *Mycobacterium* sp.: identification of ring oxidation and ring fission products. Appl Environ Microbiol 54:2556–2565

Himanshu P, Vashi RT (2010) Treatment of textile wastewater by adsorption and coagulation. Eur J Chem 7(4):1468–1476

Holliger C, Schraa G, Stams AJM, Zehnder AJB (1992) Enrichment and properties of an anaerobic mixed culture reductively dechlorinating 1,2,3- trichlorobenzene to 1,3-dichlorobenzene. Appl Environ Microbiol 58:1638–1644

Hugo N, Meyer IC, Armengaud IJ, Gaillard J, Timmis KN, Jouanneau Y (2000) Characterization of three XylT-like [2Fe-2S] ferredoxins associated with catabolism of cresols or naphthalene: evidence for their involvement in catechol dioxygenase reactivation. J Bacteriol 182:5580

Jadhav JP, Govindwar SP (2006) Biotransformation of malachite green by *Saccharomyces cerevisiae*. Yeast 23(4):315–323

Kagalkar AN, Jagtap UB, Jadhav JP, Govindwar SP, Bapat VA (2010) Studies on phytoremediation potentiality of Typhonium flagelliforme for the degradation of Brilliant Blue R. Planta 232:271–285

Kim YH, Freeman JP (2005) Effects of pH on the degradation of phenanthrene and pyrene by *Mycobacterium vanbaalenii* PYR-1. Appl Microbiol Biotechnol 67:275–285

Kot-Wasik A, Dabrowska D, Namie J (2004) Photodegradation and biodegradation study of benzo(a)pyrene in different liquid media. J Photochem Photobiol A: Chem 168:109–115

Lee SE, Seo JS, Keum YS, Lee KJ, Li QX (2007) Fluoranthene metabolism and associated proteins in *Mycobacterium* sp. JS14. Proteomics 7:2059–2069

Li Y, Li J, Wang C, Wang P (2010) Growth kinetics and phenol biodegradation of psychrotrophic *Pseudomonas putida* LY1. Bioresour Technol 101:6740–6744

Liang Y, Gardner DR, Miller CD, Chen D, Anderson AJ, Weimer BC, Sims RC (2006) Study of biochemical pathways and enzymes involved in pyrene degradation by *Micobacterium* sp. strain KMS. Appl Environ Microbiol 72:7821–7828

Liang Y, Zeng F, Qiu G, Lu X, Liu X, Gao H (2009) Co-metabolic degradation of dimethoate by Raoultella sp. X1. Biodegradation 20:363–373

Liotta LF, Gruttadauria M, Carlo GD, Perrini G, Librando V (2009) Heterogeneous catalytic degradation of phenolic substrates: catalysts activity. J Hazard Mater 162L:588–606

Mallick S, Chatterjee S, Dutta TK (2007) A novel degradation pathway in the assimilation of phenanthrene by *Staphylococcus* sp. strain PN/Y via *meta*-cleavage of 2-hydroxy-1-naphthoic acid: formation of trans-2,3-dioxo-5-(2′-hydroxyphenyl)-pent-4-enoic acid. Microbiology 153:2104–2115

Mathur N, Bathnagar P, Nagar P, Bijarnia MK (2005) Mutagenicity assessment of effluents from textile/dye industries of Sanganer, Jaipur (India): a case study. Ecotoxicol Environ Saf 61(1):105–113

Mitrowska K, Posyniak A (2004) Determination of malachite green and its metabolite, leucomalachite green, in fish muscle by liquid chromatography. Bull Vet Inst Pulawy 48:173–176

Murali O, Mehar SK (2014) Bioremediation of heavy metals using spirulina. Int J Geol Earth Environ Sci 4(1):244–249

Nadeau LJ, Menn FM, Breen A, Sayler GS (1994) Aerobic degradation of 1,1,1- trichloro-2,2-bis(4-chlorophenyl)ethane (DDT) by Alcaligenes eutrophus A5. Appl Environ Microbiol 60:51–55

Nam IH, Kim YM, Schmidt S, Chang YS (2006) Biotransformation of 1,2,3-tri- and 1,2,3,4,7,8-hexachlorodibenzo-p-dioxin by Sphingomonas wittichii strain RW1. Appl Environ Microbiol 72:112–116

Odjegba VJ, Fasidi IO (2007) Phytoremediation of heavy metals by Eichhornia crassipes. Environ Sci Technol 27:349–355

Padhye, Chakrabarti T (2015) 2, 4-dichorophenol degradation by P. chrysosporium under ligninolytic condidtions. Inter J 4(1):1–5

Patil PS, Shedbalkar UU, Kalyani DC, Jadhav JP (2008) Biodegradation of Reactive Blue 59 by isolated bacterial consortium PMB11. J Ind Microbiol Biotechnol 35:1181–1190

Patil PS, Phugare SS, Jadhav SB, Jadhav JP (2010) Communal action of microbial cultures for Red HE3B degradation. J Hazard Mater 181:263–270

Patil P, Desai N, Govindwar S, Jadhav J, Bapat V (2009) Degradation analysis of Reactive Red 198 by hairy roots of Tagetes patula L. (Marigold). Planta 230:725–735

Phugare SS, Patil PS, Govindwar SP, Jadhav JP (2010) Exploitation of yeast biomass generated as a waste product of distillery industry for remediation of textile industry effluent. Int Biodeterior Biodegrad 64:716–726

Rani K, Dhania G (2014) Bioremediation and biodegradation of pesticide from contaminated soil and water – a noval approach. Int J Curr Microbiol Appl Sci 3(10):23–33

Rao MA, Scelza R, Scotti R, Gianfreda L (2010) Role of enzymes in the remediation of polluted environments. J Soil Sci Plant Nutr 10(3):333–353

Rehmann K, Hertkorn N, Kettrup AA (2001) Fluoranthene metabolism in Mycobacterium sp. Strain KR20: identity of pathway intermediates during degradation and growth. Microbiology 147:2783–2794

Revenkar SM, Lele SS (2007) Synthetic dye decolorization by white rot fungus, *Gandoerma sp. WR-1*. Biresour Technol 98:775–780

Seo JS, Keum YS, Cho IK, Li QX (2006a) Degradation of dibenzothiophene and carbazole by Arthrobacter sp. p 1-1. Int Biodeterior Biodegrad 58:36–43

Seo JS, Keum YS, Hu Y, Lee SE, Li QX (2006b) Phenanthrene degradation in *Arthrobacter* sp. P1-1: initial 1,2-, 3,4- and 9,10-dioxygenation, and *meta*- and *ortho*-cleavages of naphthalene-1,2-diol after its formation from naphthalene-1,2-dicarboxylic acid and hydroxyl naphthoic acids. Chemosphere 65:2388–2394

Singh RL, Khanna SK, Singh GB (1990) Detection of trace quantities of aromatic amines in finished dyestuffs. Bev Food World 17:16–18

Singh RL, Khanna SK, Singh GB (1991) Metabolic disposition of (14-C)metanil yellow in guinea pigs. Vet Hum Toxicol (USA) 33:220–223

Singh RP, Singh P, Singh RL (2014) Bacterial decolorization of azo textile dye acid orange by *Staphylococcus hominis* RMLRT03. Toxicol Int 21(2):160–166

Singh RL, Singh PK, Singh RP (2015) Enzymatic decolorization and degradation of azo dyes- a review. Inter Biodeterior Biodegrad 104:21–31

Telke AA, Kadam AA, Jagtap SS, Jadhav JP, Govindwar SP (2010) Biochemical characterization and potential for textile dye degradation of blue laccase from *Aspergillus ochraceus* NCIM-1146. Biotechnol Bioprocess Eng 15:696–703

Tiwari S, Gaur R (2014) Decolorization of distillery spentwash (Melanoidin) by immobilized consortium (bacterium and yeast) cell: entrapped into sodium alginate bead. J Environ Sci Technol 7(3):137–153

Tiwari S, Gaur R, Singh A (2014) Distillery spentwash decolorization by a noval consortium of *Pediococcus acidilactici* and *Candida tropicalis* under static condition. Pak J Biol Sci 17(6):780–791

Yang CF, Lee CM (2007) Enrichment, isolation, and characterization of phenol degrading *Pseudomonas resinovorans* strain P-1 and *Brevi bacillus sp.* strain P-6. Inter Biodeterior Biodegrad 59:206–210

Yang Z, Xia C, Zhang Q, Chen J, Liang X (2007) Catalytic detoxification of polychlorinated dibenzo-p-dioxins and polychlorinated dibenzofurans in fly ash. Waste Manag 27:588–592

Yao FX, Jiang X, Yu GF, Wang F, Bian YR (2006) Evaluation of accelerated dechlorination of p, p'-DDT in acidic paddy soil. Chemosphere 64:628–633

Zhuang X, Chen J, Shim H, Bai Z (2007) New advances in plant growth- promoting rhizobacteria for bioremediation. Environ Int 33:406–413

Ziarati P, Alaedini S (2014) The phytoremediation technique for cleaning up contaminated soil By Amaranthus sp. Environ Anal Toxicol 4(2):208

Suggested Further Readings

Abdellatif S (2012) Industrial dye decolorization using membrane bioreactor. LAP Lambert Academic Publishing, Los Angeles, p 56

KubilayYE (2009) Dye treatment with Fungi: Azo dye decolorization by *Phanerochaete chrysosporium*. VDM Verlag Dr. Müller, p 184

Sankar CVR (2015) Study on phenol degradation using biological & irradiation techniques. LAP Lambert Academic Publishing, Los Angeles, p 88

Singh M (2010) Microbial decolorization and degradation of Azo dyes: an approach to biotreatment of textile and dye manufacturing industries wastewaters. LAP LAMBERT Academic Publishing, Los Angeles, p 184

Singh SN (2011) Microbial degradation of xenobiotics, Environmental science and engineering. Springer, Heidelberg, pp 1–486

Thummar V, Ramani V (2014) Microbial decolorization and degradation of textile dye. LAP Lambert Academic Publishing, Los Angeles, p 56

Management and Remediation of Problem Soils, Solid Waste and Soil Pollution

5

Shiv Shankar and Shikha

Contents

Abstract

Soils with serious constraints to cultivation and that need special management techniques and practices are called as problem soils. These constraints may be physical such as dryness, wetness, steepness and extreme textures and chemical such as acidity, salinity, sodicity and lack of fertility. Reversing the degradation

S. Shankar • Shikha (✉)
Department of Environmental Science, School for Environmental Sciences, Babasaheb Bhim Rao Ambedkar University (A Central University),
Vidya Vihar, Raebareli Road, Lucknow 226025, India
e-mail: dr_shikha2003@yahoo.co.in

R.L. Singh (ed.), *Principles and Applications of Environmental Biotechnology for a Sustainable Future*, Applied Environmental Science and Engineering for a Sustainable Future, DOI 10.1007/978-981-10-1866-4_5

of soil, water and biological resources and enhancing crop production through appropriate management and remediation are essential components in achieving food and livelihood security.

Due to rapid increase in the production and consumption processes, societies generate as well as reject solid materials regularly from various sectors—agricultural, commercial, domestic, industrial and institutional. Solid wastes are the wastes generated from anthropogenic activities that are generally solid and are refuge as useless or unwanted. Generation of solid wastes exerts pressure on natural resources and seriously undermines sustainable development. One of the best ways to solve the situation is to manage solid waste efficiently.

Presence of xenobiotic (man-made) compounds or other alterations in the natural soil environment culminates into soil pollution. The principal sources of soil pollution include industrial activity, agricultural chemicals or improper disposal of waste. The most common chemicals involved are petroleum hydrocarbons, polynuclear aromatic hydrocarbons (such as naphthalene and benzo(a) pyrene), solvents, pesticides, lead and other heavy metals. Soil pollution causes health and ecological risks. In this chapter, we shall discuss different soil contaminants, including solid wastes and problem soils, their health and ecological risks and overall management practices to control soil pollution.

Keywords
Tons per day • Ministry of Environment and Forests • Common Bio-Medical Waste Treatment and Disposal Facilities

5.1 Solid Waste Management

It has been observed that with increase in the global population and the rising demand for food and other essentials, there has been a rise in the amount of waste being generated every day by each household, apart from the wastes generated from other activities (Minghua et al. 2009). Unmanaged waste, especially excreta and other liquid and solid wastes from households and the community, causes health hazards leading to spread of infectious diseases (CPCB 2010).

Due to increase in production and consumption processes, urban population generates solid materials from diverse fields, viz. commercial, agricultural, domestic, institutional and industrial settings. The significant quantity of wastes thus generated is known as solid waste. In other words, wastes generated from human and animal activities that are generally solid and discarded as unwanted can be termed as solid wastes. Generation of solid waste mounts pressure on natural environment and compromises the pace of sustainable development. By choosing efficient solid waste management system, the above situation can be salvaged (Biswas et al. 2010).

In India, 133,760 metric tons of solid waste is generated by municipal areas per day. Major proportion of solid waste goes uncollected and untreated. Only 91,152 tons per day (TPD) waste is collected, and 25,884 TPD is treated (MOEF 2014).

The Ministry of Environment and Forests has enacted Municipal Solid Wastes (Management and Handling) Rules, 2000, for the sustainable management of municipal solid waste in India. Municipal authorities in India are in compliance with these rules within their jurisdiction for the management of solid waste. About 4.16 tons of medical waste is generated in India every day. In India, there are 190 Common Bio-Medical Waste Treatment and Disposal Facilities (CBMWTDFs) under operation, and 29 CBMWTDFs are being established. The Ministry of Environment and Forests has amended Biomedical Waste (Management and Handling) Rules for management of medical waste.

A huge amount of solid waste is generated in India from thermal power plants in the form of fly ash. To deal with the safe disposal of fly ash, a notification on utilization of fly ash has been issued by the Ministry of Environment, Forest and Climate Change in India in 1999.

In India, landfill space is hard to find in and around urban centres (DEA 2009). Dumpsites in almost all cities are already handling more waste than their capacity. Finding new landfills sites near cities has now become nearly impossible due to the sheer lack of space (DEA 2009). For the next coming 20 years, the only way India's large quantities of post-recycled mixed municipal waste can be treated is through a combination of mechanical biological treatment (MBT), waste to energy (WTE) and sanitary landfilling (SLF). City municipalities in India are involved in collection of waste primarily (Sujauddin et al. 2008). However, the collection process does not include any gradation of waste product, e.g. glasses, polybags, biodegradable, paper shreds, etc. The municipality dumps these wastes to the city outskirts. Local raddiwala/kabadiwala collects small iron pieces by magnets; similarly glass bottles, news paper, etc. are also collected for recycling.

In Mumbai (2005), the sewage lines were clogged due to a large number of plastic bags. Blast in the Bhusan Steel factory at Noida was caused by imported scrap from Iran. A massive reduction in the number of migratory birds has been reported due to consumption of contaminated food; animals have been found dying on streets and farmland due to consumption of plastic bags, which block the food movement in their stomach.

5.1.1 Solid Wastes

Solid wastes may be organic and inorganic waste materials including wrappings of the products, clippings of the grasses, wooden furniture, cloths, glass and plastic bottles, kitchen waste, used paper, electronic equipments, bottles and cans of paint, batteries, etc. generated in a society, devoid of any value to the first consumer (UNEP 2015). Solid wastes not only represent heterogeneous mass of wastes from the areas but also homogeneous accumulation of industrial, agricultural and mineral wastes. Awareness regarding different sources and categories of the solid wastes including information on composition and the rate at which wastes are generated/disposed is, therefore, necessary for the design and operation of the functional elements of solid waste management system. The rationale of management of solid waste is reduction, collection and disposal of solid waste in eco-incentive manner consistent with the protection of public health. Solid wastes have been categorized on the basis of source of generation and type.

5.1.2 Solid Wastes: Classification

5.1.2.1 On the Basis of the Sources

Generation of solid wastes depends on fields and activities (UNEP 2015), and these sources include:

(a) Residential: Wastes from apartments, dwellings, etc. consisting from leftover food, vegetable peels, plastic, clothes, ashes, etc.
(b) Commercial: Commercial waste comprised of glasses, leftover food, ashes, metals, etc. generated from stores, markets, restaurants, hotels, motels, medical facilities, auto repair shops, etc.
(c) Institutional: Institutional waste includes plastic, paper, glasses, etc. originated from administrative, educational and public buildings such as schools, colleges, prisons offices, etc.
(d) Municipal: Municipal waste consists of dust, building debris, leafy matter, treatment plant residual sludge, etc. originated from various municipal activities like construction and demolition, landscaping, street cleaning, etc.
(e) Agricultural: Agricultural waste comprised of vegetables, agricultural remains, spoiled food grains, litter, etc. resulting from fields, orchards, vineyards, farms, etc.
(f) Open areas: Streets, parks, vacant lots, alleys, playgrounds, beaches, recreational areas, highways, etc.

5.1.2.2 On the Basis of Type

The waste can be categorized on the basis of physical, chemical and biological characteristics (UNEP 2015):

(a) Garbage: Garbage refers to vegetable and animal wastes originating from the sale, handling and preparation, storage cooking and serving of food. Garbage contains biodegradable organic matter, which results in the production of obnoxious odour.
(b) Ashes and Residues: Burning of coal, wood, coke, charcoal and other combustible materials for household cooking and heating and industrial establishments. Ashes and residues if produced in huge quantities, as in the case of power generation plants and factories, are kept under the category of industrial wastes. Ashes are inorganic in nature and include powdery mass, containing traces of metals and glass.
(c) Combustible and Non-combustible Wastes: These wastes comprise of wastes generated from institutions, households, commercial activities, etc., excluding biodegradable waste. Combustible waste includes paper, cardboard, garden

trimmings, rubber, etc. Non-combustible waste comprised of items as crockery, glass, aluminium cans, tin and ferrous and non-ferrous material and dirt.

(d) Bulky Wastes: This waste includes large defunct household equipments like washing machines, refrigerators, crates, furniture, tyres, vehicle parts, etc. Such wastes cannot be accommodated in normal storage containers, hence, require a special collection mechanism.

(e) Street Wastes: Street waste refers to wastes that are collected from walkways, streets, alleys, parks and vacant plots and include cardboard, paper, plastics, vegetable matter and dirt. Littering in public places is widespread and an acute problem in several developing nations including India, and a concrete solid waste management system is required to combat this problem.

(f) Biodegradable and Non-biodegradable Wastes: Biodegradable wastes are the waste comprising of organic matter such as vegetable and fruit peels, leftover food, textile, paper, wood, etc., originated from households and industrial activities. Complex biodegradable organic waste is degraded into simpler compounds by microorganisms. Non-biodegradable wastes comprised of inorganic and recyclable materials such as glass, plastic, metals, cans, etc.

(g) Dead Animals: In connection with municipal waste, dead animals include animals which die naturally or accidentally on the road. Dead animals have been categorized into two groups, large and small. Large animals include cows, horses, goats, pigs, sheep, etc., and among the small ones are dogs, cats, rabbits, rats, etc.

(h) Deserted Vehicles: This class includes deserted four-wheeler and two-wheeler automobiles. These vehicles have significant scrap value towards their metal content.

(i) Construction and Demolition Wastes: These wastes are produced due to construction, repair and demolition of houses, commercial buildings and other structures. They are comprised of stones, bricks, concrete, lumber, heating systems, roofing and plumbing materials and electrical wires and parts of the general municipal waste stream.

(j) Farm Wastes: Farm wastes originate from different agricultural practices such as harvesting, planting, rearing of animals for slaughter, production of milk and the operation of feedlots.

(k) Hazardous Wastes: Such wastes can be defined as 'wastes generated from industries, institutes or consumers causing threat to human health and environment either immediately or over a longer period of time'. Hazardous nature of such wastes is due to their physicochemical and biological or radioactive properties like ignitability, corrosivity, reactivity and toxicity. Examples of hazardous wastes are paints, empty solvent and pesticide containers, merged with municipal wastes which become part of the urban waste stream. Some hazardous articles may cause explosions in incinerators and set fires at landfill sites. Pathological wastes generated from hospitals and radioactive wastes demand careful handling. Effective hazardous waste management practices must ensure that such wastes are stored, collected, transported and disposed off separately.

(l) Sewage Wastes: By-products of sewage are referred as sewage wastes. Organic matter is the chief component of sewage waste which is produced after the treatment of raw and treated sewages. During the primary waste water treatment, the inorganic fraction of raw sewage is segregated as it may entrain biodegradable organic matter with pathogenic microorganism and therefore, must be buried immediately.

5.1.3 Solid Waste Management (SWM) System

A SWM system includes a cluster of several functional elements involved in management of solid wastes (UNEP 2015). SWM system, when operational, facilitates the collection and disposal of solid wastes at minimal costs, while preserving public health and ensuring less impacts on the environment. Different components of solid waste management system include:

5.1.3.1 Generation of the Waste

Material that is thrown away because it has served its purpose or is no longer useful becomes the waste.

5.1.3.2 Storage and Processing

The practices which are associated with the handling storage and processing of solid waste at the point of generation.

5.1.3.3 Collection of the Waste

The activities concerned with the collection of solid wastes at specific locations.

5.1.3.4 Transfer and Transport

Activities involved in the transfer of wastes from the collection points to the vehicles and then transportation of wastes to the disposal sites.

5.1.3.5 Processing and Recovery

Methods and facilities that are used to recover the wastes for recycling and other treatments.

5.1.3.6 Disposal

Disposal is the ultimate stage of disposal of solid wastes to a landfill site. Solid wastes are mostly generated in the urban settlements, and hence, solid waste disposal belongs to urban areas. Huge amounts of wastes are produced by different activities, which need to be properly handled.

5.1.4 Treatment and Disposal of Solid Wastes

Different waste treatment techniques tend to transform the waste into a form which is not only more manageable, with reduced volume and toxicity thereby making the waste easier to dispose off. Waste treatment methods are chosen depending upon the form, composition and quantity of the waste material. Present-day waste treatment methods include exposing the waste to very high temperatures, landfilling and employing biological methods to treat the waste. Treatment and disposal options are, however, chosen as a final resort to the other management strategies, viz. reducing, reusing and recycling of waste (Guerrero et al. 2013).

5.1.4.1 Incineration

It is the thermal treatment process which involves combustion of waste in the presence of oxygen. Subsequent to incineration, the wastes get converted to CO_2, water vapour and ash (Saffarzadeh et al. 2015). This method may help in energy recovery which can be used in heating and/or electricity supply. In addition, incineration technologies help in reducing the waste volume, rendering it less harmful besides reducing cost of transportation as well as the production of methane, a major greenhouse gas.

5.1.4.2 Pyrolysis and Gasification

Pyrolysis and gasification are two more or less similar processes, because in both the methods, organic waste is decomposed by an exposure to high temperatures and low amounts of oxygen (Chen et al. 2014). Gasification requires low amount of oxygen, while pyrolysis takes place under no oxygen environment. Both the techniques use heat under an oxygen-starved environment for biomass conversion into other forms. The end product of these processes includes a mixture of combustible and non-combustible gases along with pyroligenous liquid. These end products have a high heat value which can be utilized for different purposes. Gasification is considered advantageous because it allows the incineration of waste accompanied with energy recovery and without any air pollution which is the major characteristic of other incineration methods.

5.1.4.3 Open Burning

It is the burning of unwanted waste materials that release smoke and other emissions directly into the air, not passing through any stack or chimney. Burning the open outdoor piles, wastes in a burn barrel or the use of incinerators without pollution control devices releases the gaseous by-products as such, directly into the atmosphere. Garbage burning is an easy, convenient and cost-effective method but imposes several negative effects on both human health and the environment. Uncontrolled garbage burning may release toxic pollutants into the atmosphere such as ash, carbon monoxide, polycyclic aromatic compounds (PACs), particulate matter, dioxins, hexachlorobenzene and other volatile organic compounds which are hazardous to human health. Dioxins can cause serious health problems; they may adversely affect reproduction and development, disrupt the hormonal systems and even cause cancer. Polycyclic aromatic compounds and the hexachlorobenzene are deemed to be carcinogenic. The particulate matter can precipitate respiratory problems such as bronchitis and asthma, while carbon monoxide may induce neurological symptoms. Open burning often releases acidic gases, e.g. halo-hydrides, along with the oxides of carbon and nitrogen. Nitrogen oxides are known to contribute acid rain, ozone depletion, smog, etc. which ultimately leads to global warming.

Besides being a greenhouse gas, carbon monoxide may react with sunlight and produce ozone which can be harmful. Particulate matter may create smoke and haze which ultimately leads to air pollution.

5.1.4.3.1 Dumps and Landfills

(a) Sanit*ary Landfills*:
Areas serving as natural buffer between environment and landfill site are ideal for the construction of sanitary landfills. The rationale behind establishment of such sites is to minimize and discount the risks likely to be caused by waste disposal on public health and environment (Dace et al. 2015). Areas selected for sanitary landfills should have low water table and high proportion of clay. Clay soil is impervious in nature and less susceptible for leaching of contaminants into the groundwater. A landfill is divided into several isolated cells which can be filled with trash as and when required. A thin sheet/cover can be employed to minimize odours and keep out pests. A landfill is sealed with clay soil particles when it is filled with the waste. In landfills as a result of anaerobic respiration, landfill gases like carbon dioxide, methane are generated which can be recovered and used for energy requirement (Dace et al. 2015).

(b) *Controlled Dumps*:
In controlled dumps, targeted volume of waste is disposed in a cell equipped with leachate management system. Controlled dumps are cost-effective, are easy to operate and significantly reduce the risk of environmental contamination.

(c) *Bioreactor Landfills*:
Bioreactor landfills are employed to facilitate the decomposition of solid waste by enhanced microbiological processes (Berge et al. 2009). During decomposition, landfill leachate is incorporated by recirculation to maintain appropriate moisture required for microbial action. Bioreactor landfill may be anaerobic and aerobic in nature or a combination of both. Waste in bioreactor landfill is rapidly reduced. Bioreactor landfill also maximizes the generation and collection of methane for energy recovery systems, and they reduce the costs associated with leachate management.

5.1.4.4 Biological Waste Treatment

(a) *Composting*:
In composting, organic matter is decomposed by microbes under controlled conditions aerobically. A wide spectrum of composting techniques like static pile composting, windrow composting, vermicomposting and in-vessel composting are being practised these days.

The composition and constituents of the biodegradable organic matter, i.e. ratio of carbon and nitrogen (C/N ratio), operating temperature, water content and proportion of air, generally control the pace of biodegradation. Composting is an eco-friendly process of disposal of solid waste which is biodegradable in nature. The product of composting can be utilized to improve fertility of soil.

(b) *Anaerobic Digestion*:

Anaerobic digestion employs anaerobic bacteria (especially methanogenic archaea) and oxygen-free environment to decompose biodegradable organic waste (Nalo et al. 2014). The end product is biogas which contains methane and can be combusted to produce fuels. The process starts with hydrolysis of the input waste by bacteria whereby carbohydrates and other insoluble polymers are broken down into soluble forms which can be easily utilized by other bacteria. Subsequently sugars and amino acids are converted into CO_2, NH_3, hydrogen and organic acids by acidogenic bacteria. Organic acids are further converted by these bacteria into acetic acid in addition to CO_2, NH_3 and hydrogen. These products are finally converted into carbon dioxide and methane by methanogenic bacteria. For proper anaerobic digestion, nutrients such as nitrogen, phosphorus and potassium should be optimum. The pH of the vessel should be around 7. Anaerobic digestion is an important component of Integrated Solid Waste Management (ISWM) favouring management of solid waste in a cost-incentive, eco-friendly and socially acceptable manner (Chandra et al. 2012). An Integrated Solid Waste Management System involves the employment of a cluster of diverse treatment methods of solid waste disposal, and key to the functioning of such a system is the proper collection and screening. Effective management schemes are needed to be operated in ways meeting contemporary social, economic and environmental needs of the society (Nalo et al. 2014).

5.1.5 Factors Affecting SWM System

Many factors affect the decision-making process towards effective implementation of an SWM system. Some of the factors that need to be considered in developing an SWM system are as follows.

5.1.5.1 Quantity and Characteristics of Wastes

The quantum of wastes generated depends on the income of a family; generally, families with higher-income category happen to generate more waste, as compared to low-income category families. The quantity of waste varies from about 0.25 to about 2.3 kg per person per day, insinuating a strong resemblance between waste generation and per capita income. Proportion of paper and packaging materials in the waste largely account for the differences. When this proportion is higher, the density of the waste is less. The wastes of high density are represented by a relatively high proportion of moisture, organic matter and lower levels of recycling.

5.1.5.2 Climate and Seasonal Variations

In temperate climatic regions, drifting snow and frozen soil interfere with landfill operations. Therefore, trenches must be dug in summer season and cover material stockpiled for winter use.

In tropical region, sharp seasonal variations from wet to dry season cause significant changes in the moisture content of solid waste, ranging from less than 50 % in dry season to greater than 65 % in wet months.

During monsoon period, the collection and disposal of waste is problematic. High temperatures and humidity result in rapid degradation of biodegradable organic matter. Therefore, the frequency of waste collection in tropical regions should be higher than in temperate regions. In desert areas, there is no considerable variation in water content of wastes (due to low rainfall) and low production of leachate from sanitary landfill. Intense winds and windblown dust and sand, however, cause problems at landfill sites. While temperature inversions can cause airborne pollutants to be trapped near ground level, landfill sites can affect groundwater by altering the thermal properties of the soil.

5.1.5.3 Physical Characteristics of an Urban Area

In towns and cities, wherein the layout of houses and streets enable the entry of collection vehicle, door-to-door collection of solid wastes goes smoothly. However, houses where access ways are narrow, unpaved and tortuous, collection vehicles are not accessible. Problems of solid waste storage and collection are most acute in these areas.

5.1.5.4 Financial and Foreign Exchange Constraints

Solid waste management demands adequate proportions of budgetary input by municipal corporations. This is allocated for capital resources, which is utilized for the purchase of vehicles, equipments, fuel and labour costs. Generally, municipalities allocate 10–40 % of the revenues towards solid waste management.

5.1.5.5 Management and Technical Resources

For effective solid waste management, adequate workforce is needed. Regions which exploit maximum indigenous crafts and professional skills can be considered strong aspect of solid waste management system. Solid waste management system works better if there is a participation of the general public in formulating plan strategies; spread of awareness regarding screening, recycling and reuse of solid waste; and process and product replacement (e.g. promotion of paper bag instead of plastic bags, consistent garbage collection). Such practices play significant role in minimizing the volume of solid waste generated (WASTE 2015).

For the better management of solid waste, the concept of four Rs, i.e. reduction, reuse, recovery and recycling, must be embraced:

(a) Reduction: It involves minimum consumption of natural resources by consumer and industry and reduction in the cost of waste management.
(b) Reuse: Defunct electronic equipments, appliances and gadgets are collected by respective companies and repaired for reuse.
(c) Recovery: It involves recovery of energy and useful products.
(d) Recycling: It involves conversion of waste into useful products.

5.1.6 Solid Waste Management in India: Legal Framework

Sustainable development, precaution (*measures should be taken to avoid environmental degradation and hazards*) and polluter are the foundation pillars of Indian waste management rules formulated by the Ministry of Environment and Forests. Such principles have been a nonintegrated part of Indian environmental law jurisprudence. The semblance of such principles is reflected in many verdicts of Supreme Court of India. Such principles make it necessary for the companies and industrial units to act in an environmentally sustainable manner. Keeping in view the problem of generation of waste due to increased urbanization, the Ministry of Environment and Forests has formulated several subsidiary laws like the Environment Protection Act of 1986 ('EPA') to regulate the generation and waste disposal practices.

5.1.6.1 Biomedical Waste (Management and Handling) Rules, 1998

The Biomedical Waste (Management and Handling) Rules ('BMW Rules') offer detailed provisions and practices to be followed for appropriate disposal of solid waste (MOEF 2014). Several institutions like hospitals, animal houses, nursing homes, blood banks, pathological labs, etc. come under the ambit of biomedical rules. It is obligatory for every stakeholder to take necessary measures to ensure that BM waste is handled in such a way that it may not cause any negative impact on human health and environment, segregated at point of origin in containers and disposed in accordance with established standard procedures. In addition, under Rule 5(2), it is necessary for the institutions to establish treatment facilities like incinerator, autoclave and microwave system.

5.1.6.2 The Batteries (Management and Handling) Rules, 2001

These rules cover manufacturers, exporters, importers, recyclers and bulk consumers of lead-acid batteries. The rationale behind formulation of the Batteries (Management and Handling) Rules ('Batteries Rules') was to impose a regulatory mechanism for safe disposal of consumed lead-acid batteries and their components (MOEF 2014).

5.1.6.3 The E-Waste (Management and Handling) Rules

With a view to establish a sound management system for electronic waste, the E-waste (Management and Handling) Rules, 2011 ('E-waste Rules'), have been formulated. These rules regulate import, disposal and recycling of e-wastes (MOEF 2014). These rules are applicable to manufacturers, consumers, sellers along with all collection centres and recyclers of electronic waste.

5.1.6.4 The Plastic Waste (Management and Handling) Rules, 2011

To put a regulatory framework on the production, use and recycling of polyethylene as well as for prevention of plastic pollution by its logical disposal, the Plastic Waste (Management and Handling) Rules, 2011 ('PWM Rules'), have been formulated (MOEF 2014). Plastic wastes cover used polythene sheets, articles such as pouches and carry bags, etc. Manufacturers, stockists, distributors, retailers and users of plastic products are covered under these rules.

5.1.6.5 The Hazardous Wastes (Management, Handling and Transboundary Movement) Rules, 2008

The Hazardous Wastes (Management, Handling and Transboundary Movement) Rules, 2008 ('HWM Rules') have been formulated to regulate production, storage, recycling, import, transportation, reuse and disposal of hazardous wastes (MOEF 2014).

India has signed the Basel Convention, 1992, dealing with transboundary movement and treatment of hazardous waste. HWM Rules, Schedule-1, have a list of practices producing hazardous waste which inter alia includes industrial units engaged in production of mines and minerals, petroleum products, copper, zinc, textiles, lead-based production, oil and gas, asbestos, steel, electronic and tannery, etc.

5.2 Problem Soils and Management

Sound soil management is a key to sustainable agricultural practices. In majority of cases, there is no conflict between good profitable farming, upholding fertility of soil (and the capital value of the farm) and good management of environment, as they all need the soils to be maintained or improved.

Patently, what a farmer can achieve by good management is totally dependent on the climate and soil of the area, but it is also true that poor soil management can very quickly decrease the value of land for agriculture leading to environmental problems.

5.2.1 Problem Soils

Soils which have severe constrain to cultivation and which require specific management practices are called problem soils (FAO 2015). These constraints may be chemical in the form of low pH, presence of high concentration of total salts, inadequate fertility and presence of high sodium contents. Such soils are formed due to different types of land degradation. Without reclamation, such soils are not fit for agricultural practices. FAO has identified following category of problems which include:

5.2.1.1 Highly Organic Soils

Soils rich in organic matter and undecomposed plant material are called histosols. Such soils occur particularly in areas where the rate of decomposition of organic matter has been reduced due to low temperature (in Boreal climates) or where the degradation is hampered by prolonged wet conditions (in the Wet Tropics).

5.2.1.2 Heavy Cracking Clays

These soils mainly belong to vertisols and vertic subgroups of other soils and found mainly in (sub) tropical areas with a marked dry season.

5.2.1.3 Gypsiferous Soils

Soils which contain more than 25 % of gypsum are called gypsiferous soils. These soils reduce plant growth. The soil material then lacks plasticity, does not stick together and becomes completely unstable in water.

5.2.1.4 Calcareous Soils

These soils contain more than 15 % $CaCO_3$ which may occur in different forms (powdery, nodules, crusts, etc.). Soils with high $CaCO_3$ belong to the calcisols and related calcic subgroups of other soils. They are relatively widespread in the drier areas of the earth.

5.2.1.5 Acid Soils

Soils with a pH value of less than 5.5 are called acid soils. They exert a number of toxicities (aluminium) as well as deficiencies (molybdenum) and other plant growth-restricting conditions. Majority of acid soils belong to acrisols, alisols, podzols and dystric subgroups of other soils.

5.2.1.6 Sandy Soils

Soils which are generally coarse textured until 50 cm depth and consequently retain few nutrients and have a low water holding capacities are called sandy soils.

5.2.1.7 Salt-Affected Soils

When salts more soluble than calcium carbonate and gypsum are present in the soil and affect crop growth and yield of most crops, these soils are considered salt affected. Majority of such soils have an electrical conductivity of more than 4 Ω/cm.

5.2.1.8 Steep Lands

Steep lands often cause specific problems to agriculture and the environment because the steep slopes and accidental relief result in shallow soils that are much prone to erosion and landslides.

5.2.2 Reclamation and Management of Problem Soils

5.2.2.1 Highly Organic Soils (Histosols) and Their Management

Highly organic soils (histosols) are soil comprising primarily of organic materials as per the classification FAO and the USDA. They are defined as soils having 40 cm (16 in.) or more of organic matter in the upper 80 cm (31 in.). Depending on the clay proportion of the soil, organic matter content in such soil is in the range of 12–18 % by weight. Histosols include mucky peat (hemic soil material), muck (sapric soil material) or peat (USDA 2015).

Most histosols occur in Scandinavia, Canada, Sumatra, the West Siberian Plain, Borneo and New Guinea. Histosols occur in patches in other parts of the Europe, the Russian Far East (chiefly in Khabarovsk Krai and Amur Oblast), Florida and other areas of permanent swampland. In histosols, the cultivation of crops is difficult due to poor drainage and low chemical fertility. Histosols, when drained, produce high-grade pasteur for dairying and get more productive. These soils can be used for fruit production if managed properly. However, organic matter in such soils may get dried and eroding under the effect of winds. Such soils have the tendency of shrinkage and compaction also. Like gelisols, histosols have scant use for civil engineering applications due to heavy structures tend to subside in the moist soil. In order to reclaim histosols, shallow drainage ditches should be built. The natural vegetation is left standing for longer period to facilitate drying of the peat. At 20–40 m intervals, 1 m-deep drains can be built.

At the start of reclamation of histosols, construction of a complex drainage system should be refrained as it may result uneven subsidence of the soil thereby disrupting the connections between sucker drains and collecting drains. Sometimes, to free nutrients and to raise the pH of the surface soil, small-scale farmers burn the peat which stimulates plant growth.

5.2.2.2 Management of Vertisols

In both the FAO and USDA soil taxonomy, a vertisol is a soil characterized by high content of expansive clay (montmorillonite) which forms deep cracks in drier seasons or years (USDA 2015). Alternate shrinking and swelling results in self-mulching, where the soil material consistently mixes itself, leading vertisols to have an extremely deep 'A' horizon and no 'B' horizon. (A soil with no B horizon is called an A/C soil). Generally, microrelief known as gilgai is created due to heaving of the underlying material to the surface. The management of soil water is a significant aspect of management of soil in the semiarid tropics. Extremely slow hydraulic conductivity and poor internal drainage lead to waterlogging and delayed planting. Utmost consistency properties of the soil permit tillage practices within a narrow soil moisture range only due to soils being sticky when moist and hard when dry. Tillage operations under moist conditions may culminate into soil sticking and the formation of large clods. The presence of heavy clays may be resembled with a microrelief called gilgai that is an outcome of the consistent churning of these soils. Infrastructure and buildings may seriously be damaged in the long run, if they are built on these soils. Besides the requirement to get as much of the rain as possible into the soil for use by the crop, there is the need to incorporate adequate surface drainage to discount plant injury or slow growth cropped due to waterlogging once the cracks have closed and infiltration rates have decreased. Creation of cambered bed is a traditional and relatively early method of vertisol reclamation. A cambered bed can be formed by ploughing up and down so that the soil is turned inwards to the centre. In Africa, cambered beds have been used successfully. Waterlogging is not a significant problem in very dry areas, and tillage plays an important role in getting every drop of water in the soil and prevents runoff and evaporative loss.

The roles of ridges and furrows are conversed: water is projected to run off the ridge, sometimes appropriately broadened, which then acts as water harvesting device designed to carry runoff into the furrow in which the crop is to be cultivated. The ridges may be 'tied' at intervals with a cross ridge in order to block the water movement in the furrow. For the management of heavy cracking clays, proper management and timing of cultivation practices are very crucial. In regions with predictable rains, the ICRISAT system of post-harvest ploughing followed by seedbed preparation and dry seeding before the rains has been found to be very effective.

5.2.2.3 Reclamation of Gypsiferous Soils

Gypsiferous soils are rich in gypsum (calcium sulphate) which interferes with the growth of the plants (USDA 2015). Such soils occur in areas with ustic, xeric and aridic moisture regimes. Gypsiferous soils also occur in dry areas where sources of the calcium sulphate are present. In majority of cases, gypsum is associated with other salts of calcium and salts of sodium and magnesium. The productivity of gypsiferous soils can be improved in under rain-fed conditions by:

(a) Practising terracing of the soil present on deep hills to reduce erosion.
(b) Harrowing the land after harvesting and prior to rainy season for improving the infiltration of water and conserve soil moisture.
(c) Supplementary irrigation where water resources are available.
(d) Replacement of fallow by tiny-grain leguminous crops in wheat-fallow rotations to increase soil organic matter content.
(e) Subsoiling to disintegrate the consolidated gypsic subsoil to promote penetration of the plant roots.
(f) The use of fertilizers, especially nitrogen and phosphorus for cereals.
(g) Leaching of gypsum is required to keep the content of the salt low in areas where irrigation water is not a problem and available in adequate amount (FAO 2015).
(h) In order to maintain a relatively low water table and soil salinity, efficient drainage system is necessary.
(i) Cavities created by leaching gypsum from the surface soil make it requisite to level the surface of the soil each year. The regular employment of nitrogen fertilizers is necessary for maximum crop yield under irrigated agriculture on gypsiferous soils in which content of organic matter and total nitrogen is less. The potential productivity of gypsiferous soils is resembled to the depth of the gypsic layer. In soils with a gypsic layer below the depth of 60 cm, the plant roots penetrate easily, and there is sufficient soil volume for nutrients. In soils with a gypsic layer near the surface, the soil volume is limited, and plants do not survive generally. Gypsiferous soils can be employed for production of small grains, cotton, alfalfa, etc. if they contain only little gypsum in the upper 30 cm soil layer. Many gypsiferous soils in (young) alluvial and colluvial deposits have relatively low proportion of gypsum. Such soils can turn very productive if appropriately irrigated. Even soils bearing 25 % powdery gypsum or more may still be utilized for the production of excellent yields of wheat, alfalfa hay,

dates, apricots, maize and grapes, if irrigated at increased rates in combination with forced drainage.

5.2.2.4 Management and Remediation of Calcareous Soils

Calcareous soils are relatively alkaline; in other words, they have a high pH. Alkalinity is due to weak acidity of carbonic acid, presence of calcium carbonate in the parent material and a layer of secondary accumulation of carbonates (usually Ca or Mg) in excess of 15 % calcium carbonate equivalent and at least 5 % more carbonate than an underlying layer (FAO 2015).

Calcareous soils are often devoid of phosphorous. Amounts of phosphorus to apply depend on how deficient the soil is and the crop requirements. Excess application of phosphorus may lead zinc or iron deficiency. Phosphorus should be applied in water-soluble form in case of calcareous soils. Application of phosphorus at the time of seeding has been proven to be most appropriate since phosphorus is needed mostly during the early growth of seedlings of the plant. Calcareous soils also suffer from deficiency of micronutrients, especially zinc and iron. Zinc deficiency is most conspicuous in maize, especially under high yield intensive cultivation systems.

Zinc sulphate, the most popular form of zinc, is an effective zinc source. A single application of zinc sulphate lasts for several years. On fruit trees, foliar applications of zinc are effective. Heavy employment of animal manure is helpful in preventing deficiency of iron and zinc.

5.2.2.5 Acid Soils and Their Management

Acid soils are those that have a pH value of less than 5.5 for most part of the year. They are associated with a number of toxicities (aluminium) as well as deficiencies (molybdenum) and other plant growth-inhibiting conditions. Majority of acid soils belong to podzols, acrisols, alisols and dystric subgroups of other soils (FAO 2015). An extreme case of an acid soil is the acid sulphate soil (thionic fluvisols and thionic cambisols).

There are two main belts of acid soils:

(a) Humid northern temperate zone, which is covered mainly by coniferous forests
(b) Humid tropics, which is covered by Savannah and tropical rain forest

Acid sulphate soils are generally left under natural vegetation or used for mangrove forestry. They can support oil palm and rice if water is managed well. Crops like rice, cassava, mango, cashew, citrus, pineapple, cowpeas, blueberries and certain grasses grow on acid soils across the world.

For the management of acid soils, an integrated approach involves the employment of acid-tolerant species, suitable crop rotations, efficient use of fertilizers and crop diversification. For the determination of lime requirement, soil testing should be performed every 2–3 years. The buffering capacity of the soil should be assessed to gauze the amount of lime needed to neutralize acidity of the soil to acceptable level. Incorporation of high organic matter partly compensates the negative impacts of soil acidity on physical and chemical properties of the soil. Acid sulphate soil management is more fragile tactic and has to be based on cautious water management in order to prevent oxidization of pyrite in the process:

(a) The very first strategy is to drain and fully oxidize the soil and then flush the acidity formed out of the soil. This strategy lessens the problem but has severe negative effects: it is costly, exerts threat to the environment (acid drain water) and results loss of useful elements together with the unsolicited ones.
(b) The second strategy is to try to restrict oxidation of pyrite by ensuring a high water table. A precondition is the sufficiency of water in the soil. This method also demands substantial input in management of water, while the potential risk of acidification is not discounted. In temperate regions and in the tropics, this strategy is widely practised with ingenious adaptations to suit local conditions and practices.

Addition of dolomite or lime into the top cultivable soil layer is an efficient strategy for reclamation of acidic soils. At the time of sowing, banding or pelleting lime onto the seed is also a common practice employed to assist the establishment of temperate pasture legumes. Application of lime acts as a preventative treatment for soil infertility and to supply calcium and magnesium to deficient soils. Liming escalates the pH of acid soil; subsequently, the activity of nitrogen-fixing bacteria becomes uninhibited, and the rate of nitrogen fixation increases. Mineralization of nitrogen from plant residues and organic matter has been proven to increase when lime is applied to acid soil.

5.2.2.6 Sandy Soils and Their Management

Soils which are generally coarse textured until the depth of 50 cm and consequently retain few nutrients and have a low water holding capacity called sandy soils. Soil management practices leading to an increase in the fine fraction help in improving physical and chemical properties of the soil and productivity of the crops:

(a) For the remediation of sandy soils, inorganic fertilization is the main practice and deemed as essential.
(b) The physicochemical properties of the sandy soils can be improved by input of organic manures which supply nutrients in slowly available forms.
(c) Surface application of organic manures to sandy soils does not last for longer time; hence, the manure should be incorporated deeper into the soil. If carpet-like layer of manure is spread, the thickness should not be less than a centimetre. It ensures improved storage of the water, biological activity, nutrient status and crop yield.
(d) Mulch can be practised to enhance storage of the water by lowering the rate of evaporation. Residues of the crop on the surface of the soil minimize evaporative loss of water, reduce the range between maximum and minimum soil temperature and reduce wind erosion of the soil.

(e) For tillage to be effective in real sense, it should be performed at the earliest possible time after irrigation or rainfall when the evaporation rate is still high.
(f) Maintenance of a cover crop, minimum tillage, crop rotations, strip cropping, control of grazing and establishment of shelter belts and windbreaks are some of the protective measures to combat the high susceptibility of sandy soils to erosion.
(g) Employment of artificial surface sealants such as synthetic rubber, petroleum, chemicals and water-soluble plastics has also been adopted for dune and drift sand stabilization besides conventional dry vegetation method.
(h) Afforestation with some selected shrubs and trees is a complementary practice which should follow stabilization of dunes.
(i) Overgrazing on coarse textured soils must be prevented. The introduction of rotational grazing helps in overcoming this hazard.

5.2.2.7 Salt-Affected Soils and Their Management

When salts more soluble than calcium carbonate and gypsum are present in the soil and affect crop growth and yield of most crops, these soils are considered salt affected (FAO 2015). Majority of salt-affected soils have an electrical conductivity of more than 4 Ω/cm. Most of the salt-affected soils are classified as solonchaks. Uptake of nutrients and the microbiological activity in the soil are affected due to presence of salts (Parida and Das 2005). Methods of removal of excess salts from the soil surface and the root zone in saline soils include the following:

(a) In order to prevent excessive accumulation of salt in the root zone, irrigation water (or rainfall) must be applied in excess of that required for the evaporation of the crop. Leaching can be timed to preface the critical growth phases at which environmental stress should be discounted. This can be timed through irrigation during drier period. Leaching at the times of low evapotranspiration demands is more effective, for example, during high humidity, in cooler weather or outside the cropping season and at night.
(b) Leaching is only effective at the time when salty drainage water is released through subsurface drains which carry the leached salts out of the area under reclamation.
(c) Salt-affected soils can be categorized into saline, saline-sodic and sodic, depending on amount of the salts present, type of salts and amount of sodium present and alkalinity of the soil. Each type of salt-affected soil possesses different characteristics, which determines the extent and nature of remediation.

A set of different circumstances results in formation of salt-affected soils. Because of this reason, there is no any specific premeditative measure for the reclamation and management of salt-affected soils to ensure desired agricultural production.

As a priority, it is necessary to discount the possibilities of the development of salt-affected soils by employing sound agricultural practices, keeping in view local conditions including soil type (and parent material from which it is formed), type of crops, terrain elevation, water quality, soil nutrient status, irrigation practices and drainage.

Once salt-affected soils have developed either through natural or anthropogenic processes, they are required to be monitored with a view to determine the type of salt-affected soil and the extent to which they are affected. Even under bio-saline agriculture, severely affected soils are often too costly to reclaim or very difficult to manage.

A combination of approaches and technologies and the consideration of socio-economic aspects under local conditions are required for the management of salt-affected soils for agricultural practices (Fig. 5.1). Agricultural productivity in salt-affected soils is heavily dependent on climatic conditions, water availability, crops and the availability of resources (capital, inputs and time). In majority of cases, salt-affected soils can be successfully managed to restore to other applications besides crop production, e.g. recreation areas, grassland restoration, restoration of biodiversity, cultivation of medicinal plants, etc.:

(a) *Hydraulic Practices*
   1. *Leaching*: For the prevention of excessive accumulation of salt in rhizosphere, irrigation water (or rainfall) must be applied for prolonged period in excess and must pass through the root zone. This amount in fractional terminology is called as leaching requirement. In order to restrict raising the groundwater and minimize the total load to the drainage system, leaching

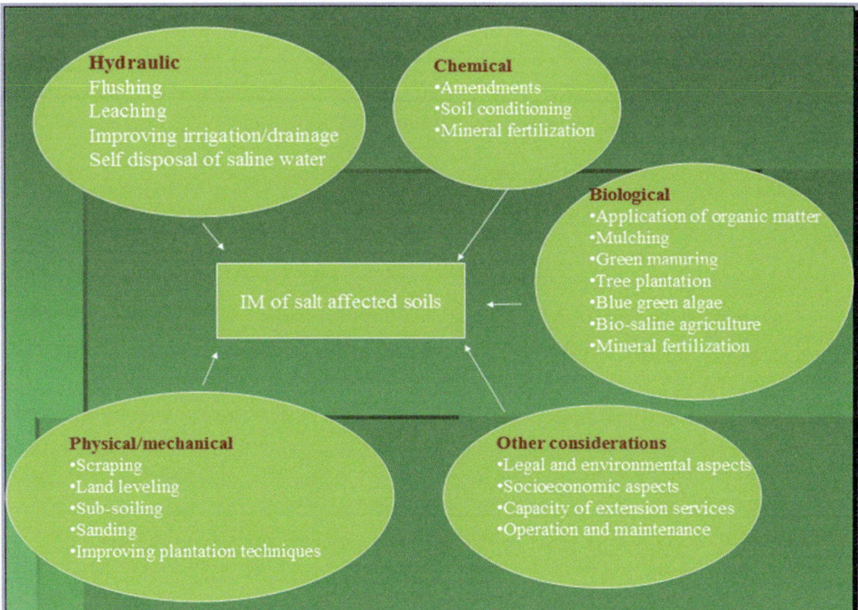

Fig. 5.1 Integrated management (IM) of salt-affected soils (Adopted from @ FAO 2015)

requirements should be minimized as far as possible. Depending on the degree of existing salinity, a leaching requirement of 10–20 % can be used.

2. *Drainage*: Natural drainage may work well if underlying layers are pervious in nature and relief is decent. Diverse drainage methods are practised globally. In surface drainage, ditches are made to outflow excess water before it joins the soil. Subsurface drainage involves restriction of the water table at a safe designated depth. In mole drainage, shallower channels are left by a bullet-shaped device pulled through the soil acting as a complementary drainage system connected to the main drainage system (open or closed). Vertical drainage is practised by flushing out extra water from boar wells when the deep layers have decent hydraulic conductivity.

(b) *Physical/Mechanical Management*

In order to enhance infiltration and permeability of water in topsoil and in root zone, diverse physical methods are employed to minimize saline and sodic conditions. Some of them include:

1. *Land levelling* is performed for uniform application of water for improved leaching and control of salts.
2. *Tillage* is practised for the preparation of seedbed and improvement in soil permeability.
3. *Deep ploughing* loosens the soil aggregate in stratified soils possessing impervious layer. It also improves soil atmosphere and hydraulic conductivity of soil.
4. *Planting procedures* reduce the accumulation of salt around seeds. Raised furrows in single or double rows and planting on sloping beds help in getting decent stands under high salt conditions.

(c) *Chemical Practices*

Chemical practices include employment of chemical fertilizers and chemical treatments. Chemical amendments balance sodic soil conditions, followed by leaching for the removal of salts formed due to the reaction of the amendments with sodic soils. Sulphuric acid, sulphur and gypsum are commonly used in chemical amendments.

(d) *The Biological Practices*

Biological practices for treatment of salt-affected soils include application of organic matter, growing legumes, farm manure, mulching, crop residue and selection of salt-tolerant crops:

1. *Crop Residue Application*: Crop residue application is the simplest method generally employed to enhance water infiltration in soil by small farmers. Application of green manure crops like *Sesbania aculeata* (dhaincha) for reclamation of alkali soils has been a practice for a very long time. Decomposition of organic residue of biological amendments releases carbon dioxide and other organic acids which enhance the solubility of native $CaCO_3$, and this provides calcium for the removal of exchangeable sodium. *Leptochloa fusca* (karnal grass), *Chloris gayana* (rhodes grass) and *Panicum maximum* (gatton panic) have been found tolerant and better than other grasses for improving sodic soils.

2. *Salt-tolerant Crops*: Salt-tolerant crops like barley, wheat, sugar beet, millet, rice and grasses can be grown in moderately saline or sodic conditions. In India, different types of land degradation have resulted the formation of wasteland covering more than 175 m ha area of the country. With the development of wastelands, there will be an improvement in the microclimate and soil productivity rendering enhanced agricultural production. Wastelands can be utilized for the cultivation of agricultural crops.

5.3 Soil Pollution and Management

Soil is the thin layer of organic and inorganic materials covering the earth's rocky surface. The organic component of the soil, which is derived from the plant and animal biomass, is concentrated in the dark upper most topsoil. The inorganic component of the soil comprised of rock components formed over thousands of years by physical and chemical weathering of parent rock material. Productive soils are necessary for good agricultural productivity.

Pollution may be defined as 'an unsolicited alterations in the physicochemical and biological properties of environment and its components causing threat to lives of human, useful living plants and animals, industrial progress, living conditions and cultural assets'.

A pollutant is something which adversely interferes with health, comfort, property or environment of the people. Generally most pollutants are released into the environment by sewage, waste and accidental discharge. Due to environmental pollution, the soil, the basis of agriculture producing crops for human food and animal feed, is under tremendous pressure.

Soil pollution is 'fortification of persistent toxic compounds, salts, chemicals or disease-causing agents and radioactive materials causing negative impact on growth of the plants and health of animals.' The most common chemicals involved in soil pollution are petroleum hydrocarbons, polynuclear aromatic hydrocarbons (such as naphthalene and benzo(a)pyrene), pesticides, solvents, lead and other heavy metals. Contamination can be correlated with the degree of industrialization and intensity of chemical usage.

5.3.1 Causes of Soil Pollution

5.3.1.1 Coal Ash

Areas that were industrialized before 1960, deposition of coal ash employed for commercial, residential and industrial heating, as well as for industrial processes such as ore smelting, were a common source of contamination. Burning of coal concentrates lead, zinc and other heavy metals in ash.

Coal ash and slag may contain meagre quantities of lead to qualify as a 'characteristic hazardous waste', defined in the USA as containing more than 5 mg/L of extractable lead using the TCLP procedure. In addition to lead, coal ash typically contains variable but significant proportion of polynuclear aromatic hydrocarbons (PAHs; e.g. benzo(b)fluoranthene, benzo(a)anthracene, indeno(cd)pyrene, benzo(k) fluoranthene, benzo(a)pyrene, anthracene, phenanthrene and others). These PAHs have been known as human carcinogens, and the permissible limit of PAHs in soils are typically around 1 mg/kg.

5.3.1.2 Sewage

Unrestricted disposal of sewage and other liquid wastes generating from household applications, agricultural runoff, effluents from animal husbandry, urban runoff and industrial wastes containing a variety of pollutants often cause soil pollution.

Irrigation of crops with sewage water causes significant changes in physical (changes in humus content, leaching and porosity, etc.) and chemical (base exchange status, soil reaction, salinity, quantity and availability of nutrients like nitrogen, potash, phosphorus, etc.) properties of the soil.

Application of treated sewage sludge (biosolid) as a fertilizer has sparked much controversy. Being the by-product of sewage treatment, it is generally rich in pollutants such as pathogenic microorganisms, pesticides and heavy metals than other soils (Snyder 2005; Olawoyin et al. 2012).

Urban Waste Water Treatment Directive in the European Union allows sewage sludge to be sprayed onto land. There is a need to control this so that pathogenic microorganisms do not get into water courses and to ensure that there is no accumulation of heavy metals in the topsoil.

5.3.1.3 Pesticides and Herbicides

Substance or mixtures of substances used to kill a pest are called pesticides. A pesticide can be a chemical substance, biological agent (such as a virus or bacteria), antimicrobial agent and disinfectant.

Plant pathogens, molluscs, birds, mammals, fish, nematodes (roundworms) and microbes that contend with humans for food, destroy property, spread pathogenic diseases or cause a nuisance for public health are called pests.

However, there are several benefits of using pesticides, but they have been notoriously known to cause potential toxicity to humans and other organisms. Herbicides are used to eliminate weeds, growing especially on pavements and railways. Herbicides are biodegradable by soil bacteria and are similar to auxins. A group of herbicides derived from trinitrotoluene (2,4-D and 2,4,5-T) contain dioxin as impurity which is highly toxic and may cause potential threat to the environment and public health even in low concentrations. Paraquat herbicide is also very toxic, but it is rapidly degraded by soil microflora. Insecticides are employed to kill pests of farms causing substantial damage to the crops. There are now two main classes of synthetic insecticides: organochlorine (DDT, aldrin, dieldrin) and BHC. DDT was extensively used on a massive scale from the 1930s, with a peak of 72,000 tonnes used in 1970. Then application of DDT declined due to harmful environmental effects. DDT is slightly soluble in water but highly soluble in bloodstream. It exerts harmful effects on nervous and endocrine systems and results the eggshells of birds to lack calcium making them very fragile. DDT has been held responsible for the decline of bird's population. Application of DDT has been banned in the UK and the

USA to prevent its concentration in food chain (Apitz 2008). However, US manufacturers sell DDT in developing countries, which cannot afford the expensive substitute of chemicals and do not have appropriate stringent legal frameworks governing the application of pesticides.

5.3.1.4 Heavy Metals

Heavy metals are elements with a density more than 5. Heavy metals are generally found in absorption sites in the soil where they persist for a very long time either on the organic and inorganic colloids. Heavy metals are widely distributed in the environment like soils, plants and animals and in their tissues. For the growth of plants and animals, heavy metals are required in traces. Heavy metal pollution is caused by combustion of fuels, urban and industrial aerosols, mining wastes, liquid and solid from animals and human beings, industrial and agricultural chemicals, etc. Weathering of rocks may result in presence of heavy metals in uncontaminated soils. In soil sediments and plants, the concentration of heavy metals has been mentioned in Table 5.1.

Applications of chemicals, farm slurries, sewage sludge and increased doses of fertilizers, pesticides or agricultural chemicals over longer period increase the concentration of heavy metals in agricultural soils. Some phosphorus fertilizers frequently contain trace amounts of cadmium which may accumulate in these soils. Likewise, some fertilizers when added to the soils contribute heavy metals (Table 5.2) and sludge (Table 5.3). Different physical, chemical and biological processes taking place in soil determine the fate of heavy metals (Urzelai et al. 2000).

5.3.1.5 Agents of War

The manufacture and disposal of munitions caused by the urgent production can pollute soil for longer periods. However this type of contamination is not well documented because of restrictions imposed by governments of several countries on the publication of material related to war effort.

Table 5.1 Heavy metal concentration in the lithosphere, soils and plants (μg/g dry matter)

S.N.	Heavy metal	Lithosphere	Soil range	Plants
1	Cadmium (Cd)	0.2	0.01–0.7	0.2–0.8
2	Cobalt (Co)	40	1–40	0.05–0.5
3	Chromium (Cr)	200	5–30,000	0.2–1.0
4	Copper (Cu)	70	2–100	4–15
5	Iron (Fe)	50,000	7000–5,50,000	140
6	Mercury (Hg)	0.5	0.01–0.3	0.015
7	Manganese (Mn)	1000	100–4000	15–100
8	Molybdenum (Mo)	2.3	0.2–5	1–10
9	Nickel (Ni)	100	10–1000	1
10	Lead (Pb)	16	2–200	0.1–10
11	Tin (Sn)	40	2–100	0.3
12	Zinc (Zn)	80	10–300	8–100

Adopted from Ashraf et al. (2014)

Table 5.2 Heavy metal content of different fertilizers (μg/g)

S.N.	Fertilizer	Co	Cr	Cu	Mn	Mo	Ni	Pb	Zn
1	Nitro chalk	–	–	22	24	–	2	–	15
2	Calcium	0.1	Traces	Traces	Traces	–	–	–	1
3	Nitrate	–	–	10	–	–	–	–	–
4	Ammonium sulphate	<5	<5	0.800	0.80	<0.05–0.22	<5	200	0.800
5	Super Phosphate	0.02–13	0.–1000	1000	2842	35	32	92	70–3000
6	Potassium chloride	0.001	–	0–10	8	<0.05	<1	<1	0.3
7	Sulphate	<5	<5	0–300	80	0.09	<5	<50	<50

Adopted from Ashraf et al. (2014)

Table 5.3 Heavy metal contents in sludge (ppm)

S.N.	Heavy metals	Range (ppm)
1	Cadmium	<60–1500
2	Cobalt	2–260
3	Chromium	40–8800
4	Copper	200–8000
5	Iron	6000–62,000
6	Manganese	150–2500
7	Molybdenum	2–30
8	Nickel	20–5300
9	Lead	120–3000
10	Zinc	700–49,000

Adopted from Ashraf et al. (2014)

Mustard gas stored during World War II has contaminated some sites for up to 50 years, and the testing of anthrax as a potential biological weapon contaminated the whole island of Gruinard.

5.3.2 Impact of Soil Pollution

5.3.2.1 Health Effects

Soil pollution directly causes threat to human health via direct contact with soil or via inhalation of soil contaminants which have vaporized. Infiltration of soil contamination into groundwater aquifers causes threat to public health. Health effects cropped because of exposure to soil contamination differ markedly and depend on type of pollutants, routes of attack and susceptibility of the exposed population. Chronic exposure to lead, chromium and other metals, solvents, petroleum and many herbicide and pesticide formulations can be carcinogenic and may result in congenital disorders. Nitrate and ammonia released due to natural and anthropogenic activities have also been identified as health hazards in soil and groundwater.

Increased cases of leukaemia have been reported due to chronic exposure to benzene at sufficient concentration. Mercury and cyclodienes have been reported to induce higher cases of kidney damage and some irreversible diseases. PCBs and cyclodienes are associated with hepatotoxicity.

Organophosphates and carbamates cause neuromuscular blockage. Many chlorinated solvents cause changes in the liver and kidney and depression of the central nervous system. There is a wide spectrum of other health effects such as nausea, headache, fatigue, eye irritation and skin rash for aforesaid chemical.

5.3.2.2 Ecosystem Effects

Soil pollutants may have adverse effect on soil ecosystem in terms of radical changes in soil chemistry thereby manifesting biochemical activities of soil flora and fauna. Majority of harmful effects are well known today, such as the accumulation of persistent DDT materials for avian consumers, leading to thinning of eggshells, increased mortality of chick and extinction of animal species. Soil affects biochemical activities of plant cell leading to reduction in crop yield.

5.3.3 Integrated Soil Management Options for Polluted Soils

Cleanup or remediation of soil pollution is assessed by environmental scientists who carry out field measurement of baseline concentration of soil pollutants and also apply computer models (GIS in Environmental Contamination) for analyzing transport and fate of soil chemicals. Various technologies have been practised for remediation of oil-contaminated soil/sediments. There are several principal strategies for remediation.

5.3.3.1 Regulation of Pollutant Entry into the Soil

Entry of air pollutants in soil cannot be controlled, but farmers and farm input manufacturers can adopt several measures to reduce extent of soil pollution. Employment of low heavy metal containing mineral materials for the production of phosphatic fertilizer can reduce cadmium contamination in soil.

Composts prepared from municipal solid waste in India generally contain high amount of heavy metals (Saha et al. 2010). On the other hand, preparation of composts from source separated biodegradable municipal solid wastes can significantly decrease heavy metal entry and enhance productivity of agricultural land.

5.3.3.2 Decontamination Using Physical and Chemical Means

For the remediation of heavy metal-contaminated soil, enumerable technologies have been developed, i.e. excavation of soil, thermal extraction for volatile metals (e.g. arsenic, mercury and cadmium as well as their derivative compounds which can be evaporated at 800 °C), solidification/stabilization, electrokinetics, chemical oxidation, vitrification and soil flushing with chemical extracts (Mulligan et al. 2001). Aeration of soils contaminated with PAHs has been found as effective practice of degrading these organic pollutants. The specific mitigation practices chosen for the treatment of a contaminated soil depend on the speciation of the contaminants and other site-specific conditions.

One or more of aforesaid approaches are often clubbed together for efficient and cost-incentive treatment. Majority of abovementioned chemical and physical treatments, however, are costly and irreversibly affect properties of the soil and destroy biodiversity, and many render the soil unfit as a cultivation medium plant growth.

5.3.3.3 Bioremediation

Bioremediation has emerged as a strong tool for the remediation of soils contaminated with pollutants. It has drawn significant attention of the researchers in past few years as it offers eco-incentive and sustainable approach to decontaminate soil from pollutants utilizing plants and microbes.

Bioremediation techniques endeavour to use the astonishing, natural ability of microbial xenobiotic metabolism to degrade, transform or accumulate a wide spectrum of pollutants including hydrocarbons, e.g. oil, polynuclear aromatic hydrocarbons (PAHs), polychlorinated biphenyls (PCBs), heterocyclic compounds (such as pyridine or quinoline), radionuclides, pharmaceutical substances and metals. Major methodological advancements in past few decades have enabled detailed genomic, proteomic, metagenomic, bioinformatic and other high throughput analysis of environmentally relevant microbes offering unprecedented insights into key biodegradative metabolic pathways and the efficiency of microorganism to adapt to changing environmental conditions. Environmental cleanup of heavy metals from soils and sediments and biosurfactants produced by bacteria and yeasts could be used efficiently (Mulligan et al. 2001). Biosurfactants augment the mobility of heavy metals by lowering the interfacial tension by forming micelles between the metals and soil. Biosurfactants are biodegradable and produced naturally by bacteria and are effective and nontoxic agents for the remediation of the contaminated sites. Surfactant foam technology has been investigated for the removal of hydrophobic organic compounds from the soil, such as polynuclear aromatic hydrocarbons (PAHs), polychlorinated biphenyls (PCBs), pentachlorophenol (PCP) and other chlorinated hydrocarbons from contaminated soils.

5.3.3.4 Phytoremediation of Contaminated Soil

In past few years, phytoremediation has gained importance as a promising ecoremediation technology, exclusively for the decontamination of soil and water. In phytoremediation, plants and their associated rhizospheric microflora are used to remove various pollutants from contaminated soil sediments, surface water and groundwater. There are some specific plants which can be utilized to treat different classes of pollutants including chlorinated solvents, petroleum hydrocarbons, metals, pesticides, radionuclides, explosives and inorganic nutrients like nitrates and phosphates. The plants intended for phytoremediation/phytodegradation are basically chosen on the basis of their growth rate and biomass, their potential to evapotranspirate groundwater, the depth of their root zone and their capacity to tolerate and bioaccumulate particular contaminants. Plants after phytoremediation can then be harvested, and the incinerated plant ash of these parts can be disposed of as hazardous matter in specialized dumps. The application of phytoremediation techniques is restricted by type of the soil and climatic and agro-ecological conditions of the site to be remediated. Phytoremediation takes more time than other technologies, and the development of vegetation may be limited by extremes of environmental toxicity.

Selection of plants for phytoremediation based on soil type, extent of soil contamination climatic conditions must be taken into account for effective integrated soil management and remediation of contaminated soils.

5.3.3.5 Crop/Cropping System Modification

Some vegetable crops, particularly lettuce and spinach, have high metal uptake capacity. Such crops should be avoided in metal-polluted soils. Growing of nonedible crops like fibre crops, flowering plants and agroforestry may prevent contamination of environment.

5.3.3.6 Other Techniques

(a) Excavation of soil and disposal of excavated soil away from ready pathways in contact with human or sensitive ecosystem.
(b) Thermal remediation by introduction of heat to raise subsurface temperatures sufficiently high to volatize chemical contaminants out of the soil for vapour extraction. Technologies include ISTD, electrical resistance heating (ERH) and ET-DSP.
(c) Extraction of groundwater or soil vapour with an active electromechanical system, with subsequent stripping of the contaminants from the extract.
(d) Containment of the soil contaminants (such as by capping or paving over in place).

Acknowledgement The authors acknowledge the support extended by Council of Science & Technology, UP (Sanction No. CST/242/2015) to Shiv Shankar, under Young Scientist Scheme.

References

Apitz SE (2008) Is risk based, sustainable sediment management consistent with European policy. J Soil Sediment 8:461–466

Ashraf MA, Maah MJ, Yusof I (2014) Soil contamination, risk assessment and remediation. In: Hernandez-Soriano (ed) Environmental risk assessment of soil contamination. INTECH, Croatia, pp 3–56

Berge ND, Reinhart DR, Batarseh ES (2009) An assessment of bioreactor landfill costs and benefits. Waste Manag 29(5):1558–1567

Biswas AK, Kumar S, Babu et al (2010) Studies on environmental quality in and around municipal solid waste dumpsite. Resour Conserv Recycl 55(2):129–134

Chandra R, Takeuchi H, Hasegawa T (2012) Methane production from lignocellulosic agricultural crop wastes: a review of second generation of biofuel production. Renew Sustain Energy Rev 16:1462–1476

Chen D, Yin L, Wang H, He P (2014) Pyrolysis technologies for municipal solid waste: a review. Waste Manag 34(12):2466–2486

CPCB (2010) Air quality assessment, emissions inventory and source apportionment studies: Mumbai. New Delhi. http://cpcb.nic.in/Mumbai-report.pdf. Accessed on 3 Dec 2015

Dace E, Blumberga D, Kuplais et al (2015) Optimization of landfill gas use in municipal solid waste landfills in Latvia. Energy Procedia 72:293–299

DEA (2009) Soild waste management sector in India. Department of Economic Affairs Ministry of Finance Government of India. http://www.pppinindia.com/pdf/ppp_position_paper_solid_waste_mgmt_112k9.pdf. Accessed on 4 Dec 2015

FAO (2015) Management of calcarious soils. http://www.fao.org/soils-portal/soil-management/management-of-some-problem-soils/calcareous-soils/en/. Accessed on 3 Dec 2015

Guerrero LA, Maas G, Hogland W (2013) Solid waste management challenges for cities in developing countries. Waste Manag 33:220–232

Minghua Z, Xiumin F, Rovetta et al (2009) Municipal solid waste management in Pudong New Area, China. Waste Manag J 29:1227–1233

MOEF (2014) Annual report 2014–2015 Ministry of Environment and Forest Government of India. www.envfor.nic.in. Accessed on 4 Dec 2015

Mulligan CN, Yong RN, Gibbs BF (2001) Heavy metal removal from sediments by biosurfactants. J Hazard Mater 85(1–2):111–125

Nalo T, Tasing K, Kumar S, Bharti A (2014) Anaerobic digestion of municipal solid waste: a critical analysis. Int J Innov Res Sci Eng Technol 3(4):2347–6710

Olawoyin R, Oyewole S, Grayson R (2012) Potential risk effect from elevated levels of soil heavy metals on human health in the Niger delta. Ecotoxicol Environ Saf 1:120–130

Parida AK, Das AB (2005) Salt tolerance and salinity effects on plants: a review. Ecotoxicol Environ Saf 60:324–349

Saffarzadeh A, Arumugam A, Shimaoka T (2015) Aluminum and aluminum alloys in municipal solid waste incineration (MSWI) bottom ash: a potential source for the production of hydrogen gas. Int J Hydrogen Energy. http://dx.doi.org/10.1016/j.ijhydene.2015.11.059

Saha JK, Panwar N, Singh MV (2010) An assessment of municipal solid waste compost quality produced in different cities of India in the perspective of developing quality control indices. Waste Manag 30:192–201

Snyder C (2005) The dirty work of promoting recycling of Americas sewage sludge. Int J Occup Environ Health 11(4):415–427

Sujauddin M, Huda MS, Rafiqul Hoque ATM (2008) Household solid waste characteristics and management in Chittagong, Bangladesh. Waste Manag J 28:1688–1695

UNEP (2015)http://www.unep.or.jp/ietc/publications/spc/State_of_waste_Management/2.asp. Accessed on 3 Dec 2015

Urzelai A, Vega M, Angulo E (2000) Deriving ecological risk-based soil quality values in the Basque Country. Sci Total Environ 247:279–284

USDA (2015) Vertisols http://www.nrcs.usda.gov/wps/portal/.nrcs/detail/soil/scientists/?cid-nrcs. Accessed on 2 Dec 2015

Waste (2015) The organic waste flow in integrated sustainable waste management. http://www.waste.nl/en/product/the-organic-waste-flow-in-integrated-sustainable-waste-management. Accessed on 3 Dec 2015

Suggested Further Readings

Helmut M (2013) Soil remediation and rehabilitation (treatment of contaminated and disturbed land), ISBN: 978-94-007-5751-6, Springer, Dordrecht

Mirsal IA (2008) Soil pollution: origin, monitoring & remediation, ISBN: 978-3-540-70775-2 (Print) 978-3-540-70777-6 (Online), Springer, Berlin

Osman KT (2014) Soil degradation, conservation and remediation, ISBN: 978-94-007-7589-3 (Print) 978-94-007-7590-9 (Online), Springer Science, Business Media, Dordrecht

Tchobanoglous G, Kreith F (2002) Handbook of solid waste management, ISBN-13: 978-0071356237, McGraw-Hill Professional, New York

United Nations Environmental Programme (2005) Solid waste management, ISBN: 92-807-2676-5, United Nations Environment Programme, Paris

Wastewater Treatment

6

Monika Asthana, Avnish Kumar, and B.S. Sharma

Contents

M. Asthana (✉) • A. Kumar
Department of Biotechnology, School of Life Sciences, Dr. Bhim Rao Ambedkar University, Khandari Campus, Agra 282004, UP, India
e-mail: mailtomonikasaxena@gmail.com

B.S. Sharma
Department of Environmental Studies, School of Life Sciences, Dr. Bhim Rao Ambedkar University, Khandari Campus, Agra 282004, UP, India

R.L. Singh (ed.), *Principles and Applications of Environmental Biotechnology for a Sustainable Future*, Applied Environmental Science and Engineering for a Sustainable Future, DOI 10.1007/978-981-10-1866-4_6

Abstract

Water is a universal solvent and a vital constituent of living organisms. It recirculates in the environment through the hydrologic cycle which can be hampered due to human activities in terms of pollution. Polluted water, known as wastewater or effluent, should not be drained without treatment, as it constitutes a serious threat to living beings. Parameters like biochemical oxygen demand (BOD), chemical oxygen demand (COD), total dissolved solids (TDS), etc. are used to determine water quality. Therefore, wastewater treatment methods are targeted to get these parameters at optimum level. Such processes are either physico-chemical or biotechnological in nature and may be categorised as preliminary, primary, secondary and tertiary treatments. All of these must be followed by disinfection to obtain potable water. Major objectives of preliminary and primary treatment include removal of coarse and fine particle by screening, filtration, sedimentation, equalisation and flotation. Secondary treatment consists of biological treatment, i.e. aerobic, anaerobic and specialised reactors. Tertiary treatment entails chemical processes to purify wastewater. In the future, as the world's population continues to grow, such research priorities will become increasingly paramount. At present, a change in research priorities can be observed, and new technologies that meet the requirements of sustainable development and multidisciplinary approach are being applied. Therefore, modified methods must be developed to be used in systemic combination to fulfil the demand of potable and reusable water.

Keywords

Activated sludge • Trickling filters • Rotating biological contactor • Anaerobic process/digesters • Aerobic processes

6.1 Introduction

Water is indispensable in the living world for many reasons. It is an essential constituent of photosynthetic process that is important to animal and plant ecosystems. It establishes itself as a means of nutrition for plants and forms the natural living conditions for many living species and acts as a solvent of organic and inorganic materials (universal solvent). It is also a vital constituent of living organisms. The amount of water contained in the human body represents about 65–70 % of the body weight. The protoplasm of most living cells contains approximately 80 % water, and any reduction of this amount can have damaging effects and may even be fatal. Thus, water plays an important role in metabolism and exists as necessary component of body fluids. Most biochemical reactions take place in the presence of water. It thus supports the entire system of life on our planet, constituting crucial elements for all ecosystems. In fact, the availability of water has governed the establishment of civilisations and the development and progress of man's economic activities.

6.1.1 The Earth's Water Resources

Water is in recirculation in the environment through the hydrologic cycle. This signifies the movement of water evaporated from surface waterbodies or evapotranspirated from plants to the atmosphere where it condenses and precipitates to the earth as rain, snow or in some other forms. On the surface of the earth, some of the precipitated water then runs off into streams, lakes, ponds and the sea. The rest percolates through soil strata to form groundwater aquifers that ultimately flow into surface waterbodies.

Although 70 % of the world's surface area is covered with water, only 0.00192 % of the total stock is available for human use because 98 % water occurs in oceans and seas and 1.998 % is locked up in arctic regions, glaciers, mountains and clouds and, thus, remains unavailable.

6.2 Wastewater

Wastewater may be defined as 'a combination of liquid or water – carried wastes removed from residences, institutions and commercial and industrial establishment together with such groundwater, surface water and storm water as may be present'. It is the used water supply of a community and consists of domestic waterborne wastes, called sewage, which include human excrement and wash water as well as everything that goes down the drain of a home and into a sewage system.

Most industries are water based and release a considerable volume of wastewater which is generally discharged into water courses either untreated or inadequately treated and causes water pollution (Noorjahan 2014). Industrial waterborne wastes usually contain acids, alkaline materials, oils, greases and animal and vegetable matters discharged by factories. Industrial growth encompasses setting up of new industries, producing new chemicals and biochemicals. Varied industries such as distilleries, tanneries, textiles, antibiotics, drugs and pharmaceuticals, pulp and paper, dairies, oil refineries, petrochemicals, fertilisers, organic chemicals, etc. in the process of manufacturing contribute to water pollution. Thus, all small-, medium- and large-scale industries have their own role in water pollution. Water pollution is also caused by solid and hazardous wastes dumped on land.

6.2.1 Problems Associated with Improper Wastewater Discharge

Generally, problems associated with improper wastewater discharge include the following:

(a) Wastewater affects natural water quality through the production of taste, odour and malodorous gases. The gases that may be produced include CO_2, H_2S, CH_4, NH_3 and other trace gases such as H_2 and N_2.

(b) Wastewater contains pathogenic microorganisms that cause many diseases.
(c) Wastewater sludge may introduce highly persistent detergents, pesticides and other toxic wastes and compound.
(d) Massive quantity of solids may produce objectionable and dangerous levels of sludge on the bottom of waterbodies or along their banks. These solids add to the chemical, biological and physical degradation of natural water courses.
(e) Wastewaters containing grease and oils render bathing sites unusable, present extra problems for treatment works, produce unsightly conditions and interfere with the process of biodegradation.
(f) Wastewater may produce eutrophication or the enrichment of water by plant nutrients, biomass of phytoplankton, attached algae, macrophytes, etc.

For the aforementioned reasons, the immediate and nuisance-free removal of wastewater from its sources of generation, succeeded by proper treatment and final disposal, is not only desirable but also necessary for appropriate environmental sanitation.

6.3 Classification of Wastewater

The sources of origin and generation of wastewater may be grouped as follows:

(a) Domestic (Sanitary) Wastewater – Domestic wastewater includes discharges from residences and commercial, institutional and similar facilities.
(b) Industrial Wastewater – Industrial wastewater signifies the industrial waste generated from industrial localities. It varies with the type and size of the industry and other factors affecting production and processes.
(c) Infiltration/Inflow – Infiltration as a class of wastewater addresses extraneous water that enters the sewer system from the ground through various means. Likewise, it includes storm water that is discharged from sources such as roof leaders, foundation drains and storm sewers.
(d) Storm Water – Storm water refers to the water resulting from precipitation runoff.

6.4 Properties and Characteristics of Water and Wastewater

6.4.1 Physical Characteristics

6.4.1.1 Temperature

Temperature is one of the physical parameters that reveal a great deal of information about the water source and its state. Changes in temperature may be due to seasonal or daily variation or disposal of hot water or due to disposal of wastes from industrial processes and power stations (thermal pollution). Notable effects of temperature increase include:

(a) Speeding up and motivation of chemical reaction and reaction rate. The temperature may affect the reaction rate of microorganisms to the extent of doubling it for each 10 °C increase. Higher temperature can support the growth of undesirable aquatic plant and wastewater fungus.
(b) Reduction of dissolved oxygen concentration.
(c) Reduction of solubility of gases.
(d) An increase of biochemical oxygen demand (BOD).
(e) An increase in the rate of corrosion of substances.
(f) An increase in sensitivity of aquatic animals for toxic dissolved substances in the aquatic environment.
(g) An increase in malodour.

The change in temperature depends on the latitude and altitude of the earth too.

6.4.1.2 Turbidity

Turbidity is the phenomena whereby a specific portion of a light beam passing through a liquid medium is deflected from undissolved particles. Organic particles, oil, algae, colloidal and soluble substances present in water often initiate turbidity or colour in wastewaters. Small particles with density close to water may never settle and stabilise in the water. However, coagulation and flocculation of these particles into larger flocs are the necessary steps for their removal followed by sedimentation. Measurement of turbidity can be made easily and rapidly by using turbidimeters. Turbidity is an expression of optical property that causes light to be scattered and absorbed rather than transmitted in straight line through the sample. Measurements of turbidity in water can be defined in two ways: the turbidity resulting from 1 mg per litre of fuller's earth suspended in water or the depth of the column of water that just obscures the image of burning standard candle viewed vertically through the sample (Jackson candle unit, JCU)

1 nephelometric turbidity unit (NTU) $\simeq$ JCU

It is also measured by Beer-Lambert's law

$$A = \log_{10}\ [\mathrm{Io}/\mathrm{I}] = \varepsilon.c.l$$

Here,

ε is the molar absorbance coefficient.
c is the concentration of the solute in Moll^{-1}.
l is the light path of liquid in cell or cuvette in cm.

6.4.1.3 Taste

Usually, drinking water must be almost tasteless to please the consumer. The taste and odour are subjective properties which are rather difficult to measure. The presence of taste may be due to dissolved impurities organic or inorganic in origin. Examples of organic substances are phenols, chlorophenols, oils, fats, grease and unsaturated hydrocarbons. Inorganic substances include dissolved salts, iron, manganese, chlorides and gaseous substances such as hydrogen sulphide that are produced by decomposition of organic matter by microorganism such as fungi, algae, protozoa, bacteria, etc. These chemicals may be originated from municipal and industrial waste discharges, from natural sources such as decomposition of vegetables matters or from associated microbial activity. Taste and odour may also result from decaying aquatic vegetation as well as decaying leaves, weeds, grasses, etc.

Measurement of taste is rather difficult and classification generally included salty, bitter, sour or sweet. While it is relatively easy and safe for a person to evaluate the taste of drinking water supply, no one would be anxious to taste wastewater before or after conventional treatment. Production of potable water from unchlorinated effluent needs lime treatment, recarbonation, filtration, ion exchange, carbon absorption, ozonation, another carbon absorption treatment, reverse osmosis and finally chlorine dioxide treatment which make it costly.

6.4.1.4 Odour

A wholesale supply of water is odour-free. Existence of odour in water may be due to a number of reasons such as:

(a) Biodegradation of organic and inorganic compounds of nitrogen, phosphorus and sulphur
(b) Decomposition of algae and other microorganisms
(c) Generation of substances such as ammonia sulphide, chlorine, cyanide and hydrogen sulphide

It is easy to check for an odour associated with a drinking water supply. Offensive odours are generally associated with the raw wastewater that has been in the sewage system too long and aerobic conditions have developed. The sewage turns black and gives off hydrogen sulphide. The effects of this unpleasant odour include psychological stress, headache, nausea, vomiting, mental depression and blurred vision, fatigue or loss of appetite, impaired respiration, irritation of eyes, loss of sleep as well as reduction in production and work efficiency.

6.4.1.5 Colour

Sources of water colour include:

(a) Natural sources such as extracts from organic debris (leaves, wood and peats)
(b) Industrial origins such as mine waste, textile industry, paper industry and dye industry
(c) Domestic sewage

Pure water is colourless. True colour in natural water is caused by large organic molecules. Colour in water may also result from the presence of natural metallic ions such as iron oxide (causes red colour) and manganese oxide (causes brown and black colour). Other sources are humus and peat material, plankton, weed and industrial waste (e.g. textile and dye operation, paper and pulp production, food processing, chemical manufacturing, mining, refining and slaughterhouse operation). The greatest contribution of colour by plants is humic acid which produces a yellow brownish colour together with tanning and humate from the decomposition of lignin. These lignin derivatives are highly coloured and resist biological degradation.

6.4.1.6 Conductivity

Conductivity is a numerical expression of the tendency of an aqueous solution to carry an electric current. This ability depends on:

(a) First presence and type of ion
(b) Total concentration of ions
(c) Mobility, balance and relative concentration of ions
(d) Temperature of solution

Solutions of most organic acids, bases and salts are relatively good conductors. Electrical resistance, R (in ohms), of conductor may be derived as

$$R = R_s \times L / A$$

where

Rs = resistivity of conductor
R = resistance
L = length of conductor
A = cross-sectional area of conductor

The reciprocal of resistance is conductance expressed in reciprocal ohms or mhos.

$$R_d = 1 / R_s = \mathrm{Ac} / \mathrm{Rm}$$

where

Rs = specific resistance
Ac = cell constant
Rm = measured resistance

Conductivity may be defined as the electrical conductance of a conductor of unit length and unit cross-sectional area and commonly expressed in μmhoscm^{-1}.

Freshly distilled water has a conductivity of 0.5–2.0 μmhoscm^{-1}, increasing after a few weeks of storage to 2–4 μmhoscm^{-1}. The increase is mainly due to absorption of atmospheric CO_2 and to a lesser extent NH_3. Pure water is thus normally not a good conductor of electricity. Increase of dissolved salts in water increases its conductivity. As such, the conductivity of water is sometimes used for indicating the degree of its purification or pollution. The conductivity is proportional to the concentration of dissolved solids.

$$a.\text{EC} = \text{TDS}$$

where

EC = electrical conductivity
TDS = total dissolved solids
a = constant

6.4.1.7 Salinity

Salinity is the total dissolved solids in water after all carbonates have been converted to oxides, all bromides and iodides have been replaced by chlorides and all organic matter has been oxidised.

$$\text{Salinity}(\text{gm}/\text{kg}) = 0.03 + 1.805.\text{chlorinity}(\text{gm}/\text{kg})$$

The sources of chlorides in natural water can be from:

(a) Leaching of chloride containing rocks and soils
(b) Salt water intrusion (coastal areas)
(c) Agricultural, industrial and domestic wastewater

6.4.1.8 Solid Content

Solid content is defined as the matter that remains as residue upon evaporation and drying at 103–105 °C. According to particle size, wastewater solids have been classified as indicated in Table 6.1.

6.4.1.8.1 Dissolved Solids

In potable water, these consist of inorganic salts and a small concentration of organic matter. Water with high dissolved solids is generally of inferior palatability and may induce an unfavourable physiological reaction in the consumer. Highly mineralised water is also unsuitable for many industrial applications.

Table 6.1 Solids classification system

Solids	Size ($\times 10^{-6}$m)
Settleable	>100
Super colloidal	1–100
True colloidal	0.001–1
Dissolved	<0.001

6.4.1.8.2 Suspended Solids

These may be inorganic particles such as clay, silt and other soil constitutes, or they may be of organic origin such as plant fibres and biological solids like algae, bacteria, etc. These are the solids that can be filtered out by a fine filter paper. Water high in suspended solids may be aesthetically unsatisfactory for such purposes as bathing. Suspended solids also provide adsorption sites for chemical and biological agents.

6.4.1.8.3 Volatile and Fixed Solids

They give a measure of the amount of organic matter present in a sample. The test is carried out by burning organic matter to convert it to carbon dioxide and water, at a controlled temperature of 550 °C, to prevent the decomposition and volatilisation of inorganic substances.

6.4.1.8.4 Settleable Solids

These are solids in suspension that can settle in quiescent conditions under the influence of gravitational attraction.

6.4.1.9 Density

The density of fluid is defined as its mass per unit volume.

$$\rho = \mathrm{m} / \mathrm{V}$$

For water at standard pressure 760 mmHg and at 4 °C the density is 1000 $\mathrm{Kgm^{-3}}$

The reciprocal of the density is termed as specific volume. It is defined as the volume of a fluid occupied by a unit mass of the fluid. The ratio of the weight (density, ρ_s) of a substance to the weight (density, ρ_w) of an equal volume of water at standard conditions is denoted as the specific gravity (s.g.)

$$\mathrm{s.g.} = \rho_s / \rho_w$$

Since molecular activity and spacing increase with temperature, fewer molecules exist in a given volume of fluid as the temperature is increased. Therefore, density decreases with an increase in temperature. The application of pressure forces a larger number of molecules into a given volume. This results in an increase in density.

6.4.1.10 Radioactivity

Radiation is a characteristic feature of unstable atoms. Therefore, the approach to understanding radioactivity should begin at the level of atom. Certain nuclides spontaneously emit particles or gamma radiation or X radiation following orbital electron capture or undergo spontaneous fission.

Radioactivity may be of artificial, induced and natural types. Artificial radioactivity is a man-made radioactivity produced by particle bombardment or electromagnetic irradiation as opposed to natural radioactivity. Induced radioactivity is produced in substance after bombardment with neutrons or other particles. The resulting activity is naturally induced radioactivity if formed by nuclear reaction occurring in nature and artificially induced radioactivity if the reactions are caused by man. The property of radioactivity exhibited naturally by more than 50 radionuclides such as uranium, thorium, radium, polonium, etc. is known as natural radioactivity.

6.4.2 Chemical Characteristics

6.4.2.1 Hydrogen Ion Concentration (pH)

pH is a measure of the acidic or alkaline nature of a solution and affects the quality of a water or wastewater.

$$\mathrm{pH} = -\log[\mathrm{H}^+]$$

where $[H^+]$ is the concentration of hydrogen ions.

pH ranges from 0 to 14, with 7 as neutral, <7 being acidic and >7 being alkaline. It is an important parameter for both natural water and wastewater. The concentration range suitable for the existence of most biological life is narrow and critical. Wastewater with an adverse concentration of pH is difficult to treat by biological means, and if the concentration is not altered before discharge, the wastewater effluent may change the pH in natural water (Frobisher et al. 1974). Most microorganisms cannot survive below pH 4, but sulphate-oxidising bacteria can exist at pH 0.1. In practice, pH control is the most significant economic control the sanitary microbiologist has over the growth and death of microorganism.

6.4.2.2 Alkalinity

Alkalinity is the measure of buffering capacity of water. Alkalinity is caused primarily by chemical compounds dissolved from rocks and soil and is mainly due to the presence of hydroxyl, carbonate and bicarbonate ions. These compounds are mostly carbonates and bicarbonates of sodium, potassium, magnesium and calcium. Normally, wastewater is alkaline. In the anaerobic digestion process, sufficient alkalinity has to be present to ensure that the pH will not drop below 6.2, because the methane bacteria cannot function below that point. When digestion is proceeding satisfactorily, the alkalinity will normally range from 1000 to 5000 mg/l as $CaCO_3$.

Alkalinity in water is determined by titrating a sample of water with 0.02 N, H_2SO_4 solution. Total alkalinity is found by titrating to pH 4.5 (methyl orange end point) with a colour change from orange to pink.

$$\text{Alkalinity mg / l as CaCO}_3 = (A - B) \times N \times 50{,}000 / \text{ml sample}$$

where

A = ml standard acid used for sample
B = ml standard acid used for blank
N = normality of acid (0.02 N)

6.4.2.3 Acidity

Acidity is usually attributed to sample with a pH below the value of 7. In unpolluted water, acidity comes from dissolved CO_2 or organic acids leached from the soil. Atmospheric pollution may cause acidity. Acid water corrodes metal or concrete.

The acidity of water is determined by titrating a water sample with 0.02 N NaOH to pH 8.3.

Acidity as $\mathrm{mg\,CaCO_3/l} = [(A - B) \times C] - [(D \times E)] \times 50{,}000/\mathrm{ml\ sample}$
where

A = ml NaOH titrant used for sample
B = ml standard NaOH titrant used for blank
C = actual normality of standard NaOH titrant (0.02 N)
D = ml standard H_2SO_4 used
E = actual normality of standard H_2SO_4 (0.02 N)

6.4.2.4 Hardness

Hardness in water will prevent the formation of soap lather and is usually due to divalent cations such as Ca^{2+}, Mg^{2+}, Sr^{2+}, Fe^{2+} and Mn^{2+}. When hardness is numerically greater than the sum of carbonate and bicarbonate alkalinity, that amount of hardness equivalent to the total alkalinity is called carbonate hardness. The amount of hardness in excess of this is called non-carbonate hardness. When the hardness is less than or equals total alkalinity, all hardness is carbonate hardness and non-carbonate hardness is absent.

Hardness, $[\mathrm{mg\,equivalent\,CaCO_3/l}] = 2.497\mathrm{Ca^{2+}}[\mathrm{mg/l}] + 4.118\mathrm{Mg^{2+}}[\mathrm{mg/l}]$ (Rice et al. 2012).

When alkalinity < total hardness, carbonate hardness = alkalinity [mg/l].

Alkalinity > total hardness, carbonate hardness = total hardness [mg/l].

Impact of hardness includes:

(a) Economic losses to water uses through consumption of soap.
(b) Formation of precipitates on hot water appliances, boilers, kettles and domestic appliances, bath tubs, sinks, dishwashers and washbasins.
(c) Staining of clothes, dishes and other household utensils.
(d) Residues of the soap precipitate may remain in pores of skin making it feel rough and uncomfortable.
(e) Development of laxative effect on new consumers, especially due to the presence of $MgSO_4$.

The total hardness of water sample can be easily determined by the EDTA titrimetric method.

6.4.2.5 Dissolved Oxygen

Oxygen dissolved in sewage or water is needed for the maintenance of aerobic conditions, but the solubility of the oxygen in water is low. Drinking water saturated with oxygen has a pleasant taste, while water lacking dissolved oxygen has an insipid taste.

$$C_g = P_g.MW / R_u T$$

where

Cg = gas concentration in gas phase (g/m^3)
Pg = partial pressure of respective gas in gas phase [Pa=N/m^2]
MW = molecular weight of gas
Ru = universal gas constant (8.3143 J/K.mol)
T = absolute temperature (K)

Oxygen is slightly soluble in water. The actual quantity of oxygen that can be present in solution is governed by solubility of gas, partial pressure of gas in atmosphere, temperature and purity of water.

6.4.2.6 Oxygen Demand

Oxygen demand is the amount of oxygen needed to stabilise organic water.

(a) Biochemical oxygen demand (BOD) is a measure of amount of pollution by organic substances in water.
(b) Permanganate value (PV) is the chemical oxidation of water sample using a potassium permanganate solution.
(c) Chemical oxygen demand (COD) is the chemical oxidation of water sample using a mixture of concentrated H_2SO_4 and $K_2Cr_2O_7$.

$$PV < BOD < COD$$

6.4.2.7 Dissolved Gases

Natural water contains dissolved gases with varying concentration depending upon their solubility in water. When water is anaerobic and microbial activity exists, free ammonia, hydrogen sulphide and methane may exist. In the latter case the water needs to be oxygenated before use. From the point of view of water purity, the most important gases are oxygen and CO_2.

6.4.2.8 Chloride

Sources of chloride in natural water include leaching of chloride from rocks and soils; salt water intrusion (in coastal areas); agricultural, industrial, domestic and human wastewaters; and infiltration of groundwater into sewers adjacent to salt water. Chloride in the form of Cl^- ions is one of the major inorganic anions in water and wastewater. In potable water, the salty taste produced by chloride concentration is variable and depends on the chemical composition of water. Some waters containing 250 mg/l chloride have a detectable salty taste if the cation involved is Na^+ ion. On the other hand, the typical salty taste may be absent in water containing as much as 1000 mg/l Cl^- when predominant cations are calcium and magnesium (Rice et al. 2012).

The chloride concentration is higher in wastewater than in raw water because NaCl is a common part of diet and passes unchanged through the digestive system. When chlorine dioxide is used in water treatment (disinfection), chlorite ion is formed as a by-product. Chlorite is known to cause methaemoglobinaemia [a condition in which haemoglobin of the blood is oxidised to a metabolically inactive (ferric) state].

6.4.2.9 Nitrogen

In waters and wastewaters, nitrogen exists in four main forms, and biological treatment cannot proceed unless some of these forms are present:

(a) Organic nitrogen N is organically bound in the tri-negative oxidation state. Organic nitrogen includes such natural materials as proteins, peptides, nucleic acids, urea and numerous synthetic materials.

$$\text{Total oxidised nitrogen} = \text{nitrite nitrogen} + \text{nitrate nitrogen}$$

(b) Ammonia NH_3-N is present naturally in surface and wastewaters. Its concentration generally is low in groundwater because it adsorbs to soil particle and clays and is not leached readily from soil. It is produced largely by deaeration of organic nitrogen-containing compound and by hydrolysis of urea.
(c) Nitrite NO_2-N is an intermediate oxidation state of nitrogen. It can enter a water supply system through use as a corrosion inhibitor in industrial purpose water. Nitrite is the actual etiologic agent of methaemoglobinaemia. Nitrous acid, which is also formed from nitrite under acidic conditions, can react with secondary amine to form nitrosamines, many of which are known to be carcinogens.
(d) Nitrate NO_3-N is derived from the oxidation of ammonia. High concentration of nitrate (>10 mg/l) in water can cause cyanosis in infants. Nitrate is an essential nutrient of many photosynthetic photoautotrophs and in some cases has been identified as a growth-limiting nutrient.

6.4.2.10 Toxic Metals

Toxicity is the adverse effect a substance has on a test organism exposed to that substance. Toxicity is the result of a concentration and time exposure test, modified by variables such as temperature, chemical form and availability. Toxicity may be:

(a) Acute (short-term lethal)
(b) Chronic (long-term effects that may be related to changes in appetite, growth, metabolism, reproduction and even death or mutations)

The degree of toxicity depends upon the element involved such as copper, lead, silver, chromium, arsenic and boron. These metal(loid)s have to be taken into consideration when designing biological treatment system. The presence of other trace metals such as nickel, manganese and mercury at high concentrations also interferes with wastewater treatment processes. Toxic anions such as cyanide and chromates, often found in industrial wastewater, also hinder biological treatment and should be removed by pretreatment at the source before discharge to the municipal sewage system.

6.4.2.11 Nutrients

Nitrogen and phosphorus are essential growth factors together with other trace elements like iron, potassium, magnesium, calcium, cobalt, copper, sulphur and zinc. If wastewaters are to be treated by biological processes, the nutrient balance has to be considered in order to establish optimum operating conditions.

6.4.2.12 Proteins

Proteins are nitrogenous organic substances of higher molecular weight found in the animal kingdom and to a lesser extent in the plant kingdom. The amount present varies from a small percentage in watery fluids (e.g. tomatoes) and in the fatty tissue of meat to quite a high percentage in beans and lean meats. Protein consists wholly or partially of very large numbers of amino acids united by peptide links. They contain carbon, hydrogen, oxygen, nitrogen, sulphur and sometimes phosphorus.

It has been shown that proteinaceous materials constitute a large part of the wastewater sludges and that the sludge particles, if they do not consist of pure protein, are covered by a layer of protein which governs their physical and chemical behaviour. Under the influence of microorganisms, proteins undergo decomposition, giving end products which often have objectionable foul odours.

6.4.2.13 Oil and Greases

Oil and grease compounds are insoluble in water but dissolve in such organic substances as petroleum, chloroform, ether, etc. These are esters of alcohol or glycerol and fatty acids. Fats are among the more stable organic compounds and are not easily decomposed by bacteria. However, they can be attacked by mineral acids resulting in the formation of glycerin and fatty acids.

When grease is encountered in sufficient quantities, it causes clogging of filters, nozzles and sand beds. It coats the walls of sedimentation tanks and on decomposing increases the amount of scum. If grease is not removed before discharge of wastewater, it can interfere with the biological processes in the surface waters and create unsightly floating matter. Both trickling filters and activated sludge process are adversely affected by grease which can coat the biological forms sufficiently to interfere with oxygen transfer from the liquid to the interior of living cells.

6.4.2.14 Carbohydrates

Carbohydrates are organic substances that include starch, cellulose and sugars. They contain carbon, hydrogen and oxygen. Carbohydrates may be grouped as simple sugar (monosaccharides) or complex sugars (disaccharides and polysaccharides).

Bacteria utilise carbohydrates for the synthesis of fats and proteins as well as for energy. The majority of carbohydrates in wastewater are in the form of large molecules that cannot penetrate the cell membrane of microorganisms. It should be noted that formation of organic acids (anaerobic respiration) in large quantities can overtax the buffering capacity of wastewater resulting in a drop in pH and a cessation of biological activity.

6.4.2.15 Phenols

Phenols are a group of aromatic compounds with one or more hydroxyl group attached to the benzene ring. Phenols can be recovered from coal tar while greater amounts are manufactured synthetically. Phenols in wastewater may be industrial in origin, such as from coal, gas or petroleum operations. Phenols cause taste problem in drinking water, particularly when the water is chlorinated. This is due to the formation of chlorophenol.

6.4.2.16 Detergents

Detergents are large organic molecules. They are slightly soluble in water and may cause foaming in wastewater treatment plants and in the surface water into which the wastewater effluent is discharged. They can also seriously reduce the oxygen uptake in biological treatment processes. Synthetic detergents are classified as anionic, cationic or nonionic due to their electrical charge or lack of one when they dissolve in water (Frobisher et al. 1974). Synthetic detergents are used in households and industry. Detergents affect the wastewater treatment adversely as they lower the surface or interfacial tension of water and increase its ability to wet surfaces with which they come in contact; emulsify grease and oils and deflocculate colloids; induce floatation of solids and give rise to foams; and may kill useful bacteria and other living organisms.

6.4.2.17 Biochemical Oxygen Demand

BOD determination involves the measurement of the dissolved oxygen consumed by microorganisms in the biochemical oxidation of organic matter. The test determines the appropriate quantity of oxygen that will be required to biologically stabilise the organic matter present. Advantages of the test include:

(a) Determination of the size of waste treatment facilities
(b) Measurement of the efficiency of some treatment processes
(c) Determination of the approximate quantity of oxygen needed for the stabilisation of organic matter present

Biological oxidation is a slow process and theoretically takes an infinite time to go to completion. Within a 20-day period, the oxidation is about 95–99 % complete, and in the 5-day period used for the BOD test, oxidation is 60–70 % complete. The 20 °C temperature used is an average value for slow-moving streams in temperate climates and is easily duplicated in an incubator. Different results would be obtained at different temperature because biochemical reaction rates are temperature dependent. The test requires exclusion of light during the incubation period to prevent oxygen formation by alga in the sample.

6.4.2.17.1 BOD Kinetics

Waste management studies are usually done using calibrated and verified water quality models. Dissolved oxygen (DO) in rivers results from the combined effect of aeration and oxidation of organic matter. A commonly used one-dimensional steady-state mathematical model to predict DO level in the rivers receiving organic matter can be written as (Thomann and Mueller 1987)

$$\begin{aligned} D = \mathrm{DO}e^{-\mathrm{kat}} &+ [\{K_d \mathrm{Lo} / (K_a - \mathrm{K}_r)\}(e^{-\mathrm{Krt}} - e^{-\mathrm{Kat}})] \\ &+ [\{K_n \mathrm{Lno} / (K_a - \mathrm{Ln})\}(e^{-\mathrm{Knt}} - e^{-\mathrm{Kat}})] \end{aligned} \tag{6.1}$$

where DO is the initial oxygen deficit, Lo is the ultimate carbonaceous biochemical oxygen demand (CBOD) in the river after mixing, Lno is the ultimate nitrogenous biochemical oxygen demand (NBOD) in the river after mixing, *Ka* is the re-aeration rate coefficient, *Kr* is the BOD removal rate coefficient and *Kd* is the river CBOD deoxygenation rate coefficient, *Kn* is the NBOD deoxygenation rate coefficient and '*t*' is the travel time in the river. *Ka* can be determined by using different empirical relationships.

These calibrated and verified DO models are used to determine the required degree of wastewater treatment to maintain DO standards to meet the specific use of the waterbody. The models can then be used to formulate river water quality management strategies.

The rate coefficients *Kr*, *Kd* and *Kn* are related to the oxygen sink and depend upon the nature of the wastewater and other physical, chemical and biological factors particular to the river. *Kr* is the removal rate of carbonaceous organic matter and is determined from river surveys and is equal to

$$K_r = K_d + K_s \tag{6.2}$$

where *Ks* is the removal rate due to settling. *Kd* may be considered to consist of a component (*K*), characteristic of the type of wastewater, and can be determined from the analysis of long-term BOD measurements. Significant portion of particulate BOD is removed up to the secondary level treatment (i.e. suspended solids <30 mg/L); therefore, for such effluents, *Ks* in Eq. (6.2) may be neglected. The other component is 'ϕ', the characteristic of the conditions in the river, and may include factors that are not included in long-term BOD analysis. These components can be related to each other as

$$K_d = K + \phi \tag{6.3}$$

Wastewaters from urban areas are a mixture of carbohydrates, proteins and fats and vary in nature. With respect to biodegradation, their value changes with the level of treatment as readily biodegradable organic matter is first consumed. As such, the *Kr*, *Kd* and *Kn* which represent the biokinetic rates in rivers also change with the level of treatment. Bhargava (Bhargava 2008) developed a composite model considering the effect of settleable BOD for a river receiving wastewater from multiple outfalls by relating the rate constants with discrete and flocculent settling types.

The exertion of BOD is a first-order reaction kinetics and may be expressed as

$$\text{Rate} = \text{dBODr} / dt = -K_1\text{BOD}_\text{r}$$

where

t = time, days
BODr = amount of BOD remaining at time t
$k1$ = first-order reaction rate, 1/day

Integrating the above expression between the limits of UBOD and BODt and t = 0 and $t = t$ yields

$$\text{BODr} = \text{UBOD}(e^{-k1t})$$

where UBOD = total or ultimate carbonaceous BOD, mg/L

The amount of BOD exerted at time t (what gets regulated) is

$$\text{BODt} = \text{UBOD} - \text{BODr} = \text{UBOD} - \text{UBOD}(e^{-k1t})$$
$$= \text{UBOD}(1 - e^{-k1t})r\,r\,k\,\text{BOD}\,dt\,\text{dBOD}\,\text{r}$$

The higher the concentration of waste matter in wastewater, the stronger it is said to be. Wastewater strength is most often judged by its BOD or COD.

Limitations of the BOD test include:

(a) A high concentration of active acclimated seed bacteria is required.
(b) Pretreatment is needed when dealing with toxic wastes, and effects of nitrifying organisms must be reduced.
(c) Only the biodegradable organics are measured.
(d) The test does not have stoichiometric validity after the soluble organic matter present in solution has been used.
(e) An arbitrary long period of time is required to obtain results.

Perhaps the most serious limitation is that the 5-day period may or may not correspond to the point where the soluble organic matter that is present has been reduced. This reduces the usefulness of the test results.

6.4.2.18 Chemical Oxygen Demand

The COD test involves an acid oxidation with potassium dichromate. A measured amount of dichromate is added, the acidified samples is boiled for 2 h, cooled and the amount of dichromate remaining is measured by titration with a 0.25 N solution of ferrous ammonium sulphate, using ferroin indicator for end point determination. COD results are generally higher than BOD values since the test oxidises material such as fats and lignin which are only slowly biodegradable.

6.4.3 Biological and Bacteriological Characteristics

6.4.3.1 Environmental Microbiology

Environmental microbiology is a growing field of biology which often brings together issues of concern to engineers, geologists, hydrologists, microbiologists and public health officials.

Microbes are nature's decomposers. Drinking water is obtained either from surface sources such as rivers and lakes or from underground water. Such natural waters are likely to be polluted with domestic and industrial wastes. Although water purification systems envisage protection from pollution, sometimes the water supply can become a potential carrier of pathogenic organisms and endanger public health. A number of diseases such as cholera, typhoid, viral hepatitis, etc. are known to be waterborne. These pathogens are commonly transmitted through drinking water and cause infection of the intestinal tract. It is, therefore, necessary to employ treatment facilities to purify water and to provide safe drinking water (potable water).

Water can be perfectly clear in appearance and free from odour and taste and yet be contaminated by microorganisms. Pathogenic organisms enter into water through sewage contamination or discharges from animals or humans into the reservoirs. The coliforms (*E. coli* and related organisms), *Streptococcus faecalis* and *Clostridium perfringens* which are normal inhabitants of the large intestine of animals and humans enter water supplies through faecal contamination. The presence of any of these bacterial species in water is evidence of sewage or faecal pollution. Techniques are available by which the presence of these specific groups can be easily identified. The routine bacteriological examination consists of (i) plate count to determine the number of bacteria present and (ii) biochemical test to reveal the presence of coliform bacteria since these are indicator organisms for faecal contamination.

A variety of other bacteria and organisms which may not be serious pathogens including faecal streptococci, slime-forming bacteria, sulphur bacteria, algae, etc. may also cause problems of odour, colour and taste, and it is essential that these are eliminated from the drinking water.

It is important therefore that the fundamental activity of microorganisms and their metabolic and biochemical control should be more fully understood by those who are involved.

6.4.3.2 Bacteria

The word bacteria comes from a Greek word meaning rod or staff, a shape characteristic of many bacteria. Bacteria are single-celled microscopic organisms that multiply by binary fission. In order to multiply, they need carbon obtained from carbon dioxide if they are autotrophs or from organic compounds (dead vegetation, sewage, meat) if they are heterotrophs. Their energy comes either from sunlight if they are photosynthetic or from chemical reaction if they are chemosynthetic. Bacteria are present in air, water, earth, rotting vegetation and the intestines of the animal. Bacteria are fundamental to all biological processes, especially in the degradation of organic matter which takes place in trickling filter, activated sludge processes and sludge digestion.

The bacterial communities found in wastewater treatment plants are complex. The bacterial flora of all aerobic treatment systems is basically the same and includes *Zoogloea*, *Pseudomonas*, *Chromobacter*, *Achromobacter*, *Alcaligenes* and *Flavobacterium*.

6.4.3.2.1 Facultative Bacteria

Most of the bacteria that absorb the organic material in a wastewater treatment system are facultative in nature. This means they are adaptable to survive and multiply in either anaerobic or aerobic conditions. The nature of individual bacteria is dependent upon the environment in which they live. Usually, facultative bacteria will be anaerobic unless there is some type of mechanical or biochemical process used to add oxygen to the wastewater. When bacteria are in the process of being transferred from one environment to the other, the metamorphosis from anaerobic to aerobic state (and vice versa) takes place within a couple of hours. When mixed cultures are present, aerobic and facultative bacteria first decompose organic matter, gradually depleting the dissolved oxygen. When dissolved oxygen is exhausted, facultative bacteria continue to use O_2 in the form of SO_4 and NO_3, while some facultative and anaerobic organisms produce organic acids, alcohol, methane, etc.

6.4.3.2.2 Anaerobic Bacteria

Anaerobic bacteria (*Methylococcus*, *Desulfovibrio*, *Clostridium*, *Thiobacillus denitrificans*, *Enterobacter*, etc.) live and reproduce in the absence of free oxygen. These carry on fermentation activity using organic substances as electron donors and acceptors. They produce organic acids, alcohol and least amount of energy.

$$\text{Organics} \rightarrow \text{Organic acid} + H_2O + CO_2 + \text{Energy}$$

If methane-producing bacteria are present, then the anaerobic digestion process is completed by converting organic acids into methane and CO_2.

$$\text{Organic Acids} \rightarrow CH_4 + CO_2 + \text{Energy}$$

Some anaerobic bacteria use nitrate and sulphate as an electron acceptor.

$$\text{Organics} + NO_3 \rightarrow (\text{Microorganism})N_2 + CO_2 + \text{Energy}$$
$$\text{Organics} + SO_4 \rightarrow (\text{Microorganism})H_2S + CO_2 + \text{Energy}$$

H_2S is given out by sulphate-reducing bacteria if the wastewater becomes anaerobic. Slightly acidic gas is absorbed in water. Sulphur bacteria can tolerate even pH 1.0 and oxidise H_2S to H_2SO_4.

$$H_2S + O_2 \rightarrow H_2SO_4 + \text{Energy}(\textit{Thiobacillus})$$

H_2SO_4 reacts with lime in concrete to form calcium sulphate which lacks structural strength.

In order to remove a given amount of organic material in an anaerobic treatment system, the organic material must be exposed to a significantly higher quantity of bacteria and/or detained for a much longer period of time. A typical use for anaerobic bacteria would be in a septic tank. The slower metabolism of the anaerobic bacteria dictates that the wastewater should be held several days in order to achieve even a nominal 50 % reduction in organic material. That is why septic tanks are always followed by some type of effluent treatment and disposal process. The advantage of using the anaerobic process is that electromechanical equipment is not required. Anaerobic bacteria release hydrogen sulphide and methane gas, both of which can create hazardous conditions. Even as the anaerobic action begins in the collection lines of a sewer system, deadly hydrogen sulphide or explosive methane gas can accumulate and be life-threatening.

6.4.3.2.3 Aerobic Bacteria

Aerobic bacteria (*Azotobacter*, *Pseudomonas*, *Nitrosomonas*, *Nitrobacter*, hydrogen-oxidising bacteria, etc.) live and multiply in the presence of free oxygen. Facultative bacteria always achieve an aerobic state when oxygen is present. While the name 'aerobic' implies breathing air, dissolved oxygen is the primary source of energy for aerobic bacteria. The metabolism of aerobes is much higher than that of anaerobes. This increase means that up to 90 % fewer organisms are needed compared to the anaerobic process or that treatment is accomplished in 90 % less time.

$$\text{Organic matter} + O_2 \rightarrow (\text{Aerobic condition})H_2O + CO_2 + \text{Biomass}$$
$$+ \text{Energy}(\text{e.g. } \textit{Escherichia coli} \text{ reaction in wastewater})$$

Aerobic procedures are biochemically efficient and rapid. Their by-products are chemically simple and highly oxidised like CO_2 and H_2O. Iron bacteria occur normally in mine wastewater which is iron-rich. They can also occur in wastewater. They also contribute to corrosion and clogging of iron pipes.

$$Fe^{2++}O_2 \rightarrow Fe^{3++}\text{Energy}(\textit{Leptothrix})$$

Concrete pipes can collapse if they become too weak due to corrosion problem. Vitrified clay or PVC may be used to overcome the corrosion problem. Sewers can be lined with corrosion-resistant coating.

Aerobic procedures provide a number of advantages including a higher percentage of organic removal. The by-products of aerobic bacteria are carbon dioxide and water. Aerobic bacteria live in colonial structures called floc and are kept in suspension by the mechanical action used to introduce oxygen into the wastewater. This mechanical action exposes the floc to the organic material while treatment takes place. Following digestion, a gravity clarifier separates and settles out the floc. Because of the mechanical nature of the aerobic digestion process, maintenance and operator oversight are required.

Autotrophic Aerobes Autotrophs do not use organic matter. They oxidise inorganic compounds for energy and use CO_2 or CO_3 as carbon source. In wastewater treatment, autotrophs are relatively less important if high organic matter is the problem. Nitrifying bacteria, sulphur bacteria and iron bacteria are particularly important. Nitrifying bacteria oxidise ammonia.

$$NH_4^+ + O_2 \rightarrow NO_2^- + \text{Energy}(\textit{Nitrosomonas})$$
$$NO_2 + O_2 \rightarrow NO_3^- + \text{Energy}(\textit{Nitrobacter})$$

Conventional biological wastewater treatment generates large amounts of low-value bacterial biomass. The treatment and disposal of this excess bacterial biomass, also known as waste activated sludge, account for about 40–60 % of the wastewater treatment plant operation cost.

6.4.3.3 Fungi

Fungi are tiny aerobic, heterotrophic Protista containing no chlorophyll. They can tolerate dryer and more acidic conditions than most bacteria. They live in the earth, freshwater and sea water. Many grow as filaments and may be seen in polluted rivers, trickling filters and activated sludge. The optimum pH for most types is between 2 and 9. Because fungi are wholly aerobic, they can, in animals, exist only on the skin or in the bloodstream or lungs. Consequently, there are relatively very few fungi that cause disease in humans. Many organic substances can be attacked by fungi such as cellulose, phenols and hydrocarbons. Attacked organic compounds are converted into simple compounds which can be used as nutrients by other organisms. They produce aflatoxins that are harmful to man and animals. Fungi produce biomass with a higher value that can significantly change the economics of wastewater treatment. The biomass produced during fungal wastewater treatment has, potentially, a much higher value than that from the bacterial activated sludge process. The fungi can be used to derive valuable biochemicals and can also be used as a protein source. Various high-value biochemicals are produced by commercial cultivation of fungi under aseptic conditions using expensive substrates. Food-processing wastewater is an attractive alternative as a source of low-cost organic matter and nutrients to produce fungi with concomitant wastewater purification. Some species of fungi *Bjerkandera adusta*, *Aspergillus niger*, *Penicillium*, *Rhizopus arrhizus*, *Rhizopus oryzae*, etc. are useful in various sewage treatments.

6.4.3.4 Algae

Algae form a large group in the Protista. Being photosynthetic, they are classified as plants. Algae are single-celled or multicellular autotrophs. At night, some algae consume oxygen by chemosynthesis. Thus, water containing algae has a diurnal variation in dissolved oxygen. However, during sunlight, the carbon dioxide concentration falls. The CO_2 originated from the symbiotic relation with bacteria or from bicarbonates releases hydroxyl ion which tends to raise the pH of water: $HCO^{3-} \rightarrow CO_2 + OH^-$. Photosynthetic reaction during the day can be represented by the following reaction: $CO_2 + H_2O + \text{Sunlight} \rightarrow CH_2O + O_2$. This process utilises CO_2. The utilisation of CO_2 during the day may lead to a considerable rise of pH and results in a softening of water due to precipitation of $CaCO_3$, as represented by the following reaction: $Ca(HCO_3)_2 \rightarrow CaCO_3 + H_2O + CO_2$. For most waste stabilisation ponds to function properly, algae such as *Chlamydomonas*, *Chlorella* and *Euglena* are needed to supply oxygen to aerobic heterotrophic bacteria that consume and oxidise the organic matter in sewage. The history of the commercial use of algal cultures spans about 75 years with application to wastewater treatment and mass production of different strains such as *Chlorella* and *Dunaliella*. The most important classes of freshwater algae are *Chlorophyta*, *Euglenophyta*, *Chrysophyta* and *Cyanophyta*. Biotreatment with microalgae is particularly attractive because of their photosynthetic capabilities, converting solar energy into useful biomasses and incorporating nutrients such as nitrogen and phosphorus causing eutrophication (Abdel-Raoufa et al. 2012). The other frequently found algae are *Ankistrodesmus*, *Scenedesmus*, *Oscillatoria*, *Micractinium* and *Golenkinia*.

In drinking water, algae are troublesome because they clog filters, may leave a taste when they die and produce toxin that can poison cattle. Algae in reservoirs can be reduced by oxygenating the water and reducing its CO_2 content or by adding an algicide such as copper sulphate or potassium paramagnet or by destratification of lake or reservoir. Algae growth in the catchment area of water supply system may lead to obnoxious taste and other odour problem that are hard to remove and may require treatment with activated carbon.

6.4.3.5 Protozoa

Protozoa are single-celled eukaryotic organisms. Some protozoa are parasitic organisms that are present in municipal sewage so are present in the treatment tanks. Such organisms are disease-causing agents, so part of the treatment of sewage is aimed to kill these organisms, e.g. *Cryptosporidia* and *Giardia*. Some protozoa scavenge and digest bacteria, and these are important for controlling bacterial pathogens as well as the overgrowth of other bacteria in a treatment system, e.g. *Paramecium* and *Vorticella*. Protozoa may indicate, by their type, the condition of activated sludge. They are also important in the operation of trickling filters. They feed on the bacteria and some utilise alga. Most protozoa are harmless and only a few cause illness in humans. *Entamoeba histolytica* (amoebiasis), *Giardia lamblia* (giardiasis), *Trypanosoma* (trypanosomiasis: sleeping sickness) and *Plasmodium* (malaria).

6.4.3.6 Rotifers

Rotifers are tiny aerobic creatures ranging from 50 to 250 μm in length; rotifers are the simplest of the multicelled invertebrate animals. They have cilia around their mouth and can swallow bacteria or other organic matter. Their presence in an effluent indicates highly efficient aerobic biological treatment. These organisms are important to wastewater treatment because their main source of food is bacteria. It is believed that rotifers scavenge free-floating bacteria, thus reducing BOD and also the numbers of pathogenic organisms in the water.

6.4.3.7 Crustaceans

Crustaceans mainly are water animals that use O_2, consume organic substances and have a hard body or crust. They are an important food for fish. Crustaceans are not normally found in the biological treatment processes. Usually they are an indicator of clean water. The metabolic complexity of the crustaceans limits their growth to relatively stable streams and lakes.

6.4.3.8 Worms and Larvae

Worms and larvae are the normal inhabitant in organic mud and biological slime. They have aerobic requirements but can metabolise solid organic matter not readily degraded by other microorganisms. The common organism used in stream pollution studies as indicators of pollution are the worm, tubifex and the midge fly larva of Chironomidae.

6.4.3.9 Viruses

A virus is an entity that carries the information for its replication but does not possess the machinery for such replication. Thus, all viruses are obligatory parasite and they are unable to reproduce outside a host cell. Viruses are disease-causing agents that are present in sewage and are removed during sewage treatment by adsorption to the floc. Some kind of floc or support is necessary to prevent washout of the active cells. A rich carbon source is a benefit in that it allows the formation of extracellular polymers which provide a glue-like substance to allow organisms to stick together (floc) or to a solid support. Viruses of particular interest in drinking water are hepatitis A, Norwalk-type viruses, rotaviruses, adenoviruses, enteroviruses and reoviruses.

6.5 Impact of Pollutants on Biotreatment

Industrial effluents may have a different impact on microorganisms depending on the nature and composition of effluent pollutants. This can be as follows.

6.5.1 Biodegradation

Biodegradation is nature's way of recycling wastes or breaking down organic matter into nutrients that can be used and reused by other organisms. Biodegradation is a process of biological catalytic reduction in complexity of chemical compounds using living microbial organisms (Asthana et al. 2014; Gupta et al. 2014; Kumar et al. 2015; Marinescu et al. 2009). There is, therefore, complete compatibility between the bacterial floral and oxo-biodegradable degradation of compounds. The term is often used in relation with ecology, waste management, environmental remediation (bioremediation) and for plastic materials, due to their long life span.

6.5.1.1 Inhibition of Biodegradation

Heavy metals like Cd, Cr, Cu, Hg, Zn, Ni, Pb, etc. are often present in a variety of industrial effluents and will inhibit biological activity. The presence of metals and metalloids will not allow otherwise degradable organic matter also to get degraded. Among organic solvents, chlorinated organics and alcohols are toxic to biological processes. Also phenols, pesticides and surfactant are inhibitory to biological activities in treatment plants.

Anaerobic processes are susceptible to sulphides (H_2S). High sulphides are present in the effluents of molasses fermentation industry, tanneries, petrochemical industry and paper mills. Anaerobic processes are also sensitive to pH outside the range of 6.5–8.2. Volatile fatty acids in excess of 2000 mg l^{-1}, NH_4^+ concentration in excess of 3000 mgl^{-1} and free ammonia at a concentration of 150mgl^{-1} are also very toxic.

6.5.2 Incidental Removal

Incidental removal of pollutant may occur by absorption, adsorption, precipitation and consequential concentration into sludge produced. Activated sludge has the capacity to bind heavy metals by polysaccharides of microbial flocs. Incidental removal of organic compounds by association with settleable solids or floatable scum or grease also takes place. This may occur if the compounds are insoluble or slightly soluble in water or hydrophobic.

6.5.3 Co-metabolism

When a particular compound is altered chemically by microbial metabolism without that compound serving as a source of carbon or energy, that compound is said to be co-metabolised or co-oxidised. A co-metabolite does not support the growth of organism concerned, and end products of transformation are accumulated stoichiometrically. The transformation does not require energy consumption. Co-metabolism is thought to occur because some enzymes produced by organisms for metabolism of their major carbon source are not substrate specific and can act on other compounds (Table 6.2). The rate of co-metabolism is often quite slow.

Table 6.2 Growth substrate and co-metabolite for some microorganisms

Organism	Growth substrate	Co-metabolite
Methylococcus	Methane	Ethane
Achromobacter	Hexadecane	Ethylbenzene
	Benzoic acid	3-chlorobenzoate
Nocardia	Hexadecane	Toluene
Corynebacterium	Hexadecane	Glucose
	Naphthalene	Anthracene

In pure culture, co-metabolism can be considered as a dead end transformation without benefit to that microorganism. Synergistic transformation and substrate utilisation may lead to recycling of relatively recalcitrant compound.

Co-metabolism appears to be an important route for the degradation of hydrocarbon, especially the more recalcitrant alicyclic compound in the environment. It is also an important pathway for pesticide degradation.

6.6 Wastewater Treatment Process

Selection of a treatment process depends on the characteristics of wastewater (i.e. form of pollutants like suspended, colloidal or dissolved; biodegradability; toxicity of the components), required effluent quality, cost, availability of treatment devices and area (land). Wastewater treatment units can be classified according to their capacity.

Small Wastewater Treatment Plants Small wastewater treatment units address wastewater treatments as applied to individual household or small community. Usually they are on-site treatment and disposal units.

Large Wastewater Treatment Units Large wastewater treatment plants are wastewater works that govern the discharge and treatment of large population sector. Sewage is collected from different localities and diverted to a central treatment plant.

Reasons for Treatment

The major reasons for the wastewater treatment may be as follows:

- Reduction in the spread of communicable diseases – to be achieved through the elimination or reduction of pathogens in the sewage
- Prevention or reduction of pollution that may enter the surface or groundwater sources
- Stabilisation of sewage without causing any odours or nuisances and without endangering the public health
- Water reuse aspects or for waste by-product recovery

Typical Strategies of Wastewater Treatment Processes

Wastewater may be treated on site or as sewage treatment works (STW) by the following methods:

Preliminary treatment
Physical (primary) treatment
Secondary treatment
Tertiary treatment (chemical)
Disinfection of wastewater

6.6.1 Preliminary Treatment

Preliminary treatment is used essentially to prepare wastewater for treatment. The objective of preliminary treatment is the removal of coarse particles and other large materials often found in raw wastewater, which could obstruct flow through the plant or damage equipment. Removal of these materials is necessary to enhance the operation and maintenance of subsequent treatment units. These materials are composed of floating objects such as rags, wood, faecal material and heavier grit particles.

Elements that could damage treatment units are removed, and usually the flow is equalised, reducing maximum flow conditions and allowing a smaller plant treat the wastewater flow. A typical pretreatment unit contains the following.

6.6.1.1 Bar Racks

A bar rack (or bar screen) traps debris as wastewater influent passes through. Typically, a bar screen consists of a series of parallel, evenly spaced bars or a perforated screen placed in a channel used to remove large objects that could damage other plant devices (Singh et al. 2012). Large floating objects can be removed by passing the sewage through bars spaced at 20–60 mm. Retained material is raked from the bars at regular intervals. Bar screens with relatively large openings of 75–150 mm are provided ahead of pumps, while those ahead of sedimentation tanks have smaller openings of 50 mm.

6.6.1.2 Grit Chamber

These are designed to remove grit (inert dense material, such as sand, broken glass, silt, pebbles) from wastewater, to keep it from eroding and damaging pumps and other mechanisms. Preliminary treatment operations typically include coarse screening and grit removal of large objects. Sufficiently high velocity of water flow is maintained in grit chambers, or air is used, so as to prevent the settling of most organic solids.

6.6.1.3 Equalisation Basin

Most treatment devices must be designed with specific conditions of maximal and minimal flow, but normally the wastewater flow from a community is far from being constant. So, in order to keep the wastewater flow entering the primary treatment unit constant, an equalisation basin is used to collect and store wastewater flow. From this basin the wastewater is pumped at a constant rate into the primary treatment unit.

6.6.2 Primary Treatment

The objective of primary treatment is the removal of settleable organic and inorganic solids and the removal of materials that will float (scum) by skimming. Approximately 25–50 % of the incoming biochemical oxygen demand (BOD_5),

50–70 % of the total suspended solids and 65 % of the oil and grease are removed during primary treatment. Some organic nitrogen, organic phosphorus and heavy metals associated with solids are also removed during primary treatment, but colloidal and dissolved constituents are not affected. Pathogen removal during primary treatment is highly varied with various removal rates reported for different organisms.

6.6.2.1 Screening

Screening is the first operation at any wastewater treatment works. This process essentially involves the removal of large non-biodegradable and floating solids that frequently enter a wastewater works, such as rags, papers, plastics, tins, containers and wood.

Efficient removal of these constituents will protect the downstream plant and equipment from any possible damage, unnecessary wear and tear, pipe blockages and the accumulation of unwanted material that will interfere with the required wastewater treatment processes. The screening equipment is designed for efficient and cost-effective solution with durable and dependable operations.

Wastewater screening is generally classified into either coarse screening or fine screening.

6.6.2.1.1 Coarse Screens

Coarse screens remove large solids, rags and debris from wastewater and typically have openings of 6 mm (0.25 in.) or larger. Types of coarse screens include mechanically and manually cleaned bar screens, including trash racks (Fig. 6.1).

6.6.2.1.2 Fine Screens

Fine screens are typically used to remove material that may create operation and maintenance problems in downstream processes, particularly in systems that lack primary treatment. Typical opening sizes for fine screens are 1.5–6 mm (0.06–0.25 in.). Very fine screens with openings of 0.2–1.5 mm (0.01–0.06 in.) placed after coarse or fine screens can reduce suspended solids to levels near those achieved by primary clarification.

Most modern wastewater treatment plants utilise a combination of coarse and fine screening (i.e. upstream coarse screens providing protection to downstream fine screens).

6.6.2.2 Equalisation

For sewage treatment plants of small community, where wastewater flow rates vary considerably with time, and for industrial wastewater treatment plants, where wastewater flow and characteristic vary with time, equalisation becomes essential to obtain proper performance of the treatment plant by avoiding shock loading (hydraulic and organic) to the systems. Due to possibility of variation in flow rate received at treatment plant, there may be deterioration in performance of the treatment plant. To facilitate maintenance of uniform flow rate in the treatment units, flow equalisation is used. Equalisation can also be used to provide continuous

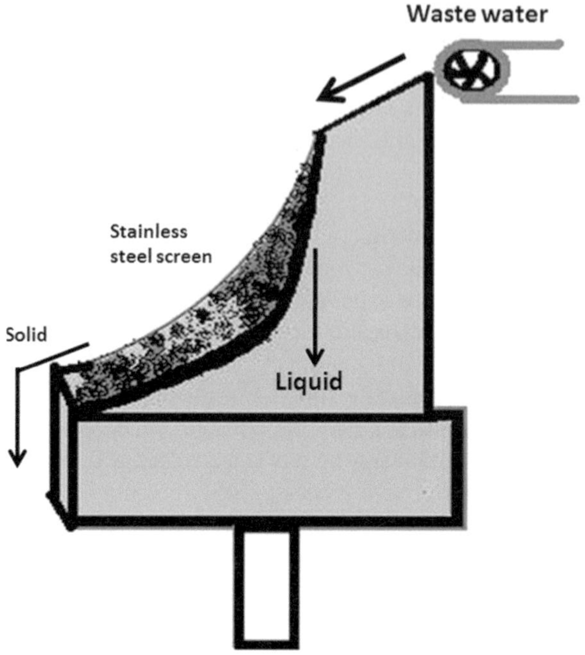

Fig. 6.1 Diagram of an inclined or tangential screen

feeding to the treatment system when the wastewater generation is intermittent, to control pH fluctuations or to control toxic concentration in the feed to the biological reactor, and this can also be used to control the discharge of industrial effluent in to the sanitary sewers.

Equalisation can be of two types:

(a) Inline: where all flow passes through equalisation basin.
(b) Offline: where the flow above average daily flow is diverted to equalisation basin. The pumping is minimised in this case, but the amount of pollutant concentration damping is considerably reduced.

Location of equalisation: Location of equalisation basin after primary treatment and before biological treatment is appropriate. This arrangement considerably reduces problem of sludge and scum in the equalisation basin. If the equalisation basin is placed before primary treatment, it must be provided with sufficient mixing to prevent deposition of solids and concentration variations and aeration to prevent odour problem. Most commonly used submerged or surface aerators have power levels of approximately 0.003–0.004 KW/m^3. In diffused air mixing, air requirement of 3.74 m^3/m^3 (air flow rate to water flow rate) is used.

6.6.2.3 Sedimentation

After removal of the coarse materials, sewage passes to sedimentation tanks, which aims to remove the settleable solids (represent up to 70 % of the total settleable solids) by gravity. A well-designed sedimentation tank can remove 40 % of the

BOD in the form of settleable solids. Sedimentation is used to remove suspended solids present in wastewaters. In fishery wastewaters, these include fish scales, portions of fish muscle and offal (relative proportions varying with the particular process being used).

Sedimentation is based on the difference in density between the bulk of the liquid and the solid particles, which results in the settling of the solids present. The terms sedimentation and settling are often used interchangeably. This operation is conducted as part not only of the primary treatment but also in the secondary treatment for separation of solids generated in the biological treatments such as activated sludge or trickling filters. Depending on the properties of solids present in the wastewater, sedimentation can proceed as:

- Discrete settling, if the wastewater is relatively diluted and the particles do not interact.
- Flocculent settling, if the particles that coalesce or flocculate are living particles of larger mass and faster settling rate. This is typical of untreated wastewater and is encountered in primary settling facilities.
- Zone settling, which is also called hindered settling and occurs when the particles adhere together and settle as a blanket, forming a distinguishable interface with the liquid above it. This reaction occurs in secondary clarifiers for sludges of biological treatments.

Each case has different characteristics. For discrete settling, calculations can be made on the settling velocity of individual particles. In a settling tank, these move both downwards (settling) and towards the outlet zone with the waterflow (Fig. 6.2). The particles that reach the bottom before the outlet will be separated from the effluent while the rest will pass through the settling tank. The critical velocity (*vc*) below in which a particle will be carried out of the tank is given by the depth of liquid (*d*), the volume of the tank (*V*) and the flow rate (*Q*):

$$v_c = d / (V / Q)$$

The ratio of *V*/*Q* is also known as the residence time of the liquid in the tank. It is called the overflow rate when *vc* is expressed in terms of volume of effluent per unit

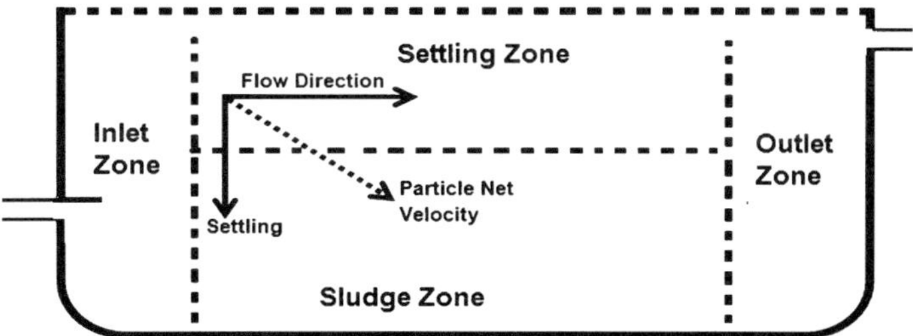

Fig. 6.2 Schematics of discrete settling

surface area of the tank per unit of time. This case may be present in fishery wastewaters but is not the most common.

In the case of a flocculant suspension, the formation of larger particles due to coalescence depends on several factors, e.g. the nature of the particles and the rate of coalescence. A theoretical analysis is not feasible due to the interaction of particles which depends among other factors on the overflow rate, the concentration of particles and the depth of the tank.

A settling column is used to evaluate the settling characteristics of a flocculant suspension (Fig. 6.3). The same kind of column using only one sampling port can be used to study discrete settling.

The zone (or hindered) settling, which occurs when the particles do not settle independently, is also studied by batch tests. In this case, an effluent that is initially uniform in solid concentration (Fig. 6.4), if allowed to settle, will do so in zones, the first of which is that of clarified water (1) and below is the interfacial zone (2) in which the solid concentration is considered uniform. In the bottom, a compact sludge develops in the so-called compaction zone (4). Between (2) and (4), a transition zone (3) generally exists.

In some cases, further compaction may occur. The detailed design procedures for all these cases are beyond the scope of this document and can be found elsewhere. The actual configuration of a sedimentation tank can be either rectangular or circular. Rectangular settling tanks (Fig. 6.5) are generally used when several tanks are required and there is space constraint, since they occupy less space than several circular tanks.

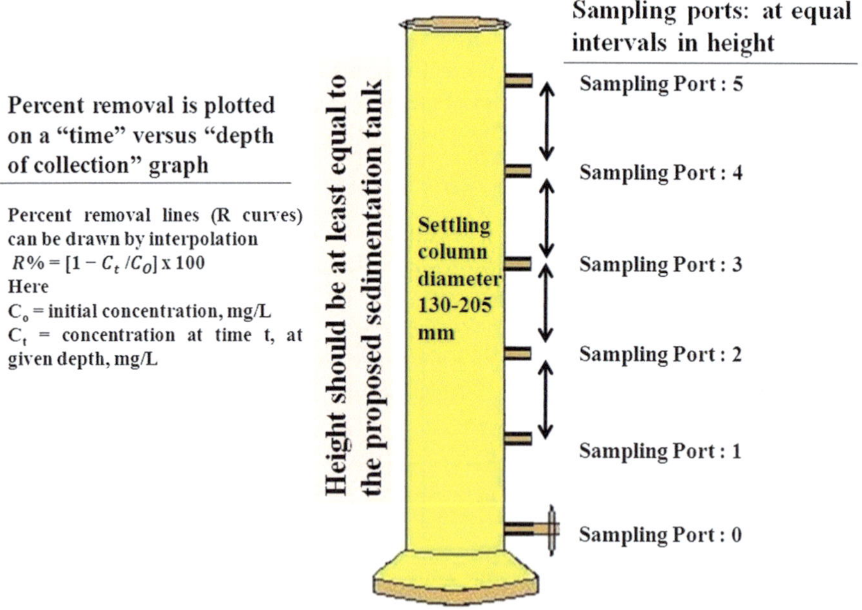

Fig. 6.3 Laboratory settling column

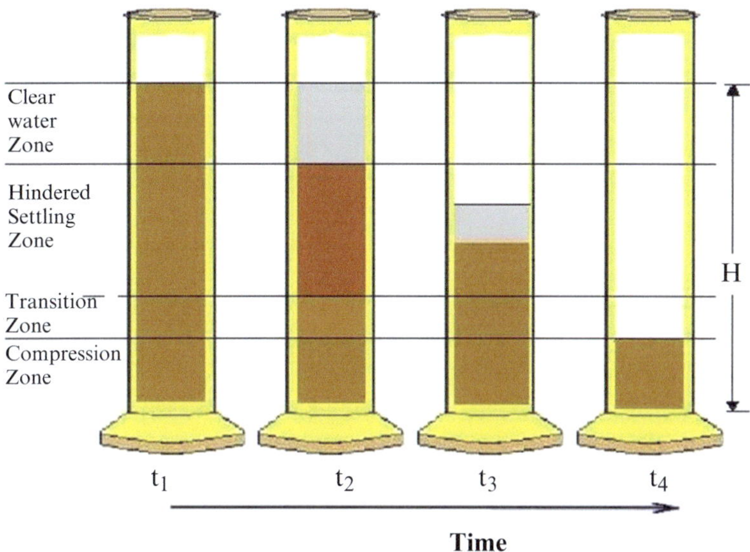

Fig. 6.4 Diagram of a zone settling process: Zone (Type III) Settling. In the bottom, a compact sludge develops in the so-called compaction zone (*4*). Between (*2*) and (*4*), a transition zone (*3*) generally exists. As time proceeds, the clarified effluent and compaction zones will increase in size, while the two intermediates will eventually disappear

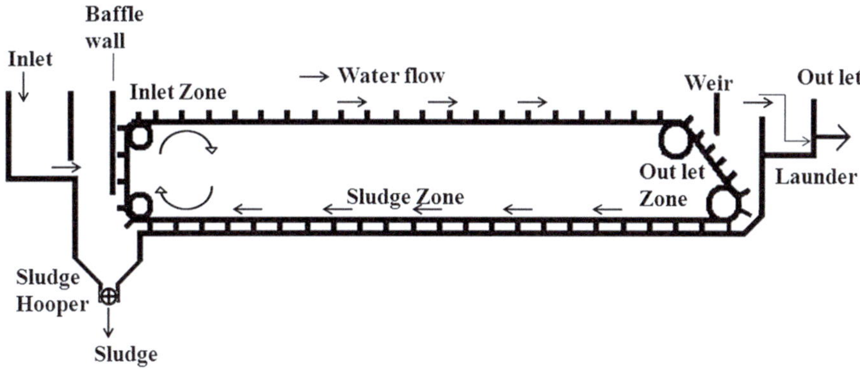

Fig. 6.5 Diagram of a rectangular clarifier

For removal of solids, a series of chain-driven scrapers are used: these span the width of the tank, are regularly spaced and move at 0.5–1 m/min. The sludge is collected in a hopper in the end of the tank, where it may be removed by screw conveyors or pumped out. Configurations exist in which the sludge is forced opposite to the flow, as shown here, but concurrent flow of the liquid and solids is also used.

Circular tanks are reported to be more effective. In these, the effluent circulates radially, the water being introduced at the periphery or from the centre. The solids are generally removed from near the centre, for which a slope of 10 % is required in the bottom of the tank. The sludge is forced to the outlet by two or four arms provided with scrapers which span the radius of the tank. For both types of flow, a means of distributing the flow in all directions is provided: for centre feed tanks there is a circular well, while for the rim-fed tanks a baffle is usually installed and the effluent enters tangentially. An even distribution of inlet and outlet flows is important to avoid short-circuiting in the tank that would reduce the separation efficiency.

A critical factor for selection of tank size is the so-called surface-loading rate, generally expressed as volume of wastewater per unit time and unit area of settler (m^3/m^2d). This loading rate depends on the characteristics of the effluent and the solids content and can be determined from the settling tests described above. The retention time in the settlers is generally in the order of 1–2 h, but the capacity of the tanks must be determined taking into account the peak flow rates so that good separation is also obtained in these cases.

In cases of small or elementary settling basins, the sludges can be removed using an arrangement of perforated piping placed in the bottom of the settling tank. The pipes must be regularly spaced, they must be of a diameter wide enough to be cleaned easily in case of clogging, and the flow velocities should also be high enough to prevent sedimentation. These last two requisites are somewhat contradictory, and a compromise is usually reached, using pipes of 5 cm in diameter, perforated with holes of 1–1.5 cm in diameter, 1 m apart. Flow in individual pipes may be regulated by valves. This configuration is best used after screening and is also found in biological treatment tanks for sludge removal. An alternative to the above configurations for settling tanks is that of the inclined tube separators. These separators consist of tubes (although there are alternate designs that use plates close to each other) which are tilted.

The concept is that, when a settling particle reaches the wall of the tube or the lower plate, it coalesces with another particle to give one of larger mass and higher settling rate.

The media are usually inclined at 45–60°. They are also commonly used to upgrade existing settling tanks since they have a higher separation rate.

6.6.2.4 Floatation

Flotation is an operation that removes not only oil and grease but also suspended solids. The most common procedure is that of dissolved air floatation (DAF), in which the waste stream is first pressurised with air in a closed tank. After passing through a pressure-reduction valve, the wastewater enters the floatation tank where, due to the sudden reduction in pressure, minute air bubbles in the order of 50–100 μm in diameter are formed. As the bubbles rise to the surface, the suspended solids and oil or grease particles adhere to them and are carried upwards. It is common practice to use chemicals to enhance floatation performance. As with coagulants (discussed later), these aids should preferably be innocuous, since these recovered solids are frequently used in animal feed formulations.

One alternate design involves the recycling of part (10–30 %) of the treated water. All systems contain a mechanism for removing the solids that may settle to the bottom of the flotation tanks, usually by a helical conveyor placed in the conical bottom. The main advantage claimed by DAF systems is the faster rate at which very small or light suspended solids can be removed in comparison with settling.

In one case, oil removal was reported to be 90 %. In tuna-processing wastewaters, the DAF removed 80 % of oil and grease and 74.8 % of suspended solids in one case and 64.3 % of oil and grease and 48.2 % of suspended solids in another case.

Another floatation system exists in which air is not dissolved but forced through the wastewater by surface aerators. This system generates air bubbles of larger sizes than DAF systems, and no report exists about its application to fishery wastewaters.

Prior to the design or selection of a DAF system, it is advisable to carry out laboratory experiments to evaluate its applicability and critical operating factors such as the air-to-solid ratio, the effectiveness of flocculants and the proper pH. This can be conveniently done in laboratory units. In these devices, water with or without chemicals and pH adjustment is introduced, and the pressure is increased to the desired value. After mixing to saturate the liquid with air, pressure is released and the liquid flows to a graduated cylinder where time is allowed for separation. The detailed procedures for conducting the evaluations are available elsewhere.

6.6.3 Secondary Treatment

The main purpose of the primary sedimentation stage is to produce both a generally homogeneous liquid capable of being treated biologically and a sludge that can be separately treated or processed. The secondary treatment process aims to reduce the BOD exerted by reducing organic matter. This is mediated, primarily, by a mixed population of heterotrophic bacteria that utilise the organic constituent for energy and growth.

Secondary treatment includes biological treatment: aerobic processes, anaerobic processes and specialised (Table 6.3). The objective of secondary treatment is the further treatment of the effluent from primary treatment to remove the residual organics and suspended solids.

6.6.3.1 Aerobic Processes

In most cases, secondary treatment follows primary treatment and involves the removal of biodegradable dissolved and colloidal organic matter using aerobic biological treatment processes. Aerobic biological treatment is performed in the presence of oxygen by aerobic microorganisms (principally bacteria) that metabolise the organic matter in the wastewater, thereby producing more microorganisms and inorganic end products (principally CO_2, NH_3 and H_2O). Several aerobic biological processes are used for secondary treatment differing primarily in the manner in which oxygen is supplied to the microorganisms and in the rate at which organisms metabolise the organic matter.

Common high-rate processes include the activated sludge processes, trickling filters or biofilters, rotating biological contactors (RBC), fluidised bed reactor,

Table 6.3 Flow chart of waste water treatment systems

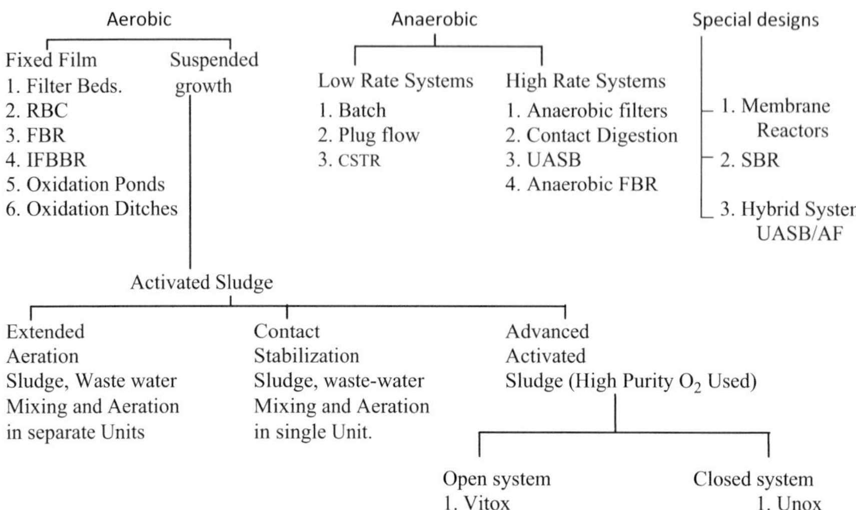

IFBBR, oxidation pond, oxidation ditches, rotating drums and discs. A combination of two of these processes in series (e.g. biofilter followed by activated sludge) is sometimes used to treat municipal wastewater containing a high concentration of organic material from industrial sources.

Secondary treatment is designed to substantially degrade the biological content of the sewage such as those derived from human waste, food waste, soaps and detergent. It is normally done through aerobic processes, so elements needed include availability of microorganisms, oxygen, contact between these microorganisms and the organic material and finally, favourable environmental conditions. These requisites can be satisfied by several approaches, being the most commonly used activated sludge, trickling filters and oxidation ponds. Lastly, the rotating biological contactor is a process that does not fit into any of the previous categories, but employs principles common to trickling filters and activated sludge.

A large number of biological unit operations are available to achieve the aerobic oxidation of BOD. All operations can be classified on the basis of their microbial population, into either fixed-film or dispersed growth processes. Fixed-film reactors have biofilms attached to a fixed surface where organic compounds are adsorbed into the biofilm and aerobically degraded. In suspended (e.g. activated sludge) growth reactors, the microorganisms mix freely with the wastewater and are kept in suspension by mechanical agitation or mixing by air diffusers.

Several investigators have pointed out that biological oxidation systems can remove over 90 % of pathogenic bacteria from sewage; however, the removal of viruses is much more varied. The major mechanism of viral removal is thought to be adsorption. In suspended growth reactors, the intimate mixing of solid flocs and sewage gives 90 % removal, while the smaller surface areas of biological adsorption sites in film reactors give varied reductions.

6.6.3.1.1 Biological Filters: Fixed-Film Systems

Fixed-film systems use a medium to retain and grow microorganisms. Fixed-film systems (FFS) are biological treatment processes that employ a medium such as rock, plastic, wood or other natural or synthetic solid material that will support biomass on its surface and within its porous structure.

At least two types of fixed-film systems may be considered – those in which the medium is held in place and is stationary relative to fluid flow (trickling filter) and those in which the medium is in motion relative to the wastewater (e.g. rotating biological disc).

Fixed Bed: Trickling Filters

Trickling filter systems are typically constructed as beds of media through which wastewater flows. Oxygen is normally provided by natural or forced ventilation. Flow distributors or sprayers distribute the wastewater evenly onto the surface of the medium. As the wastewater moves by gravity through the medium, soluble and colloidal organic matter is metabolised by the biofilm that forms on the medium. Excess biomass sloughs from the medium and is carried with the treated wastewater to the clarifier, where the solids settle and separate from the treated effluent. At this point the treated wastewater may be discharged or recycled back to the filter medium for further treatment.

The trickling filter consists of a bed of coarse material, such as stones, slats, rocks, gravel, slag, sphagnum, peat moss, plastic materials or another highly permeable media, over which wastewater is applied. As the wastewater trickles through the bed, a microbial growth establishes itself on the surface of the stone or packing in a fixed film. The wastewater passes over the stationary microbial population, providing contact between the microorganism and the organic contaminants, allowing degradation processes.

The trickling filter works through the formation of a biofilm on the media which degrades the organic compounds and ammonium in the wastewater. The wastewater is fed to the top of the trickling filter using a pump and collected at the bottom in a sump. Oxygen is made accessible to the bacteria through the use of a fan or solely through natural convection. As the biofilm continues to grow, an anaerobic section of the filter is formed, since the oxygen cannot diffuse through the thickening biofilm well. Parts of the biofilm slough off as it grows in size, and this slough is clarified or settled out after being collected at the sump. Trickling filters are advantageous for small to medium communities that do not have a lot of land to dedicate to wastewater treatment.

Rotating Biological Contactors (RBC)

These mechanical secondary treatment systems employ rotating discs that move within the wastewater are referred to as a rotating biological contactor (RBC). This is a fixed-film biological treatment process that is robust and capable of withstanding surges in organic load. Developed in the late 1960s, the RBC employs a plastic medium configured as discs and mounted on a horizontal shaft. Rotating biological contactors (RBCs) consist of closely spaced, slowly rotating, plastic discs on a shaft which is moved by a motor.

The shafts are rotated slowly (1–2 rpm) by mechanical or compressed air drive. For a typical aerobic RBC, approximately 40 % of the medium is immersed in the wastewater. Anoxic or anaerobic RBCs (far less common) are fully immersed in the wastewater. Wastewater flows through the medium by simple displacement and gravity. Biomass continuously sloughs from the discs, and some suspended biomass develops within the wastewater channels through which the discs rotate, making the addition of a secondary clarifier necessary. The rotation of the discs exposes the attached biomass to atmospheric air and wastewater. Oxygen is supplied by natural surface transfer to the biomass. The biofilm size is controlled by the shear forces of rotation. This is advantageous, because it prevents overgrowth of the biofilm which could limit oxygen to the aerobic bacteria. Sheared off excess biomass is removed through a final settler or clarifier where the microorganisms in suspension settle as sludge.

Some oxygenation of the wastewater is also created by turbulence at the disc-water interface. The use of exposed and submerged stages in multiple tanks to create aerobic and anoxic conditions may be employed where nitrogen removal is required.

RBCs are simpler to operate than activated sludge systems and have longer contact times between the bacteria and the pollutants than trickling filters. The start-up cost for RBCs is high, but the maintenance and energy costs are low.

Fluidised Bed Reactor (FBR)

A fluidised bed reactor (FBR) is a type of reactor device that can be used to carry out a variety of multiphase chemical reactions. In this type of reactor (Fig. 6.6), a fluid (gas or liquid) is passed through a granular solid material (usually a catalyst possibly shaped as tiny spheres) at high enough velocities to suspend the solid and cause it to behave as though it were a fluid. The solid substrate (the catalytic material upon which chemical species react) material in the fluidised bed reactor is typically supported by a porous plate, known as a distributor. The fluid is then forced through the distributor up through the solid material. At lower fluid velocities, the solids remain in place as the fluid passes through the voids in the material. This is known as a packed reactor. As the fluid velocity is increased, the reactor reaches a stage where the force of the fluid on the solids is enough to balance the weight of the solid material. This stage is known as incipient fluidisation and occurs at this minimum fluidisation velocity. Once this minimum velocity is surpassed, the contents of the reactor bed begin to expand and swirl around much like an agitated tank or boiling pot of water. The reactor is now a fluidised bed. Depending on the operating conditions and properties of solid phase, various flow regimes can be observed in this reactor. In the 1930s, the fluidised bed reactor (FBR) appeared as a new alternative for biological treatment of wastewaters. In this type of reactor, a high

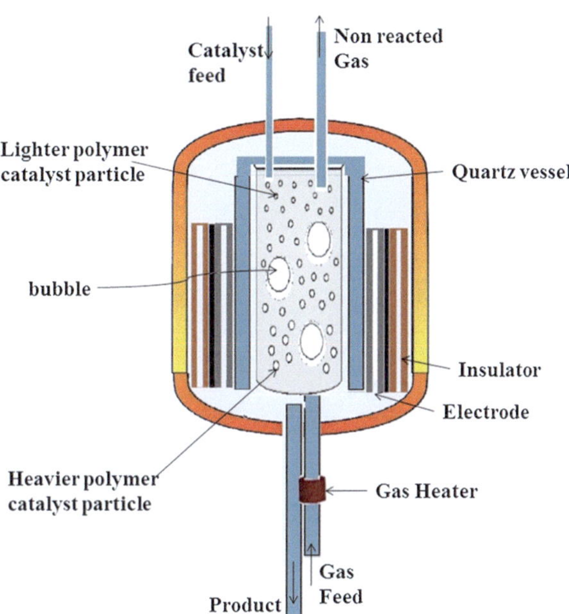

Fig. 6.6 Basic diagram of a fluidised bed reactor

concentration of biomass is maintained inside because microorganisms are attached to support particles.

The main advantages of using FBR are the lower hydraulic retention time (HRT) and the small size of equipment. In FBRs, it is possible to achieve a high concentration of biomass depending on the operational conditions used in the process and the type of support used to immobilise the microorganisms, which are found within a complex structure of cells and their extracellular products, referred to as a biofilm. The cell volume inside the biofilm is only a small part of its total volume. Dense particles were traditionally used in FBRs as a support for the biofilm. Entrainment of support particles from the reactor was a problem that appeared as the biofilm grew, decreasing particle density.

FBRs have many advantages over other processes.

1. High specific area is available to microorganisms.
2. Minimum problems related to channelling, plugging and gas hold-up.
3. Biofilm thickness is controlled by particle motion and liquid upflow velocities minimising the diffusion problems through biofilms.
4. Uniform liquid flow distribution.
5. Small installation is required.

Inverse Fluidised Bed Biofilm Reactors (IFBBR)

In inverse fluidised bed biofilm reactors (IFBBR) containing low-density particles, fluidisation can be conducted either by an upward co-current flow of gas and liquid through a bed (Fig. 6.7) (Bandaru et al. 2007) or by a downward flow of liquid and

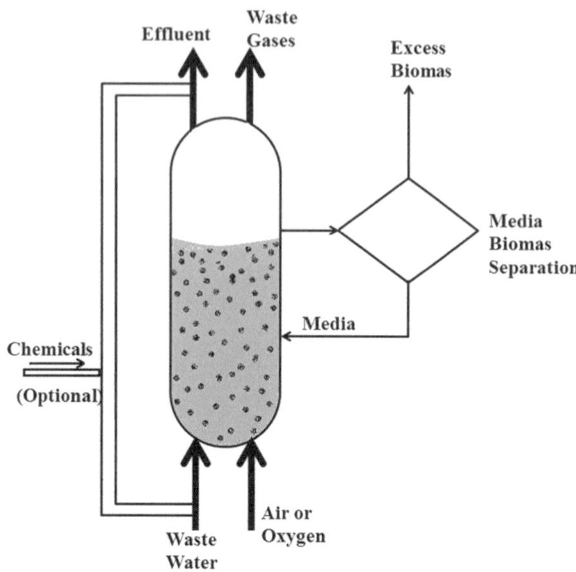

Fig. 6.7 Inverse fluidised bed biofilm reactor scheme

countercurrent upward flow of gas. In the former, fluidisation is achieved by an upward flow of gas whereby the gas bubbles make the bed expanding downwards into the less dense mixture of gas and liquid. In the latter, the bed is fluidised by a downward flow of a liquid counter to the net buoyancy force of the particles. At a small flow of the liquid, not sufficient to counter to the net buoyancy force, fluidisation can be achieved by an adequate upward flow of the gas. Fluidisation where fluidised bed expands downwards is termed as inverse fluidisation.

Three advantages of IFBBR are:

1. Effective and simple control of biofilm thickness
2. Large specific support surface area
3. Fast biofilm formation

Oxidation Ponds

The term oxidation ponds has been used lately as a collective name for all kinds of treatment ponds. Essentially, all work on the same principle: the use of ponds and basins designed taking wastewater flow conditions in consideration so that wastewater remains for sufficient time in the basin to allow degradation of organic matter take place. There are five basic types, namely, aerobic, anaerobic, facultative, maturation or tertiary ponds and aerated lagoons.

Oxidation Ditches

It is particular type of extended aeration process, where an aeration tank is constructed in the shape of a ditch (oval shape). An aeration tank consists of a ring-shaped channel 1.0–1.5 m deep and of suitable width forming a trapezoidal or rectangular channel cross section. An aeration rotor, consisting of Kessener brush, is placed across the ditch to provide aeration and wastewater circulation at velocity of about 0.3–0.6 m/s. The oxidation ditch can be operated as intermittent with fill and draw cycles consisting of (a) closing inlet valve and aerating the wastewater for duration equal to design detention time, (b) stopping aeration and circulation device and allowing the sludge to settle down in the ditch itself and (c) opening the inlet and outlet valve allowing the incoming wastewater to displace the clarified effluent. In case of continuous operation, called as Carrousel process, it is operated as a flow-through system where wastewater is continuously admitted. The vertically mounted mechanical aerators are used to provide oxygen supply and at the same time to provide sufficient horizontal velocity for not allowing the cells to settle at the bottom of the ditch. Separate sedimentation tank is used to settle the sludge, and the settled sludge is recirculated to maintain necessary MLVSS in the oxidation ditch. The excess sludge generation in oxidation ditch is less than the conventional ASP and can be directly applied to the sand-bed for drying.

6.6.3.1.2 Suspended Growth

In a suspended growth system, the waste flows around and through free-floating microorganisms, gathering into biological flocs that settle out of the wastewater. The settled flocs retain the microorganisms, meaning they can be recycled for further treatment.

Activated Sludge Process (ASP)

In the activated sludge process, atmospheric air or pure oxygen is bubbled through primary wastewater combined with organisms (a suspension of the wastewater and microorganisms in the mixed liquor) in an aeration tank to develop a biological floc which reduces the organic content of the wastewater. The contents of the aeration tank are mixed vigorously by aeration devices which also supply oxygen to the biological suspension. Aeration devices commonly used include submerged diffusers that release compressed air and mechanical surface aerators that introduce air by agitating the liquid surface. Hydraulic retention time in the aeration tanks usually ranges from 3 to 8 h but can be higher with high BOD_5 wastewaters. Following the aeration step, the microorganisms are separated from the liquid by sedimentation, and the clarified liquid is secondary effluent. A portion of the biological sludge is recycled to the aeration basin to maintain a high mixed liquor suspended solids (MLSS) level. The remainder is removed from the process and sent to sludge processing to maintain a relatively constant concentration of microorganisms in the system. Once the wastewater has received sufficient treatment, it is discharged into settling tanks, and the treated supernatant is run off to undergo further treatment before discharge. To achieve specific effluent goals for BOD, nitrogen and phosphorus, different adaptations and modifications have been made to the basic activated sludge design.

(a) Extended aeration sludge

The extended aeration process is one of the modifications of the ASP. It is a complete mix system and provides biological treatment for the removal of biodegradable organic wastes under aerobic conditions. Air may be supplied by mechanical or diffused aeration to provide the oxygen required to sustain the aerobic biological process. Mixing must be provided by aeration to maintain the microbial organisms in contact with the dissolved organics. Since a complete stabilisation occurred in the aeration tank, there is no need for a separate sludge digester. Further primary settling tank is also omitted, and settleable organic solids are also allowed to settle in the aeration tank due to long detention time in the aeration tank.

(b) Contact stabilisation sludge

It is developed to take advantage of the absorptive properties of activated sludge (Fig. 6.8). The BOD removal in ASP occurs in two phases: in the first phase absorption and second phase of oxidation. The absorptive phase requires 30–40 min, and during this phase most of the colloidal, finely divided, suspended solids and dissolved organic matter get absorbed on the activated sludge. Oxidation of organic matter then occurs. In contact stabilisation, these two phases are separated out and they occur in two separate tanks.

The settled wastewater is mixed with re-aerated activated sludge and aerated in the contact tank for 30–90 min. During this period the organic matter is absorbed on the sludge flocs. The sludge with absorbed organic matter is separated from the wastewater in the SST. A portion of the sludge is wasted to maintain requisite MLVSS concentration in the aeration tank. The return sludge is aerated before sending it to aeration tank for 3–6 h in sludge aeration tank, where the absorbed organic matter is oxidised to produce energy and new cells. Contact stabilisation is effective for treatment of sewage; however, its use to the industrial wastewater may be limited when the organic matter present in the wastewater is mostly in the dissolved form. Existing treatment plant can be upgraded by changing the piping and providing partition in the aeration tank. This modification will enhance the capacity of the existing plant. This is effective for sewage treatment because of the presence of organic matter in colloidal form in the sewage. Contact stabilisation may not be that effective for the treatment of wastewater when the organic matter is present only in soluble form.

(c) Advanced activated sludge

Advanced sludge treatment (AST) processes have been developed to improve sludge dewatering and facilitate the ultimate disposal. These processes include

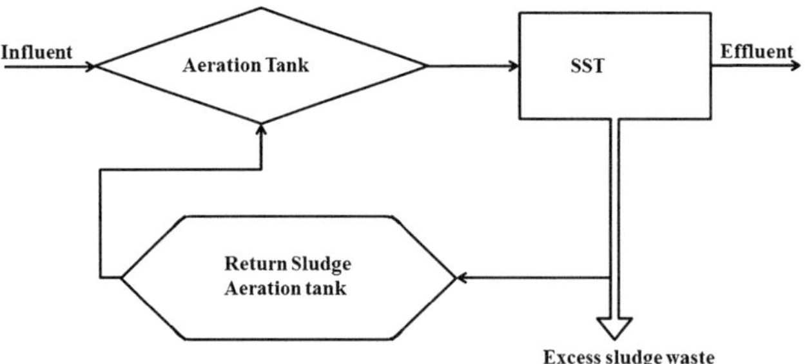

Fig. 6.8 Contact stabilisation activated sludge process

thermal hydrolysis (neutral, acid, alkaline) or chemical oxidation by H_2O_2, O_3, O_2, etc.. Peroxidation (H_2O_2) can thicken the sludge (5–6 % dry solids) from a municipal sewage treatment plant. Effective dewatering requires the attack of the microorganisms. The vast majority of these microorganisms live in aggregates such as films, flocs and sludges.

Advance activated sludge processes are classified into two types of system: 'open' and 'closed'. The closed system was firstly developed by the Union Carbide Corporation, USA, while the 'open' system was developed at the British Oxygen Company, UK. Further modifications were developed elsewhere.

In the closed system, oxygen is dispersed into covered tanks, the content of which are mixed by Verticle spindle vane impellers. In open system, oxygen is injected into the throat of a Venturi tube in a side stream through which mixed liquor is continuously recycled.

Closed Systems

UNOX

The UNOX system has been developed to improve upon the conventional activated sludge process. The use of enriched oxygen in a simple and economical multistage gas-liquid contacting device allows oxygen to be transferred to wastewater at increased rates with significant decreases in power requirements over those required when using air as the oxygen supply media. The elimination of the mass transfer restriction allows operation at solid levels of 4500–7000 mg/l while maintaining a dissolved oxygen level of 8–10 mg/l in the mixed liquor. Retention times for the process can be correspondingly decreased to 1–2 h. A highly flocculant sludge is obtained which has excellent settling and dewatering characteristics and is produced in less quantities than normally produced by a conventional air activated sludge process. The process has been demonstrated in a 2.5 mgd activated sludge plant at Batavia, New York. During the Federal Water Quality Administration (FWQA) contract, the UNOX process was able to demonstrate consistent BOD and suspended solid removals in excess of 90 %. A number of pilot plant programmes in municipal waste applications continue to verify and confirm the excellent treatment effectiveness and decreased power requirements achieved with the system.

OASES

The OASES is an oxygen activated sludge process for secondary treatment of wastewater. The system includes parallel trains consisting of a number of covered, separated stages. Influent wastewater, recycled activated sludge (RAS) and high purity oxygen are introduced to the first stage of each train of operation and flow through succeeding stages. The mixed liquor is subsequently sent to additional processes for further biological treatment and/or clarification. The process produces a good settling sludge and can be operated at high organic loading rates.

This is developed by Air Products and Chemicals Inc., USA. Many features are similar to UNOX system. It has surface aerators in enclosed aeration chamber as in UNOX but also has additional bottom turbines for enhanced mixing. Gas and liquid flow is concurrent through a series of contacting stages, and oxygen supply is controlled by a pressure demand system. As the oxygen is consumed, the head space pressure drops and more oxygen is admitted through the pressure control valves.

Forced Free Fall (F_3O)

This has been developed by Airco Inc. in the USA. In this process, mixed liquor falls through an enclosed oxygen-rich atmosphere contained in a precast module. The oxygenation module is completely submerged in the aeration basin, from which mixed liquor is pumped into the module. After oxygenation, the mixed liquor is returned to the aeration basin, and the dissolved oxygen concentration in the mixed liquor is used to control the oxygen supply rate and height of liquid fall.

SIMPLEX

The simplex system has been developed at Bury Sewage Works by Joshua Bolton. This is a low-cost modified conventional air activated sludge system. It has a plastic cover inside the basin. O_2 is introduced into specially installed fine double diffusers. O_2 in the head space is recirculated throughout the existing coarse bubble sparger and partially purged using a gas blower.

Open Systems

MAROX

This has been developed by the FMC Corporation in the USA. Here, rotating diffusers are used, and 92–95 % oxygen utilisation occurs, making it two to six times more efficient than the conventional process.

VITOX

This has been developed by the BOC Ltd., and VITOX 1 and VITOX 2 are the two systems used. In VITOX 1, liquid is pumped at a pressure and oxygen is injected into it. O_2 is trapped in the form of fine bubbles. This gas-liquid flow is then discharged into a mixed liquor oxygenation tank through an expansion nozzle. High velocity discharge and subsequent turbulence in the tank give very fine bubbles, and oxygen utilisation achieved is 95 %. The VITOX 2 system has a bell diffuser. Here, O_2 is injected with the liquid stream. This then falls at very low velocity into the vessel, and only small bubbles fall while larger bubbles are retained in contactor vessels.

Oxygen transfer can be achieved by using air or pure oxygen. However, air contains only ~21 % oxygen in it. So when wastewater treatment demands vary and performance standards become stringent, pure oxygen-based delivery system can be utilised to significantly enhance the process.

Vitox systems are extremely efficient and powerful dissolution systems for upgrading existing plants and for purpose-built systems. These systems are ideal solutions for permanent or intermittent effluent overloading or oxygenation shortfall.

MEGOX

This has also been developed by BOC Ltd. The wastewater is mixed with sludge recycle stream, injected with oxygen and then passed into the central reaction zone of a treatment tank. No additional agitation is required, and sludge settles with the aid of a slow speed rotating consolidator. The treated liquor goes to a clarifying zone and then discharged. The O_2 injection is controlled by a sensor in the sludge recycle system. The process is particularly suitable for nutrient-rich wastewater as in food-processing industries, and up to 86 % COD is removed from the wastewater feed with 3800–4400 g/m^3 COD and 2000 g/m^3 suspended solids.

PRIMOX

This is used as a part of sewage system. Anaerobic conditions and the associated problem of corrosive H_2S production are avoided by the application of this process.

6.6.3.2 Anaerobic Process

Anaerobic treatment (Fig. 6.9) is a biological process carried out in the absence of O_2 for the stabilisation of organic materials by conversion to CH_4 and inorganic end products such as CO_2 and NH_3.

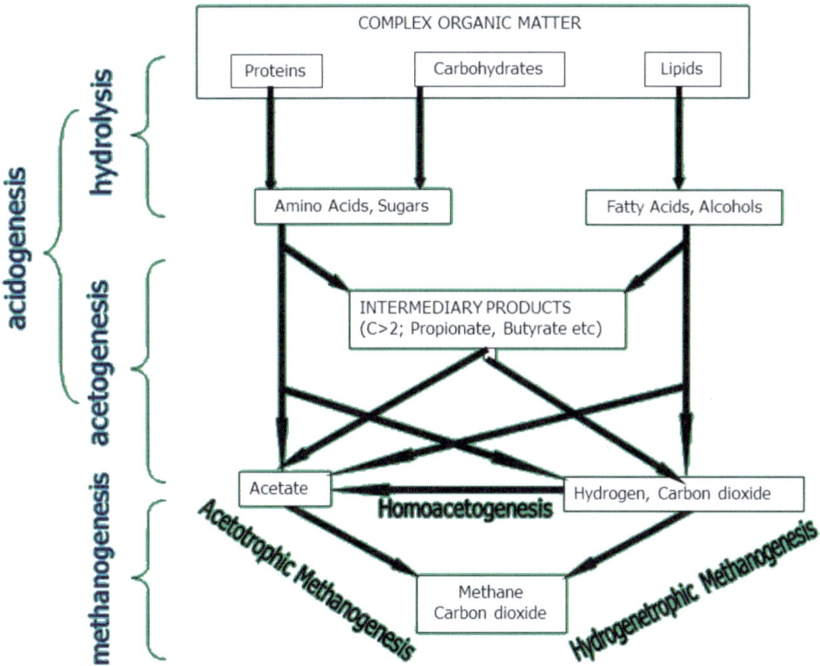

Fig. 6.9 General scheme for organic conversion in anaerobic system S_o and Q can be measured easily and known upfront VOLR can be selected

$$\text{Organic materials} + \text{Nutrients} \xrightarrow[\text{Anaerobic microbes}]{} CH_4 + CO_2 + NH_3 + \text{Biomass}$$

Anaerobic treatment is mainly used for reducing mass of high solid wastes, e.g. human waste, animal manure, agricultural waste and sludge. The treatment is divided in 'low-rate' systems, in which long hydraulic retention times are applied, and 'high-rate' systems, in which hydraulic retention time is relatively short.

Design based on volumetric organic loading rate (VOLR):

$$\text{VOLR} = \frac{S_o \cdot Q}{V}$$

VOLR: volumetric organic loading rate (kg COD/m^3-day)
So: wastewater biodegradable COD (mg/L)
Q: wastewater flow rate (m^3/day)
V: bioreactor volume (m^3)

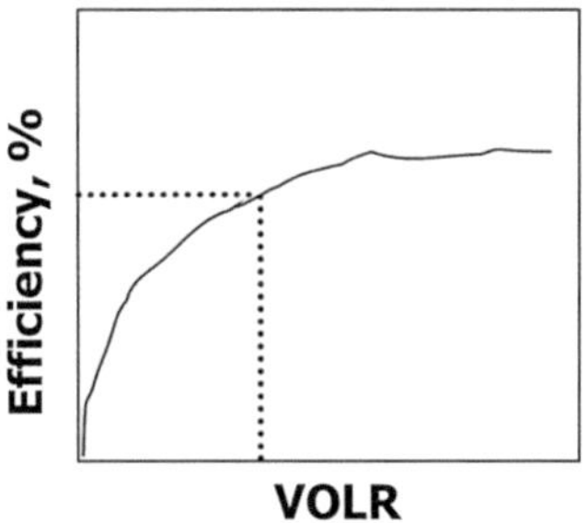

Conduct a pilot scale studies.
Find out removal efficiency at different VOLRs.
Select VOLR based on desired efficiency.

Design based on hydraulic loading rate:

$$V = \theta a \cdot Q$$

$$A = \frac{\theta_a \cdot Q}{H}$$

H = reactor height (m)
θ_a = allowable hydraulic retention time (hr)
Q = wastewater flow rate (m^3/h)
A = surface area of the reactor (m^2)

Permissible superficial velocity (*Va*)

$$V_a = H / \theta \text{ For dilute wastewater with COD} < 1000\text{mg / L}$$

Low-rate systems are used mainly for waste streams such as slurries and solid waste, which require a long time for sufficient anaerobic degradation. High-rate systems are used mainly for wastewater. The retention time of sludge in a low-rate system (solid retention time; SRT) is equal to the hydraulic retention time (HRT). In high-rate systems, however, the sludge retention time should be much higher than the hydraulic retention time.

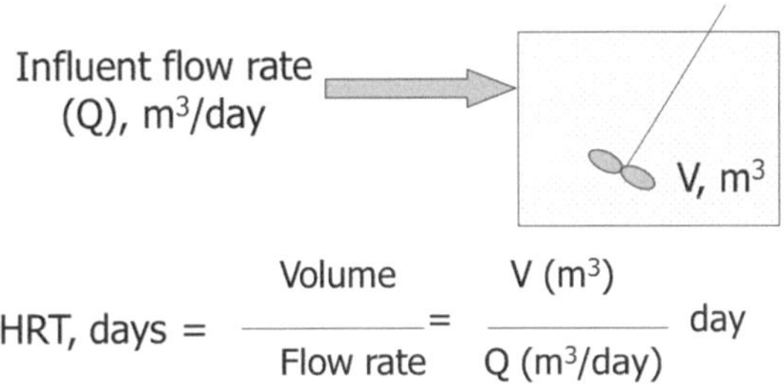

For a given SRT (HRT), the size of reactor can be easily determined since flow rate (Q) is known to us

Digester volume, V (m^3) = Flow rate (Q) x SRT (θ_C)

Examples of low-rate systems are batch, accumulation, plug flow and continuous stirred tank reactor (CSTR) systems. Examples of high-rate systems are contact process, anaerobic filter, fluidised bed and upflow anaerobic sludge bed (UASB)/ expanded granular sludge bed (EGSB).

6.6.3.2.1 Low-Rate Systems

Batch System

This is a process by which a reactor is filled with feedstock in one sequence, then processed and finally emptied in one instance.

Some of the first dry digesters were envisioned as modified landfills. In batch-fed digesters, the reactors are filled with a feedstock, closed and left for a period of time (i.e. the retention time) and then opened again and emptied (Khalid et al. 2011). Vandevivere et al. (2003) state that batch systems represent the lowest technology of all systems and are also the cheapest. Due to their simple design and lower investments costs, batch systems are recommended for application in developing countries. However, experience shows that these reactors have some serious limitations. Each batch, once closed, undergoes the whole start-up phase of the methanogenic process. This implies that there will be high fluctuations in gas production until the system operates in a stable way. Variations are also observed in gas quality. The height of the reactor is limited to ensure good infiltration of the percolate. Furthermore gastight sealing of inlet/outlet can be challenging especially as the doors are regularly closed and opened after each batch sequence. This may result in biogas losses and the risk of explosion when emptying a residual methane in the reactor mixes with air (Vandevivere et al. 2003).

Plug Flow Reactor

Plug flow digesters use slurries, e.g. almost undiluted manure and have a total suspended solid concentration of 10–12% TS. The basic digester design is a long trough (Fig. 6.10), often built below ground level with a gastight but expandable cover. At low TS concentration problems with floating and settling layers can appear. This problem can be solved using vertical mixing inside the pipe. In this particular process, anaerobic stages such as hydrolysis and methanogenesis are separated over the length of the pipe. At first, hydrolysis mainly occurs, whereas later in the process methanogenesis takes place at full velocity.

Using this system, the SRT is equal to the HRT. These systems are frequently used to treat slurries with a high fraction of suspended solids, as the hydrolysis of particulate matter is rate limiting; hence, only low loading rates can be applied.

Continuous Stirred Tank Reactor (CSTR)

The most common form of low solid reactor is the continuous stirred tank reactor (CSTR).

Feed is introduced into the reactor, which is stirred continuously to ensure complete mixing of the reactor contents. At the same time an equal quantity of effluent is removed from the reactor.

Retention time within the reactor can be varied according to the nature of the feedstock and process temperature applied, which is typically in the range of 2–4 weeks. Such systems have a low operating expenditure.

The CSTR is generally used for treatment of slurries with a TS percentage of approximately 2–10%. The influent concentration range applicable for CSTRs is determined by:

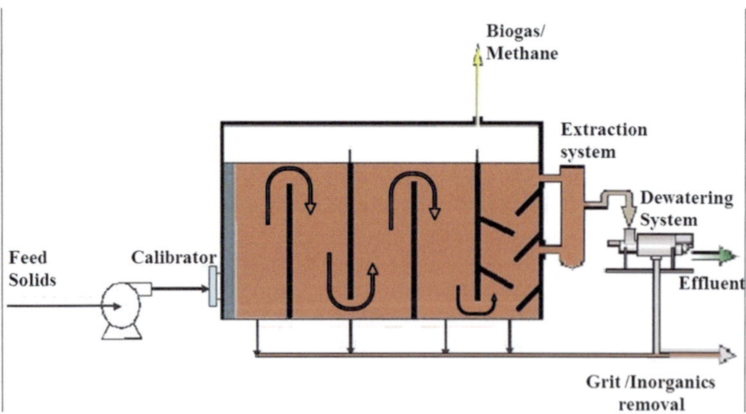

Fig. 6.10 Schematic diagram of a plug flow digester

- Gas yield in relation to the energy requirement for heating
- Possibility of mixing the reactor content

CSTR systems are applied in practice for treating animal manure, sewage sludge, household waste, agricultural wastes, faeces, urine and kitchen waste or mixtures of these substrates. Mixing creates a homogeneous substrate, preventing stratification and formation of a surface crust, and ensures solids remain in suspension. Bacteria, substrates and liquid consequently have an equal retention time resulting in SRT which is equal to HRT.

Digester volume ranges from around 100 m^3 to several thousand cubic metres, often with retention times of 10–20 days, resulting in daily capacities of 6–400 m^3. Examples of CSTR digesters with different mixing and heating systems are shown in Fig. 6.11.

6.6.3.2.2 High-Rate Systems: Anaerobic Filters

An anaerobic filter is a fixed-bed biological reactor with one or more filtration chambers in series (Fig. 6.12). As wastewater flows through the filter, particles are trapped and organic matter is degraded by the active biomass that is attached to the surface of the filter material. Developed by Young and McCarty in the late 1960s to treat dilute soluble organic wastes, the filter was filled with rocks similar to the trickling filter. Wastewater was distributed across the bottom, and the flow was in the upward direction through a bed of rocks. The filter is submerged completely. Anaerobic microorganisms accumulate within voids of media (rocks or other plastic media). The media retain or hold the active biomass within the filter. Non-attached biomass within the interstices forms bigger flocs of granular shape due to rising gas bubble/liquid. Non-attached biomass contributes significantly to waste treatment. Anaerobic filters are widely used as secondary treatment in household black or grey water systems and improve the solid removal compared to septic tanks or anaerobic baffled reactors. Since anaerobic filters work by anaerobic digestion, they can be designed as anaerobic digesters to recover the produced biogas. An anaerobic filter is an attached biofilm system (fixed-bed or fixed-film reactor; see also fixed-film activated sludge) that aims at removing non-settleable and dissolved solids (Morel and Diener 2006). As septic tanks or anaerobic baffled reactors, anaerobic filters are

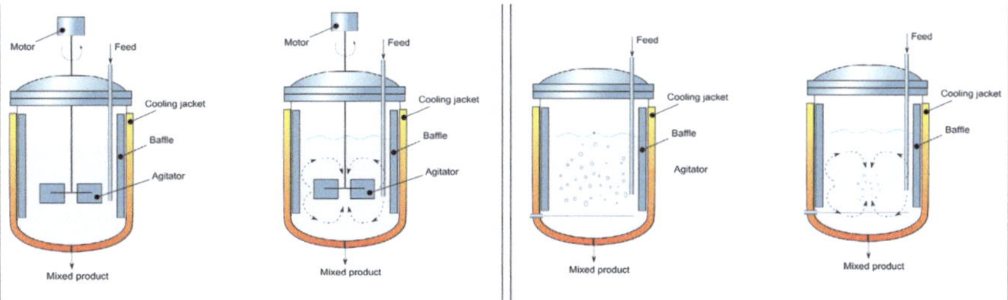

Fig. 6.11 Schematic diagram of a CSTR system, mechanically stirred (*left*) and stirred by biogas recirculation (*right*)

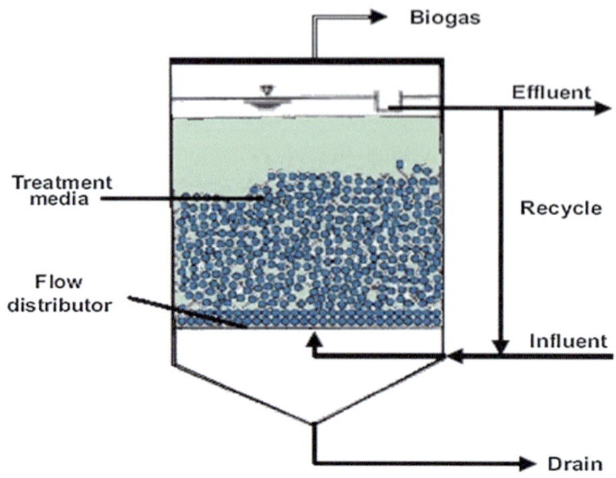

Fig. 6.12 Schematic diagram of an anaerobic filter

based on the combination of a physical treatment (settling) and a biological treatment (see also anaerobic digestion).

Contact Digestion

Anaerobic contact process (ACP) is essentially an anaerobic activated sludge process. It consists of a completely mixed reactor followed by a settling tank (Fig. 6.13). The settled biomass is recycled back to the reactor. Hence, ACP is able to maintain high concentration of biomass in the reactor and thus high solid retention time (SRT) irrespective of hydraulic retention time (HRT). A degasifier allows the removal of biogas bubbles (CO_2, CH_4) attached to a sludge which may otherwise float to the surface. ACP was initially developed for the treatment of dilute wastewater such as meat packing plant which had tendency to form settleable flocs. An ACP is suitable for the treatment of wastewater containing suspended solids which render the microorganisms to attach and form settleable flocs.

Upflow Anaerobic Sludge Blanket Reactor (USAB)

UASB was developed in 1970s by Lettinga in the Netherlands. It is essentially a suspended growth system in which proper hydraulic and organic loading rate is maintained in order to facilitate the dense biomass aggregation known as granulation. The size of granules is about 1–3 mm diameter. Since granules are bigger in size and heavier, they will settle down and be retained within the reactor. The concentration of biomass in the reactor may become as high as 50 g/L. Thus a very high SRT can be achieved even at a very low HRT of 4 h. The granules consist of hydrolytic bacteria, acidogen/acetogens and methanogens. Carbohydrate-degrading granules show layered structure with a surface layer of hydrolytic/fermentative acidogens. The mid-layer is comprised of syntrophic colonies and an interior with acetogenic methanogens (Fig. 6.14).

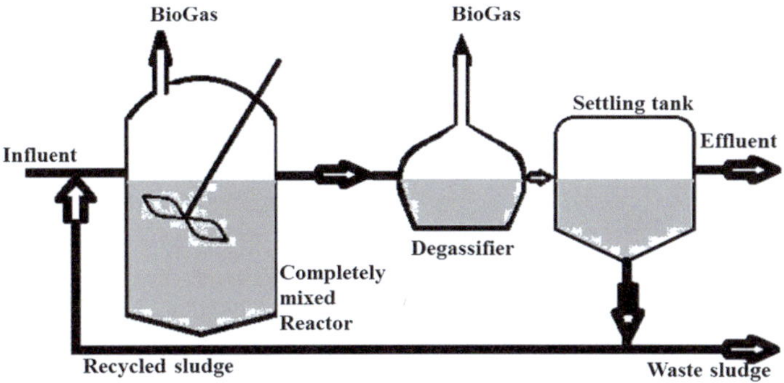

Fig. 6.13 Anaerobic contact process

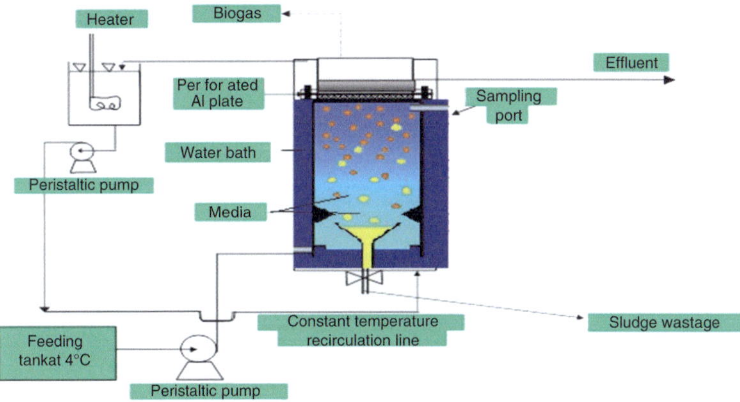

Fig. 6.14 Upflow anaerobic sludge blanket reactor

Anaerobic Fluidised Bed Reactor (FBR)

FBR is truly a fixed-film reactor as suspended biomass is washed out due to high upflow velocity. The bed expansion is 25–300 % of the settled bed volume, which requires much higher upflow velocity (10–25 m/h). The biomass gets attached on bio-carriers such as sandman, pulverised polyvinyl chloride and shredded tyre beads. The bio-carriers are supported entirely by the upflow liquid velocity and therefore able to move freely in the bed. The fluidised bed reactor is free from clogging problem short-circuiting and better substrate diffusion within the biofilm.

Expanded bed reactor (EBR) is an attached growth system with some suspended biomass that is similar to FBR.

6.6.3.3 Special Design Reactors

Membrane Reactors

Membrane bioreactors (MBR) are treatment processes (Fig. 6.15), which integrate a permselective or semipermeable membrane with a biological process. It is the combination of a membrane process like microfiltration or ultrafiltration with a

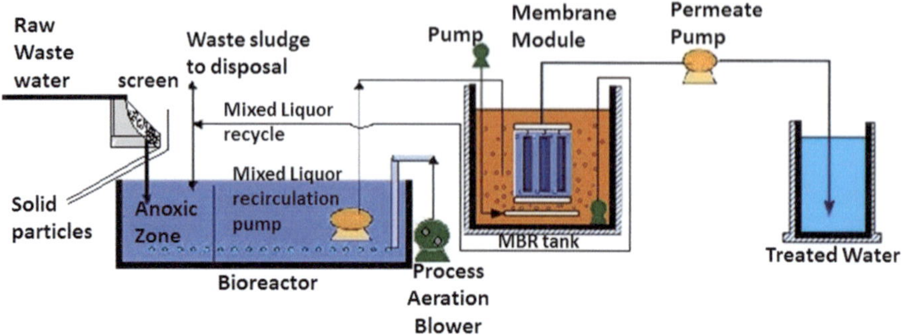

Fig. 6.15 Typical schematic for membrane bioreactor system

suspended growth bioreactor and is now widely used for municipal and industrial wastewater treatment with plant sizes up to 80,000 population equivalents. Due to it being a very technical solution, it needs expert design and skilled workers. Furthermore, it is a costly but efficient treatment possibility. With the MBR technology, it is possible to upgrade old wastewater plants.

Membrane bioreactors combine conventional biological treatment (e.g. activated sludge) processes with membrane filtration to provide an advanced level of organic and suspended solid removal. When designed accordingly, these systems can also provide an advanced level of nutrient removal. In an MBR system, the membranes are submerged in an aerated biological reactor.

The MBR process involves a suspended growth activated sludge system that utilises microporous membranes for solid/liquid separation in lieu of secondary clarifiers. This very compact arrangement produces a MF/UF quality effluent suitable for reuse applications or as a high-quality feed water source for reverse osmosis treatment. Indicative output quality of MF/UF systems includes SS <1 mg/L, turbidity <0.2 NTU and up to four log removal of virus (depending on the membrane nominal pore size). In addition, it provides a barrier to certain chlorine-resistant pathogens such as *Cryptosporidium* and *Giardia*.

The MBR process is an emerging advanced wastewater treatment technology that has been successfully applied at an ever increasing number of locations around the world. In addition to their steady increase in number, MBR installations are also increasing in terms of scale. A number of plants with a treatment capacity of around 5–10 ML/day have been in operation for several years now while the next generation (presently undergoing commissioning or under contract) have design capacities up to 45 ML/day.

Sequence Batch Reactor (SBR)

Sequence batch reactor is a special type of activated sludge process in which all the treatment takes place in the reactor tanks and clarifiers are not required. These processes treat the wastewater in batch mode, and each batch is sequenced through a series of stages.

Wastewater treatment is achieved by a timed sequence of operations which occurs in the same SBR tank, consisting of filling, reaction (aeration), settling, decanting, idling and sludge wasting (Fig. 6.16).

Secondary Clarifier the SBR tank acts as a secondary clarifier during the settling and decanting stages where the mixed liquor is allowed to settle under quiescent conditions and the overflow is discharged to the next stage of treatment.

Sludge Return System the activated sludge, following settling in the SBR tank, is mixed with the influent similar to the sludge return system, except that the feed is

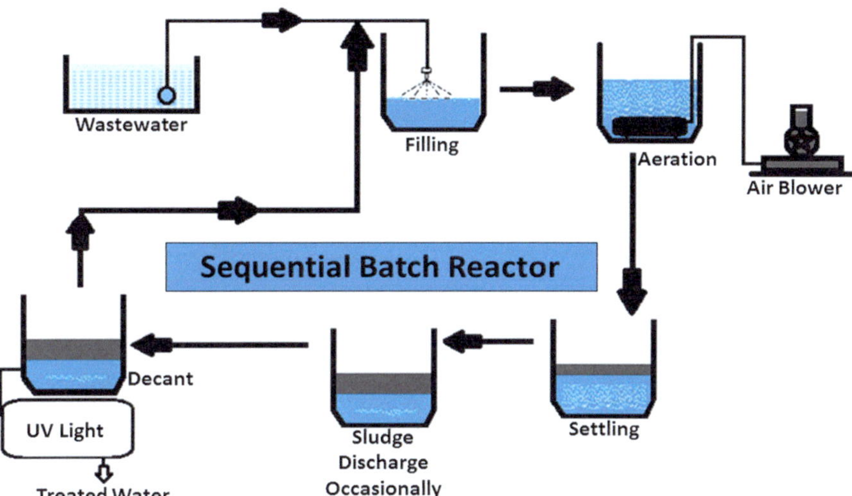

Fig. 6.16 Sequence batch reactor having various stages. ***Stage* 1**: **Filling**. During this stage, the SBR tank is filled with the influent wastewater. In order to maintain suitable F/M (food-to-microorganism) ratios, the wastewater should be admitted into the tank in a rapid, controlled manner. This method functions similarly to a selector, which encourages the growth of certain microorganisms with better settling characteristics. ***Stage*2**: **Reaction**. This stage involves the utilisation of biochemical oxygen demand (BOD) and ammonia nitrogen, where applicable, by microorganisms. The length of the aeration period and the sludge mass determines the degree of treatment. The length of the aeration period depends on the strength of the wastewater and the degree of nitrification (conversion of the ammonia to a less toxic form of nitrate or nitrite) provided for in the treatment. ***Stage* 3**: **Settling**. During this stage, aeration is stopped and the sludge settles leaving clear, treated effluent above the sludge blanket. Duration for settling varies from 45 to 60 min depending on the number of cycles per day. ***Stage* 4**: **Decanting**. At this stage of the process, effluent is removed from the tank through the decanter, without disturbing the settled sludge. ***Stage* 5**: **Idling**. The SBR tank waits idle until it is time to commence a new cycle with the filling stage. ***Stage* 6**: **Sludge wasting**. Excess activated sludge is wasted periodically during the SBR operation. As with any activated sludge treatment process, sludge wasting is the main control of the effluent quality and microorganism population size. This is how the operator exerts control over the effluent quality by adjusting the mixed liquor suspended solids (MLSS) concentration and the mean cell residence time (MCRT).

transferred to the sludge rather than the sludge being transferred to the front end of the plant.

Hybrid System

Hybrid system may be any combination of two types of reactor. Hybrid system incorporates both granular sludge blanket (bottom) and anaerobic filter (top) (Fig. 6.17). Such an approach prevents washout of biomass from the reactor. Additional treatment takes place at the top bed due to the retention of sludge granules that escape from the bottom sludge bed. A UASB reactor facing a chronic sludge wash-out problem can be retrofitted using this approach.

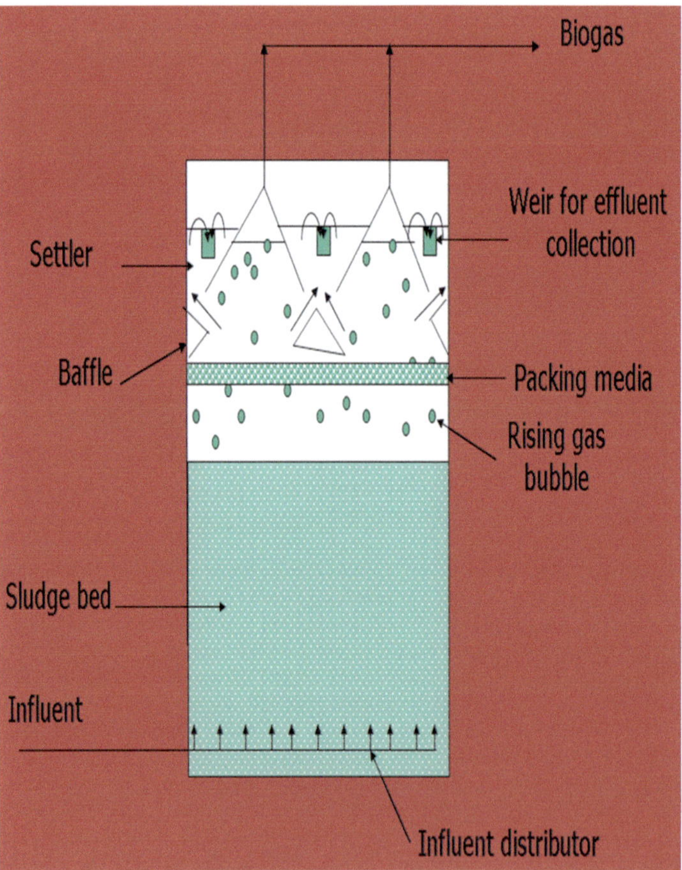

Fig. 6.17 Hybrid system having UASB (*bottom*) and AF (*top*)

6.6.4 Tertiary Treatment

Tertiary wastewater treatment is employed when specific wastewater constituents which cannot be removed by secondary treatment must be removed such as nitrogen, phosphorus and heavy metals (Fig. 6.18). Each tertiary treatment is designed to remove certain elements, so it is entirely possible that a plant employs several tertiary treatment processes, depending on the composition and characteristics of the wastewater flow that remove specific types of residuals by filtration (removal of suspended or colloidal solids), adsorption, chemical oxidation (removal of organics, nitrogen, phosphorus), ozonation, chlorination and UV radiation (disinfection). All these processes have their place in overall wastewater treatment scheme.

Effluent from primary clarifiers flows to the biological reactor, which is physically divided into five zones by baffles and weirs. In sequence, these zones are (i) anaerobic fermentation zone (characterised by very low dissolved oxygen levels and the absence of nitrates), (ii) anoxic zone (low dissolved oxygen levels but nitrates present), (iii) aerobic zone (aerated), (iv) secondary anoxic zone and (v) final aeration zone. The function of the first zone is to condition the group of bacteria responsible for phosphorus removal by stressing them under low oxidation-reduction conditions, which results in a release of phosphorus equilibrium in the cells of the bacteria. On subsequent exposure to an adequate supply of oxygen and

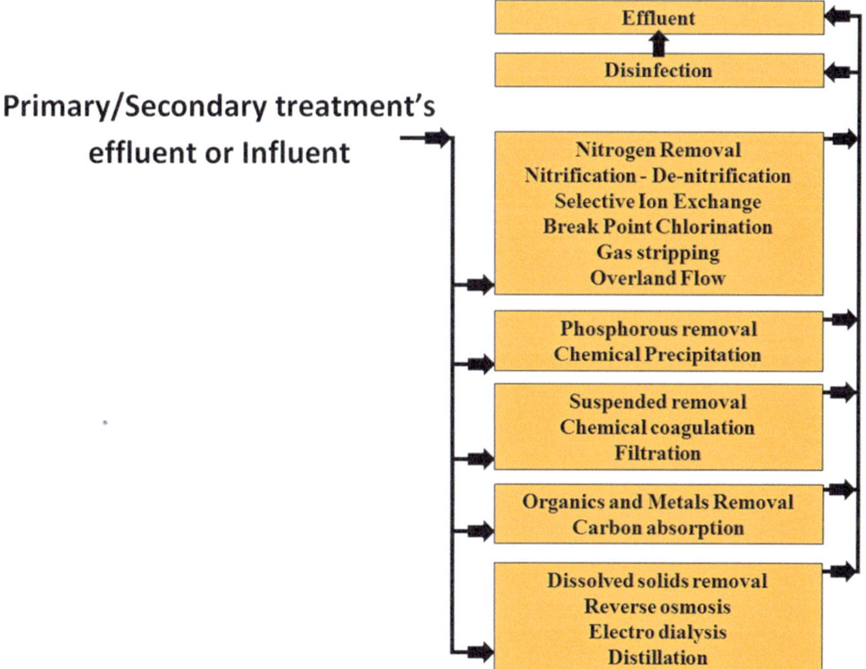

Fig. 6.18 Tertiary treatment

phosphorus in the aerated zones, these cells rapidly accumulate phosphorus considerably in excess of their normal metabolic requirements. Phosphorus is removed from the system with the waste activated sludge.

Most of the nitrogen in the influent is in the ammonia form, and this passes through the first two zones virtually unaltered. In the third aerobic zone, the sludge age is such that almost complete nitrification takes place, and the ammonia nitrogen is converted to nitrites and then to nitrates. The nitrate-rich mixed liquor is then recycled from the aerobic zone back to the first anoxic zone. Here denitrification occurs, where the recycled nitrates, in the absence of dissolved oxygen, are reduced by facultative bacteria to nitrogen gas, using the influent organic carbon compounds as hydrogen donors. The nitrogen gas merely escapes to atmosphere. In the second anoxic zone, those nitrates which were not recycled are reduced by the endogenous respiration of bacteria. In the final re-aeration zone, dissolved oxygen levels are again raised to prevent further denitrification, which would impair settling in the secondary clarifiers to which the mixed liquor then flows.

In many situations, where the risk of public exposure to the reclaimed water or residual constituents is high, the intent of the treatment is to minimise the probability of human exposure to enteric viruses and other pathogens (Table 6.4). Effective disinfection of viruses is believed to be inhibited by suspended and colloidal solids in the water; therefore, these solids must be removed by advanced treatment before the disinfection step.

Because advanced treatment usually follows high-rate secondary treatment, it is sometimes referred to as tertiary treatment. However, advanced treatment processes are sometimes combined with primary or secondary treatment (e.g. chemical

Table 6.4 Infectious agents potentially present in untreated domestic wastewater

S. No.	Pathogen	Disease
Bacteria		
1.	*Escherichia coli* (enterotoxigenic)	Gastroenteritis
2.	*Leptospira* (spp.)	Leptospirosis
3.	*Salmonella typhi*	Typhoid fever
4.	*Vibrio cholera*	Cholera
Protozoan		
1.	*Balantidium coli*	Balantidiasis
2.	*Entamoeba histolytica*	Amoebiasis (amoebic dysentery)
Helminths		
1.	*Ascaris lumbricoides*	Ascariasis
2.	*Taenia solium*	Taeniasis
Viruses		
1.	Enteroviruses (72 types, e.g. polio, echo, and Coxsackie viruses)	Gastroenteritis
		Heart anomalies
		Meningitis
2.	Hepatitis A virus	Infectious hepatitis

addition to primary clarifiers or aeration basins to remove phosphorus) or used in place of secondary treatment (e.g. overland flow treatment of primary effluent).

6.6.4.1 Filtration

This process removes residual unsettled microorganisms flocs and any other suspended solids. Two main types are used: granular filtration, using sand filters or multimedia filters, and membrane filtration.

6.6.4.2 Carbon Adsorption

Several soluble organic materials can resist a full secondary treatment and filtration persisting in the effluent. These materials, known as refractory organics, are removed by adsorbing them on activated carbon.

6.6.4.3 Chemical Oxidation

The chemical oxidation processes utilising activation of H_2O_2 by Fe(II) salts are referred to as Fenton's reactions. Fenton's reaction causes the dissociation of the oxidant and the formation of highly reactive hydroxyl radicals that attack and destroy organic pollutants. Fenton's reagent is a mixture of H_2O_2 and ferrous iron, which catalyses the formation of hydroxyl radicals according to the reaction

$$Fe^{2+} + H_2O_2 \rightarrow Fe^{3+} + OH^{\bullet} + OH^{-}$$

6.6.4.4 Phosphorus Removal

Phosphorus is typically found as monohydrogen phosphate (HPO_4^{2-}) in wastewater. In order to prevent or reduce eutrophication in wastewater, phosphorus is chemically precipitated. This reaction is usually produced using one of these three compounds: ferric chloride, alum or lime. The process itself requires a reaction basin and a settling tank. In case of using ferric chloride, the aeration tank serves as reaction basin, and the secondary clarifier can be used as settling tank, so performing this process as the secondary treatment takes place.

6.6.4.5 Nitrogen Control

In order to control algal growth in the effluent receiver, it is necessary on many occasions to remove nitrogen from wastewater. This process can be achieved biologically or chemically. The biological process is known as nitrification/denitrification, while the chemical process is called ammonia stripping.

6.6.5 Disinfection

Primary, secondary and even tertiary treatment cannot be expected to remove 100 % of the incoming waste load, and as a result, many organisms still remain in the waste stream. To prevent the spread of waterborne diseases and also to minimise public health problems, regulatory agencies may require the destruction of pathogenic organisms in wastewaters. While most of these microorganisms are not pathogens,

pathogens must be assumed to be potentially present. Thus, whenever wastewater effluents are discharged into receiving waters which may be used for water supply, swimming or shellfishing, the reduction of bacterial numbers to minimise health hazards is a very desirable goal. Disinfection is the treatment of the effluent for the destruction of all pathogens.

The effectiveness of disinfection depends on the quality of the water being treated, the type of disinfection being used, the disinfectant dosage and other environmental variables. Cloudy water will be treated less successfully since solid matter can shield organisms, especially from ultraviolet light or if contact times are low. Common methods of disinfection include ozone, chlorine or ultraviolet light.

6.6.5.1 Chlorination

Disinfection normally involves the injection of a chlorine solution at the head end of a chlorine contact basin. Chlorine contact basins are usually rectangular channels, with baffles to prevent short-circuiting, designed to provide a contact time of about 30 min. However, to meet advanced wastewater treatment requirements, a chlorine contact time of as long as 120 min is sometimes required for specific irrigation uses of reclaimed wastewater. The chlorine dosage depends upon the strength of the wastewater and other factors, but dosages of 5–15 mg/l are common. The treatment of wastewaters for the destruction of pathogens demands the use of practical measures that can be used economically and efficiently at all times on large quantities of wastewaters which have been treated to various degrees. The bactericidal effects of chlorine and other disinfectants are dependent upon pH, contact time, organic content and effluent temperature. The prevalent use of chlorine has come about because chlorine is an excellent disinfecting chemical and, until recently, has been available at a reasonable cost. However, the rising cost of chlorine coupled with the fact that chlorine even at low concentrations is toxic to fish and other biota as well as the possibility that potentially harmful chlorinated hydrocarbons may be formed has made chlorination less favoured as the disinfectant of choice in wastewater treatment.

6.6.5.2 Ozonation

Ozone is produced when oxygen (O_2) molecules are dissociated by an energy source into oxygen atoms and subsequently collide with an oxygen molecule to form an unstable gas, ozone (O_3), which is used to disinfect wastewater. Most wastewater treatment plants generate ozone by imposing a high-voltage alternating current (6–20 kV) across a dielectric discharge gap that contains an oxygen-bearing gas. Ozone is generated on site because it is unstable and decomposes to elemental oxygen in a short period of time after generation. Ozone is a very strong oxidant and virucide.

When ozone decomposes in water, the free radicals hydrogen peroxy (HO_2) and hydroxyl (OH) that are formed have great oxidising capacity and play an active role in the disinfection process. It is generally believed that the bacteria are destroyed because of protoplasmic oxidation resulting in cell wall disintegration (cell lysis). The effectiveness of disinfection depends on the susceptibility of the target organisms, the contact time and the concentration of the ozone.

Ozone is considered to be safer than chlorine because unlike chlorine, which has to be stored on site, ozone is generated on site as needed. Ozone also produces less disinfection by-product than chlorination.

6.6.5.3 UV Radiation

UV light can be used instead of other chemical compounds. Because no chemicals are used, the treated water has no adverse effect on organism that later consume it, as may be the case with other methods. UV radiation causes damage to the genetic structure of bacteria, viruses and other pathogens, making them incapable of reproduction. Ultraviolet light has recently undergone studies to determine its effectiveness and cost when used at large wastewater treatment plants.

Both ozone and ultraviolet light, as well as being an effective disinfecting agent, leave no toxic residual. Ozone will additionally raise the dissolved oxygen level of water. However, ozone must be generated and has only recently begun to compete favourably with chlorination in terms of economics. Therefore, the increased use of ozone (ozonation) or ultraviolet light as a disinfectant in the future is a distinct possibility in wastewater disinfection.

6.7 Effluent Storage

Although not considered a step in the treatment process, a storage facility is, in most cases, a critical link between the wastewater treatment plant and the irrigation system. Storage is needed for the following reasons:

(i) To equalise daily variations in flow from the treatment plant and to store excess when average wastewater flow exceeds irrigation demands; includes winter storage.
(ii) To meet peak irrigation demands in excess of the average wastewater flow.
(iii) To minimise the effects of disruptions in the operations of the treatment plant and irrigation system. Storage is used to provide insurance against the possibility of unsuitable reclaimed wastewater entering the irrigation system and to provide additional time to resolve temporary water quality problems.

6.8 Reliability of Conventional and Advanced Wastewater Treatment

Wastewater reclamation and reuse systems should contain both design and operational requirements necessary to ensure reliability of treatment. Reliability features such as alarm systems, standby power supplies, treatment process duplications, emergency storage or disposal of inadequately treated wastewater, monitoring devices and automatic controllers are important. From a public health standpoint, provisions for adequate and reliable disinfection are the most essential features of the advanced wastewater treatment process. Where disinfection is required, several reliability features must be incorporated into the system to ensure uninterrupted chlorine feed.

6.9 Consideration for Future Use

Research activities to improve wastewater handling have been focused on optimising the built systems and the development of new methods (Table 6.5). A change in research priorities has gradually occurred in order to comply with sustainability principles. At present new technologies should meet requirements of sustainable development and multidisciplinary approach (Fig. 6.19). Further, the following approaches are being applied:

(a) Production technology: safe personal care products, pharmaceuticals and drugs
(b) Communication: interactions between consumers, water companies, etc.
(c) Technology development: membrane, removal of microorganics, deammonification, phosphorus recovery, etc.
(d) Recipient effects: new threats
(e) Resources recovery: phosphorus recovery

Future goals may be categorised as follows:

1. Development and optimisation of new methods and process configurations for resource-effective wastewater treatment
2. Tests for development of equipment for wastewater treatment and separation technology
3. Development of new methods/process configurations for drinking water production from wastewater

Table 6.5 Changes in priorities of wastewater treatment with time

Time	Main problem	Remedies
From 1930	Visible pollutants	Mechanical treatment
From 1950	Low oxygen contents in recipient	Secondary/biological treatment
From 1970	Eutrophication of lakes	Tertiary/chemical treatment
From 1990	Marine eutrophication	Removal of nitrogen/phosphorus
Present and future	Recovery of resources (as phosphorus and energy)	Eco-cycling
From 2025 (predicted)	Deposition of sludge	Implementation of Agenda 21
	'Unwanted substances' in aquatic environment (pharmaceuticals, synthetic chemicals, hormones) New unknown??	Sustainable technologies
		Modified sludge handling
		Increased public participation/responsibility
		Development of new wastewater treatment technologies (Anammox)

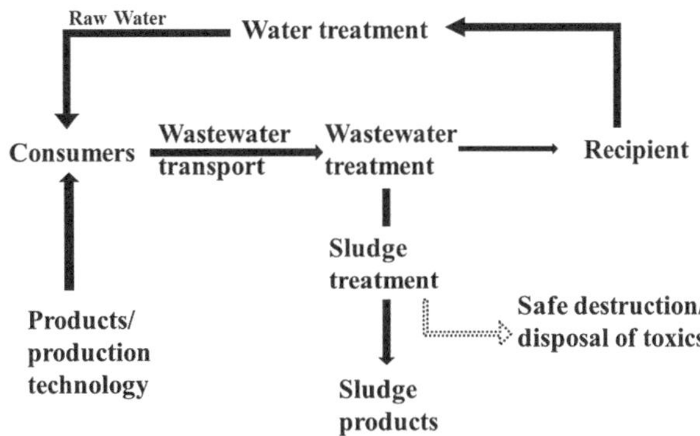

Fig. 6.19 Multidisciplinary approach in water and wastewater treatment

Wastewater treatment research which has the potential to advance safe, reliable and cost-effective technologies to reuse effluents should be an international priority. Emerging technologies such as on-line sensors with real-time feedback will certainly play a major role in water reuse in the near future. Advances in membrane technologies will also be critical in lowering energy needs and increasing water recovery rates. These technological advances should be prioritised comprehensively for achieving the correct water quality for the application needed. Creating ultra-high purity water for irrigation, toilet flushing and washing laundry does not make good sense. Therefore, the future will lie squarely in fit-for-purpose treatment and more distributed systems that can be interlinked and autonomously controlled.

References

Abdel-Raoufa N, Al-Homaidanb AA, Ibraheem IBM (2012) Microalgae and wastewater treatment. Saudi J Biol Sci 19(3):257–275

Asthana M, Kumar A, Vikrant P, Gupta P (2014) Tannery effluents de-colourization efficiency of bacterial isolates from River Yamuna and industrial effluents. Int J Curr Microbiol Appl Sci 3(5):869–880

Bandaru KSVSR, Murthy DVS, Krishnaiah K (2007) Some hydrodynamic aspects of 3-phase inverse fluidized bed. China Particuol 5(5):351–356

Bhargava DS (2008) Dissolved oxygen Sag analysis for a settling fields overlapping type multiwastewater-outfall. Environ Springer Sci Bus Media 28:128–136

Frobisher M, Hinsdill RD, Crabtree KT, Goodheart CR (1974) Fundamental of microbiology, 9th edn. W.B. Saunders, Philadelphia, pp 120–152

Gupta P, Kumar A, Khatri A, Asthana M (2014) Acid green dye decolourizing bacteria from Yamuna water and textile effluents. Int J Environ Sci 5(3):482–490

Khalid A, Arshad M, Anjum M, Mahmood T, Dawson L (2011) The anaerobic digestion of solid organic waste-review. Waste Manag 31(8):1737–1744

Kumar A, Asthana M, Gupta P, Yadav S, Sharma D, Singh KN, Kumar S (2015) 16SrRNA sequencing of Dye decolourizing bacteria isolated from soil. Bioinformation 11(1):1–5

Marinescu M, Dumitru M, Lacatusu A (2009) Biodegradation of petroleum hydrocarbons in an artificial polluted soil. Res J Agric Sci 41(2):157–167

Morel A, Diener S (2006) Greywater management in low and middle-income countries, review of different treatment systems for households or neighbourhoods (= SANDEC News No.14/06). Swiss Federal Institute of Aquatic Science (EAWAG), Department of Water and Sanitation in

Developing Countries (SANDEC), Duebendorf, http://www.sswm.info/sites/default/files/reference_attachments/MOREL%20and%20DIENER%202006%20Greywater%20Management.pdf. Accessed: 29 Dec 2015
Neyens E, Baeyens J, Dewil R, De Heyder B (2004) Advanced sludge treatment affects extracellular polymeric substances to improve activated sludge dewatering. J Hazard Mater 106B:83–92
Noorjahan CM (2014) Physicochemical characteristics, identification of bacteria and biodegradation of industrial effluent. J Bioremed Biodegrad 5(3):1–5
Rice EW, Baird RB, Eaton AD, Clesceri LS (2012) Standard methods for the examination of water and wastewater, 22nd edn. American public health association, Washington, DC
Singh U, Maheshwari R, Swami VK, Chauhan V (2012) Integrated effluent treatment in textile industries-feasibility study. Int J Chem Life Sci 01(01):1018–1025
Thomann RV, Mueller JA (1987) Principles of surface water quality modeling and control. Harper International Edition, New York
Vandevivere P, De Baere L, Verstraete W (2003) Types of anaerobic digesters for solid wastes. In: Mata-Alvarez J (ed) Biomethanization of the organic fraction of municipal solid wastes. IWA Publishing, Barcelona, pp 111–140

Suggested Further Readings

Hammer MJ, Hammer MJ Jr (2000) Water and wastewater technology, 3rd edn. Prentice Hall of India Pvt Ltd, New Delhi
Hill MK (1997) Understanding environmental pollution. Cambridge University Press, New York
Jogdand SN (2003) Environmental biotechnology (industrial pollution management), 2nd edn. Himalaya Publishing House Mumbai, India
Olguin EJ, Sanchez G, Hernandez E (2000) Environmental biotechnology and cleaner bioprocesses, 1st edn. Taylor and Francis, London
Rowe DR, Abdel-Magid IM (1995) Handbook of wastewater reclamation and reuse. CRC Lewis Publisher, Boca Raton

Bioreactors

7

Rajeeva Gaur, Anurag Singh, Ashutosh Tripathi, and Ranjan Singh

Contents

R. Gaur (✉) • A. Singh • A. Tripathi • R. Singh
Department of Microbiology (Centre of Excellence), Dr. Ram Manohar Lohia Avadh University, Faizabad 224001, UP, India
e-mail: rajeevagaur@rediffmail.com

R.L. Singh (ed.), *Principles and Applications of Environmental Biotechnology for a Sustainable Future*, Applied Environmental Science and Engineering for a Sustainable Future, DOI 10.1007/978-981-10-1866-4_7

Abstract

Bioprocess technology/engineering deals with the development and applications of competent strains of microorganisms for their optimum metabolite production in a specialized bioreactor, depending on the growth kinetics and nature of metabolite production at their suitable physicochemical and nutritional levels. However, three important components are taken under consideration involving several microbiological and biochemical engineering skills, also known as biomolecular engineering along with architecture and design of bioreactor systems, because the design and architecture of a bioreactor is solely based on the nature of the microorganisms, growth, and metabolite production along with the elimination of toxic substances during fermentation. In the current scenario, the development of bioreactor technology can change any process parameters economically with greater productivity and quality of microbial products, therefore, the design and architecture of a bioreactor is important, and can make a new revolution in bioprocess engineering. The mutational and recombinant DNA technology has developed several beneficial microorganisms, and their large-scale economical production requires fermentation technology mainly in the development of suitable bioreactor systems and process parameters. Much work has been done by microbiologists and biochemists for the production of various microbial metabolites including enzymes, hormones, polysaccharides, organic acids, vitamins, and so on at laboratory scale through flask culture experiments, whereas commercial-scale production requires several correction factors as well as a specialized bioreactor for continuous production of microbial metabolites with minimum energy consumption. The importance of bioreactor systems and their evaluation for the architecture requires optimization of fermentation parameters by a benchtop fermentor, and then a bioreactor should be fabricated on the guidelines of the microbiologist adopting the corrective measures of several growth parameters for the optimum production of microbial metabolites.

Keywords

Bioreactor • Microbial metabolites • Solid-state fermentation • Immobilized cells

7.1 Introduction

The bioreactor is one of the important components of industrial microbiology where microorganisms are being cultivated in such a way to their optimal physicochemical and nutritional levels for the production of variable microbial products via a specific fermentation process. The variety of microorganisms of different nutritional classes having a different mode of metabolic process along with the requirement of air, pH, and temperature change during fermentation is regulated through bioreactor systems depending on the nature of the microorganisms. In the last ten decades, microorganisms have been employed in the production of numerous chemical feedstocks, energy sources, enzymes, pharmaceutical hormones, vitamins, polysaccharides, protein, oil, and vaccines among others. The nature of primary and secondary microbial metabolite production at different stages of growth of microorganisms at changing substrate concentration along with the status of biomass and metabolite production initiates the option of a specific fermentation process that further initiates the architecture and design of bioreactor systems. The upstream process parameters, mainly inoculum production with changing nutrient levels and also requirements of physicochemical conditions, and ratio of inoculums into the culture vessel fermentation to the main bioreactor, require the mathematical norm as well as devices to maintain optimal conditions.

The traditional models and practices are modified with the advent of mathematical as well as statistical basis to achieve higher productivity. Microbial systems require special care for proper proliferation into the initial vessel or series of vessels, if required. Therefore, live biomass recycling is the only option requiring the centrifugation and recycling into the required vessel as per the optimization norms, and the continuous fermentation approach can be followed for even those microbes that show their tendency in batch fermentation. Batch fermentation is not economical over a continuous approach, therefore nutritional manipulative conditions and modifications in the bioreactor depending on the microbes alone or through mix culture at different stages may solve the problems for the microbiologists/biotechnologists.

Mathematical concepts are required, and the principles of thermodynamics, miscibility of O_2/CO_2 into the solution, viscosity, temperature control system, pH effects, and solute translocation at a particular osmotic and redox-potential, should be learned by biologists along with the basic knowledge of instrumentation principles that can easily be studied. However, the responsibility of microbiologists/biotechnologists is more to develop this area for future research and development (Shuler and Kargi 2002).

The strain improvement either through the mutational approach or modern recombinant DNA technology has initiated the change of various designs and models of the fermentor/bioreactor with modern automatic devices for measurement of biomass, substrate, and metabolites during fermentation, therefore such improvements in the design and architecture through all types of fermentation, and diversified microbial systems as well as their specific requirements of growth and metabolite, can be facilitated with minimum energy and higher productivity. In this chapter, the basic designs and process parameters for different microbial systems are discussed.

7.2 Basic Concepts in the Development of Bioreactors

The initial steps in the development of bioreactor status with the knowledge of microbial growth kinetics along with nutritional categories and the rate of their uptake are essential. The nutrient from the environment must be transported across the cell membrane into the cell. This is often the rate limiting step in the conversion of raw materials to the products and therefore is important in assessment of a fermentation process. The uptake of a few nutrients is by passive diffusion, which does not require carriers. Such nutrients are usually soluble in lipids and can enter in the hydrophobic membranes, for example, glycerol and urea. The mechanism is not efficient as the rate of uptake is dependent on the magnitude of the concentration gradient across the membrane. Most solute must be transported via active mechanisms, because membranes are selectively permeable. During fermentation, the substrate and metabolite concentration change quickly and may create either favorable or adverse conditions that must be noticed for a particular microorganism and its fermentation process, therefore the type of fermentation process with other parameters such as the retention period concentration of the solute and the like, finally initiate the change in bioreactor conditions. Microorganisms usually inhibited in the natural environment where nutrient concentration is low and consequently it is essential that they can accumulate solute against concentration gradients as intracellular concentrations of compounds generally higher than the environmental levels.

Most solute uptake involves carrier proteins, permeases; the resulting facilitated diffusion requires the direct input of energy, therefore, it is run solely by the concentration gradient across the membrane and is reversible. However, nutrient uptake into the cell continues because intracellular concentration does not increase, as the nutrients are immediately metabolized on entry. This mechanism rarely occurs in prokaryotes but frequently occurs for the transport of sugar and amino acids, and their selective carriers usually function for a group of related solutes. Such phenomena increase the diffusion rate at least up to concentration at which point the carrier protein becomes saturated with its nutrients (Mercille et al. 2000). Knowledge of the nutrient uptake and the behavior by the microbial strain for a particular fermentation process must be assessed to achieve optimum metabolite production in the specialized design of a bioreactor. The active transport mechanism allows the accumulation of nutrients against a concentration gradient which is important in an environment where nutrient levels are low. Some systems allow accumulation to 100–1000 times greater than the external concentration. However, these mechanisms require the direct input of substantial amounts of metabolic energy, ATP, or proton gradients for transportation.

In facilitated diffusion, protein carriers are involved; many are highly specific, whereas this functions with a group of related compounds. Sometimes the proton gradient transport amino acid sugar and/or organic acid molecule is simultaneously transported (known as symport); such a phenomenon in specific microorganisms in a bioreactor is essential. A proton gradient may also be used to establish a sodium ion gradient across the membrane: the sodium ions leave the cell in exchange for the entry of protons, which is termed antiport. Some compounds may be modified during uptake such as sugar, which is phosphorylated using phosphorenol pyruvate as the phosphate donor through a process referred to as group translocation and is performed by many prokaryotic cells. These aspects must be evaluated while using consortia of different nutritional classes as well as their specific metabolic process for energy yielding, synthesis of biomolecules, and extra- and intracellular status of metabolites. The membrane permeability and the nature of extracellular compounds affects production which has a specific design and process requirement especially for the rate of aeration and temperature increase/decrease in the bioreactor at the point of nutrient uptake as well as metabolite production. Certain initial ions that increase/decrease the membrane permeability and enzyme stability should also be taken care of during optimization of parameters as well as the design of the bioreactor to combat the process parameters (Blanch and Clark 1996; Mercille et al. 2000; Singh et al. 2011).

Utilization of polymeric substrates (viz. polysaccharides, proteins, and lipids) require additional activities such as phagocytosis by organisms having a membrane-bound food vacuole with the capacity of producing hydrolytic enzymes. The organisms with a rigid cell wall secrete extracellular hydrolytic enzymes including proteases, amylases, cellulases etc., and then uptake the hydrolyzed product. The nature of the constituents and induced enzymes is also an important phenomenon that should be evaluated at different changing concentrations at different phases of growth and stages of either primary or secondary metabolite production.

7.3 Growth Kinetic Parameters in Batch and Continuous Bioreactors

Microbial kinetics must be assessed for selection of fermentation parameters and design of the bioreactor. Mechanical growth is an orderly increase in cellular components resulting in cell enlargement and finally cell division. The consequence of growth is always an increase in cell numbers. In certain conditions, growth can occur without cell division; for example, when cells are a synthesized storage compound such as glycogen or poly beta hydroxyl butyrate, the cell concentration becomes constant, but the concentration of biomass continues to increase. This is true for coenocytic organisms such as some fungi that are not divided into separate cells resulting only in increased size. The theory of suspension used for growth of the microorganism is one of the important components of microbial kinetics and should be considered during the design development of a bioreactor. Growth kinetics of a homogeneous unicellular suspension culture can be developed using a differential equation in a continuous culture, whereas heterogeneous cell aggregates and assemblages, particularly bio films, colonies, flocks, and pellicles are more difficult. In fact, heterogeneous systems require a very different approach using cellular automation with solution flow theory, especially viscosity of the medium, sedimentation rate of the biomass along with the morphology of the organisms, and filamentous organisms.

The growth model for bacteria is considered by their binary fission in a homogeneous suspension culture; where cell division produces identical daughter cells, such cells generate with minimum generation time or doubling time, the time required for a microbial population to double. Theoretically, after one generation, both the microbial cell population and biomass concentration have doubled. Under certain conditions, growth can be associated with an increase in biomass and not cell numbers. Moreover, the generation time recorded during microbial growth is in reality an average value, as cells will not be dividing at exactly the same rate. At a particular time there are cells at different stages of their cell cycle called asynchronous growth. However, under certain conditions, synchronous growth can be induced to all cells dividing simultaneously which is a useful research aspect in the study of microbial physiology and bimolecular synthesis at certain energy yielding pathways and the role of physicochemical and nutritional inhibitors for metabolite production. These aspects can be minimized and controlled through the architectural manipulations in the bioreactors and addition of nutrients, and retention time of cells at a particular stage or maintaining a constant supply or disrupting the supply of a particular concentration of nutrients into the bioreactor. Different devices with their automation can be operated to achieve an ideal situation for growth and metabolite production (Kargi and Moo-young 1985).

Microbial fermentation in liquid media (submerged system) can be operated under different conditions: batch, fed batch, and continuous growth. Batch growth involves a closed system where all nutrients are present at the start of the fermentation within a fixed volume, whereas other physicochemical parameters are maintained in accord with their manipulation at any stage of the fermentation process. In a fed batch system, fresh medium or medium components either in the same nutrient conditions or different concentrations according to the requirements of the microorganisms on the basis of their physiology and tolerance limits are operated. The continuous system maintains the chemostatic as well as turbidostatic conditions depending on the design of the bioreactor either on a continuous basis or through the transfer of whole components in different vessels having different nutritional setups, known as a multi-vessel continuous system, for example, for production of single cells and some of the secondary metabolite production. Throughout the continuous system the volume of the medium at all the stages should be constant as the spent medium and cells are removed at the same rate, therefore recycling of cells (dead/viable) through a large cup centrifuge is controlled by optical devices. In this process, the live and dead cells are recycled, and the cost involved for such a process added to the cost of the products. In order to control such expenses, again the design of a bioreactor can be utilized. The next approach is to use the immobilization of microbial cells/enzymes for microbial metabolite production. This method is more cost effective for microbial metabolite production compared to a simple continuous system. Every method has its own limitations which are finally corrected through the manipulation in the cultural, nutritional, and modifications in the bioreactor design. The scope in this area for a microbiologist and biochemist is to work together to solve such problems to achieve quality products at economical cost. For this, the yield coefficient is an important parameter determined on the basis of the quantity of rate limiting nutrients, normally the carbohydrate sources converted into the microbial products and cellular mass. In the case of biomass production, the yield coefficient relates to the quantity of biomass produced per gram of utilized substrate. Therefore, the greater the percentage of the original substrate converted into microbial biomass or microbial metabolite products, the yield coefficient is related to the quantity of metabolite produced in relation to the quantity of substrate used. Therefore, determination of the yield coefficient is very important because the carbon source and quantity assesment is important in the overall cost of the production (Shuler 1985).

For optimization, in several experiments under different operating conditions with varying medium constituents and component concentration including pH, temperature, and aeration, among others, optimum growth and production condition can be established. It is also important to determine the maximum specific growth rate of the organism used for production of any specific metabolites. This is particularly true for primary metabolites where product fermentation is related to growth. For optimization of overall productivity of the system, the microorganism must usually be grown at its maximum specific growth rate where the operating substrate concentration has a major effect on the growth rate of a microorganism. These can be easily determined by the mathematical growth equations (Shuler 1985, 2002).

The continuous process is more economical than the batch, if microbial growth kinetics favor the process, otherwise the option is batch. Therefore the cell growth, physiology, and biochemistry should be studied. The continuous bioreactors initially start as a batch culture but exponential growth is extended indefinitely until or unless the metabolite production rates decrease. The continuous addition of fresh fermentation medium in constant volume is maintained by an overflow through some specific devices, therefore chemo- and turbidostatic conditions are created in the bioreactor, and a steady-state condition prevails. There are several advantages of continuous fermentation over batch along with the limitations where cell growth and product formation are a complex process reflecting the overall kinetics, as thousands of intracellular reactions run within a cell. For many process calculations, potential substrates in terms of cell mass yield, product yield, and evolution of heat related to thermodynamic are being under considerations. If a system is close to its thermodynamic limits, the organism need not undergo mutation or genetic engineering approach (Bailey 1998).

7.3.1 Advantages and Disadvantages of Batch and Continuous Fermentation

The main concepts of microbial growth and changing temperature, pH, and O_2 concentration along with substrate and product concentration (especially in batch), the specific growth rate possesses the sequence of growth phases (log, stationary, and death phases), and consequently the system never achieves steady-state condition. Several distinct practical stages are associated with the operation batch fermentation.

1. Changing of the bioreactor with fresh medium after every batch.
2. Sterilization of bioreactor and medium after every batch, therefore there is more change of contaminants.
3. For every batch, fresh inoculum production and inoculation of the bioreactor involve more time, energy, and cost.

4. Cleaning the substrate feeding vessels and pipeline.
5. Loss of products through evaporation.
6. Major economic implication for industrial process along with less productivity compared to other fermentation processes.
7. Several benefits of continuous fermentation over batch process but both have some limitations.

7.3.1.1 Advantages of Submerged Fermentation (SmF)

- All the substrate is uniformly available to the microbes as it is completely dissolved.
- In SmF, aeration is normally not a problem.
- In liquid phase, there is better heat transfer.
- In liquid substrate, mixing is much easier.
- In liquid broth, microbial growth is faster.
- In mixed SmF, control of pH is easier.

7.3.1.2 Disadvantages of Submerged Fermentation

- Water consumption is high.
- Formation of foam.
- Wastewater output is high.
- High water activity, augmented possibility of contamination.
- Cost of media is high, as substrates are often costly compared to SSF.
- High demand for sterility because of higher water activity.

7.4 Basic Concepts of Solid-State and Solid-Substrate Fermentation

Bioreactor system categorization at a basic level is mainly based on the process approach, that is, submerged as well as solid-state/solid-substrate fermentation. Both approaches are dependent upon the nature of the microorganism mainly regarding their water activity (a_w). Filamentous fungi mostly grow at low water activity except for a few such as *Aspergillus*, *Penicillium*, and *Fusarium* spp. Some strains of these fungi have been used in submerged fermentation for production of citric acid and antibiotics as well as alkaloids. These fungi grow at low as well as higher water activity just like yeast and bacteria. Most of the yeast and bacteria prefer higher water activity for metabolite production, and a few strains may grow even at lower water activity. Furthermore, the bioreactors have been generalized for bioethanol (or any alcoholic group), biogas (from industrial effluent), antibiotic production, SCP, and various other metabolites. Surface fermentation is another submerged fermentation approach. Various subgroups of fermentation have been adopted depending on the fermentation kinetics; that is, on the basis of the substrate utilization rate, biomass and metabolite production in the time scale may give a complete idea regarding the selection of a suitable fermentation process to achieve maximum yield in an economical way (Rejsman 1993; Lee and Papoutsakis 1999; Singh et al. 2011).

These selection procedures may further be manipulated according to the nature of the microorganisms, requirements of nutrients, and the bioreactor system to achieve quality products. This chapter discusses several technical aspects of fermentation of microbial products especially for the quality of microorganisms, the bioreactor system as a factor affecting fermentation, as well as downstream processing of the product along with quality control. The engineering approach, especially the architecture and design of bioreactor and equipment related to downstream processing (Midgett 1986), is also discussed.

Solid-substrate fermentation in which moisture content is not the barrier may therefore be separated even in a submerged system as well as at a low moisture level depending on the specificity of the microorganisms. The main difference from solid state is that the solid support is utilized as a carbon and energy source by the microorganism and the solid substrate to be exhausted after a certain period of time, requiring continuous feeding of the solid substrate; the rate of supplementation depends upon the utilization rate of the substrate, therefore microbial leaching of metals from ores or desulphurization of coal and some of the cellulose particles for conversion into glucose. The Kogi process in SSF requires *Aspergillus*, *Penicillium*, *Rhizopus*, *Mucor*, or growth on rice, wheat, or other cereals for food fermentation. Various fermented food such as meso, temph, sofu, tofu, and soya sauce are examples of solid-substrate fermentation. Various bioreactor models have a large flat surface with a rotating surface in close plates, keeping the aeration and turning, if operated in low moisture while in a submerged system. The common submerged bioreactor with proper aeration has been used for proper solid-substrate utilization either for the production of biomass or for the utilization of substrate . The operation may be operated either in batch or continuous process (Perez-Guerra et al. 2003).

7.4.1 Advantages and Disadvantages of Solid-State Fermentation

Fermentation techniques can be classified into two main categories: solid-state fermentation (SSF) and submerged fermentation (SmF). The dissimilarity between these two techniques consists in the amount of free-flowing liquid present in the system. SSF has the growth of microorganisms on solid materials in the absence or near absence of free-flowing water, whereas in SmF, the microorganisms cultivate on a continuous liquid phase. Because of the small quantity of water present in SSF, the formation of definite products, which are not formed under SmF, can take place. For the products that can be obtained using both techniques, SSF presents elevated volumetric productivity and improved performance over SmF (Moo-Young et al. 1983).

SSF processes have shown to be generally appropriate for the production of enzymes by filamentous fungi (Moo-Young et al. 1983) because they reproduce the natural living state of such microorganisms.

One of the advantages frequently cited for SSF processes is that enzyme titers are higher than in SmF, when comparing similar strains as well as fermentation broth (Viniegra-Gonzalez et al. 2003). They differentiated the productivity of three fungal enzymes, invertase, pectinase, and tannase, via SSF and SmF techniques. They reported that the elevated titers found in SSF over those in SmF were because of SSF cultivation in fed batch culture with oxygenation through a slow sugar supply. Castilho et al. (2000) performed a relative economic study of solid-state and submerged processes for the production of lipases by *Penicillium restrictum*. They found that for a plant producing 100 m^3 lipase concentrate per year, the procedure based on SmF needed an entire capital investment of 78% more than SSF, and its product had a unitary cost 68% more than the product market price. These outcomes showed the large advantage of SSF because of its low cost, but not with all the filamentous fungi.

7.4.1.1 Advantages of Solid-State Fermentation

- Minimum water consumption.
- Foaming has no problems.
- Output of wastewater is low.
- Low water activity used in SSF places lower demand on sterility.
- Low cost media, as substrate are often agricultural by-products or waste.
- Fermentation productivity is higher.
- Higher end-concentration of products.
- Product stability is higher.
- Catabolic repression is lower.

7.4.1.2 Disadvantages of Solid-State Fermentation

- Partial conversion of substrate by microbe.
- In the three-phase system there is a problem of aeration.
- Product recovery and purification are more expensive processes.
- Pretreatment of the substrate is normally needed (grinding, chopping, homogenization, physical, chemical, or enzymatic hydrolysis, cooking, or vapor treatment).
- In a large amount of solid substrate there is the problem of mixing.
- Slower microbial growth on solids.
- Difficult to control process variables such as heat, pH, mass transfer, oxygen.

There are numerous contributions to the field of laccase production under SmF by different microorganisms, at diverse scale with the potential use of immobilization supports with the addition of inducers. Several of the most remarkable consequences in terms of laccase activity were obtained by the *Trametes genus*, *T. pubescens* (Galhaup et al. 2002), *T. versicolor* (Font et al. 2003; Tavares et al. 2006), and *T. hirsuta*. In nearly all cases, the cultures were supplemented with laccase-inducing compounds. Galhaup et al. (2002) got a highest laccase activity of 740,000 U/L in *T. pubescens* cultured within a 20-L stirred-tank reactor (STR) with a stirring speed of 100 rpm and with 2 mM concentration of Cu^{+2}. Font et al. (2003) showed a highest laccase activity of 16,000 U/L by free pellets of *T. versicolor* in a 0.5-L pulsed-bed reactor. Tavares et al. (2006) reported a highest laccase activity of 11,403 U/L culturing the same fungus on a STR of 1 l, supplementing the medium with 30 μM of xylidine. A highest laccase activity of 19,400 U/L was recorded by culturing *T. hirsuta* within a 6-L airlift reactor (ALR), moreover supplementing the medium with glycerol and Cu^{+2}. Laccase has also been formed in SSF, especially during the last decades. The use of natural solid substrates, particularly lignocellulose agricultural residues, as augmentation substrates for fungi has been considered for laccase production in current years. Moreover, such residues have cellulose, which serves as an inducer of laccase activity. The insufficiency of bioreactor designs to carry out solid-state processes among the advantages presented by such processes encourages the need for developing new bioreactor configurations or modifying the designs that previously were present. These bioreactor designs should be capable of working in a nonstop manner by way of high enzyme productivity for extended periods of time with no operational harms and permit the scale-up of the process. Rivela et al. (2000) developed a novel bioreactor design used for the production of ligninolytic enzymes in SSF conditions called an immersion bioreactor. They attained high ligninolytic activities and the bioreactor was able to work in a nonstop manner. Dominguez et al. (2001) designed a rotating drum reactor (RDR) for the formation of ligninolytic enzymes in SSF by the white-rot fungus *Trametes pubescens* used for their possible relevance to synthetic dye treatment. This bioreactor was capable of working in batch and continuous methods. Bohmer et al. (2006) further reported the advantages of adapting the provisional immersion RITA®-System as a bioreactor for laccase production by white-rot fungi along with its relevance to synthetic dye discoloration. Rodriguez Couto et al. (2003) experienced three bioreactor configurations (immersion, expanded bed, and tray) by means of diverse agitation systems (mechanical, pneumatic, and static, respectively) for laccase production by *T. versicolor* in a SSF situation with an inert (nylon sponge) and a noninert (barley bran) support on seperate specillized designed tray system for maximum laccase production. They differentiated two bioreactor configurations (immersion and tray) for laccase production by *T. hirsuta* with grape seeds as support-substrate, and found much superior laccase behavior in the tray bioreactor (Rodriguez et al. 2006a). In addition, they reported much superior laccase activities in a tray bioreactor than in a fixed-bed one for *T. hirsuta* grown on ground orange peelings (Rosales et al. 2007).

7.5 Bioreactor Design Components: Temperature and Aeration

7.5.1 Basic Design of Solid-State and Submerged Fermentation in Relation to Temperature and Aeration

Microbial growth generally generates about 40–50% of the energy stored in a carbon and energy source is converted to ATP by aerobic metabolism and rest released as heat. The heat balance in microbial growth depends upon the rate of utilization of substrate and production of cellular materials which indicate the enthalpy balance. The heat of combustion of the substrate is equal to the sum of the metabolic heat and the heat combustion of the cellular material. In aerobic fermentation, the rate of metabolic heat evolution can roughly be correlated to the oxygen uptake because oxygen is the final electron acceptor, and for anaerobic there are several other inorganic sources may be as electron acceptor. Metabolic heat is released during fermentation, therefore cooling devices such as circulating cool water either through a cooling coil or cooling jacket in the bioreactor with several designs are used. The cooling coils inside the bioreactors create problems in the batch process due to favoring contamination, and a higher cost is involved in sterilization, however, continuous fermentation should be adopted for such a process. The adequate heat removal from the bioreactor is an important limitation on reactor design. The costing parameter during design of a bioreactor and adaptation of a set of process is always essential (Kossen 1985; Singh et al. 2011).

Aerobic fermentation is most prevalent for microbial metabolite production as fungi and yeast are aerobic and grow well when proper O_2 is available in the fermentation. Moreover, in some cases, anaerobic fermentation of microbial metabolism also requires O_2 at the initial stage for production of cell mass. Therefore, aeration and temperature control initiate the technologists in designing a bioreactor, which may prove both the conditions according to the nature of microbes in the bioreactor for optimum metabolite production. The design of a bioreactor is based on the economics of almost all the growth parameters and the release phase of the product and temperature in the bioreactor. The removal of heat from the bioreactor/fermentor is through the various other mechanisms instead of cooling coils or through heat exchanger jackets. The recycling of the medium which is known as hydrodynamic conditions has been designed for cooling as well as proper missing of the substrate along with proper aeration, as sterile air in large bioreactors is very costly and has some limitations. The sterile compressed air generates high heat resulting in the change of temperature of the bioreactor, therefore it is just not possible to supply huge amount of the air into bioreactor.

Several growth inhibitors are produced at high concentration of substrate by the fermentative microorganisms and the non-uptake substrate may favor another group of microorganisms (mainly bacteria) to multiply there even though in stress conditions. The growth is directly dependent on the inhibitor concentration.

The enzymes are more sensitive to the inhibitors, and various types of inhibitors are identified:

1. Substrate inhibition: The low and high both inhibit growth as well as metabolite production.
2. Product inhibition: High concentration of product can be inhibitory for microbial growth. The product inhibition may be competitive or uncompetitive and sometimes other physiological factors may inhibit noncompetitive; an example is ethanol production from glucose by yeast.
3. Inhibition by toxic compounds: This may be due to competitive, noncompetitive, and uncompetitive which only occurs in enzyme inhibition.

All three factors of inhibition are based on logistic equations, which are a set of equations that characterize growth in terms of carrying capacity. The usual approach is based on a formulation in which the specific growth rate is related to the amount of unused carrying capacity. Mathematical modeling is essential for bacterial growth, whereas fungi and algae require different growth models with high cell density in suspension culture. The growth model of molds should include the simultaneous diffusion and consumption of nutrients. The exact modeling is difficult with the fungus as well as bacteria and yeast when they are trapped in spherical gel particles. In solid-state or substrate fermentation, when fungi grow on the moist solid surface, growth assessment is a complicated process. In this system, the biomass and rate of nutrient consumption is not essential to estimate the metabolite production. It is the only way to assess the fermentation efficiency in term of the carbon source supplied initially which will be an accurate way to assess the productivity and cost effectiveness. Several intrinsic parameters affect the cell simulations, mainly aeration, temperature, pressure, and movement of the fluid. The large cells of filamentous fungi, mainly the sporulating/nonsporulating nature and growth rate with nutrient uptake, may affect those cells with proper growth conditions and may multiply fast and divide little but late, especially for fungi, whereas bacteria and yeast have shown their doubling period in the early exponential phase, and in the later stage, may be slightly affected if the conditions are improper. In the batch fermentation, one should analyze the kinetics of the organisms as to whether the organism is fit for such process. Mostly bacteria and yeast can follow all types of fermentation systems, therefore for economic reasons may be used in continuous fermentation of production. The concentration of the microbial product against the growth parameters for single culture as well as in consortium may also be discussed in the similar way when growth is not affected by the other microorganism. Sometimes the consortium utilize the same substrate; then the condition of the fermentation as well as the intermediate compounds must be evaluated to assess the other physicochemical parameters alongwith different substrate concentration taken in the account for development of bioreactor in different design and architecture accordingly.

Another modeling approach has been used to predict growth under several available substrates. These substrates may be complementary (e.g., carbon and nitrogen) or substitutable. For example, glucose and lactose are substitutable as these compounds supply carbon and energy both through a diauxic phenomenon for sequential use of glucose and lactose by regulation of lac-operon and catabolite repression. The metabolite regulation was necessary for the transition from one primary pathway to another. The modeling of growth on multiple substrates is a cybernetic approach meaning that a process is approached for the maximization of the growth rate by assessing the cellular metabolism. In this way, an easy approach is to assess all the microorganism growth on a dual basis as the physiology of the organism at individual bases may differ, therefore, the exact DNA sequencing and strain approach must be followed prior to use at the individual level. In bioprocess technology, the growth rate and growth yield are important to assess as environmental conditions may vary these, but it is very difficult to assess all the constant conditions completely. There are several factors simultaneously acting on a growing microorganism in different fermentation processes, therefore online monitoring can only be operated in set of processes for optimum growth and metabolite production (Flickinger and Drew 1999; Midgett 1986).

Continuous culture is entirely different from batch, as the growth kinetics that have been discussed; continuous culture gets fresh nutrients continuously supplied to a well-stirred culture and products and cells are simultaneously withdrawn. Growth and product formation can be maintained for prolonged periods in a continuous culture. After a certain period of time the system attains steady state where all products and substrate concentrations remain constant. This system provides constant environmental conditions for growth and product formation and produces an informed quality product. The microorganisms produce their maximum ability to transform substrate to product. The product degradation into the bioreactor is checked through this process, and the chemostatic and turbidostatic conditions are maintained through the equipment by recycling of the cells or addition of more cells at a particular stage through an auto-suspensor based on optical instrumentation. All three technologies based on optical methods, chromatographic and electroanalytical tools have been used for the assessment of nutrient and microbial products during/ after fermentation for proper maintance of online monitoring.

Several designs have been adopted to maintain chemostatic as well as turbidostatic conditions. The stirred-tank bioreactor is the basic bioreactor structure which is differentiated from the non-stirred one. The basic unit of a benchtop bioreactor model is the main one adapted for slight modification.

7.6 Classification and Configurations of Bioreactors

The classification of bioreactors mainly confers:

1. Stirred-tank bioreactor by internal mechanical agitation: The design and place may vary depending on the size at industrial scale.
2. A bubble column which is through gas sparging for agitation: it is also known as an airlift bioreactor, which is available in different designs depending on the process requirement. The bubble column reactor is also run in loop bioreactors in which liquid circulation is induced by the motion of an injected gas or by a mechanical pump.
3. Hydrodynamic bioreactor: Circulation of medium by the pump is either through swarming or by simple serration by pipeline from bottom to top feeding.

All three types are mainly concerned with the aeration mixing of the substrate as well as temperature control within minimum energy. The control parameters of the aeration and temperature depend on the nature of the microorganism as well as process parameters. The traditional fermentation is the stirred-tank reactor with internal mechanical agitation. The 450–500 m^3 bioreactors are used for antibiotic production. The air pressure is supplied to the sparger either through a ring with holes or a tube with a single hole of different size. The size of the gas bubble and dispersion throughout the tank should be in small bubbles which is only possible through sparger, but single discharge points is often better for media having a high level of suspended solids.

The gas distribution through the impeller increases the residence time of bubbles in the liquid. Some large bioreactors having side impellers attached to the side wall of the reactor may be considered for better aeration with a varying design of impellers and size of the blade with up and down flow of liquid. These designs are not suggested for fungal, algal, or animal cells. The hard baffles are also designed with four to five baffles of varying width especially for solid-state fermentation, whereas in submerged fermentation baffles cause shear damage.

The flow of medium current is an important aspect, therefore impeller speed and size with specific design in a large reactor have the two main limitations i.e., requirement of aeration and release of heat removal. The bioreactor is always filled to less than the full working capacity due to foaming which creates a hindrance to aerobic growth and also favors contaminants. It also loses product, therefore mechanical devices have been used for foam breaking but they are energy consuming. Chemical agents are more efficient and economical for the control of foam, but in some cases chemicals affect the cells. The microbial protein favors foaming, however, the working volume in a bioreactor is typically about 75% of the total bioreactor volume. The bioreactor should have minimum openings to avoid contamination. The main ports are at least three to four at the maximum. A port on the top of the bioreactor is for exhaust gas with some control devices. As bioreactor exhaust may be used for the desired pressure, which is required for some facultative anaerobic microorganisms, and pressure is also required at cooler countries to maintain temperature and gas stay into the bioreactor (Hammer and Hammer 2001).

Two inlet ports on the top are used for the culture and medium supply into the bioreactor separately. The pipelines should be stainless steel which does not react with the nutrient solution at any pH range and can be sterilized by live steam after every fed, in batch fermentation. In continuous culture, there is no need of such process. The culture pipeline is used once a year or wherever required otherwise; this pipeline does not require steaming and saves revenue. The fourth port is used for sampling as well as for controlling pH or addition of some compounds such as antibiotics or other micronutrients whenever required during fermentation. These openings are made leakproof with an o-ring and a large opening from the gaskets. The shifts and valves should be designed to be leakproof. Bioreactor cleaning is very essential. Hard detergent and chlorinated water are used for washing.

The bubble column reactor is generally not suitable when the fermentation medium creates huge foaming; the slow movement of the fermentation medium is required. The range of appropriate gas flow varies with the nature of the broth, therefore multistage columns along with a perforated plate which disrupts the pressure and maintains proper flow rate. A bioreactor with immobilized microbial cells with column aeration along with multi-segmentation has been worked out by Singh et al. (2011) in which bubble column and loop (air lift) bioreactor principles were integrated in its design with slight modification in order to maintain, the aeration, agitation and heat transfer with single action reducing the cost of production.

At industrial-scale design of a bioreactor, oxygen supply and heat transfer are the main parameters; the large amount of oxygen is essential at the commercial level. The specific rate of oxygen (mol oxygen/g/h) and the presence of salt and surfactants will alter the bubble size and liquid film resistance around the gas bubble, and affect oxygen solubility, whereas temperature and pressure also affect oxygen solubility. The bacteria require a specific oxygen uptake rate (respiration rate) 2–90 m mol oxygen/g dry wt. per hour for example- *E. coli* 10–12, *Azotobactor* 30–90, *Streptomyces* 2–4, *Yeast*, *Saccharomycaecerevisiae* 8, *mold*, *Penicillium* sp. 3–4, *Aspergillus*-3, plant cells 0.2–5, *saccharum* 1–3, animal cell He La 0.4/106 cells/ml, diploid embryo w138–0.15/106 cell/ml. The optimization of oxygen uptake at approximately the amount in pilot scale may give a correct value for industrial scale. The benchtop bioreactor has been used to evaluate oxygen uptake at laboratory scale and should also be followed before pilot scale to know the corrective measures for this.

Despite the knowledge of different types of bioreactors based on the stirred principle, aerodynamic as well as hydrodynamic in different models and modifications with the integration of each can make the bioreactor system efficient regarding the proper and uniform mixing of substrate and effective heat transfer as well as aeration as per the requirement of the microorganisms. The thermodynamic principles, law of viscosity, Reynolds theory, Bernoulli principles, and diffusion of gases in liquid, as well as the law of mass action act on the fermentation system accordingly. Furthermore, biomass maintenance especially for simple continuous production in a submerged system or through immobilized microbial cells/enzymes may be considered for the development of new designs to achieve optimum fermentation conditions.

Active inoculum production either in a batch or continuous system is another approach, thus knowledge of active inoculums of a specific group of microorganisms is also very essential to achieve proper fermentation. Therefore, some of the important aspects of microorganism growth are important to know, such as generation time, and optimum nutritional and physicochemical growth conditions along with water activity (a_w).

7.7 Kinetic Parameters for Development of Bioreactors

Appropriate bioreactor conditions and architecture provide proper growth of microorganisms depending upon their nature, nutritional types and concentration, microbial cell size, rate of uptake of nutrients and change of temperature, pH, and requirement of air for particular periods. Microorganisms can grow under a variety of physical, chemical, and nutritional conditions, in a suitable nutrient medium and other optimum physicochemical conditions; organisms extract nutrient at the maximum capacity from the medium and convert it into biological compounds after a series of metabolic levels for utilization of carbohydrates, protein, lipids, and minerals from simple to complex form depending on their genetic and physiological states. A part of nutrients are used for energy production and some parts are used for biosynthesis and product formation either of primary or secondary metabolites. Primary metabolites are used for the dispensable function of cell metabolism whereas secondary metabolites are not used for a dispensable function; thereby both forms of metabolites are secreted out.

Primary and secondary metabolites are useful for human welfare. The production of antibiotics, toxins, alkaloids, and color compounds are examples of secondary metabolites. It has been explained that some of the deleterious compounds generated inside the cell through various metabolic processes as the result of their metabolite defects constituted inside the cell through the metabolic process therefore must be secreted out as they are toxic to the cell. However, the substance produced in excess by certain gene regulations either induced or constituted enzymes that are released from the cell in the form of primary metabolites, therefore release of such metabolites either in an increased or decreased amount by several ways such as by an appropriate mutational approach, depends upon proper growth through autocatalytic reaction. The rate of growth is directly related to cell concentration and cellular production is the normal outcome of this consequence.

The rate of growth is directly related to cell concentration, and cellular reproduction is the normal outcome of this reaction. The microbial growth is characterized by the net specific growth rate, that is, the cell mass concentration (g/l) in a specific time (h). The net specific growth is the difference between a gross specific growth rate and rate of loss of cell mass due to cell death or endogenous metabolisms. Moreover, microbial growth can also be defined in terms of cell number concentration which shows the net replication rate per hour. In this way, the cell death is not taken into account, therefore the growth rate changes with its environment may be discussed when the microorganism is grown in a specialized vessel/vessels called a fermentor or bioreactor. The literal meanings of both have their own definitions. A fermentor is the vessel where a single type of microorganism is cultivated for obtaining single product fermentation with a specific substrate and product, whereas in a bioreactor a series of different substrates are released by different groups of microorganisms having different physiology and reaction to various products produced in a single vessel.

The behavior of the microorganisms and the nature of uptake of nutrients (substrate) into the rate and stage of metabolite production have been categorized in different types of fermentation processes. The main categories are batch, fed batch, and continuous, as well as continuous through immobilization of microbial cells/enzymes. All these stages generally depend upon the nature of the rate of substrate utilization, biomass, and metabolite production, but such conclusions may be modified by the microbiologist or biotechnologist by optimization of the process, as well as change in the devices, mainly using multivessel systems with evaluation of the proper retention period of the biomass in a particular period by maintaining specific flow rate of the medium and biomass, as the stage of biomass with relation to the metabolite production period especially at the early or late log as well as stationary phases. Therefore, mathematical calculations regarding the time cell mass of a particular stage as well as requirements of other supplements at different stages must be optimized to achieve proper design of fermentors and/or bioreactors.

In the light of this, selection of a proper fermentation process in a specialized vessel to achieve maximum productivity is the main objective of a microbiologist. The basic understanding for the quantification of cell concentration either through biomass/cell number density along with growth and kinetics in a set of processes either batch, fed batch, or continuous with their mathematical and substrate standards is essential. Furthermore, the evaluation of how environmental conditions affect the growth kinetics must be evaluated prior to the fermentation process option.

7.7.1 Kinetic Parameters Other Than Biomass and Substrate Utilization

It has been observed that the slight modification in the culture conditions and manipulation/assessment of cell mass transfer into another vessel either with the same or changing conditions may give a better option of another fermentation mode to achieve more economical microbial products with maximum productivity along with the above factors; the heat generation by microbial growth and aeration required for a set of microbial biomass is necessary. It is also necessary to evaluate the microbial dependency over the nutrients, which has not been taken into account in the assessment, as some microorganisms are nutrient dependent and some nutrient independent for a set period. Nature may affect the fermentation assessment and stage of microorganisms in the continuous fermentation, as well as continuous through immobilization of the microbial cell. The nutrient-dependent microorganisms therefore just reduce their growth as nutrient exhaust, whereas the nutrient-independent microbes having a larger accumulation of cell reserve materials in the form of volutin granules, sulfur, phosphorus, PHA and PHB, glycerol, and the like, may survive for a longer period after exhaustion of nutrient, and therefore may be managed in a multivessel continuous system without supplementation of nutrient. Likewise, there are some other parameters, including the requirement of O_2 and temperature and pH tolerance capability affecting the release of microbial metabolite. Cell membrane bound metabolites released at the specific stage should be evaluated for smooth functioning of the metabolite production at industrial scale.

In the same way, the specific growth rate should be assessed with substrate limiting growth, growth inhibitor-like substances, and product inhibitor along with the inhibition by toxic compounds, therefore in time delay of such an effect, and how to make up such a gap by addition of specific compounds may be the same substrate or any nutrients and nontoxic chemicals may give a better way to control such change to achieve maximum productivity. Almost all the factors depend upon the type of fermentation process.

Batch growth of microbial cells occurs in a vessel with initial feeding of nutrients only, and is not altered by further nutrient addition or removal. This process is very simple and widely used in industry with some limitations.

7.8 Assessment of Fermentation by Measurement of Biomass and Other Bioreactor Products

To work out a specific fermentation process at industrial scale, some basic knowledge of quantifying cell concentration is essential that comprises direct and indirect methods. In some cases, the direct method is not feasible due to the presence of suspended solids or other interfering compounds in the medium. The cell mass or cell number depending on the requirement may be assessed. Cell mass concentration is often preferred to the measurement of cell number density, but in some cases at a particular stage when a viable number of cells are essential, the culture medium standard plate count is the only alternative as per the requirement from pour, spread, as well as sandwich agar plate depending on the nature of the microorganism from aerobic, facultative anaerobic, to anaerobic levels.

The cell number density determination can be determined by Petroff–Hausser slide or a hemocytometer. In this method, a calibrated grid is placed over the culture chamber, and the number of cells per grid square is counted under a microscope; counting at least 20 square grids is essential. In these methods, stains may be used for determination of viable and dead cells but not for the entire microorganism. Suspension should be clean without suspended particles. It is difficult to count filamentous fungi, the best result for bacterial and yeast counts.

Several other methods such as miniature culture on a microscopic slide are examined under the microscope. Another method is based on the relatively high electrical resistance of cells. This method is suitable for discrete cells in a particle-free medium and cannot be used for mycelia organisms. The number of particles in solution can be determined from the measurement of scattered light intensity with the aid of a phototube. Light passes through the culture sample and a phototube measures the light scattered by the cells in the sample. The intensity of scattered light is proportional to cell concentration. This method is good for dilute cell and particle suspensions.

To determine cell mass concentration, the direct method of determination of cellular dry wt. is the common approach and only for cells grown in solid free medium. For the noncellular solids, such as molasses, cellulose, and corn steep liquor, the dry wt. method will be inaccurate; therefore centrifugation, filtration, washing at several steps with either buffer or water, and then drying the cell mass at 80 °C for 24 h before measurement is necessary. For a rough estimate, the packed cell volume is also determined for cell concentration in fermentation broth, for example, antibiotic fermentation. Some other methods comprise the measurement of turbidity or optical density of culture medium providing a fast, inexpensive, and simple method of estimating cell density: the instrument is a spectrophotometer. The medium should be particle free and of proper wavelength to minimize absorption by medium components (600–700 nm are often used) along with the blank having a calibration curve, which relates optical density (O.D.) to dry wt. measurement. Such a calibration curve can become nonlinear at high O.D. values (>0.3) and depend to some extent on the physiological state of the cells which are generally known as direct method for determination of cell no./mass. There are several indirect methods for determination of microbial component like estimation of microbial metabolites, nucleic acids and proteins.

In batch culture, these intracellular components will change with time during a batch growth cycle. Therefore, RNA content/cell weight will vary and DNA and protein content will be constant. Thus in a complex medium, DNA concentration can be used for the measurement of microbial growth. Cellular protein content can be determined by various methods, including total amino acids, biurate, Lowery (folin-reagent), and Kjeldahl nitrogen measurement. Total amino acid and the Lowery method are more reliable in some cases; microbes utilize protein as substrate which limits the use of such a method. Therefore, another approach is used, that is, ATP concentration (mg ATP/mg cells), which is almost constant for a given organism. The ATP concentration determination in a fermented broth for estimation of cell concentration is through luciferase activity, which catalyzes oxidation of luciferin at the expense of O_2 and ATP with emission of light.

$$\text{Luciferin} + O_2 + \text{ATP} \xrightarrow{\text{luciferase enzyme}} \text{light}$$

A bacterial cell generally contains mg ATP/g dry wet cell.

Approximately, the utilization of a carbon source or uptake of O_2 can be measured to main cellular growth rate when cell mass is a product. For aerobic fermentation, CO_2 is a common product that can be related to microbial growth. In some cases, change in the pH or acid base addition to control pH can be used to monitor nutrient uptake and microbial growth. For example, the utilization of ammonium results in the release of hydrogen ion (H^+) and therefore drops of pH. The amount of base added to neutralize the H^+ released is proportional to ammonium uptake and growth. Similarly, when nitrate is used as the nitrogen source, hydrogen ions are removed from the medium, resulting in an increase in the pH. In this case, the amount of acid added is proportional to nitrate uptake and therefore to microbial growth; sometimes viscosity of a polysaccharide before and after degradation can be used for the determination of cells as well as product concentration.

7.9 Current Concepts for Optimization of Large Bioreactors

Growth kinetic parameters in batch, fed batch, and continuous culture can be discussed in the length of optimization and scaling up of the process; that is, from laboratory, pilot, to industrial scale is essential. There are two major categories of fermentation process: liquid or submerged fermentation and solid or semisolid fermentation. The liquid medium is uncalculated with a bacterium or bacteria in the consortium and will selectively take up dissolved nutrients from the medium and increase their cell numbers as biomass or biomass with some or more products depending upon the nature of the microorganisms. Batch culture will show different phases of growth: lag, log or exponential, stationary, and death phases. The lag phase occurs immediately after inoculation, and is a period of adaptation of cells to a new environment. At industrial scale, the inoculum is transferred into the same medium with a slightly low concentration of carbon source, therefore the lag phase is not expected which shows the correct approach of transfer of an inoculum to the wort, the medium ready for fermentation. Microorganisms reorganize their molecular constituents when they are transferred into a new medium; new enzymes are synthesized depending on the composition of nutrients and are repressed. The internal macromolecules adopt the new environmental condition, and the changes reflect the intracellular mechanism for the regulation of the metabolic process.

The growth phases always depend upon the types and concentration of nutrients like carbon and nitrogen sources alongwith micronutrients. The lag phase of the growth depends upon the state of microorganism and presence of nutrients along with other stresses. Sometimes Mg^{+2}, which is an activator of the enzyme phosphatase, affects the growth rate. The age of the inoculum may affect the lag phase. The age is the period that refers to how long a culture is grown in a batch culture. The lag period increases with the age of the inoculum, and is minimized with the growth medium and conditions before the inoculum, therefore the middle of the exponential phase rather than late exponential to stationary phase should be better for inoculum growth. The cell should be youngest and active and generally should be 5–10% by volume, therefore nutrient medium and cell size optimization for a particular fermentation must be optimized. Multiple lag phases may be observed when the medium contains more than one carbon source, therefore the organism utilizes a simple carbon source and then will regulate its enzyme for utilization of a complex carbon source showing a diauxic growth, which is caused by a shift in metabolic pathways in the middle of a growth cycle and may give further release of the beneficial metabolites. Some metabolites including antibiotics, ethanol, and toxins may be inhibitory to the growth at higher concentrations, but cell resistance against the product may give a better result; ethanol production by yeast is an example of a fermentation in which the product is inhibitory to growth, therefore in such condition, the simultaneous removal of product either through the membrane, however, simple continuous fermentation may be better option to overcome this event, where cell death may start in the stationary phase. It is clear that the dead cell will release intracellular nutrients into the medium and is used by the viable cells during the stationary phase. Most microorganisms follow the same trend with slight changes, but the above conditions are truly followed by bacteria and yeast, whereas filamentous fungi may grow through their vegetative mycelia cells along with spore germination, therefore, both may simultaneosly impart for metabolite production extacellularly and intracellularly or both.

To know the details of growth, we must know how the cell works. The exponential growth phase is also known as the logarithmic phase. This is a period of balanced growth in which the entire composition of a single cell remains approximately constant. In this growth phase, the specific growth rate determined from either cell number or cell mass will be the same. Small sufficient nutrient is present there; thus the growth rate is dependent on nutrient concentration. The calculation of doubling time or generation time $g = \ln 2 / \mu$, μ can be calculated from another equation in which the value of the initial and final population is put over zero to the final time. In the exponential growth phase, the nutrient depletion and generation of toxic metabolites lead to the stationary phase and further the resistance capacity toward the accumulated toxic compound and ability of microorganisms to survive (not to multiply) due to endogenous reserve as well as the death of some cells due to the immediate susceptibility of such factors, and the ability of nutrient-independent cells may multiply there, and therefore maintain the same cell number during the stationary phase for a longer period and thereafter growth decreases to the death phase.

The nature of primary and secondary metabolites may simultaneously produce late in the exponential growth phase. Primary metabolites are growth-related products and secondary metabolites are nongrowth related. The correct relationship of the cell number and metabolite production may further be explained with some of the facts: the total cell mass concentration may stay constant, but the number of viable cells may decrease. It may be possible that cell lyses occur and cells get nutrients from the dead cell debris (i.e., cryptic growth), and cells may not be growing but may be active for production of secondary metabolites and furthermore such live cells may be more active for production of more metabolites. The evaluation should follow of such metabolites and their role in the physiology of other cells in the basic unit of any live system with proper outer structure in the form of cell wall membrane and internal structure of macromolecules and organelles that are present for a specific function under proto- or cytoplasm in which enzyme functions and cellular metabolism with the uptake of carbon, nitrogen, and mineral sources depending on the requirement. Moreover, there is restriction of the membrane capability depending on the structure variability for specific organisms as well as different mechanisms of solute uptake as well as extracellular release of some of the compound produced in excess via various metabolic pathways. These types of metabolites may be primary or secondary metabolites of the cells. The metabolic regulations are the main phenomena for cell functions that maintain flow and control of information to regulate the metabolite prodcution at certain physicochemical and nutritional levels.

The central dogma is the main concept of the cell: DNA stress information that is replicated directly to form a second identical molecule. Further segments of information on the molecule can be transcribed to yield RNAs. Therefore, there is DNA replication then transcription to translation for production of proteins, and some proteins bind to DNA to regulate transcription. This approach regulates the production of various metabolic products and extracellular secretion. In a bioprocess, the microbial segment is very important along with cultivation, that is, a bioreactor system where optimum physicochemical and nutritional conditions are provided to achieve higher metabolite production within a minimum time period with suitable and efficient downstream processing. Microorganisms are a very diverse group from bacteria, yeast, and filamentous fungi as well as algae. All groups of microorganisms may be selected for use depending on their water activity (a_w), which shows whether organisms can be grown in the submerged, solid, or semi-solid state fermentation. Submerged fermentation is the economical way for growth and production as well as ease of downstream processing, but some members, mainly filamentous fungi and some bacteria of xerophytic level, may require less a_w and require solid-state fermentation rather than submerged fermentation, therefore the fermentation system categories can be further classified into three main phenomena: batch, fed batch, and continuous fermentation kinetics.

7.10 Microbial Constraints in Fermentation

Microorganism quality for an industrial application comprises several characteristics. They should have a short generation time, meaning fast multiplying and producing metabolites at a faster rate. The metabolite should be produced from 20 to 48 h in general, but some metabolites are produced at the late exponential phase; then growth is slow. For example, some filamentous fungi produce a higher rate of metabolites after just 3 days, and some fastidious fungi can attain higher biomass and metabolite production at 2 days of incubation. Fungi in general prefer solid-state fermentation rather than submerged fermentation but some filamentous fungi are fast multiplying and their growth appears in the nutrient medium within 24–48 h of incubation, and they also grow well in submerged systems only through surface fermentation; that is, fungi grow on the upper surface and a thick mat is formed, metabolites are released into the medium, therefore surface fermentation has been used with fastidious fungi such as *Aspergillus*, *Penicillium*, *Mucor*, *Rhizopus*, and so on. Of these, *Aspergillus* and *Penicillium* are most potent for surface fermentation.

Another quality of microorganisms is that they utilize a variety of high substrate concentration as well as crude substrate for economical production of products. The organisms must utilize that substrate in the presence of other types of carbon source presence. The catabolic repression should not affect the growth of microorganisms at higher concentrations of the product, but some products' concentration inhibits growth, therefore microorganisms should tolerate at high concentration of their metabolites, for example, ethanol, antibodies, alkaloids, and the like.

Microbial products can be categorized in three groups.

1. Growth-associated products: ethanol
2. Nongrowth-associated products: antibiotics
3. Mixed-growth–associated products

Lactic acid products, xanthan gum, and some secondary metabolites from cell culture, for example, ethanol fermentation yeast are grown in glucose 100 g/l. The calculation of maximum net specific growth rate is by the equation

$$\mu_{\text{net}} = \ln 2 - \ln x_1 / t_2 - t_1,$$

and apparent growth yield can be calculated by $y = " X / " S$. Furthermore, the cell concentration after 200 g of glucose X $\max = X_0 + YS_0$ (Table 7.1).

Table 7.1 Growth kinetics of microorganism

Time (h)	Cell Conc. (g/l)	Glucose Conc. (g/l)
0	0.8	100
5	2.2	95
10	8.4	80
15	18	25
20	45	6.0
25	49	0.4

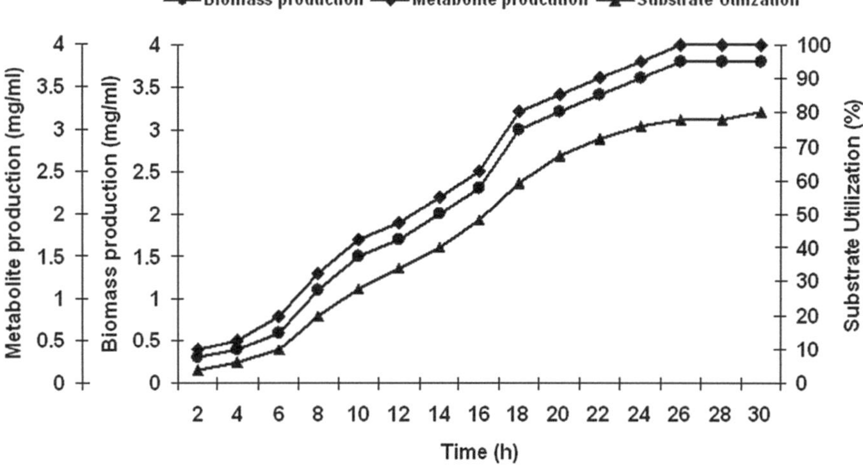

Fig. 7.1 Fermentation kinetics of continuous growth system

$$\mu_{net} = \ln X_2 - \ln X_1 / t_2 - t_1 = \ln 45.0 - \ln 8.4 / 20 - 10 = 4.5 - 0.8 / 10 = 3.6 / 10 = 0.36\text{h}^{-1}$$

$$Y = \text{"} X / \text{"} S = 49 - 0.8 / 0.4 - 100 = 48.2 / 99.6 = 48.2 / 99.6 = 0.4\text{g cell / g}$$

$$X_{max} = X_0 + YS_0 = 0.8 + 0.4(200 \text{ g glucose})$$

$$= 0.8 + 80 = 80.8 \text{ g cell / l}$$

The growth kinetic parameters in terms of the substrate utilization rate and metabolite and biomass production rate in the specific time period data can give an idea of the selection of a fermentation process. The data of substrate utilization and biomass and metabolite production are plotted on the *Y*-axis with time on the *X*-axis (Fig. 7.1). Three conditions will ideally show, if substrate utilization and biomass and product form in the same phase with time variations, the continuous process will be adopted, and the product formed by the organism from an entirely different phase of growth and substrate utilization, clearly indicating that the microorganism is producing secondary metabolites which are generally produced in the late exponential phase, therefore, only the batch fermentation process can be facilitated for metabolite production. Moreover, this process can be adopted in continuous production after manipulation in the retention period through certain modification/s in bioreactor.

It is well documented that continuous fermentation is more economical and gives higher production over the batch process. The microorganism showing the above trend of growth and metabolite production purely justifies the continuous system without any modification. Batch fermentation requires culture production for every batch, which involves huge energy for autoclaving and cooling of the medium; therefore large molecular production is a very combustive process, and will increase the cost of production as well as purity of the product, as the maintenance of each batch fermentation may deviate the parameters for optimum fermentation. However, every process has its own limitations because some microorganisms do not follow such fermentation limits, therefore, the only option is batch fermentation. Continuous fermentation may be adopted with the modifications in the fermentation system with few of the fermentation levels which are just intermediate to the kinetics of the continuous process; if the parameters are levitating especially for late metabolite production from the growth phase, then batch process is the only option rather than fed batch or continuous process.

The process selection and optimization of conditions from laboratory to pilot or industrial scale are known as the scaling-up of a process. Fermentation kinetics and parameters for any fermentation process require several parameters; almost all parameters can easily be maintained such as nutrients, pH, temperature, and aeration, out of which two parameters are critical and must be evaluated for the architecture and design of a fermentation for a set of microbial processes either for submerged or solid state. In solid state, a major change in the design of the fermentor/bioreactor is required. The solid-state system is differentiated from the solid substrate with a slight difference. The solid-state system is generally used when microorganisms produce higher metabolites at lower water activity, as the submerged condition is unfavorable; in a solid-state system, microorganisms are grown on the porous substrate and microorganisms should not utilize nutrients from the substrate. The substrate is meant for a support material to absorb different levels of moisture depending on the requirement of the microorganisms. Solid-state fermentation is useful for fungi, actinomycetes, and some bacteria of xerophytic nature.

There is another phenomenon, solid-substrate fermentation for microbial metabolite production. The modifications in the process as well as use of a multivessel system or a single bioreactor with segmentation are suggested. There are several factors, but temperature is one of the important factors affecting the performance of cells and depends upon the nature of microorganisms from psychrophillic, mesophillic, to thermophillic. For industrial application, the tolerance of temperature in the range of 2–5 °C is essential as during large-scale production of metabolites, the temperature of the fermentor increases from 2 to 8 °C depending upon the process. The fast multiplying microorganisms such as ethanol-producing yeast, where the fermentation process is completed within 24 h, the temperature may go up to 8 °C during summers, and such increase is detrimental for growth of the yeast having optima 30 °C. Therefore, temperature-tolerant *Saccharomyces cerevisiae* is the requirement, and the cooling device is one of the components, thus design and architecture are constructed in such a way to overcome such problems otherwise production will go down due to thermal death. At higher temperatures, the thermal death rate exceeds the growth rate, which causes a net decrease in the concentration of viable cells. It is established that the activation energy for growth is typically 10–20 kcal/mol, and for thermal death 60–80 kcal/mol. Temperature also affects product formation, therefore, growth as well as product formation optima may certainly vary. The yield coefficient is also affected by temperature. In some cases, the cell has the capability to maintain cell reactions by modifying cell lipids and proteins immediately, which is the requirement for the selection of a microbial strain for industrial use. Temperature also affects the rate limiting step in a fermentation process. At higher temperature, the rate of the bioreactor might become higher than the diffusion, which is a limiting factor. The activation energy of molecular diffusion is approximately 6 kcal/mol. The activation energy for most bioreactors is more than 10 kcal/mol, so diffusion limitations must be considered at high temperature.

7.10.1 Mix Culture/Consortia of Several Groups of Microorganisms with Diversified Physiological and Nutritional Categories

Mix cultures are being used for large-scale treatment/production of metabolites of different natures. The dynamics of mix culture are an important consideration in some commercial fermentations. The use of mutants and organisms with recombinant DNA has created interest regarding the interaction of two populations with each other. Many food fermentations, such as cheese, require multiple interacting microbial systems. The biological treatment of wastewater is through different groups of microorganisms. The ratio of various species in the treatment process is critical, therefore microbial shifting may damage the fermentation, however, a specialized design of a bioreactor is essential. This is true not when several populations existing together may interact either without affecting negative and positive impacts on a particular group of the population through competition, antagonism, ammensalism, commensalism, protocooperation, mutualism, parasitism predation, or symbiosis . Such competition can be controlled by a selective group of microorganisms. If the population is competing with a similar substrate by producing different metabolites then it can be managed easily. The metabolites of both microorganisms may be further utilized by another group of microorganisms as the sole source of carbon and energy that may be downstreamed simultaneously. The population dynamics and rate of metabolite produced by one group will be utilized by another group of microorganisms for a new product either in the form of gas or volatile substances or oil or alcoholic substances or biomass production and hydrolyzing to $CO_2 + H_2O$ as the final product. Several bioreactor designs have been developed for such processes, which are being used for the treatment of various industrial effluents and production of fuel in the form of methane or hydrogen.

7.11 Physical and Mathematical Concepts for Aeration and Temperature Control Fluid Dynamics for Development of Bioreactors

Selection of a bioreactor and fermentation process purely depends upon the fermentation kinetic parameters and modification in both can give higher productivity and continuous production over the batch process. Furthermore, simple continuous fermentation may be improved by a specialized continuous system that is immobilized cells or an enzyme system in multistage vessels known as a heterogeneous model. The types of bioreactor mainly based on the temperature control and air requirement of the microorganism are generally based on fluid viscosity, dissolved solid, and suspended solid; therefore air dispersion rate and movement can be controlled by intermediate sieve plates. The distance of such a plate's number and size of the perforation can be optimized. Temperature control through higher circulation may also be facilitated in aero- and hydrodynamic fermenters in homogeneous and heterogeneous systems.

The substrate status in a continuous system may be supplied in several ways. The substrate can be continuously mixed in the vessel through stirring and product released, maintaining chemostasis in some cases where the nutrient is continuously added to the bioreactor and simultaneously or intermittently added along with product using multivessels or the biomass is recycled to maintain turbidostatic conditions. In some cases, the design of the bioreactor may be in tubular form maintaining volume along with the retention period of the biomass to complete the fermentation, and release of product without backflow generally operated in a plug flow reactor. In this, the rate of product fermentation is dependent on the flow rate, but directly affected by the nature of metabolite production from primary to secondary metabolite. The idiophase (product formation phase) is separated from the trophophase (growth phase); then batch fermentation is the only option, however, this kinetics can be overcome by the changes in cultural parameters and bioreactor design maintaining the retention period into a multivessel system.

The transfer of energy, substrate, and metabolite within the bioreactor requires a suitable device as some microbial cells are damaged by stirring devices, therefore the only option is either through aerodynamic to hydrodynamic levels with suitable designs to minimize shearing stress. Fungi and algal cells are restricted to such a process, whereas bacteria and yeast require stirring without effecting mechanical damage. Stirring in mathematical notation is expressed in Reynolds numbers which is affected by the stirrer diameter, speed, dynamic viscosity, and density. Stirring ensures the transport of nutrients, mixing of gases and substrate, as well as decreasing the temperature of the bioreactor. The relative velocity between the nutrient solution and individual cell should be about 0.5 m/s. The fluid mechanics clearly indicates the dimensionless Reynold's number

$$\mathrm{Re} = D_1^{\,2}.N.\rho / n$$

D_1 = stirrer diameter in cm
N = stirrer speed sec^{-1}
ρ = density
η = dynamic viscosity r/cm.sec

The Reynold's number describes only the flow at the periphery of the stirrer to distribute the turbulence homogeneously within the whole reactor, and an impeller of appropriate shape and diameter is used.

Power number $N\rho$ = Newton's number = Ne has been defined as dimensionless parameters
Power number ($N\rho$) = Imposed force/inertial force

Therefore, several types of stirrer can be used depending on the size and nature of the microorganism. Viscosity of a normal solution containing a carbon and nitrogen sources alongwith the concentration of metabolite/s but the carbon and nitrogen sources are utilized by the microorganism during the course of fermentation, therefore, viscosity of the metabolite is the main component taken an account for the calculation of Newton's law of friction

$$T = nr$$

Newtonian fluid $n = T / r =$ Constant

T = Shear stress (kp/m^2)
R = Shear rate (sec^{-1})

Three stages of velocity have been identified: pseudo plastic, Bingham, and Newton, similarly.

The solution behavior and rate of mixing solely depend on both the flow and direction of air current in the bioreactor. Moreover, both depend on the type of bioreactor from aerodynamic to hydrodynamic along with an agitator facilitated by impellers of various designs. The impeller has a critical job in mixing throughout fermentation because it maintains the most favorable substrate and biomass concentration in the bioreactor during the entire practice. In addition, it keeps the solids suspended, disperses oxygen to maintain the maximum whole bubble surface area, and captures air bubbles to avoid air escaping prior to all the oxygen being dissolved (Freitas et al. 2000). Here, two types of agitator are used, the top entry stirrer and the bottom entry stirrer. The top entry stirrer is usually used as the process is easier to handle, and is consistent and forceful, whereas the bottom entry form is not often used. The benchscale fermenters are frequently prepared of borosilicate glass by means of a stainless steel top. Within the laboratory, a top entry stirrer can be used, which consists of a motor attached just before the shaft and simultaneously through impellers (Bloch and Soares 2007).

7.11.1 Power Requirements in Airlift Bioreactor

During aeration, if the gas liquid mixer is reduced, the power requirement decreases, therefore the aerated and nonaerated bioreactor are expressed in mathematical terms for Newtonian and non-Newtonian fermentation systems.

Power requirement of aerobic bioreactor $(Pg) = K(P_1^2.N.D_1)0.45 / Q^{0.56}$
K = constant (function of bioreactor geometry)
P_1 = power requirement of the nonaerated bioreactor
N = stirring rate (rpm)
D_1 = stirrer diameter (cm)
Q = aeration rate (vvm)

This equation is applicable for both Newtonian and non-Newtonian fermentation systems of capacity 20 l to more than 30,000 l.

The gas exchange and mass transfer is a very important component and critical in operation in an industrial-scale bioreactor, where adequate gas exchange is very essential. O_2 is an important substrate for microbial metabolism and release of CO_2 by most aerobic microorganisms. This metabolic product must be released otherwise the O_2 limiting factor will be more affected by the substrate and decrease production of the desired metabolites. CO_2 is less soluble in the fermentation medium due to the presence of various substrates, therefore only 0.3 mm of O_2 (equivalent to a ppm) dissolves in 1 l of water at 20 °C in an air bottle mixer in pure condition. The solubility of gases follows Hennery's law in the gas pressure range over which bioreactor is operated.

7.12 Basic Concepts of Immobilization of Microbial Cells/Enzymes

Immobilization of microbial cells and enzymes has the same mechanisms and process parameters. The carrier for immobilization is based on the mechanism of microorganism attachment and product specificity along with the factors affecting such a process. The cell immobilization can be seen in the submerged system with several advantages. It provides a high cell concentration with retention of viable cells with recycling approach something cost effective metabolite production. The chance of contamination is less as healthy cells are always present in the primary bioreactor, the substrate is not spared of contaminants and the wash out rate of viable cells from the fermentor is avoided to achieve higher product rate and yield. The shear stress is overcome by this process.

Every process has the same limitation. The product of interest should be excreted by the cell. The diffusion limit, growth, and gas distribution at the site of fixation is critical and may damage the support material and micro colony formation. In some cases enzyme immobilization is more feasible than cell immobilization. The primary advantage of immobilized cells over immobilized enzymes is that the immobilized cell can perform multistep cofactors requiring biosynthetic reactions that are not easily possible using purified enzyme preparations.

7.12.1 Immobilization Supports and Stabilization of Microbial Cells

Effective immobilization is the entrapment for binding cells by physical and chemical forces. The important methods are entrapment and binding. The entrapment/encapsulation within porous material/membranes can be used for the immobilization of cells. Several carrier materials have been used for such purposes. The porous polymers include agar, alginate, k-carrageenan, polyacrylamide, chitosum, gelatin/collagen, the porous metal screens, polyurethane foam of different porosity, silica gel polystyrene, L-alanine, and so on. The polymer beads should be porous; the size of the bead and porosity should also be optimized according to the nature of the microorganisms and metabolites. There are several methods for the immobilization of microbial cells for specific microorganisms and processes. Cells are entrapped/encapsulated or adsorbed/absorbed in porous material enough to allow the transport of substrate and product in and out.

Several methods have been used for immobilization of microbial cells/enzymes having certain limitations and benefits. Gelatin agar beads are used for the preparation of beads by mixing the liquid form of these polymers with microbial cell suspension and temperature gradients. Further, higher porous beads entraps more cells than less porous beads after solidification of the polymers. Hard porous plastic materials have also been used inasmuch as agar and gelatin beads are fragile and restricted to low-temperature fermentation conditions. A large bioreactor sometimes does not maintain such condition to sustain these materials. The porous beads sometimes absorb the product, therefore fluid chemistry of the medium affects such a process and optimized accordingly by manipulative conditions that is use of multivessel system with different concentration solution or immobilization to without immobilization process, moreover, thermodynamics and fluid nature of the compound must be studied. Therefore, bioreactor design is again necessary to solve such problems when the option is restricted to a certain process; several other processes such as preparation of polymers using different solvent systems, especially water miscible solvent and solutes, are used but this approach is also not effective due to direct contact. Another approach is ion exchange gelatin which takes place when a water soluble electrolyte is mixed with a salt solution and a solid gel is formed; calcium alginate and k-carrageenan are used equivalently. Ca alginate solution is mixed with $CaCl_2$ solution; similarly other ion exchange gelatin compounds are calcium alginate (CA)-carboxymethyl cellulose (CMC), Mg pectinate, k-carrageenan; and elution a polyphosphate. These ionic gels can be further sterilized by covalent cross-linking. The other mechanism that is also good for some microbial immobilization is polycondensation. The epoxy resins are prepared by polycondensation and can be used for cell immobilization but it does not have high chemical and mechanical stability. The main functional groups are –OH, $–NH_2$, epoxy, and isocyanate groups. The epoxy glutaraldehyde is very effective for this, but in a bioreactor where fluctuating conditions of high temperature, low or high pH, and toxic functional group may adversely affect the activity of cells.

Polymerization of compounds through cross-linking copolymers of a vinyl group is also used. The polyacrylamide gel beads are most commonly used. Several other monomers such as acrylamide, methacrylamide, and 2-hydroxyethyl meta-acrylate can be used for polymer formation. The polymer solution is mixed with the cell suspension and polymerization takes place to form a polymeric block that is pressed through a sieve plate to obtain regular-shaped particles for entrapment of the cells. Encapsulation is another method of cell entrapment. The hollow capsules of different dimension are in a spherical shape wrapped with semipermeable membrane for the transport of nutrients/products in/out through selective barriers. Microcapsules have certain advantages over gel beads inasmuch as more cells can be packed per unit volume of support materials and their size may increase to several times higher than the gel beads. Various polymers can be used as the capsule membrane: such as nylon, polystyrene, acrylate, polylysine, alginate hydro gel, cellulose acetate, ethyl cellulose, and polyester membrane.

Another form of entrapment is the use of macroscopic membrane-based reactors. The simple model is the hollow fiber reactor. The design has a device to maintain mass transfer through membranes of various porosity in different layers just like heat exchanger units for cooling of a bioreactor in which a tube and outer covering have a semipermeable membrane. The different design may be suggested in order to develop the efficiency of such bioreactors. The large tubing is restricted for cell suspension along with the nutrient supply through a peristaltic pump. The nutrient uptake by the cell is through diffusion from the flowing nutrient over the surface of the membrane.

The metabolic products of the cells will be back into the flowing nutrient streams but again to maintain diffusional limit is difficult in this system with the maintenance of living cells, therefore, modification in the use of several types of membrane according to the concentration gradient of solute as well as the metabolic products, which are to be separated during downstream processing. Several other problems related to removal of gases and their concentration during fermentation at the peak or moderate stage may be managed only with transfer of the whole material in another vessel/s. Several commercial reactors for animal cell cultivation use membrane entrapment. Therefore, entrapment or encapsulation of cells shows different trends from support through adsorption or covalent binding.

High cell loading can be obtained using microporous support materials. The support material may cause intraparticle pore diffusion, and nutrient availability may be a limiting factor, therefore size and number are responsible for bioconversion potentials. Adsorption capacity sometimes is not effective in the tubular or bubble column reactor. The adsorption on porous silica gel is always less than on wood shavings; there are several other materials including porous glass beads that may also have adsorption capacity in the range of 109–1010 cells/ml. The binding forces between the cell and support surface may vary, depending on the surface properties of the support material and the type of cells. Electrostatic forces may also work depending on the charge of the support materials. Covalent and hydrogen bonding may occur.

There is passive immobilization through biological film; the multilayer growth of cells on a solid support surface is known as biological film. This support material may be biologically active or inert; bio film is formed in any place where nutrients are present among the various groups of microorganisms, mainly at the site of biological wastewater treatment. The bio film contains both live and dead cells and live microbial attachment to dead cells. The distribution of microorganisms from inner, middle, and outer surfaces may vary with different physiology and production of specific metabolites, therefore several nutritional categories may work depending on the process and use of microbial system for specific fermentation in mixed cultures.

The thickness of bio film for the utilization of the same type of material may be discussed in the light of aerobic, facultative anaerobic to anaerobic, and mesophilic to thermophilic. This system is generated in trickling filters for transformation of various organic matter to their various metabolic products. Several symbiotic and other associations may help in the efficiency of microbial systems. *Lactobacillus* sp. is used for lactic acid production from glucose using gelatin as adsorption material. *Clostridium acetobutylicum* is used for acetone and butanol production from glucose using ion exchange resin as support material through the adsorption process. An exhaustive list regarding the microorganisms and nature of support materials for various substrates and products is shown in Table 7.2.

The bio film cultures efficiently work in the stagnant surface fermentation or in the immobilization in suspension culture rather than solid state to other types of continuous submerged process. The bio film thickness and diffusion of the products in the medium continuously diluting the product protect from end-product inhibition, therefore after a certain concentration, the medium flow is fixed accordingly with the adhesion of nutrients. The thickness of a bio film is an important factor affecting the performance of the biotic phase. Their bio film has a low rate of conversion due to less biomass concentration. The thick film sometimes may affect the aerobic condition, therefore facultative microbes work efficiently in this system. The film may be uniform or breaking in several segments may also give a better result as the dwell culture of some metabolite products by different strains (mesophilic to thermophilic); aerobic to facultative aerobic may selectively be used for higher yield in the batch system. Film optimization is necessary for a better result but in a practical approach it is difficult, therefore the selective cultures of different tolerance to temperature, pH, or O_2 requirements are the only right and economical approach. The mathematical calculations in bioconversion and different limits between the surface support and fluid, and the rate of limiting factors in the immobilized cell have been established, but the flat bio film surface culture and submerged cells have different values. Therefore during scaling-up of a bioreactor process for bio film having flocculated cells there are different mathematical calculations. Without applying mathematical calculations, microbiologists have optimized several bioreactors at the commercial level for various metabolite production based on direct observation of the physiological consequences and growth rate of the microorganism through multivessel continuous

Table 7.2 Cell immobilization of microbial cells by entrapment and surface attachment for various commercial productions of metabolites

Microorganism	Support Matrix	Process	Substrate and Product Conversion
Saccharomyces cerevisiae (yeast)	k- carrageenan	Entrapment	Glucose to ethanol
E.aerogens (bacteria)	k-carrageenan		Glucose to 2,3-butane di-ol
E.coli	k-carrageenan		Fumeric acid to aspartic acid
Zymomonasmobilis (bacteria)	Ca-alginate		Glucose to gluconic acid
Trichodermaressei (fungus)	k-carrageenan		Cellulose to cellulase production
Candida tropicalis (yeast)	Ca-alginate		Phenol degradation
Acetobacter (bacteria)	Ca-alginate		Glucose to gluconic acid
Nocardia (actinomycetes)	Polyurethane		Production of testosterone
E.coli	Polyurethane		Penicillin G to G.penicillic acid
Rhodotorulaminuta (yeast)	Polyurethane		Methyl succinate to methanol
Aureobasidiumpullulans (polymorphic fungus)	Polyurethane	Entrapment	Sucrose/glucose to pullulan
Lactobacillus spp. (bacteria)	Gelatin (adsorption)	Adsorption	Glucose to lactic acid
Clostridium acetobutylicum (bacteria)	Ion-exchange resin	Ion-exchanger	Glucose to acetone, butanol
Streptomyces	Sephadex	Adsorption	Sucrose/glucose to streptomycin
E.coli	Ti (I^{v}) oxide	Covalent bonding	Hormones
Bacillus substillis	Agrose-carbodi imides	Covalent bonding	Hormones
Animal cells	DEAE-sephadex	Adsorption	Hormones

system. Still mathematical concepts provide an idea in advance regarding the tentative outline for a group of microorganisms but microbial systems are very diverse having different biotic and abiotic adaptability and require practical orientation and correction for a set of processes especially for retention of microbial cells.

7.12.2 Bioreactor Models for Immobilized Cell System

Various reactor configurations can be used for immobilized cell systems as various microbial systems require different support materials with different mechanisms. The support materials are not very stable in hydrodynamic shear. The immobilized cells can be run in packed column, fluidized bed, or airlift bioreactors. The packed column with continuous release of cells and nutrients contacted in the main bioreactor and then operated through a multivessel system may give better results either through a batch or continuous system.

Several designs for immobilized microbial cell bioreactors have been used for production of various microbial metabolites as well as transformation of compounds. Each bioreactor has a specialized design for entrapment/attachment of cells for continuous production either in the initial bioreactor or in a multistage system. Singh et al. (2011) have developed a novel fermentor system having *Aureobasidium pullulans*, a polymorphic fungus, immobilized through polyurethane foam which was designed to achieve continuous production of pullulan, a polysaccharide consisting of maltotriose/tetrose. *A. pullulans* cells were immobilized in 10–20 g polyurethane foam with pore size of 1000 Å. The system has a specialized aeration provision with 80 perforations of 4 mm. Pullulan production achieved was approximately 37 g/l in 18 cycles at 42 °C, pH 5.5 and at aeration rate of 0.5 vvm. The concentration of sucrose and ammonium sulphate was 3% and 0.5%, respectively. This novel design could serve as an excellent fermentor system in industries for large-scale continuous production of pullulan. They have also optimized all the parameters for a bioreactor at different aeration rates and designs of a column packed with polyurethane foam for continuous production of polysaccharide from *Aureobasidium pullulans*. The immobilization of wild-type *A. pullulans* cells was done in a self-designed multistage continuous fermentor having an immobilization unit. The polyurethane foam quality and quantity were also optimized in the fermentor. The optimal conditions for pullulan production by *A. pullulans* were 42 °C, pH 5.5 in 48 h having 3% sucrose and 0.5% ammonium sulphate for production of pullulan (37 ± 1.0 g/l), therefore similar parameters were considered for pullulan production through immobilized cells of *A. pullulans* on polyurethane foam in a self-designed fermentor model with several specifications as described later. The grade of polyurethane foam was also optimized for better attachment of the cells for a longer period. The grade is based on the pore size mainly classified as macropore, mesopore, and micropore. Macropore (1000 Å) was selected for this study.

7.12.3 Design of the Novel Fermentor and Its Operational Parameters

The fermentor was fabricated indigenously in the laboratory by the author's group. The design consisted basically of five units: A, B, C, D, and E (Fig. 7.2). Unit A is of 10 L capacity with working volume of 7 L, fabricated in a mild steel vessel. The main fermentor (unit A), has a replaceable lid on the top having three ports, an exhaust for gas (CO_2), a pressure gauge, and an extra port. In addition to this, unit A comprises an aeration unit directly connected to the bottom of unit B. Sterile air is supplied only to unit A and B as per the design in the fermentor. All the fermentation vessels have an outlet in the bottom for total decantation during washing or for any other purpose including release of dead yeast sludge from the fermentor. A

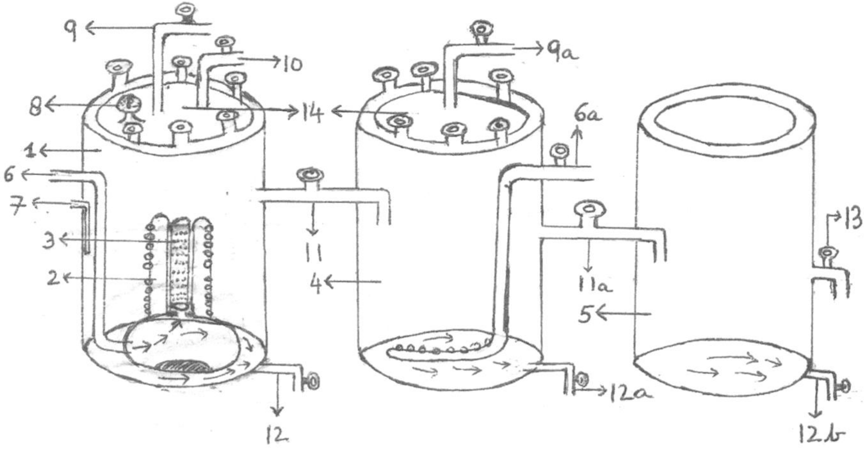

Fig. 7.2 Schematic representation of multistage continuous fermentor (*1* unit A; *2* unit B; *3* unit C; *4* unit D; *5* unit E; *6* aeration unit; *6a* aeration unit; *7* thermostatic heating device; *8* pressure gauge; *9* exhaust; *9a* exhaust; *10* extra port for medium inlet; *11* connecting pipe with regulator; *11a* connecting pipe with regulator; *12* outlet for dead yeast cells and washing; *12a* outlet for dead yeast cells and washing; *13* drain pipe for pullulan outlet; *14* replaceable lid)

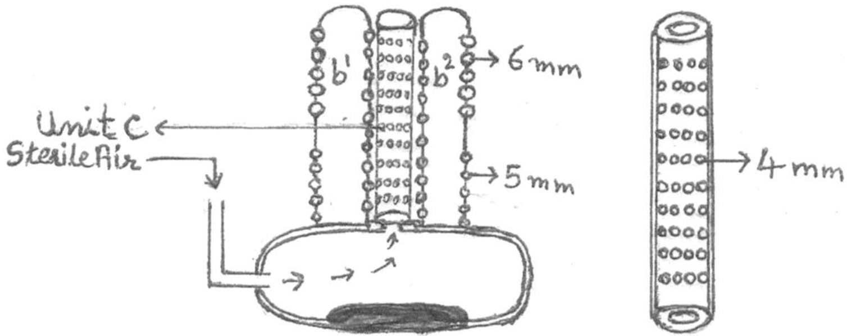

Fig. 7.3 Unit B with b1 and b2 and unit C

provision of a thermostatic heating device was made in unit A to control the desirable temperature (42 °C) for thermostable *A. pullulans* used in this investigation. Unit B is centrally fitted in the middle of the main fermentor (unit A) and has extended b1 and b2 segments with perforations on the outer and inner walls having a removable socket, immobilization unit C, which exists in the middle of b1 and b2 segments. The inner walls of unit b1 and b2 segments are in close contact with unit C. The outer walls of units b1 and b2 have five perforations of 5 mm in size on the lower and five perforations of 6 mm on the upper side (Fig. 7.3). These pores help in the circulation of air from units B, C and finally to A continuously. The upper portion of the unit B is dome shaped without perforations to reflect back the air into the lower segments of b1 and b2, helping the medium for proper mixing. The design of the flat upper portion of b1 and b2 (nondomed) was also compared with the dome-shaped structure. However, dome-shaped design was found to be better for the yeast phase of growth (data not shown). Unit C is barrel shaped, 10 in. long having a 50 mm radius, made up of mild steel, a removable unit, packed with polyurethane foam along with *A. pullulan* culture (50 × 106 CFU/g) for immobilization. Therefore, unit C is the main immobilization unit. The top of unit C is opened for regular contact with the medium and also releases the air and respired CO_2 to avoid back pressure in it. The design of unit C is especially made for polyurethane having macropores (1000 Å). Unit C has 80 perforations of 4 mm each throughout the walls. These numbers and size were optimized for pullulan production, particularly for this novel model of fermentor, especially for unit C. This design is presented in Fig. 7.3. The main fermentor unit A is interconnected with units D and E having 15-L capacity to achieve a multistage continuous fermentation system, and the flow of the medium from unit A to D and E is through gravitational force as well as water level principles to avoid back flow. Unit D is closed on the top with an air exhaust outlet having an aeration unit with sparging of air on the bottom, regulated to any capacity depending on the requirement. Both units D and E have a bottom outlet for decantation of sludge and washing materials. The number of fermented cycles was calculated on the basis of the capacity of the fermentor, that is, 7 l effluent of the fermented medium from unit D was considered as one cycle. The results of the biomass and pullulan production having an almost similar count of the yeast cells were considered as efficient process parameters for pullulan production under scale 1. The changes in production and other parameters by 5–10% variation were considered as a significant change in production under scale 2, whereas scale 3 meant poor production showing 10–15% loss in productivity. In this experiment, the immobilization efficiency was evaluated on the basis of no change in continuous production of pullulan, biomass, and yeast-like cells in the main fermentor. The design of the fermentor basic unit was totally standardized on the basis of biomass in the yeast phase essential for pullulan production.

7.12.3.1 Support Materials

The polyurethane foam having micropores (less than 20 Å pore width), mesopores (between 20 and 500 Å widths), and macropores (more than 500 Å widths) as solid support material for immobilization of *A. pullulans* cells for unit C was optimized for pullulan production. The macropore polyurethane foam is designated as polyurethane grade B (1000 Å). Different grades of polyurethane foam were purchased from Sigma Chemicals, St. Louis, MO, USA.

Fermentation kinetics of pullulan production, especially for biomass and yeast phase of growth were recorded at different time intervals in different designs of the fermentor along with the immobilization unit with polyurethane foam quality and quantity as support material for wild-type *A. pullulans* cells. The polyurethane foam having pore size of 1000 Å in 15 g with aeration rate of 0.5 vvm in the immobilization unit C having 50 × 106 cfu/g with 80 perforations of 4 mm was found to be the best in the multistage continuous fermentation process for pullulan production. Furthermore, unit B having b1 and b2 segments having five perforations of 5 mm size on the lower side and five perforations of 6 mm size on the upper side with temperature 42 °C, pH 5.5, sucrose concentration 3%, ammonium sulphate concentration 0.5% were found to be better for continuous production of pullulan (37 ± 1.0 g/l) in 18 cycles.

Solution and development of the bioreactor purely depend upon the microbial kinetics for metabolite production in the form of primary or secondary metabolites in either batch or fed batch and continuous systems. Every process can be developed with the use of selective design of a bioreactor run either in submerged or solid-state to solid-substrate level. After selection of a process and bioreactor system, scale-up, operation, and control of bioreactors are required. These parameters will be in the homogeneous bioreactor or heterogeneous bioreactor either in batch, fed batch, or continuous mode of operation. The heterogeneous reactors are immobilized cell system and solid-state fermentation. The reactor configuration and operational method are solid-state fermentation.

7.13 Bioreactors for Stem Cells and Hematopoietic and Other Cell Cultures

Mass production of animal cells for transplantation or the use of a bioreactor as artificial hybrid organs has created a new dimension in medicine. The specialized bioreactor for cultivation of stem cells is used. Some animal cells are capable of extensive replication and self-renewal. Some are highly differentiated and perform specific functions. A stem cell is an undifferentiated cell capable of continuous self-renewal that can also produce large numbers of differentiated progeny, depending on extracellular factors. The best studied system is the hematopoietic system. These are eight major types of fully mature blood cells in the human circulatory system. Hematopoietic stem cells give rise to two types of progenitor cells. These are capable of self-replication with a restricted range of cells into which they may differentiate.

There are many challenges to generate commercial-scale system hematopoiesis. Co-culture with stroma cells from the bone marrow is necessary to generate necessary growth factors, so the bioreactor must accommodate adherent cell growth. The three basic reactor types are under development. Fluidized bed reactors have macroporous support just like that in bone. However, the bioreactor technology may solve the problems. The flatbed reactors (modified T-flasks with continuous flow) can carry stroma and facilitate analysis and design because the geometry is well defined. Direct microscopic observation of cells is possible. Automated flatbed systems have been used to generate cells for human clinical trials. Membrane-based units such as hollow fiber reactors are used. The cell observation and harvesting are considered problematic. The suspension type of bioreactor has also been used. The development of a membrane system combined with other approaches and modification may be developed in the future.

A bioreactor system for such work requires more research and development, which is not only a matter of engineering, but requires biosystem physiology and biochemistry along with the evaluation of fermentation kinetics of the specific microbial system. There is far less expense with bioreactors for other stem cells/ tissues. The knowledge of microbial handling during fermentation may provide a solution to this challenge (Bronzino 2000; Naughton 1999).

7.13.1 Bioreactor for Animal Cell Cultures

Mammalian cells are large, ranging from 10 to 20 μm diameter, slow growing having doubling time from 10 to 50 h, sensitive to shearing stress, and anchorage dependent, but not all, and can be grown on glass surface treated with either plastic material polymers (collagen) or some other support medium. Some cells are not anchorage dependent and can grow in the submerged system. The product concentration is generally low (μg/ml), and some toxic substances such as lactate and ammonium are produced during growth, affecting the cell growth, therefore design of an animal cell bioreactor requires certain architectural and cultural modifications. Certain common features of these reactors are:

1. The reactor should not be stirred quickly, but with smooth aeration. Agitation should be less than 20 rpm, therefore the bubble column and airlift reactor are not recommended. Shear damage to the cell is avoided.
2. The homogeneous environment for temperature, pH, dissolved O_2, redox-potential, along with CO_2 requirement should be maintained.
3. A large support material surface-volume.
4. The removal of toxic products of metabolism such as lactic acid and ammonium.

For cultivation of mammalian cells in small-scale production in laboratory fermenters may be maintained by adopting gentle shaking either through reduced (1–3 rpm) shallow liquid suspension in trays, and incubating in CO_2 incubator at 37 °C (5–6% CO_2 containing air). An optimum condition should be evaluated for anchorage-dependent and suspension cells. On various carriers the high surface volume can be created on a microcarrier system, ceramic matrix system, hollow fiber reactors, large porous bead, for better anchorage-dependent cells. The modification in the bubble column and air lift bioreactor can be used for suspension cultures. Membrane bioreactors and microencapsulation have been developed for simultaneous cell cultivation, product concentration, and removal of toxic compounds. Such conditions can be facilitated through the retention of cells in the reactor and period at which the toxic metabolite is producing; the fresh medium may be added to the reactor continuously or semicontinuously.

The animal cell culture can be best grown commercially on a large glass rolling tubular bioreactor which will be the new design for such a process where slow rolling and CO_2 incubation can be easily maintained especially for anchorage dependent cells. In this process a large surface area with gentle rolling in packed large tubular reactors easily maintain the CO_2 concentration throughout the bioreactor, and simultaneous rolling under a temperature-controlled water bath having dimensions exactly fitted under the bathtub. Therefore, a large rolling tubular reactor combined under a water bath can be used for production of some vaccines, erythroprotein, and several other therapeutic proteins.

Several carrier materials have been used for cultivation of anchorage-dependent mammalian cells. It just works on the principle of immobilization of microbial cells/enzymes that are attached specifically on different carrier materials. Therefore, selection of carrier materials of anchorage-dependent cells should be optimized up to the level of covering a large surface area and requirement of CO_2 and stage for removal of toxic metabolites.

References

Bailey JE (1998) Mathematical modeling and analysis in biochemical engineering: past accomplishment and future opportunities. Biotechnol Prog 14:8

Blanch HW, Clark DS (1996) Biochemical engineering. Marcel Dekker, New York

Bloch HP, Soares C (2007) Mixers and agitators. In: Process plant machinery, 2nd edn. Butterworth-Heinemann, Boston, pp 617–631

Bohmer U, Suhardi SH, Bley T (2006) Decolorizing reactive textile dyes with white-rot fungi by temporary immersion cultivation. Eng Life Sci 6:417–420

Bronzino JD (2000) Editor in chief. The biomedical engineering handbook, 2nd edn. CRC press LLC, Boca Raton

Castilho LR, Polato CMS, Baruque EA et al (2000) Economic analysis of lipase production by *Penicilliumrestrictum* in solid-state and submerged fermentations. Biochem Eng J 4:239–247

Dominguez A, Rivela I, Rodríguez Couto S et al (2001) Design of a new rotating drum bioreactor for ligninolytic enzyme production by *Phanerochaetechrysosporium* grown on an inert support. Process Biochem 37:549–554

Flickinger MC, Drew SW (1999) Encyclopedia of bioprocess technology: fermentation, biocatalysis and bioseperation. Wiley, New York

Font X, Caminal G, Gabarrell X et al (2003) Black liquor detoxification by laccase of *Trametesversicolor* pellets. J Chem Technol Biotechnol 78:548–554

Freitas C, Fialov M, Zahradnik J et al (2000) Hydrodynamics of a three-phase external-loop airlift bioreactor. J Chem Eng Sci 55:4961–4972

Galhaup C, Wagner H, Hinterstoisser B et al (2002) Increased production of laccase by the wood-degrading basidiomycete *Trametespubescens*. Enzym Microb Technol 30:529–536

Hammer MJ, Hammer MJJR (2001) Water and waste water technology, 4th edn. Prentice hall, Upper Saddle River

Kargi F, Moo-young M (1985) Transport phenomena in bioprocess. In: Moo-young M (ed) Comprehensive biotechnology, vol 2. Pergamon press, Oxford, pp 5–55

Kossen NWF (1985) Problems in the design of large scale bioreactors. In: Ghose TK (ed) Biotechnology and bioprocess engineering. Unit India Press, New Delhi, pp 365–385

Lee SY, Papoutsakis ET (1999) Metabolic engineering. Marcel dekker, New York

Mercille S, Johnson M, Louthier S et al (2000) Understanding factors that limit the productivity of suspension- based perfusion cultures operated at high medium renewal rates. Biotechnol Bioeng 67:435–450

Midgett RE (1986) Solid state fermentation. In: Demain AL, Solomon N (eds) A manual of industrial microbiology and technology. American Society for Microbiology, Washington, DC, pp 66–83

Moo-Young M, Moreira AR, Tengerdy RP (1983) Principles of solid sate fermentation. In: Smith JE, Berry DR, Kristiansen B (eds) The filamentous fungi. Edward Arnold Publishers, London, pp 117–144

Naughton GK (1999) Skin: the first tissue engineered products. The advanced tissue science story. Sci Am 4:8485

Pérez-Guerra N, Torrado-Agrasar A, Lopez-Macias C et al (2003) Main characteristics and applications of solid substrate fermentation. Electron J Environ Agric Food Chem 2:343–350

Rejsman HB (1993) Problems in scale up of biotechnology production processes. Crit Rev Biotechnol 13:195–253

Rivela I, Rodríguez Couto S, Sanroman A (2000) Extracellular ligninolytic enzymes production by *Phanerochaetechrysosporium* in a new solid-state bioreactor. Biotechnol Lett 22:1443–1447

Rodriguez Couto S, Moldes D, Liebanas A et al (2003) Investigation of several bioreactor configurations for laccase production by *Trametesversicolor* operating in solid-state conditions. Biochem Eng J 15:21–26

Rodriguez Couto S, Lopez E, Sanroman MA (2006) Utilisation of grape seeds for laccase production in solid-state fermentors. J Food Eng 74:263–267

Rosales E, Rodriguez Couto, Sanroman MA (2007) Increased laccase production by *Trameteshirsuta* grown on crushed orange peelings. Enzym Microb Technol 40:1286–1290

Shuler ML (1985) On the use of chemically structured models for bioreactor. Chem Eng Commun 36:161–189

Shuler Michael L, Kargi F (2002) Bioprocess engineering basic concepts, 2nd edn. Prentice-Hall, New York

Singh R, Gaur R, Jamal F et al (2011) A novel fermentor system optimized for continuousproduction of pullulan. Afr J Biotechnol 10:9839–9846

Tavares APM, Coelho MAZ, Agapito MSM et al (2006) Optimization and modeling of laccase production by *Trametesversicolor* in a bioreactor using statistical experimental design. Appl Biochem Biotechnol 134:233–248

Viniegra-Gonzalez G, Favela-Torres E, Noe Aguilar C et al (2003) Advantages of fungal enzyme production in solid state over liquid fermentation systems. Biochem Eng J 13:157–167

Suggested Further Readings

Marques BC (2005) Aprimoramento de abordagem para modelar o balanço de agua embiorreactores de fermentaçaoem estado solido. Universidade Federal do Parana (1)

Rodriguez Couto S, Sanroman MA (2005) Application of solid-state fermentation to ligninolytic enzyme production. Biochem Eng J 22:211–219

Rodríguez Couto S, Rosales E, Sanromán MA (2006) Decolourization of synthetic dyes by *Trameteshirsuta* in expanded-bed reactors. Chemosphere 62:1558–1563

Singhania RR, Patel AK, Soccolc CR et al (2009) Recent advances in solid-state fermentation. Biochem Eng J 44:13–18

Biopesticides

8

Sudhir Mehrotra, Shiv Kumar, Mohd Zahid, and Minal Garg

Contents

Abstract

Biopesticides, an alternative to chemical pesticides, are typically derived from living organisms, microorganisms, and other natural sources. Microorganisms such as bacteria, fungi, viruses, or protozoans as an active ingredient can control varieties of pests and exhibit specificity for their target pests. Certain weeds can be controlled by some fungi, whereas other fungi can target and kill specific insects. None of microbial, biochemical pesticides and plant-incorporated protectants can be obtained from various microorganisms which confer protection against pest damage. Some of the additional benefits of biopesticides include complex and novel modes of action against their target pests and efficient

S. Mehrotra (✉) • S. Kumar • M. Zahid • M. Garg
Department of Biochemistry, Lucknow University, Lucknow 226007, UP, India
e-mail: sudhirankush@yahoo.com

R.L. Singh (ed.), *Principles and Applications of Environmental Biotechnology for a Sustainable Future*, Applied Environmental Science and Engineering for a Sustainable Future, DOI 10.1007/978-981-10-1866-4_8

resistance management to extend the product life of conventional pesticides. Biopesticides pose less risk to people and the environment as compared to synthetic pesticides and hence gain global attention as a new tool to kill insects and plant pathogens and suppress growth of weeds. In this chapter, the global scenario of biopesticides is discussed with special reference to its current demand, use, constraints, and remedies.

Keywords
Biopesticides • Active ingredients • Constraint and remedies • Genetic incorporation • Agricultural commodities

8.1 Introduction

Biopesticides are made up of living organisms which are found in nature including microbes, bacteria, plant extracts, as well as cellular products like fatty acids or pheromones. Biopesticides can be either living organism or act as natural enemies or are derived from animals (e.g., nematodes), microorganisms (e.g., *Bacillus thuringiensis* (*Bt*), *Trichoderma*, etc.), and plants (*Chrysanthemum*, *Capsicum annuum*, *Allium sativum*). They can be the products including phytochemicals, microbial products, or by-products and control pests by nontoxic mechanisms in an eco-friendly manner (Gupta and Dikshit 2010; Mazid et al. 2011). The demand and production of biopesticides are rapidly increasing worldwide. India although has a vast potential for biopesticides; however, lack of proper education to farmers in India limits its wide-scale applications. Consumers' demand for residue-free commodities warrants increased use of biopesticides by the farmers, and this requires awareness to farmers for maximizing the benefits. Since biopesticides are more eco-friendly and pose less risk to environment and people as compared to synthetic/conventional pesticides, they are gaining lot of global attentions as an important tool to control pest populations (Steinwand 2008).

Globally, the use of biopesticides has been steadily increased annually by approximately 10 % every year (Bailey and Mupondwa 2006). Continual accumulations of chemical fertilizers and pesticides in the environment not only harm the ecosystem and cause pollution but also inflict diseases at alarming levels (Arora et al. 2010). Biopesticides have gained widespread attention as a tool of next generation for sustainable agriculture products (Mazid et al. 2011). A number of advantages are associated with the use of biopesticides which include reduced risk of exposure to chemicals, reduced water pollution through fertilizer runoff, less number of applications, narrow range of targeted pests, and biodegradability and thus allow crops to provide more and better nutritional content as well as quality. Keeping check on pest populations and thereby protecting crops for sustainable annual agricultural production are potential benefits of the biopesticides (Mazid et al. 2011). Conventional or chemical pesticides, on the other hand, are normally synthetic materials with wide host specificity. They work by directly killing or inactivating the pests. One of the examples to further explain this includes the presence of chitosan which induces systemic resistance (ISR) and allows the plants to protect themselves against the damaging effects of pathogens and pests.

In the year 2014, the US Environmental Protection Agency (EPA) registered more than 430 active biopesticide ingredients and more than 1320 products. Biopesticides are known to be more complex than chemical pesticides. Commercial success of the use of biopesticides depends upon their mode of action, type of pests, extent of pest control, consistency of response, number of applications, host target, and nontarget susceptibility (Saxena and Pandey 2001).

Trichogramma is used as biocontrol agent and acts against an array of important soilborne plant pathogens which cause serious diseases of crops. Cotton wool worm is one of the examples of the pests in some parts of South India and in Punjab, which are successfully eliminated or controlled by *Trichogramma* (Bailey and Gilligan 2004).

Microbial biopesticides are one of the classes of biopesticides. Three broad modes of action are identified for them, namely, (i) biological or ecological means, (ii) physical means, or (iii) chemical or biochemical processes. Biological modes of action include predation or competition. Physical means of protection may involve creating barriers or occupying space. Chemical or biochemical approaches control the pest populations by altering biochemical compound or hormone and as a result disrupt the normal life cycle of pests. Advent of recombinant DNA technology facilitates genetic incorporation of DNA from microorganisms for consistent production of gene products, toxic to pests, into agricultural commodities. This approach is gaining recent widespread attention for successfully controlling pests. A single biopesticide may produce more than one secondary function in the targeted pest. Commercially successful biopesticides rely on utilization of the products/compounds produced by the microorganisms, rather than depending only on their infection. It is reported that toxic proteins produced from *Bt* and agrastatins from *Bacillus subtilis*, respectively, are toxic to pests. The biodegradable property of biopesticides makes them an eco-friendly alternative to chemical pesticides.

Biopesticides based on secondary metabolites may be more akin to chemical pesticides. Pest resistance is a major concern associated with synthetic pesticides which restricts its use. Single mode of action of pesticides is one of the probable reasons for the development of resistance. Biopesticides utilize multiple modes of action, greatly reduce the risk of resistant compounds, and work synergistically by establishing the physiological interaction of the specific microbes with the narrow-range hosts (Strange 2007). The total annual worldwide production of biopesticides is over 3000 tons, which is increasing at a rapid rate. India has a vast potential for biopesticides. The market share of biopesticides is only 2.5 % of the total pesticide market. Therefore, an eco-friendly alternative is the need of the hour. Biopesticides or biological pesticides based on pathogenic microorganisms specific to a target pest offer an eco-friendly and effective solution to pest problems.

A complete global look at today's scenario indicates that the farmers and particularly the industries based on them are still in an insecure position in comparison to the chemicals which rule the agriculture process (Glare et al. 2012). Cost incurred in the production of biopesticides and lack of proper awareness to farmers regarding the potential advantages of biopesticides still limit their use.

8.2 Types of Biopesticides

Based on the active ingredient, biopesticides are categorized into three types, viz., (i) microbial pesticides, (ii) plant-incorporated protectants (PIPs), and (iii) biochemical pesticides (Table 8.1).

8.2.1 Microbial Pesticides

Active ingredients of microbial pesticides are derived from microorganisms including bacteria, fungi, viruses, protozoans, or algae. Each active ingredient has been shown to be relatively specific for its target pest. For example, varieties of fungal organisms can control their target pests, but some of them are active to certain weeds, while other fungi can kill specific insects. One of the examples of the most widely known microbial pesticides is varieties of the *Bt* bacterium which can control certain insects by producing harmful toxins, in cabbage, potatoes, and other agricultural crops (Copping and Menn 2000; Glare and O'Callaghan 2000).

Certain other microbial pesticides function by outcompeting pests. Table 8.2 describes the species-specific pathogenic effects of microbial pesticides on the target pests. Pesticidal effects by microbial entomopathogens have been shown via their invasion through the integument or gut of the insect, followed by their multiplication and finally death of the insect host. Toxins produced by microbial pathogens vary greatly in terms of structure, toxicity, and specificity; however, most of them have been identified as peptides. These microbial pesticides are recognized as potential alternative to chemical insecticides which confer increased target specificity and ecological safety so that they can be used either alone or in combination with other pest management programs.

8.2.1.1 Fungal Biopesticides

Many important field and horticultural crops are very prone to infection by fungal pathogens. They develop diseases in the crop plants and thereby result in severe

Table 8.1 List of active ingredients of different biopesticides along with their representative examples

S. no.	Types	Active ingredient/substances	Representative examples
1	Microbial pesticides	Bacterium, fungus, viruses, or protozoan	Strains of Bt
2	Biochemical pesticides/herbal pesticides	Naturally occurring substances	Scented plant extracts
3	Plant-incorporated protectants (PIPs)	Genetic material/toxins	Plant-producing Bt toxin

Table 8.2 Some commonly employed pathogens as biopesticides and their applications

Pathogens	Host range	Applications
Bacteria		
Bt var. *kurstaki* (Bt)	Caterpillars (larvae of moths and butterflies)	Mainly effective for foliage-feeding caterpillars and Indian meal moth in stored grain. Undergo biodegradation, deactivated rapidly in sunlight, applied on crops in the evening or on overcast days and sometimes directly in the form of spray to lower surfaces or leaves
Bt var. *israelensis* (Bt)	Larvae of *Aedes* and *Psorophora* mosquitoes, black flies, and fungus gnats	Potentially effective against larvae only. Active only if ingested. *Culex* and *Anopheles* mosquitoes are not controlled at moderate/normal concentrations. Show biodegradability
Bt var. *tenebrionis*	Larvae of Colorado potato beetle, elm leaf beetle adults	Mainly effective against Colorado potato beetle larvae and the elm leaf beetle. Like other Bt toxins, its ingestion is required. It is subjected to breakdown when exposed to ultraviolet radiations and does not cycle extensively in the environment
Bt var. *aizawai*	Wax moth caterpillars	Mainly used only for killing moth infestations in honeybee hives
Bacillus popilliae and *Bacillus lentimorbus*	Larvae (grubs) of Japanese beetle	The main Illinois lawn grub (the annual white grub, *Cyclocephala* species) resistant to milky spore disease
Bacillus sphaericus	Larvae of *Culex*, *Psorophora*, and *Culiseta* mosquitoes, larvae of some *Aedes* species	Active only if ingested like Bt toxins, for use against *Culex*, *Psorophora*, and *Culiseta* species; also effective against *Aedes vexans*. Potentially active in stagnant or turbid water
Fungi		
Beauveria bassiana	Aphids, fungus gnats, mealy bugs, mites, thrips, whiteflies	Effective against several pests. Require high moisture for their growth; lack of storage longevity and competition with other soil microbial organisms are limitations associated with them
Lagenidium giganteum	Larvae of most pest mosquito species	Larvae of most pest mosquito species can be effectively controlled; active in the environment through dry periods. Main inability to survive high summer time temperatures is its main drawback

plant yield losses (Khandelwal et al. 2012). Potentially hazardous toxic compounds to humans can accumulate in the environment and develop resistance to pathogens as a result of intensified use of fungicides. Fungal biopesticides can be used to control insects and thus save plants from various diseases. Their mode of action is specific for a particular pesticidal fungus infecting the target pest. Fungal biopesticides are not required to be eaten up to show their pathogenic effects. They are the living organisms and often require a narrow range of host specificity. Besides this, moist soil and cool temperatures are necessary for them to proliferate. Biocontrol fungal agents like *Trichoderma* are acclaimed as effective, eco-friendly, biodegradable, and cost-effective. Therefore, of late, these pesticides are identified to act against an array of important soilborne plant pathogens which are responsible for crop destruction (Bailey and Gilligan 2004).

Trichoderma harzianum (*T. harzianum*) is an example of fungal biopesticide which is used against multiple plant pathogens including *Rhizoctonia*, *Pythium*, *Fusarium*, and other soilborne pathogens (Harman 2005). *Trichoderma* is recognized as a fungal antagonist that grows into the main tissues of a disease-causing fungus and secretes enzymes that degrade the cell walls of the other fungus and multiplies its own spores. *Trichoderma* is one of the common fungi with pesticidal effects and used worldwide for suitable management of various foliar and soilborne plant pathogens including various species of *Ceratobasidium*, *Fusarium*, *Rhizoctonia*, *Macrophomina*, *Sclerotium*, *Pythium*, and *Phytophthora* spp. (Anand and Reddy 2009). Research studies on *Trichoderma viride* observe its promising pesticidal effects on soilborne plant parasitic fungi (Khandelwal et al. 2012). *Beauveria bassiana* (*B. bassiana*) (Balsamo), *Vuillemin*, and *Metarhizium anisopliae* (Metchnikoff) are naturally occurring entomopathogenic fungi and are shown to have pathogenic effects against insect-sucking pests including *Nezara viridula* (L) (green vegetable bug) and *Creontiades* species (green and brown mirids). The pathogenic effects of fungi are demonstrated by their unique ability to attack insects by penetrating through the cuticle and, thereby, kill sucking pests. *B. bassiana* is commercially registered in the USA as Mycotrol ES® (Mycotech, Butte) and Naturalis L® (Troy Biosciences). A number of sucking pests such as whitefly, aphids, thrips, mealybugs, leafhoppers, and weevils are known to be killed by these registered fungal products. Studies further report the virulent effects of *B. bassiana* against *Lygus hesperus* Knight (Hemiptera, Miridae), a major pest of alfalfa and cotton in the USA.

8.2.1.2 Viral Biopesticides

Like bacteria and fungi, viral biopesticide is another subclass of microbial pesticides that play a significant role in antagonizing pathogens. Viruses are host specific, infect only one or a few closely related species, and thus offer minimal off-target impacts (Hewson et al. 2011). Bacteriophages, also known as bacteria eaters, are the type of viruses that infect bacterial microorganisms. Baculoviruses when ingested by the host insect infect the host cell and take over its replication machinery. Once they enter larval gut, the viral protein overcoat quickly disintegrates, and the viral DNA proceeds to infect digestive cells. New infectious multiple viral particles are liberated and are ready to infect another host (Fig. 8.1).

As a result of colonization, the host larvae are unable to digest food and die within few days of infection. Baculoviruses particularly are gaining attraction as biopesticides due to their high host specificity. Figure 8.2 enlists few examples of pests which are attacked by specific types of viruses. They have been shown to have no negative impacts on plants, mammals, birds, fish, or nontarget insects.

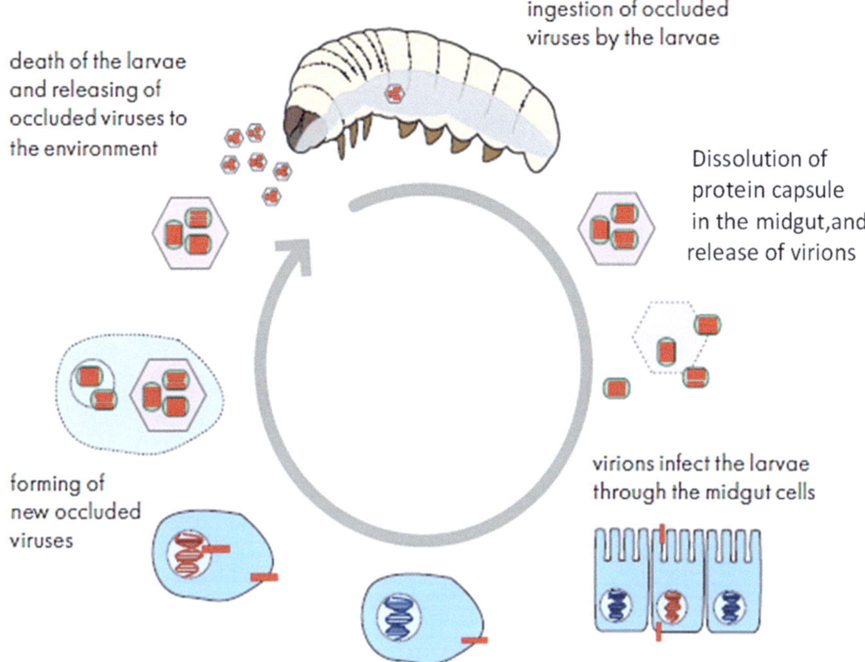

Fig. 8.1 Replication inside the host and the infection of new host cell

Sudden and severe outbreak of viral particles, employed for complete control of the disease, within the host population is one of the serious limitations with the use of baculoviruses can be.

8.2.2 Plant-Incorporated Protectants (PIPs)

Plant-incorporated protectants (PIPs) represent another class of biopesticides. These are natural substances produced by genetically modified plants where the genetic material from the microbial organism has been added to the plant. The plants can be made transgenic, and their nuclear material can be stably integrated with the genetic material obtained from the source microorganism. Biological or physical means of transformation can be employed to make plants transgenic plants. These foreign gene plant products are expressed inside the plants irrespective of outside environmental conditions. Although PIPs are very effective in destroying pests, however, acute reactions, such as toxicity, allergenicity, and skin and eye irritation, as well as long-term effects including cancer, birth defects, and reproductive and neurological system disorders are few potential side effects associated with the use of PIPs. Application of PIPs may be more useful and economical in the developing countries to enhance safe food, feed, and forage production (Koundal and Rajendran 2003).

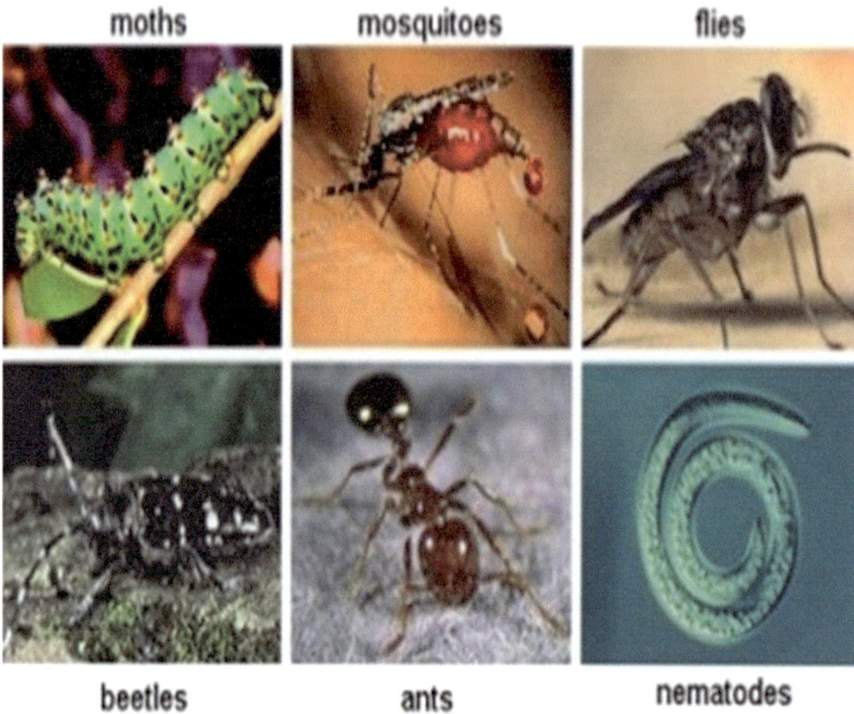

Fig. 8.2 Examples of pest which are attacked by specific viral biopesticides

One of the popular examples of PIPs includes transgenic plants with cry genes from Bt expressing Crystals of proteinaceous insecticidal δ-endotoxins (Crystal proteins or cry proteins) during sporulation. Cry toxins obtained from different varieties of Bt have specific activities against insect species of the orders Lepidoptera (moths and butterflies), Coleoptera, and Hymenoptera and nematodes. Insects when feed on transgenic plants ingest expressed Crystal toxins. Alkaline digestive tracts of insects denature the insoluble Crystals, making them soluble and thus amenable to being cut with proteases. Active form of toxin is liberated from the Crystal (Dean 1984). The cry toxin is then inserted into the insect gut cell membrane, paralyzes the digestive tract, and forms a pore. As a result, the insect stops feeding and starves to death. Figure 8.3 shows the mode of action of cry protein produced by genetically modified plant.

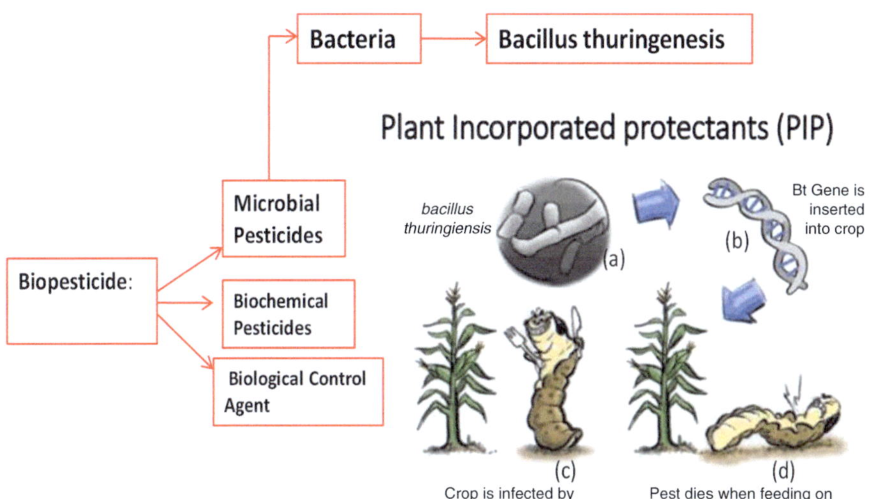

Fig. 8.3 Mode of action of cry toxin protein produced by genetically modified plant (**a**) Crystal of insecticidal toxin (cry protein) produced during sporulation of *Bacillus thuringiensis* (Bt). (**b**) Bt gene inserted into the crop. (**c**) Alkaline digestive tracts of insects ingesting the toxins; denature the insoluble Crystals, and solubilize it. This is followed by its cut with proteases found in the insect gut. As a result, active part of toxin is liberated. The toxin then binds to insect gut cell membrane, where it forms a pore and paralyzes the digestive tract. (**d**) The insect stops eating and starves to death

8.2.3 Biochemical Pesticides

Biochemical pesticides are naturally occurring substances that control pests by non-toxic mechanisms. Biochemical pesticides are different from conventional pesticides in their structure (source) and mode of action (mechanism employed to kill or control pests) (O'Brien et al. 2009). Biochemical pesticides are the noble class of biopesticides which work by interfering with normal growth or mating cycles in insects. Examples of biochemical biopesticides may include plant growth regulators or substances that repel or attract pests, such as insect sex pheromones, and interfere with their mating and population buildup, various scented extracts (that attract insect pests to traps), and some vegetable oils (Mazid et al. 2011; Singh et al. 2012).

8.3 Potential Advantages and Disadvantages of Biopesticides

Selection of potential biopesticides may require training and knowledge about pests/pathogens for effective elimination of pest populations. There are many factors which limit their use. One of them includes the cost, and hence, it is important to maintain a balance between the costs incurred and the benefits of using biopesticides over chemical pesticides. Appropriate number of application times and their moderate concentration are additional factors to consider while ensuring their efficacy. Multiple regulatory issues and economic challenges associated with newly introduced biopesticides must be addressed jointly by the social and natural scientists, the policy makers, and the industry. Lack of trust between regulators, producers, and consumers, limited resources and capabilities, and relative immaturity of the policy network are some of the serious issues to ponder.

Better understanding of the host specificity, mode of action, biodegradable properties, and the side effects to human and other animal populations associated with the biopesticide use is important to work upon before commercializing any particular type of biopesticide.

Biopesticides are usually better and safer than conventional pesticides. They often are effective in very small quantities and owing to biodegradability, they decompose quickly, thereby lowering down the accumulation levels and thus largely avoiding the pollution problems caused by conventional pesticides. Biopesticides can greatly decrease the use of conventional pesticides, while maintaining higher crop yields, when used as a component of integrated pest management (IPM) programs. Biopesticides are not only eco-friendly and renewable and can be handled safely but at the same time make it difficult for insects to develop resistance against them. Lack of persistence, rapid degradation by UV light, slow residual action, slow effect, and poor water solubility are some of the limitations associated with biopesticides. Moreover they are very expensive and generally nonsystemic in nature.

8.4 Common Biopesticides Produced and Used in India

8.4.1 Neem

Neem tree, *Azadirachta indica* (*A. indica*), has been well recognized as an important and valuable medicinal plant. Several phytochemicals are known to be extracted from this tree including "azadirachtin." Azadirachtin has been studied to affect the reproductive and digestive processes of many pests. Azadirachtin has been extensively studied as one of the most potent neem-based biopesticides. It is chemically a limonoid tetranortriterpenoid and effective against a wide spectrum of insect pests. Effective formulations of neem have been recently developed and are commercially produced in India as well as abroad. Demand for the plant extract as biopesticide has been recently increasing owing to its nontoxicity to birds and mammals and its anticarcinogenic effects. Although more than 100 firms are registered to produce neem-based pesticides in India, only very few of them are actually producing it. Furthermore, most of its production is traded for export markets. Extracts of fruits, seeds, seed kernels, twigs, stem bark, and root bark of *A. indica* have been extensively studied and shown to possess multiple biological activities. The extracts are anti-inflammatory and immunostimulating and behave as insect antifeedant. These phytochemicals are examined to confer insecticidal, fungicidal, nematicidal, and bactericidal activities (Satdive et al. 2001).

8.4.1.1 Production of Azadirachtin Biopesticides

Hairy root cultures of the plant undergoing genetic transformation with *Agrobacterium rhizogenes* (*A. rhizogenes*) have been identified as an alternative and suitable source for production of phytochemicals (bioactive molecules) with biopesticidal activity. Some of the potential strategies for enhancement of production of secondary metabolites are (i) elicitation of cell cultures with biotic elicitor and signal compounds, (ii) the presence of bacterial and fungal cell wall fragments, and (iii) altering the macro- and micronutrient composition of the media during plant tissue culture (Sivakumar et al. 2005; Savitha et al. 2006). Three different ionic strength media such as Murashige and Skoog (MS) and Gamborg B5 were compared for high biomass and azadirachtin production during hairy root cultures. Similarly, addition of fungal cell wall fragments of *Claviceps purpurea* to the hairy root cultures has been shown to significantly increase the azadirachtin content by sevenfold. Besides it, production of azadirachtin can be increased by the addition of jasmonic acid and salicylic acid into the medium of *A. indica* hairy root cultures. Higher concentrations of signaling compounds are known to result in loss of cell viability, confer toxic effects, and thereby reduce the azadirachtin productivity in hairy root cultures of *A. indica*.

8.4.1.2 Plant Materials and Tissue Culture

Disinfected fresh seeds obtained from *A. indica* (A. Juss) mature can be ground and are taken for culture in Murashige and Skoog (MS) medium supplemented with 30 gl/1 sucrose required for germination. The medium can be added with gelling agent phytagel and pH adjusted to 5.8. This is followed by growing the culture at 26 ± 1 °C and maintaining it under an intermittent period of 16 h light (white fluorescent light) and 8 h dark hours.

Another method for production of phytochemicals involves infection of 3-week-old germinated seedlings with *A. rhizogenes* by making a wound with a sterile needle and smearing the bacteria on the wounded areas. This is followed by cocultivation for 2 days. During the culture, hairy roots developed at the end of 3 weeks from the infected areas and can be excised and transferred to MS medium. This is followed by subculturing of hairy roots every 3 weeks for three to four passages on MS medium with antibiotic and in later passages without antibiotic.

DNA from Ri plasmid of *A. rhizogenes* strain LBA 9402 can be isolated and subjected to polymerase chain reaction (PCR) amplification using specific primer sequences. The amplified DNA sequences can be eluted from gel and taken for cloning and subcloning, followed by its expression. The protein products, purified using various strategies, can be commercially marketed as biopesticides.

8.4.1.3 Molecular Analysis and Characterization

Production of secondary metabolites from the hairy roots of the plants induced by *A. rhizogenes* has received a widespread attention from plant biotechnologists. Ability of increased growth of hairy roots in defined basal media without supplementation of phytohormones, their differentiated nature, genetic stability, and higher production rates of secondary metabolites are few potential advantages of hairy root cultures. Factors including differential bacterial virulence, vulnerability of the explants, and their genotype play an important role in deciding and determining the frequency of hairy root induction.

Many other important phytochemicals with biopesticidal effect include sesquiterpenes produced by *Solanum tuberosum* cell cultures (Komaraiah et al. 2003); saponin by *Panax* ginseng cell cultures (Jeong et al. 2005); and Taxol and its analogues, rosmarinic acid, indole alkaloids, anthocyanins, etc. by cell cultures of *Taxus* species, *Lithospermum erythrorhizon*, *Catharanthus roseus*, and *Vaccinium pahale*.

8.4.2 *Bacillus thuringiensis* (Bt)

Bacillus thuringiensis is classified as a gram-positive spore-forming bacterium. The special and useful property of this bacterium is the production of toxic Crystal proteins called delta endotoxins during stationary phase of bacterial growth. Delta endotoxins Crystal proteins account for 20–30 % of the dry weight of the sporulated cells and are released to the environment after lysis of the cell wall (Schnepf et al. 1998). Many potential advantages associated with Bt toxin as discussed in above sections make this as globally important biopesticide.

Strains of the Bt subspecies *kurstaki*, *galeriae*, and *dendrolimus* are the main sources for the production of Bt preparations.

8.4.2.1 Cry Proteins

There are at least 50 subgroups with more than 200 members of BT cry proteins. Larger groups of cry proteins comprise of globular molecules and constitute three structural domains which are connected by single linkers. The toxic property of long protoxins resides in the C-terminal end and is believed to be necessary for the formation of Crystal within the bacterial cell. Their mode of action involves cascade pathway of multiple events and requires several hours to kill the insect after ingestion (Höfte and Whiteley 1989). Since proteolytic activation of toxin requires only alkaline midgut conditions as in insects, hence, it is regarded as safe for humans during their accidental consumption. Binding of activated toxin to receptors located on the apical microvillus membranes of epithelial midgut cells of insects facilitates its conformation change and allows its insertion into the cell membrane.

8.4.2.2 Target Organisms

Bt *cry* toxins are classified according to the target pest they attacked over the past decades. Due to various inconsistencies in the original classification, Crickmore et al. (1998) revised the nomenclature for insecticidal Crystal proteins. This classification is based on the toxic ability of a Crystal protein in a target organism, as well as obvious sequence similarities among known cry proteins. Strains producing Bt toxin or their products and their biological activity toward target pests have been a subject of patent coverage for many years.

Many insects are known to be susceptible to the toxic activity of Bt. Some of the insects among them include lepidopterans which encompass the majority of susceptible insect species infecting agriculturally important crops. Such families belonging to order Lepidopterans include Cossidae, Gelechiidae, Lymantriidae, Noctuidae, Pieridae, Pyralidae, Thaumetopoetidae, Tortricidae, and Yponomeutidae (Iriarte et al. 2001). Cry toxin isolates produced in the transgenic plants as suitable hosts elicit toxic effects to kill the infecting insects.

Discovery of novel *Bt* strains and subspecies varieties, expressing parasporal Crystal proteins with pesticidal properties against sucking insects of agronomic importance, whiteflies, aphids, and leafhoppers, revolutionized the biopesticide industry. Potential pesticidal activities associated with this noble bacterium are not limited to insects, but also to nematodes, protozoans, flukes, collembolans, mites, and worms (Iriarte et al. 2001).

Another cry protein with pesticidal effects discovered has been identified to have nematicidal activity. Truncated form of the Cry6A protein behaves as toxin and exhibits toxic activity against the corn rootworm, a coleopteran, and also has been proved to be useful for nematode control. Cry protein toxin known as PS81F, produced by a novel isolate, has been discovered to confer killing properties to protozoans and thus can be used to treat humans and animals hosting parasitic protozoans. In addition to it, the gene encoding the toxin can be isolated from the source bacterial species and can be transferred to a suitable host plant via physical or biologically mediated transformation.

Pathogenic fungi including *Fusarium solani* quite prevalent in farmed fish and shellfish are examined to be controlled by Bt toxin proteins. This way of controlling fungal infections not only satisfies the requirements for food safety but also drastically reduces operating costs. A chitinolytic isolate of *Bt* subspecies *dendrolimus* (HD-548) has been shown to exhibit in creased toxic activity against plant pathogenic fungi as well as against certain lepidopteran pests.

8.4.2.3 *Bacillus thuringiensis*-Based Formulations

Safety and the efficiency of BT-based biopesticide formulations are the foremost requirements. In addition to it, the products must be easy to use and should have a long stable shelf life. Active ingredients in commercial formulations are spore Crystal complexes. They are not only easy to produce on large scale but also cost-effective and very efficient in terms of biopesticidal effects.

Many experimental studies have revealed the inclusion of various kinds of components or factors to enhance toxicity and attraction of insects in order to increase the palatability to insects. One of such examples is the use of formulation based on the purified and activated cry toxin isolated from a novel strain of Bt along with biodegradable, environmentally sound glycoprotein as an insect attractant for killing fire ants. These biodegradable compounds can be safely used in formulations as they are inert carriers.

Furthermore, biological and physical properties of the biopesticide formulations need to be improved in order to increase their stability during storage. The use of additives in the formulations is advocated in order to reduce evaporation and avoid formulation loss, high adherence to foliage, improved dispersion, and a long residual effect. Liquid or solid carriers, surfactants, coadjuvants, dispersants, stabilizers, moisturizers, attractants, fluidity agents, adherents, and protective agents are such examples of large variety of ingredients that can be employed to prepare formulations. One such recent and interesting novel example is the use of superabsorbent starch graft copolymer, an inert ingredient, along with Bt strain-based formulations in an agricultural environment.

Addition of an antibiotic zwittermicin has been identified to exhibit a synergistic effect on the biological activity of bioinsecticide. This combinational approach has been successful in efficient killing of pests. Addition of such compounds with synergistic effects not only can enhance the pesticidal activity of a Bt-based biopesticide but also can help in reducing the amount of Bt requirement for effective pest control. This comprises of suboptimal levels, lower than what is normally needed for commercially effective *B. thuringiensis* biopesticide, and optimal concentrations of surfactant to control pests.

B. thuringiensis-based products can be classified according to their formulations. First-generation products comprise of all Bt-based formulations which contain blend of spores and Crystals from a native strain. Majority of commercial products are first-generation types. Bt products based on spores and Crystals from a Bt strain with an artificially incorporated foreign genes that code for delta endotoxins from several Bt strains, to increase the activity spectrum against other insect pests, are classified as second-generation products. These commercial Bt-based formulations obtained from strains of various subspecies can be mixed with a number of artificial agents including chemical dryers, dispersing agents, protective agents against sunlight, diluents, lubricants, binding agents, moisturizing agents, and neutralizing agents. Addition of these agents is preferable for successful field applications of Bt-based bioinsecticides to control wide variety of agriculture and forestry pests, including disease vectors.

8.4.2.4 Chimeric Crystal Proteins

Over the past recent years, biotechnological tools are used to produce hybrid delta endotoxins with enhanced toxic activity and improved physical and biological properties. Construction of recombinant fusion genes and chimeric proteins is possible with the recent advancements in molecular methodologies. Insertional mutagenesis or site-directed mutagenesis can be employed to alter the genetic sequences encoding for proteins of interest; thus, recombinant fusion proteins can be made to express. Owing to the complex nature of cry proteins, careful selection of target genetic sequences is required for the sustainable production of engineered toxins with a high insecticidal activity.

Recent advances have been made to improve the biological properties of bioinsecticides. One of the examples toward such an effort includes the chimeric production of biopesticides against lepidopteran pests. Recombinant fusion cry protein consisting of CryIF and Cry1A(c) from Bt exhibits synergistic toxic effects. Wider spectrum of toxicity against varieties of insects, particularly mosquitoes from the genera *Culex*, *Aedes*, and *Anopheles*, is one of the potential advantages of chimeric delta endotoxins.

Substitution of aspartic acid at position 454 with proline, glycine, alanine, threonine, or serine; addition of any two amino acids among glycine, alanine, valine, leucine, isoleucine, methionine, proline, threonine, and serine; or insertion of two additional amino acids immediately after position 454 can modify Cry4Ba toxin. Modified form of Cry4Ba protein has been determined to exhibit enhanced toxicity to *Culex*, as compared to natural Cry4Ba protein. Similarly modifications in the amino acid sequence of Cry19Aa have been reported to increase its activity.

Broader insect host range against coleopteran, dipteran, and lepidopteran insects and improved insecticidal activity of novel peptides of Cry35 have been studied. Modified segments, domains, and motifs of wild-type Cry35 proteins are exchanged with the cry proteins that exhibit enhanced insecticidal properties. As a result modified Cry35 protein has been identified to confer enhanced toxin activity than wild-type Cry35 protein against plant pests including rootworms.

Extraordinarily high insecticidal activity in recombinant plants has been shown by another chimeric protein formed from Cry1Ea, derived from Bt subspecies *kenyae* 4FI, and Cty1Ca delta endotoxins, obtained from *Bt* subspecies *entomocidus* 60.5. Many other examples of new chimeric cry proteins based on genomic sequences of Cry1C, Cry1B, Cry1D, CryIA, or CryIG proteins are found to be useful in conferring the protection to plants from insect damage.

Multiple strategies can be employed to construct transgenic plants where chimeric cry proteins along with proper promoter, terminator, and marker gene sequences from microbial sources can be stably incorporated into nuclear or chloroplast DNA of plant cell. Stable incorporation would result in the expression of desired protein product by the plant cell as and when required. Transgenic crop plants thus would confer enhanced insecticidal/pathogenic effects to the insect ranges infecting them.

8.4.3 Baculoviruses

Baculoviruses constitute the major group of viruses infecting arthropods and are also used as an important expression vector with multiple applications in molecular biology (Shim et al. 2013; Contreras et al. 2014). These viruses are the most extensively studied insect pathogenic viruses. Since the twentieth century, they have been in use for biological control of forestry and agriculture pests. These are target-specific viruses, particularly effective against the lepidopterous pests of *Trichoderma* and against pathogens causing soilborne diseases such as root rot. Baculoviruses provide protection to those crops which are susceptible to these diseases. Such few examples of crop plants include dry land crops like groundnut, black gram, green gram, and chickpea.

8.5 Current Knowledge on Biopesticide Market

Major share of crops and forests infected by wide variety of insect pests are known to be controlled by Bt toxins. Many important biological, physical, and insecticidal properties of Bt toxin as discussed above make it a strong candidate in bioinsecticide market since 1950s (Bailey et al. 2010; Sarwar 2015a). There are some commercially available biopesticides in global biopesticide market listed in Table 8.3. Environmental pollution due to accumulation of toxic residues as a result of application of chemical pesticides is the major driving force for increased demands of biopesticides. Biopesticides are naturally occurring substances that control pests by nontoxic mechanisms. Usage of biopesticides has been growing at an impressive pace of 20 % worldwide. They are considered as a viable and cleaner alternative for chemical pesticides in pesticide market (Sarwar 2015b). Lack of proper education to farmers and their expensive nature, however, limits its share only to 2 % of the global pesticide market worldwide, but it is expected to increase tremendously in the next 5 years as a result of recent advancements toward its production.

Owing to various regulatory and ethical issues associated with chemical pesticides as per guidelines by Environmental Protection Agency (EPA), biopesticides are gaining lot of attention as a biocontrol agent. In view of lack of awareness, there is still disillusionment over the use of biopesticides. General public and particularly agriculturists and farmers need proper education about the dangers posed by handling and the use of chemical pesticides. In addition to it, research on the development of biological pest control methods should be promoted. Government should intervene and make policy decisions for the reduced use of chemical pesticides and increased application of biopesticides as a green alternative.

8.6 Recent Advancements and Future Prospects

Over the past recent decades, biopesticides have been successfully adopted as a clean alternative to chemical pesticides. Although biopesticides and biofertilizers are known as potential substitutes of chemical pesticides for years, nevertheless, their demands have increased for sustainable crop productivity worldwide. In view of people's choice for organic farming, safe products, and healthy food, biopesticides have become the successful replacements of synthetic pesticides.

The use of biopesticides has successfully overcome the hazardous effects of chemical pesticides when incorporated into integrated crop management (ICM) practices. Pest management in an environmental friendly manner is no longer a dream now. Advent of recombinant DNA technology, various molecular and biotechnological tools, facilitates stable and safe consistent production of biopesticides in crop plants irrespective of outside environmental conditions (Kumar et al. 2010). Many research studies are undergoing on continual search for new biomolecules and in improving the pesticidal activity of known biopesticides under suboptimal concentrations in a host-specific manner. Molecular biology tools have played an important role in facilitating such kind of research in pest management programs (Kumar 2013).

Table 8.3 Commercially available biopesticides in the global market

Category of biopesticides	Product's common name or trade name	Targets
Bactericides		
B. thuringiensis subspecies *kurstaki* EG7826	Lepinox WDG	Lepidopteran larvae
Bacillus subtilis QST 713	Serenade	Fire blight, *Botrytis* species
Agrobacterium radiobacter k84	Galltrol – A	Crown gall disease
Pseudomonas agglomerans C9-1	BlightBan C9-1	Fire blight
Pseudomonas agglomerans E325	Bloomtime	Fire blight
Bacteriophage of *Pseudomonas syringae* pv. tomato	AgriPhage	Bacterial speck
Bacteriophage of *X. campestris* species *vesicatoria*	AgriPhage	Bacterial spot
Insecticides		
B. thuringiensis subspecies *aizawai* GC-91	Turex	Lepidopteran pests
B. thuringiensis subspecies *israelensis* AM65	VectoBac	Sciarids
B. thuringiensis subspecies *tenebrionis* NB 176	Novodor	Coleoptera pests
B. bassiana ATCC 74040	Naturalis L	Thrips, whitefly, mites
Cydia pomonella GV	BioTepp	Summer fruit tortrix (*Adoxophyes orana*)
Fungicides		
B. subtilis QST 713	Serenade	Fire blight, *Botrytis* species
B. licheniformis SB3086	EcoGuard	Fungal diseases
Bacillus mycoides isolate J	BacJ	*Cercospora*
B. pumilus GB 34	GB34	Seedling diseases – *Pythium* and *Rhizoctonia*
B. pumilus QST 2808	Sonata, Ballad Plus	Powdery mildew, downy mildew
B. subtilis GB03	Companion	*Fusarium*, *Pythium*, *Rhizoctonia*
B. subtilis MBI 600	HiStick N/T	Damping off
B. subtilis subspecies *amyloliquefaciens* FZB24	Taegro	*Fusarium* and *Rhizoctonia* wilt diseases
*Pseudomonas aureofaciens*Tx-1	Spot-Less	Turf fungal diseases
*Pseudomonas chlororaphis*63–28	At-Eze	Soil- and seed-borne fungi
Pseudomonas syringae ESC 10	Bio-Save 10LP	Postharvest diseases
Pseudomonas syringae ESC 11	Bio-Save 11LP	Postharvest diseases

(continued)

Table 8.3 (continued)

Category of biopesticides	Product's common name or trade name	Targets
Streptomyces griseoviridis K61	Mycostop Biofungicide	Fungi causing damping off, stem
Streptomyces lydicus WYEC108	Actinovate	Fungi causing damping off, stem
*Ampelomyces quisqualis*M10	PowderyGon	Powdery mildew
*Aspergillus flavus*AF36	*Aspergillus flavus* AF36	*Aspergillus flavus* producing aflatoxin
Aspergillus flavus NRRL 21882	Afla-Guard	*Aspergillus flavus producing aflatoxin*
C. minitans CON/M/91–08	Contans	*Sclerotinia minor*, *Sclerotinia sclerotiorum*
Gliocladium catenulatum J1446	Prestop	Seed-borne and soilborne diseases
M. albus QST 20799	Arabesque	*Sclerotinia minor*, *Sclerotinia sclerotiorum*
Pseudozyma flocculosa PF-A22 UL	Sporodex	Powdery mildew
Trichoderma asperellum ICC 012	Tenet	Soilborne diseases

Better understanding the genomic composition of microbial organisms and important crop plants has enabled successful isolation of genes in order to develop transgenics to control insect pests and diseases. Biotechnological tools have revolutionized the development of fusion proteins as next-generation biopesticides. Technology allows designing of toxins which can work only in a targeted insect hosts (Fitches et al. 2004). Industrial and commercial production of biopesticide-based formulations can be brought about as recombinant fusion proteins in a microbial system. Several innovative approaches are being worked on to scale up their production in cost-effective manner for efficient and acceptable pest control measures.

References

Anand S, Reddy J (2009) Biocontrol potential of *Trichoderma* sp against plant pathogens. Inter J Agri Sci 2:30–39

Arora NK, Khare E, Maheshwari DK (2010) Plant growth promoting rhizobacteria: constraints in bioformulation, commercialization, and future strategies. In: Maheshwari DK (ed) Plant growth and health promoting bacteria. Springer, Berlin, pp 97–116

Bailey DJ, Gilligan CA (2004) Modeling and analysis of disease induced host growth in the epidemiology of take- all. Phytopathology 94:535–540

Bailey KL, Mupondwa EK (2006) Developing microbial weed control products: commercialization, biological, and technological considerations. In: Singh HP, Batish DR, Kohli RK (eds) Handbook of sustainable weed management. The Haworth press, Binghamton, pp 431–473

Bailey KL, Pitt WM, Derby JA, Walter S, Taylor W, Falk S (2010) Efficacy of Phomamacrostoma, a bioherbicide, for control of dandelion (Taraxacum officinale) following simulated rainfall conditions. Am J Plant Sci Biotechnol 4(Special Issue 2):35–42

Contreras-Gómez A, Sánchez-Mirón F, García-Camacho et al (2014) Protein production using the baculovirus-insect cell expression system. Biotechnol Prog 30(1):1–18

Copping LG, Menn JJ (2000) Biopesticides: a review of their action, applications and efficacy. Pest Manag Sci 56:651–676

Crickmore N, Zeigler DR, Feitelson J et al (1998) Revision of the nomenclature of the *Bacillus thuringiensis* pesticidal Crystal proteins. Microbiol Mol Biol Rev 62:807–813

Dean DH (1984) Biochemical genetics of the bacterial insect-control agent *Bacillus thuringiensis*: basic principles and prospects for genetic engineering. Biotechnol Genet Eng Rev 2:34–63

Fitches E, Edwards MG, Mee C, Grishin E, Gatehouse AM et al (2004) Fusion proteins containing insect-specific toxins as pest control agents: snowdrop lectin delivers fused insecticidal spider venom toxin to insect haemolymph following oral ingestion. J Insect Physiol 50:61–71

Glare TR, O'Callaghan M (2000) *Bacillus thuringiensis*: biology, ecology and safety. Wiley, Toronto

Gupta S, Dikshit AK (2010) Biopesticides: an ecofriendly approach for pest control. J Biopest 3:186–188

Harman GE (2005) Overview of mechanisms and uses of *Trichoderma* species 648. Phytopathology 96:190–194

Hewson I, Brown JM, Gitlin SA, Doud DF (2011) Nucleopolyhedrovirus detection and distribution in terrestrial, freshwater, and marine habitats of Appledore Island, Gulf of Maine. Microbial Ecol 62:48–57

Höfte H, Whiteley HR (1989) Insecticidal Crystal proteins of *Bacillus thuringiensis*. Microbiol Rev 53:242–255

Iriarte J, Caballero P,Biología y, ecología de (2001) *Bacillus thuringiensis*. Bioinsecticidas: Fundamentos y Aplicaciones de *Bacillus thuringiensis* en el Control Integrado de Plagas. In: Caballero P, Ferré J (eds) Phytoma-España, pp 15–44

Jeong GT, Park DH, Ryu HW et al (2005) Production of antioxidant compounds by culture of Panax ginseng CA Meyer hairy roots: I Enhanced production of secondary metabolite in hairy root cultures by elicitation. Appl Biochem Biotechnol:121–124

Khandelwal M, Datta S, Mehta J et al (2012) Isolation, characterization and biomass production of *Trichoderma viride* using various agro products- A biocontrol agent. Adv Appl Sci Res 3:3950–3955

Komaraiah P, Reddy GV, Reddy PS et al (2003) Enhanced production of antimicrobial sesquiterpenes and lipoxygenase metabolites in elicitor treated hairy root cultures of *Solanum tuberosum*. Biotechnol Lett 25:593–597

Koundal KR, Rajendran P (2003) Plant insecticidal proteins and their potential for developing transgenics resistant to insect pests. Ind J Biotechnol 2:110–120

Kumar S (2013) The role of biopesticides in sustainably feeding the nine billion global populations. J Biofertil Biopestic 4:1–114

Kumar S, Arul L, Talwar D (2010) Generation of marker-free Bt transgenic indica rice and evaluation of its yellow stem borer resistance. J Appl Genet 51:243–257

Mazid S, Kalida JC, Rajkhowa RC (2011) A review on the use of biopesticides in insect pest management. Int J Sci Adv Technol 1:169–178

O'Brien KP, Franjevic S, Jones J (2009) Green chemistry and sustainable agriculture: the role of biopesticides, advancing green chemistry. http://advancinggreenchemistry.org/wp-content/uploads/Green-Chemand- Sus.-Ag.-the-Role-of-Biopesticides.pdf. Riazuddin, S.: EP0983362

Sarwar M (2015a) The killer chemicals for control of agriculture insect pests: the botanical insecticides. Int J Chem Biomol Sci 1(3):123–128

Sarwar M (2015b) Commonly available commercial insecticide formulations and their applications in the field. Int J Mater Chem Phys 1(2):116–123

Satdive RK, Fulzel DP, Eapen S (2001) Biotechnology of neem.*Azadirachta indica*. L. In: Khan IA, Khanum A (eds) Role of biotechnology in medicinal and aromatic plants. Ukaz Publication, Hyderabad, pp 168–181

Savitha BC, Timmaraju R, Bhagyalaksami N et al (2006) Different biotic and abiotic elicitors influence betalin production in hairy root cultures of *Beta vulgaris* in shake flask and bioreactor. Process Biochem 41:50–60
Saxena S, Pandey AK (2001) Microbial metabolites as eco-friendly agrochemicals for the next millennium. Appl Microbiol Biotechnol 55:395–403
Schnepf EN, Crickmore J, Van Rie D et al (1998) *Bacillus thuringiensis* and its pesticidal crystal proteins. Microbiol Mol Biol Rev 62(3):775–806
Shim HJ, Choi Y, Wang XY et al (2013) NeuroBactrus, a novel, highly effective, and environmentally friendly recombinant baculovirus insecticide. Appl Environ Microbiol 79(1):141–149
Singh A, Khare A, Singh AP (2012) Use of vegetable oils as biopesticide in grain protection-a review. J Biofertil Biopestic 3:1–114
Sivakumar G, Yu KW, Hahn EJ et al (2005) Optimisation of organic nutrients for ginseng hairy roots production in large scale bioreactors. Curr Sci 89:641–649
Steinwand B (2008) Biopesticide ombudsman (personal communication). US Environmental ProtectionAgency, Washington, DC
Strange RN (2007) Phytotoxins produced by microbial plant pathogens. Nat Prod Rep 24:127–144

Suggested Further Readings

Bravo A, Gill SS, Soberón M (2007) Mode of action of *Bacillus thuringiensis* Cry and Cyt toxins and their potential for insect control. Toxicon 49:423–435
Castagnola A, Stock SP (2014) Common virulence factors and tissue targets of entomopathogenic bacterial for biological control of Lepidopteran pests. Insects 5:139–166
Chandler D, Davidson G (2005) Evaluation of entomopathogenic fungus *Metarhizium anisopliae* against soil dwelling stages of cabbage maggot (Diptera: anthomyiidae) in glasshouse and field experiments and effect of fungicides on fungal Activity. J Econ Entomol 98:1856–1862
Glare T, Caradus J, Gelernter W et al (2012) Have biopesticides come of age? Trends Biotechnol 30:250–258
Karthikeyan A, Sivakumar N (2007) Sericulture pupal waste a new production medium for mass cultivation of *Bacillus thuringiensis*. Indian J Biotechnol 6:557–559
Kim JK, Kim NH, Jang HA et al (2013) Specific midgut region controlling the symbiont population in an insect-microbe gut symbiotic association. Appl Environ Microbiol 79:7229–7233
Mausam V, Satinder KB, Tyagi RD et al (2007) Antagonistic fungi, *Trichod* SemaWahab. Biotechnological Approaches in the management of plant pests diseases and weeds for sustainable Agriculture. J Biopestic 2(2):115–134
Zhang ZHY, Yang L, Zhou SHF et al (2011) Improvement of resistance to maize dwarf mosaic virus mediated by transgenic RNA interference. J Biotechnol 153:181–187

Renewable and Nonrenewable Energy Resources: Bioenergy and Biofuels

9

Shiv Shankar and Shikha

Contents

Abstract

Energy is at the heart of most critical economic, environmental, and development issues facing the world today. Challenges posed on global community and national governments due to energy security, climate change, health impacts, and poverty are making it exigent to make energy sector green. Shifting toward green energy is supposed to play a critical role in addressing some of the most prominent contemporary challenges the world is facing at present.

S. Shankar • Shikha (✉)
Department of Environmental Science, School for Environmental Sciences, Babasaheb Bhim Rao Ambedkar University (A Central University),
Vidya Vihar, Raebareli Road, Lucknow 226025, India
e-mail: dr_shikha2003@yahoo.co.in

R.L. Singh (ed.), *Principles and Applications of Environmental Biotechnology for a Sustainable Future*, Applied Environmental Science and Engineering for a Sustainable Future, DOI 10.1007/978-981-10-1866-4_9

Renewable energy is derived from natural processes that are replenished constantly. Renewable energy replaces conventional fuels in four distinct areas: electricity generation, air and water heating/cooling, motor fuels, and rural (off-grid) energy services. Present chapter shall deal with different renewable energy resources and technologies, their ecological implications, commercialization, challenges, and opportunities.

Nonrenewable energy resources (also called a finite resource) are resource that does not renew themselves at a sufficient rate for sustainable economic extraction in meaningful human time frames. At present, the main energy source used by humans is nonrenewable fossil fuels. The continual use of fossil fuels at the current rate is believed to increase global warming and cause more severe climate change. This segment will highlight different nonrenewable energy resources, present status, their ecological implications, and future prospectus.

Bioenergy a form of renewable energy is obtained from biomass. Biomass contains solar energy in the form of chemical energy. As a fuel it may include wood, wood waste, straw, manure, sugarcane, and many other by-products from wide spectrum of agricultural practices. This section will describe different forms of bioenergy, routes of conversion of biomass into bioenergy, present status, and future prospectus.

Keywords
Nonrenewable energy • Biofuels • Bioenergy • Ecological implications • Fossil fuels • Biomass

9.1 Renewable Energy Resources: Introduction

The Earth serves as a source of enumerable natural resources enabling us to live healthily and to harness things that we take for granted every day. Natural processes have been generating majority of fuels like coal, oil, or natural gas. Presently, global attention is centered on energy resources due to their irreversible use, but the supplies of traditional fossil fuels (oil, natural gas) are running out fast. This is why over the last decades, attention is focused on renewable energy resources and ways to increase energy efficiency. Energy resources could be broadly classified as:

(a) *Renewable Energy Resources:* A resource that is quickly replaced or recycled by natural processes in a time frame that makes it useful for human consumption or use, e.g., solar energy, hydro-energy, wind energy, geothermal energy, biomass energy, tidal energy, etc.
(b) *Nonrenewable Energy Resources:* A resource that is replaced slowly by natural Earth processes in such a way that once used by people, it will not be available again within a useful time frame, e.g., oil, natural gas, coal, nuclear energy, etc.

Since the major component of greenhouse gases (GHGs) is carbon dioxide, a global concern has emerged about reducing carbon emissions. In this regard, different policies have come into force with an objective to reduce carbon emissions, which includes enhancing renewable energy deployment along with technological innovations. Renewable energy comes from resources which are naturally replenished on a human timescale, such as sunlight, wind, rain, tides, waves, and geothermal heat (Destouni and Frank 2010). Renewable energy can replace conventional fuels in four major areas: electricity generation, air and water heating/cooling, motor fuels, and rural (off-grid) energy services. The energy consumption may be divided as coming from traditional biomass (9 %), heat energy (non-biomass, 4.2 %), hydroelectricity (3.8 %), and electricity from wind, solar, geothermal, and biomass energy (2 %). Deployment of renewable energy along with energy efficiency has been found to result in significant energy security, climate change mitigation, and economic benefits.

9.1.1 Renewable Energy Resources

9.1.1.1 Wind Energy

In the Earth's atmosphere, wind is generated due to intense convective currents driven by heat energy from the sun. The Earth's surface contains both water and land. When the sun rises up, the temperature of the air above the land escalates quicker than the temperature of the air above the water leading to upliftment of lighter air. The cooler air is denser than warm air which falls and replaces the air above the land. In the night, the phenomenon is reversed. Air above the water gets warmer and rises and is subsequently replaced by cooler air from the land. It indicates that as long as the sun continued to be shined, wind will be generated. The air (wind) in motion contains huge amount of kinetic energy which can be converted into electrical energy employing wind turbine machines (Gonzalez and Arantegui 2016).

The blowing wind sets the turbine blade in motion, rotating a shaft connected to a generator which produces electricity. The electricity generated is sent to power station via transmission and distribution and ultimately to the home, industries, and institutes. For the conversion of wind energy into electricity, a wind speed of 14 miles per hour is required.

In Ohio during the late 1800s, the world's first power-generating turbine was established which was used to charge batteries.

9.1.1.2 Hydropower

Hydropower is a traditional method of electricity generation. Water contains huge amount of kinetic energy which can be transformed into electrical energy. The density of water is 800 times more than air; because of this reason, even the slowest flow of a river and ocean currents can be used to generate electricity. Hydropower is generated by constructing a dam on a river where water is allowed to flow through turbines to generate electricity. Production of hydropower does not result in any harmful emissions, but it affects ecological integrity of water ecosystems. The water quality of river deteriorates which exerts pressure on aquatic life (Wu et al. 2012; Li et al. 2015).

The term hydroelectricity is generally referred for hydroelectric dams on large scale. The hydroelectricity is produced by the gravitational force of falling water. The production of hydroelectricity depends on the height of the falling water and available water flow. Building up behind a high dam, water accumulates potential energy which is converted into mechanical energy when the water moves down the sluice and hits the rotary blades of turbine. The rotation of the turbine spins electromagnets which produce electric current in static coils of wire. Ultimately, the electric current is put through a transformer where the voltage is increased for long-distance transmission over power lines. Run-of-the-river hydroelectricity systems derive kinetic energy from rivers without the creation of a large reservoir.

In Asia-Pacific region, 32 % of global hydropower is produced; however hydropower is being produced in 150 countries. China is the leading producer of hydroelectricity, with 721 terawatt hours of production in 2010, representing around 13.2 % of domestic electricity use (Chang et al. 2010). Wave power, capturing the energy of ocean floor waves, and tidal power, harnessing the energy of tides, are two other forms of hydropower with boundless potential. Tidal energy and wave power are not being employed commercially. A project on trial being operated by the Ocean Renewable Power Company on the coast of Maine is harnessing tidal energy from the Bay of Fundy which is the world's highest tidal flow.

9.1.1.3 Solar Energy

Heat from the sun, solar energy, and radiant light are utilized by employing a wide spectrum of ever-evolving promising technologies such as photovoltaics, solar heating, solar architecture, concentrated solar power, and artificial photosynthesis.

Solar energy can be transformed into electricity in two ways:

(a) Photovoltaic (PV devices) or *solar cells* convert sunlight directly into electricity. Individual PV cells are clubbed into panels and arrays of panels which can be employed for diverse applications varying from single small cell charging calculator and watch batteries to systems that power single home and to large power plants spread in many acres.
(b) Solar thermal/electric power plants produce electricity by condensing solar energy to heat a fluid and generate steam which is subsequently utilized to power a generator.

Solar technologies are broadly characterized as either passive solar or active solar depending on the way they capture, convert, and distribute solar energy. Passive solar techniques involve orientation of a building to the sun, selection of materials with favorable thermal mass or light-dispersing properties, and design of spaces circulating air naturally. Active solar technologies comprehend solar thermal energy by employing solar collectors for heating applications and solar power, converting sunlight into electricity using photovoltaics (PVs) or using concentrated solar power (CSP). By operating on the principle of photoelectric effect, a photovoltaic system converts light into direct current (DC). Solar photovoltaic technology has emerged out as a cost-effective, multibillion, fast-growing industry and the most potent technology of renewable energy generation together with CSP than any other technologies.

Concentrated solar power (CSP) systems focus a large area of sunlight into a small beam utilizing lenses or mirrors and tracking systems (Lorenzin and Abanades 2016). In 1980s, commercial concentrated solar power plants were developed for the first time:

(i) There are two principal merits of solar energy:
  (a) Air pollutants and carbon dioxide are not produced by solar energy systems.
  (b) Solar energy systems have scant effect on the environment when they are installed on buildings.
(ii) There are two main constraints of solar energy:
  (a) The insolation reaching on the Earth's surface is inconsistent. The amount of insolation depends on latitude, time of the day, season, and prevailing local atmospheric conditions.
  (b) Huge surface area is needed to concentrate solar energy at a specific rate since the sun doesn't deliver that much energy to any one place at any one time.

9.1.1.4 Geothermal Energy

Thermal energy generated and stored in the Earth is called geothermal energy. Thermal energy regulates the temperature of the Earth's crust. Radioactive decay of minerals and the inner core of the Earth are the sources of geothermal energy (Gando et al. 2011). The difference in temperature between the inner core of the Earth and its surface (geothermal gradient) drives a continuous conduction of thermal energy in the form of heat from the inner core to the Earth's surface. The heat trapped deep within the Earth, all the way down to the Earth's core – 4,000 miles (6,400 km) down – can be harnessed for the generation of geothermal energy. Rocks under high temperature and pressure melt and converted into magma which covets upwardly because of being lighter than consolidated rock. Magma may elevate the temperature of the rock and water in the crust, sometimes up to 700 °F (371 °C).

9.1.1.5 Heat Pumps

The hot lava from a volcano and the hot steam from a geyser originate from the heat of deep interior Earth. A heat pump is a device rendering heat energy from a source of heat to a destination called a "heat sink." Heat pumps are designed in such a way that they transfer thermal energy in opposite direction of spontaneous heat current by absorbing heat from a cold space and orienting it to a warmer one (Abdelaziz and Shen 2012). In order to execute the function of energy transfer from the heat source to the heat sink, heat pumps use some amount of external energy. The temperature of the Earth remains consistent by 55° C at 4 ft of depth below the ground, about 4 ft underground. Generally, the pipes are buried beyond 4 ft deep inside the Earth in a typical geothermal heating system.

The system pumps a liquid through the installed pipes to absorb the heat and brings it back. A geothermal system can keep the house cool during the summer, when it works in reverse by absorbing the heat from the atmosphere inside a house and transferring it back into the Earth. Geothermal heaters are very efficient wherein none of the energy is wasted. Thus, they reduce heating bills significantly during winter. Geothermal power plants employ hydrothermal resources containing both water (hydro) and heat (thermal).

Geothermal power plants harness high-temperature (300–700 °F) hydrothermal resources coming from either hot water wells or dry steam wells. Such resources can be used by drilling wells inside the Earth and then diverting steam or hot water to the surface of the Earth. For the generation of electricity using turbines, hot water or steam is generally utilized. Geothermal wells may be as deep as 2 miles.

Geothermal plants basically are of three types:

(a) *Flash steam plants* employing high-pressure hot water from deep interior of the Earth convert it to steam to set generator turbines in motion. When the temperature of the steam gets down, it condenses to water and is again transferred back into the ground for reuse. Majority of geothermal power plants are flash steam plants.
(b) *Binary cycle power plants* are involved in the transfer of heat from geothermal hot water to another liquid. The heat converts the second liquid to steam, which is subsequently used to run a generator turbine.
(c) *Dry steam plants* employ steam which is directly piped from a geothermal reservoir to drive generator turbines. The very first geothermal power plant was developed in 1904 in Tuscany, Italy, where natural steam erupted from the Earth.

9.1.2 Nonrenewable Energy Sources

For a developed industrialized nation, adequate and dependable energy resources are required for heating, cooking, transportation, and manufacturing applications. Generally, energy can be categorized as nonrenewable and renewable. Nonrenewable energy constitutes more than 85 % of the total energy used across the world. Majority of developed nations are dependent on nonrenewable energy sources (fossil fuels and nuclear power) which cannot be replenished or reproduced for energy requirements to keep pace with their use. Fossil fuels are widely preferred types of nonrenewable energy in industrialized nations. They were created when partially decomposed plant and animal matter was buried in deep interior of the Earth and converted into material which was rich in carbon usable as fuel. This conversion took place millions of years ago. Nonrenewable energy thus is an energy derived from fossil fuels (coal, crude oil, natural gas) and uranium.

Nonrenewable energy resources are as follows.

9.1.2.1 Coal

Coal (meaning: "mineral of fossilized carbon" since the thirteenth century) is a black sedimentary rock found in rocky stratum in layers in the Earth's crust. It is a form of combustible fuel. Metamorphic rocks are the source of harder form of the coal which is called as anthracite. Mainly coal consists of carbon, but hydrogen, sulfur, oxygen, and nitrogen may be present in varying proportion. Throughout the history, coal has been used as a source of energy for the production of electricity, heating, and metallurgical operations (Xie et al. 2015).

Coal formation takes place when dead biomass is transformed into peat which in turn gets converted into lignite followed by its conversion into subbituminous coal which ultimately converted into anthracite. Geological and biological processes over a long period of time converts buried biomass into coal. On global scale, coal is a widely preferred substrate for the generation of electricity. It serves as the largest contributor of carbon dioxide into the atmosphere due to anthropogenic activities. For the generation of every megawatt hour, 2,000 lb CO_2 is emitted by coal-fired power generators. However, 1,100 lb of CO_2 is emitted by natural gas-fired electric plant per megawatt hour for the production of every megawatt-hour electricity. Coal is derived from the Earth's crust by underground and surface mining operations. China was the largest producer of coal in 1993. In 2011, the USA produced 993 million tons, India 589 million tons, and the European Union 576 million tons of coal. Coal mining activities and application of coal for power generation pose several threats to the environment.

9.1.2.2 Oil

Liquid petroleum or crude oil comes under the category of fossil fuel which is transformed into gasoline, diesel fuel, jet fuel, and heating oil. Below the ground, rocks such as shale which is abundant in organic matter result in the formation of crude oil. Crude oil moves in upward direction through pores of the rocks such as limestone or sandstone and may get trapped within impervious rock and which can be collected by drilling of such oil reservoirs in several stages. More than 70 % oil reserves are located near tectonic plate boundaries. Several usable and superior products from crude oil are obtained by a set of refining process including fractional distillation. Crude oil is also the source of plastic, asphalt, and several other important industrial chemicals. More than 50 % of the total global oil reserves are located in Middle East. At present the rate of crude oil consumption is higher than the production which has led to exploitation of majority of coal reserves of the globe. New sources are being discovered. The present rate of consumption of oil may lead to exhaustion of oil reserves in the coming 30 years. Burning of oil has resulted in enumerable environmental problems. Atmospheric pollutants like nitrogen oxides, sulfur dioxide, carbon monoxide, and carbon dioxide are released due to burning of mineral oil leading to the formation of photochemical smog and global warming (ACC 2010).

9.1.2.3 Natural Gas

Within a two commonly shared underground reservoirs, natural gas is often produced as a by-product of oil recovery. Methane (CH_4), propane (C_3H_8), ethane (C_2H_5), and butane (C_4H_{10}) are the components of natural gas. Natural gas does not contain sulfur, hence, deemed as the purest fossil fuel. Liquefied petroleum gas (LPG) is produced by eliminating propane and butane from the natural gas. LPG is transported into specific pressurized tanks as a source of fuel for household and commercial applications. The transportation of natural gas, besides being an eco-friendly fuel, is inexpensive once pipelines are in place.

In developing countries, natural gas is employed primarily for heating, cooking, and powering vehicles. The same is also employed for the production of ammonia fertilizer. At present, the natural gas reserves are estimated about 100 million metric tons. At present consumption rate, this supply will last an estimated 100 years. Majority of natural gas reserves of the world are found in Eastern Europe and the Middle East.

9.1.2.4 Oil Shale

Oil shale has been scantly utilized as fossil fuel source (Akash 2003). Oil shales are the least utilized fossil fuel sources. Oil shale is a sedimentary rock with minute pores containing kerogen, a carbon-based, waxy substance. On heating shale up to 490 °C, the kerogen gets condensed after vaporization as oil shale which is a viscous and thick liquid. Oil shale is further processed to derive useful oil products. Mining and processing the oil shale require huge amount of energy. In order to extract one barrel of oil shale, about a half barrel of oil is required. Oil shale is present in abundant amount in the Earth's crust, with calculated reserves accounting three trillion barrels of recoverable oil shale (Cook 1974). These oil shale reserves are capable to meet global oil needs for about 100 years. Environmental concerns pertaining to oil shale extraction include the following: Huge amount of water is required for processing, safe disposal of toxic waste water, and disruption of large areas of surface lands.

9.1.2.5 Tar Sand

Tar sand is a kind of sedimentary rock which is impregnated with a very thick layer of crude oil. This thick crude oil does not flow easily; hence, conventional oil extraction methods cannot be utilized to mine it. Tar sand can be easily mined if it is present near the Earth's surface. But if the tar sand is present in deep layers, injection of steam is required into the reservoir to improve flow of oil and push it into the recovery well. The cost of the energy required for the production of one barrel of tar sand is almost equal to that for oil shale. In Canada, the largest tar sand deposits are present which contain sufficient material (about 500 billion barrels) to meet the demand of oil for about 15 years. Keeping in view adverse environmental impacts and high extraction cost, tar sand is not being utilized on commercial scale.

9.1.2.6 Nuclear Power

In majority of thermal power plants, water is converted into steam by heating, which drives turbine generator to generate electricity. In such plants, coal, oil, or natural gas is combusted to generate heat. In a nuclear power plant, heat is generated by the fission of uranium atoms inside a reactor for the production of steam to generate electricity. In the USA, atomic energy is being generated commercially in several nuclear reactors (Mez 2012).

The design of a typical nuclear power plant comprised of a heavy steel pressure vessel covering a reactor core. The reactor core contains uranium as a fuel, which is formed into cylindrical ceramic pellets and sealed in the tubes of long metal known as fuel rods. Several thousand fuel rods constitute the reactor core. Heat is generated inside the nuclear reactor subsequently after neutrons hit uranium atoms, resulting into their breakdown in a continuous chain reaction. Control rods, comprising of material like boron, are placed among the fuel assemblies which absorb neutrons.

Withdrawal of neutron absorbing control rods out of the core facilitates the nuclear fission by making more neutrons available releasing more energy. Insertion of rods into the core reduces the availability of neutrons for fission; as a result the rate of chain reaction is slowed down and less energy is released. Nuclear energy initially was considered as eco-incentive and eco-friendly energy on the grounds of not emitting any air pollutants and greenhouse gases. Nuclear energy was supposed to be available for longer period than traditional fuels. However, several regulations, high capital input, and inappropriate management of nuclear power plants now established nuclear energy as a costly affair (Jim 1982). Deep concerns over the future of nuclear energy as a safest fuel have been expressed after Chernobyl Nuclear Power Plant mishap. In addition, disposal of nuclear waste has also emerged as a serious problem without logical solution (Ojovan and Lee 2005). Knowing such facts, the USA has not established any new nuclear plant in the last three decades (Mez 2012).

9.2 Bioenergy

Biomass-based energy is called as bioenergy which is produced from materials derived from biological origin. Organic material which has stored sunlight in the form of chemical energy is called biomass. Feedstock for the production of bioenergy includes wood, wood waste, straw, manure, sugarcane, and many other by-products from a variety of agricultural processes.

This is a common misconception, as bioenergy is the energy harnessed from the biomass and as the biomass is the fuel and the bioenergy is the energy contained in the fuel. In Europe, the word bioenergy is prevalent, while there is a slight tendency for the word biofuel to be favored in America.

Biomass is the material obtained from recently living organisms, which cover plants, animals, and their by-products (Bridgewater 1999). Crop residues, garden waste, and manure are the sources of biomass. It is a renewable energy source based on the carbon cycle, unlike other natural resources such as petroleum, coal, and nuclear fuels.

Animal wastes, as persistent and unavoidable pollutant produced primarily by the animals housed in industrial-sized farms, serve as another source of biomass.

In the USA, agricultural products like corn, soybeans, and willow are specifically being grown for biofuel production on a pre-commercial research level; sorghum and cassava in China; sugarcane in Brazil; rapeseed, wheat, sugar beet, and willow (15,000 ha or 37,000 acres in Sweden) primarily in Europe; palm oil and *Miscanthus* in Southeast Asia; and *Jatropha* in India.

Biodegradable wastes originating from agriculture, industry, forestry, and households can be used for the generation of bioenergy, using physical and biochemical energy conversion technologies like direct combustion, anaerobic digestion for biogas production, and gasification for the production of syngas.

Straw, timber, manure, rice husks, sewage, and food waste are some examples of biodegradable organic wastes. The use of biomass fuels promotes management of waste and fuel security or reduces the extent of climate change. However, biofuel alone is not a comprehensive solution to these problems. Energy in the form of methane gas or transportation fuels like ethanol and biodiesel can be harnessed from biomass. Agricultural waste, human waste, and decaying garbage release methane gas which is known as landfill gas or biogas. Transportation fuel like ethanol can be produced by the fermentation of corn and sugarcane. Transportation fuel like biodiesel can be produced from vegetable oils, animal fats, and leftover food products. Possibilities are also being explored for the conversion of biomass to liquids (BTLs) and cellulosic ethanol.

9.2.1 Biomass to Bioenergy: Energy Conversion Technologies

Energy from biomass can be produced in a number of ways (Fig. 9.1). Different types of feedstock and technologies can be exploited to extract energy from biomass for the generation of heat and/or electricity. Energy conversion technologies widely used for the production of bioenergy from biomass include conventional combustion, gasification, pyrolysis, and anaerobic digestion.

9.2.1.1 Conventional Combustion

Direct combustion is the simple and widely preferred method of bioenergy technology used for the conversion of plant and animal biomass to heat which can be subsequently utilized for space heating or cooling, water heating, industrial processes, and generation of electricity via a steam engine or turbine. Electrical efficiency of combustion is only 20–35 %, but cogeneration technologies can meet the energy efficiencies beyond 85 %. Generally, combustion technologies are of two types:

(i) *Fixed bed combustion* – In fixed bed combustion, feedstock is allowed to burn on a fixed or rotating grate with supply of air through it.

(ii) *Fluidized bed combustion* – In this technology, feedstock is properly mixed with sand which subsequently acts like a fluid, rendering uniform burning resulting in increased energy efficiency. The level of atmospheric pollutants emitting from fluidized bed boilers is less than fixed bed boilers:

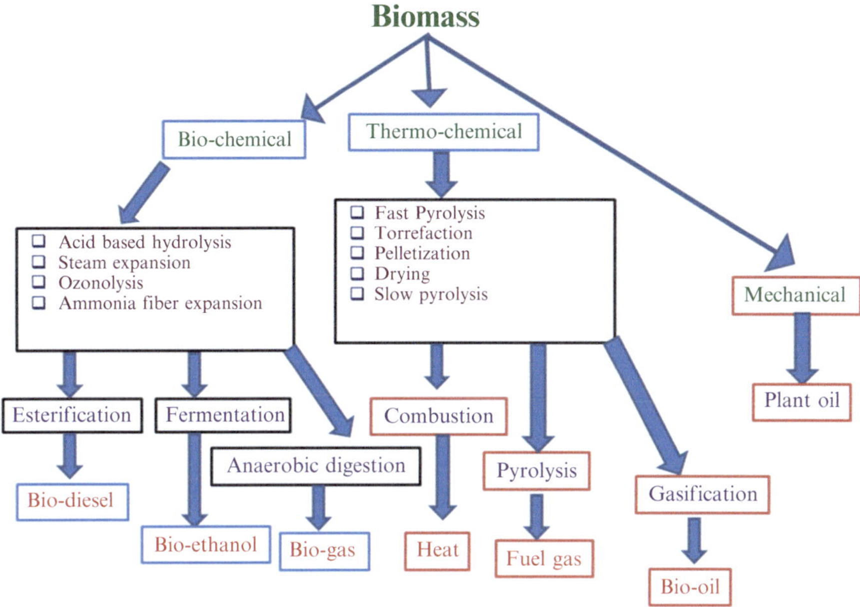

Fig. 9.1 Biomass to bioenergy: conversion technologies

(a) *Co-firing* – Biomass pellets, sawdust, or biogas is mixed and burnt in combination with another base fuel, such as LPG or coal in co-firing technology. Greenhouse gas emissions from fossil fuel power generators can be curtailed by employing co-firing which is supposed to be the cost-effective technology.

(b) *Cogeneration* – It is also known as combined heat and power (CHP). This technology offers greater energy conversion efficiencies as it enamors "waste" heat liberated during generation of electricity, which can be employed for water heating, cooling via absorption chillers, and space applications. Cogeneration is suitable for those areas where heating and cooling are required simultaneously and constantly and electricity can be used on site.

(c) *Combined cycle electricity generation* – These technologies employ a gas turbine system which captures exhaust gases and increase the temperature of water and generate steam provisioned to run a steam turbine for the production of the electricity. This technology works on the principle that the exhaust of one gas engine can be utilized as the source of heat for another, thereby extracting more useful energy from the heat, increasing the efficiency of overall system.

9.2.1.2 Gasification

In gasification, a mixture of carbon monoxide and hydrogen (syngas) is produced by heating wood and bark in minimum presence of oxygen (Neubauer 2013).

Syngas can be converted to liquid transportation fuels. At present, biochemical conversion of biomass into bioenergy needs clean wood chips (without bark), which could draw on the same wood resources as pulp mills.

Mixture of wood and bark is used to produce bioenergy from biomass in thermochemical conversion processes. Gasification, a thermochemical process, operates by heating solid biomass up to 800–1000 °C in a gasifier under controlled supply of oxygen. Under such conditions, feedstock is burnt partially and largely converted to "syngas" which is a mixture of methane, hydrogen, carbon monoxide, carbon dioxide, and nitrogen. During gasification, small traces of char are produced. Syngas can be used directly for heating or power applications, for example, to operate gas engines, gas turbines, or combined cycle power systems.

Gasification is considered more efficient than combustion for the generation of electricity. Gasification is the more preferred technology in terms of biomass specifications like particle size and moisture content. The biomass used for electricity production differs region by region. In the USA, by-products of forests, such as wood residues, are popular feedstock for gasification. In Mauritius, agricultural waste is mostly feedstock, while in Southeast Asia rice husk is used widely. Residue from animal husbandry, such as poultry litter, is common in the UK. In a mature plant, sucrose accounts for more than 30 % of the chemical energy stored; 35 % chemical energy lies in the leaves and stem tips, which are deserted in agricultural fields during harvest; and 35 % chemical energy is stored in the fibrous material (bagasse) leftover from pressing.

Plasma gasification technology uses electricity coupled with an arc gasifier to generate high temperatures which break down biomass and inorganic matter into syngas that can be used to generate electrical energy. Currently, plasma gasification is chiefly being used to treat solid waste focusing on inorganic waste streams including household rubbish.

9.2.1.3 Pyrolysis

Pyrolysis involves thermal degradation of biomass heated in the absence of air – or with very limited air or oxygen. It generates solid, liquid, and/or gaseous by-products at ratios depending on the pace and operating temperature of the pyrolysis (Goia et al. 2011). The gases and compounds in the liquids can be used to generate bioenergy:

(a) *Slow pyrolysis* involves heating of biomass up to 500 °C and produces biochar, liquid (bio-oil), and syngas in equal proportions.
(b) *Fast "flash" pyrolysis* is performed at much higher temperatures and can produce up to 80 % bio-oil which can subsequently be used in other biofuel production systems. Small quantities of syngas and biochar are produced in fast pyrolysis.
(c) *Biochar* can be produced in different ways. It is a stable form of charcoal with diverse chemical and structural properties. Biochar can be utilized for the production of heat. In addition, it can also be utilized for improvement of soil property commercially and is a valuable carbon sequestration product.

9.2.1.4 Anaerobic Digestion

Biological breakdown of biomass in conditions devoid of oxygen is called anaerobic digestion. Anaerobic digestion takes place naturally, e.g., peat swamps deprived of oxygen and in artificial environments, including effluent lagoons, landfills, and biodigesters built purposefully. Anaerobic digestion of sewage, wet agricultural residues, straw and effluents, and manure produces a mixture of methane and carbon dioxide called biogas (Saxena et al. 2007). Biogas can be combusted for the production of heat and/or power in a gas turbine. Biogas can also be converted into natural gas and can be stored in gas grid for its use in gas engine vehicles. Biodigesters are sealed systems with simple design employed for consistent production of biogas. Biodigesters are used for household applications in developing countries. In countries like Germany, larger biodigesters are used by farmers and food processors. The undigested sludge generated by biodigesters can be utilized for the production of bioenergy after burning and dehydration. It can also be used as an organic fertilizer or compost.

Biogases can be collected in sewage treatment ponds and landfills through collection pipes and flared (burnt off) or used to generate bioenergy. In Victoria, most of the larger landfills and sewage treatment plants collect biogas which is utilized for the production of heat and/or electricity for selling on site at premium rate as green power.

9.3 Biofuels

Fuels which are produced by coeval biological processes, such as anaerobic digestion and agriculture, rather than geological processes such as those involved in the formation of fossil fuels, such as petroleum and coal, from ancient biological matter are called as biofuels. Biofuel can be extracted from plants, while it can be produced indirectly from agricultural, commercial, domestic, and/or industrial wastes.

Renewable biofuels generally involve contemporary fixation of carbon, as it occurs in green plants or microalgae through the process of photosynthesis.

Other renewable biofuels can be produced from biomass of living organisms, plants, or plant-derived materials:

- The energy content of ethanol is approximately 50 % that of gasoline.
- The energy content of butanol is approximately 80 % that of gasoline.
- The energy content of biodiesel is approximately 90 % that of petroleum diesel.
- Majority of biofuels are energy denser as coal, but emit low carbon when they are burnt.
- The lower energy content of biofuels refers vehicles traveling short distances on the same amount of fuel.

9.3.1 The First-Generation Biofuels

First-generation biofuels are also known as conventional biofuels. They are produced from the feedstock like sugar, starch, or vegetable oil. Biofuels which are produced from the feedstock that can also be consumed as a food for human consumption are considered as the first-generation biofuel (Naik et al. 2010).

The most popular first-generation biofuels are as follows.

9.3.1.1 Biodiesel

Biodiesel is the widely used biofuel in European countries. Biodiesel is derived from a process called transesterification. Biodiesel is chemically similar to the mineral diesel and is also known as fatty acid methyl diesel. Biodiesel is produced by incorporating biomass with methanol and sodium hydroxide. Mustard, vegetable oils, soy, animal fats, rapeseed, flax, mahua, sunflower, hemp, palm oil, field pennycress, *Pongamia pinnata*, etc. are used as feedstock for the production of biodiesel. Carbon emissions can be reduced up to 60% by using pure biodiesel as compared to diesel second-generation B100. After mixing up with mineral diesel, biodiesel is very commonly used in various diesel engines. Nowadays, manufacturers of the diesel engine in several countries ensure that the diesel engine works well even with the biodiesel. As an oxygenated fuel, biodiesel contains scant proportion of carbon but high proportion of hydrogen and oxygen as compared to the fossil diesel. Low carbon content in biofuel reduces particulate and carbon emission. However, the use of biodiesel may increase NOx emissions. Transportation and handling of biodiesel are safe as it is nontoxic and biodegradable and has a high flash point of about 300 °F (148 °C) compared to petroleum diesel fuel, which has a flash point of 125 °F (52 °C).

9.3.1.2 Vegetable Oil

Vegetable oil can be either used as fuel or even for cooking purpose. The quality of vegetable oil determines its utility. Vegetable oil with decent quality is used for cooking. In warmer atmosphere, vegetable oil can be used as fuel in old diesel engines. In several countries, vegetable oil is principally used for the production of biodiesel.

9.3.1.3 Biogas

Biogas can be produced by anaerobic digestion of biodegradable organic matter and waste materials in digesters. The residue of anaerobic digestion can be used as organic fertilizer for agricultural application. Methane constitutes major proportion of the biogas which can be easily recovered using mechanical biological treatment systems. Landfill gas is the less cleaner form of biogas which is produced in naturally occurring anaerobic digestion. Landfill gases may cause substantial threat if they remain uncaptured and escape into the atmosphere.

9.3.1.4 Bioalcohols

Fermentation of starches and sugar by employing microbes and their enzyme systems results in the production of alcohols. Ethanol is the most common bioalcohol, while propanol and butanol are some less known alcohols. Sometimes, biobutanol, a product of ABE fermentation, is also referred to as a direct substitute of gasoline as the same can be directly applied in the various gasoline engines. Some research findings have substantiated that butanol can be directly used in the various gasoline engines since it is a more energy-efficient fuel.

9.3.1.5 Syngas

Combined process of combustion, gasification, and pyrolysis results in the production of syngas (Rauch et al. 2014). Biofuel employed in this process is transformed into carbon monoxide and subsequently into energy by pyrolysis. Very little amount of oxygen is incorporated to keep combustion under control in this process. In the ultimate step, called as gasification, the organic matter is converted into gases like carbon monoxide and hydrogen. The syngas produced in the process can be used for different applications.

9.3.1.6 Green Diesel

Hydrocracking of biological oil feedstocks, such as animal fats and vegetable oils, results in the production of green diesel. Hydrocracking, as a refinery method, uses high temperature and pressure in the presence of a catalyst to break out larger molecules, such as those found in vegetable oils, into shorter hydrocarbon chains used in diesel engines. Green diesel is also called as renewable diesel, hydrotreated vegetable oil, or hydrogen-derived renewable diesel. Green diesel has the same chemical properties as petroleum diesel. Green diesel does not require any alterations in engines, pipelines, or infrastructure for its distribution and use. However, their production is not cost incentive as compared to petroleum diesel.

9.3.2 Second-Generation (Advanced) Biofuels

Using conventional technology, sugars and oils present in arable crops can easily be extracted for the production of biofuels belonging to the first generation. Lignocellulosic agricultural waste or woody crops can be used for the production of second-generation biofuels. However, extraction of biofuels from such sources is difficult due to the presence of lignin-like compounds (Naik et al. 2010). In order to convert lignocellulosic biomass to liquid biofuels, a wide spectrum of physicochemical pretreatments may be required. Biofuels are also referred as advanced fuels which can be produced from different types of biomass of plants and animals. Biomass refers broadly as any source of organic carbon that is replenished rapidly as part of the carbon cycle.

9.3.2.1 Common Second-Generation Feedstock

To qualify as a second-generation feedstock, a source must be unsuitable for human consumption. It is not a necessity that the feedstock be grown on nonagricultural land, but it generally goes without saying that a second-generation feedstock should grow on what is known as marginal land. Marginal lands are lands which cannot be used efficiently to cultivate arable crops/edible crops. The unsaid point pertaining to second-generation feedstock is that they should not consume a great deal of water or fertilizer to grow. Enumerable grasses like *Miscanthus*, Indian grass, switchgrass, and others have been placed in the spotlight alternatively. The selection of particular grass depends on the geographical location and climate. For instance, in the USA, switchgrass is preferred, while in Southeast Asia, *Miscanthus* is preferred:

(a) *Jatropha* and Other Seed Crops:

Seed crops are promising sources for the production of biodiesel. *Jatropha* plant became extensively popular among biodiesel producers and supporters in the beginning of the twenty-first century (Demirbas 2008). The plant *Jatropha* gained popularity because of its high yield per seed, rendering the return values as high as 40 %. *Jatropha* happened to be a prolific crop as compared to soybean containing 15 % oil. Other, similar seed crops have met with the similar fate as *Jatropha*. Examples include *Camelina*, oil palm, and rapeseed.

In all cases, the initial benefits of the crops were quickly realized to be offset by the need to use cropland to achieve suitable yields.

(b) *Waste Vegetable Oil (WVO):*

Waste vegetable oil has been in practice as a fuel for more than a century (Pryde 1983). It has been used to run the earliest diesel engines. Utility of waste vegetable oil as a food has been expanded as it is deemed as second-generation biofuel. In fact, recycling of waste vegetable oil can help to improve its overall environmental impact. WVO is probably one of the best sources of biodiesel. However its collection is problematic since it is distributed throughout the world in restaurants and homes.

(c) *Municipal Solid Waste:*

Municipal solid waste means to things like human waste, landfill gas, and grass and yard clippings. Municipal solid waste is often used in cogeneration plants, where it is burned for the generation of heat and electricity. All of these sources of energy are, in majority of cases, simply being allowed to go to waste. However, such sources are not cleaner as solar and wind, but their carbon footprint is much less than that of traditionally derived fossil fuels.

(d) *Cellulosic Ethanol:*

Cellulosic ethanol is produced by fermentation, using cellulosic biomass as feedstock, derived from nonfood sources, such as trees and grasses (Brunner et al. 2015). Ethanol is generally used as a gasoline additive to enhance octane and improve vehicle emissions; however, it can be used in its pure form as a fuel in vehicles. In the USA and in Brazil, bioethanol is being used on large scale. Current plant designs do not render conversion of the lignin portion of plant raw materials to fuel components by fermentation.

Ethanol is the most common alcohol produced biologically by the action of microorganisms and enzymes through the fermentation of sugars or starch (easiest) or cellulose (which is more difficult). Propanol and butanol are less commonly produced alcohols during the fermentation. Methods used to produce ethanol include enzymatic digestion of feedstock (to release sugars from stored starch), fermentation of the sugars, distillation, and drying.

The distillation step of ethanol production needs significant input of energy for heat (sometimes unsustainable natural gas fossil fuel, but cellulosic biomass such as bagasse, is the most common fuel in Brazil, while pellets, wood chips, and also waste heat are more common in Europe).

Majority of existing car petrol engines can run on bioethanol up to 15 % with petroleum/gasoline. Ethanol possesses a lesser energy density as compared to gasoline; it means ethanol requires more fuel (volume and mass) to produce the same amount of work. Ethanol has a higher octane number than ethanol-free gasoline commercially available at roadside gas stations. Higher octane number enhances engine compression ratio for increased thermal efficiency.

Some states make it obligatory to use blend of gasoline and ethanol as a winter oxidizer to reduce atmospheric pollution emissions in high-altitude (thin air) locations.

(e) *Biohydrogen*

Biohydrogen can be produced using a set of processes such as pyrolysis, gasification, or biological fermentation like biogas (Saxena et al. 2009). Pyrolysis and gasification heat biomass for the production of hydrogen, while fermentation employs either photo-fermentation or dark fermentation for the production of hydrogen. Dark fermentation as an anaerobic process breaks down feedstock using bacteria through a series of biochemical reactions. Photo-fermentation is similar to dark fermentation, but it requires the presence of light for the reactions to take place. Production of biohydrogen has not been attempted as yet on large scale since it is a less efficient process.

9.3.2.2 Second-Generation Biofuel: Processing Technologies

An outline of processing technologies for the production of second-generation biofuel is as follows:

(a) *Thermochemical Conversion*

1. *Gasification* is the very first thermochemical conversion root. Gasification is an old technology and has been practiced extensively on conventional fossil fuels for a number of years.

   Second-generation gasification technologies have been altered slightly to accommodate the differences in biomass stock. In gasification, carbon-based materials are converted to carbon monoxide, hydrogen, and carbon dioxide. Gasification is different from combustion wherein the supply of oxygen is restricted. In gasification, synthesis gas or syngas is produced which is used to produce energy or heat. Black liquor, wood, brown liquor, and other feedstocks are used in gasification.

2. *Pyrolysis* is the second thermochemical route of conversion of biomass to bioenergy. Pyrolysis also has a prolonged history of application with fossil fuels. Pyrolysis is carried out in oxygen-free environment. Biomass after pyrolysis is generally converted into two products: tars and char. Feedstocks like wood and a number of other energy crops can be harnessed to produce bio-oil through pyrolysis (Goia et al. 2011).
3. *Torrefaction*, a third thermochemical reaction, is very similar to pyrolysis, but is carried out at low temperature. Torrefaction renders better yield of fuels for further use in gasification or combustion. Torrefaction is often used to convert biomass feedstock into fuels convenient enough to be transported and stored.

(b) *Biochemical Conversion*
Different biochemical processes are being exploited for the production of biofuel from the second-generation feedstock. Fermentation with genetically modified bacteria is exclusively popular for biochemical conversion of second-generation feedstock like landfill gas and municipal waste to bioenergy.

9.3.3 Third-Generation Biofuel

Recently, the term third-generation biofuel has been introduced in the area of bioenergy which refers to fuels derived from algae (Pribyl et al. 2014). Algae were lumped in with second-generation biofuels in the beginning. Nevertheless, when it became evident that algae may render higher yield with lower resource input than other feedstocks, several experts suggested that they may be kept under separate category.

9.3.3.1 Third-Generation Biofuel Feedstock

One of the principal benefits of using algae for the production of biofuel is that they can exploit a wide spectrum of carbon sources. Cultivation of algae, even when grown in waste water, requires huge amounts of water, nitrogen, and phosphorus. This is the main demerit of employment of algae for large-scale production of biofuels. In addition, the production of fertilizer to cater the need of algae for biofuel production would produce more greenhouse gas emissions than were saved by using algal biofuels. The cost of production of algal biofuel is much higher than other sources.

After investing more than 600 million USD into research and development of algae, Exxon Mobil, an American multinational oil and gas corporation headquartered in Irving, Texas, concluded in 2013 that algae-based biofuels will not be viable for at least 25 years. Another minor disadvantage pertaining to algal biofuel is that they tend to be less stable than biodiesel produced from other sources. This is because the oil produced in algae tends to be more unsaturated. Unsaturated oils are volatile at room temperature in general and more volatile at high temperatures and thus more prone to degradation.

9.3.3.2 Cultivation of Third-Generation Biofuels

Another merit of algae is the diversity of ways in which they can be cultivated. Algae can be grown choosing any of the following ways:

(a) *Open Ponds:*
Open ponds are the simplest system used for the cultivation of algae in open air. Cultivation of algae in open ponds does not demand high capital cost; however open ponds are less efficient than other systems. Contamination in the pond by other organisms is a matter of concern which can potentially damage or kill the algae.

(b) *Closed-Loop Systems:*
Closed-loop systems are similar to open ponds, but they are not exposed to the air and use a sterile source of carbon dioxide. Closed-loop systems are efficient systems because they may be able to be directly linked to sources of carbon dioxide (such as smokestacks) and thus they consume the gas before it is released into the atmosphere.

(c) *Photobioreactors:*
Photobioreactors are closed systems and most advanced systems. They are difficult systems to be implemented following their high capital cost. However, their advantages are unparalleled in terms of yield and control. All these three systems indicate that algae are capable to be grown almost anywhere the temperatures are warm enough. It means that no farmland will be threatened by algae. Closed-loop and photobioreactor systems have even been employed in desert conditions.

9.3.3.3 Potential of Third-Generation Biofuels

When it comes to the potential to produce fuel, algae are unmatched feedstock in terms of quantity or diversity than any other feedstock. The algal fuels are diverse in nature because of two features of the microorganism. First, oil produced by algae can easily be refined into diesel or even certain components of gasoline. Second, algae can be genetically altered to produce from ethanol and butanol to even gasoline and diesel fuel directly.

Butanol is of massive significance since alcohol is exceptionally similar to gasoline. Third-generation biofuels possess a nearly similar energy density to gasoline and an improved trend of emissions. Until the inception of genetically altered algae, researchers had a great deal of problems in butanol production.

Nowadays, enumerable facilities on commercial scale have been manufactured and are on the verge of butanol production as more popular biofuel than ethanol since butanol is not only similar in many ways to gasoline but also does not result in damage of engine or even need modification of the engine as ethanol does.

The fuels which can be produced from algae include:

- Biodiesel
- Butanol
- Gasoline

- Methane
- Ethanol
- Vegetable oil
- Jet fuel

Besides bearing potential for the production of diverse biofuels, algae render outstanding yield of biofuel. Algae have been employed to produce up to 9,000 gal of biofuel per acre, which is tenfold what the best traditional feedstock has been able to generate.

9.4 Current Research

Research is underway to explore more suitable biofuel crops and improve the oil yields of such crops. To accomplish present yields, vast amounts of agricultural land and freshwater would be required to produce significant amount of oil to fully replace existing fossil fuel applications. It would need twice the agricultural land of the USA to be used for soybean cultivation or two-thirds to be devoted for the production of rapeseed, to attain present US heating and transportation applications. Bred mustard varieties are very useful in crop rotation with cereals and can produce reasonably high oil yields. In addition, they have the added benefit that the meal that remained after the oil has been pressed out can serve as an effective and biodegradable pesticide. A new biofuel has been made from trash by a group of Spanish developers hailing from Ecofasa Company. This biofuel has been produced utilizing urban waste as a feedstock which is treated by bacteria to produce fatty acids which are used to convert biofuels.

9.5 Emerging Renewable Energy Technologies

Other renewable energy technologies like cellulosic ethanol, hot-dry-rock geothermal power, and marine energy are still under development. Such technologies are not well demonstrated and scantly commercialized. Several emerging renewable energy technologies are on the horizon and may have potential applications as compared to other renewable energy technologies, but they have not gained significant attention of researchers and manufacturers. Advanced research is being carried in the area of renewable energy by several organizations within the academic, federal, and commercial sectors. Such research endeavors span diverse areas of renewable energy. Majority of research attempts are directed toward improving energy efficiency and energy yields. In recent years, many federally funded research organizations have focused on renewable energy.

The Sandia National Laboratories and the National Renewable Energy Laboratory (NREL) are the most prominent labs in the arena funded by the US Department of Energy and supported by various corporate partners. Sandia has a total budget of \$2.4 billion, whereas NREL has a budget of \$375 million.

9.6 Advantages and Disadvantages of Renewable Energy

One major benefit of using renewable energy is that it is sustainable and will never run out because of being renewable. Generally, renewable energy facilities need lesser maintenance than traditional generators. Renewable energy fuels are derived from natural and available resources, therefore, minimizing the costs of operation. Renewable energy generates little or no waste such as carbon dioxide or other chemical pollutants, hence having minimal impact on the environment. Renewable energy projects can render monetary benefits to many regional areas, as majority of projects are established distant from large urban centers and suburbs of the capital cities. We can easily recognize the environmental benefits of renewable forms of energy, but the dark aspects of the same are equally needed to be taken into account. Demerit of the renewable energy is that it does not produce ample amount of electricity as compared to petroleum hydrocarbons. Hence, we need to curtail the amount of energy consumption or simply develop more energy facilities. It also necessitates that the best remedy to our energy problems may be to have a consistent balance of diverse energy sources. Reliability of supply of renewable energy source is another demerit of renewable energy. Production of renewable energy heavily relies on weather also. For example, production of hydro-energy requires rain to fill dams to supply flowing water. Wind turbines require wind to turn the blades, while solar collectors need transparent skies and sunshine to collect heat and make electricity. Weather sometimes may be unpredictable and inconsistent thereby affecting production of renewable energy. The existing expenditure of renewable energy technology is also far in excess of conventional fossil fuel production. This is because it is a technology which is in infancy and has extremely large capital cost.

References

Abdelaziz O, Shen B (2012) Cold climates heat pump design optimization. ASHRAE Trans 118(1):34–39

ACC (America's Climate Choices) (2010) Panel on Advancing the Science of Climate Change; National Research Council. Advancing the Science of Climate Change. The National Academies Press, Washington, DC. ISBN 0-309-14588-0

Akash BA (2003) Characterization of shale oil as compared to crude oil and some refined petroleum products. Energy Sources 25(12):1171–1182

Bridgewater AV (1999) Principles and practice of biomass fast pyrolysis processes for liquids. J Anal Appl Pyrolysis 51:3–22

Brunner A, Currie WS, Miller S (2015) Cellulosic ethanol production: landscape scale net carbon strongly affected by forest decision making. Biomass Bioenergy 83:32–41

Chang X, Liu X, Zhou W (2010) Hydropower in China at present and its further development. Energy 35:4400–4406

Cook EW (1974) Oil-shale technology in the USA. Fuel 53(3):146–151

Demirbas A (2008) Biofuels sources, biofuel policy, biofuel economy and global biofuel projections. Energy Convers Manage 49:2106–2116

Destouni J, Frank H (2010) Renew energy. AMBIO 39:18–21

Gando A, Dwyer DA, McKeown RD et al (2011) Partial radiogenic heat model for Earth revealed by geoneutrino measurements. Nat Geosci 4(9):647

Goia V, Cormos CC, Agachi PS (2011) Influence of temperature and heating rate on biomass pyrolysis in a fixed-bed reactor. Stud Univ Babes-Bolyai Chem 56:49–56

Gonzalez JS, Arántegui RL (2016) A review of regulatory framework for wind energy in European Union countries: current state and expected developments. Renew Sustain Energy Rev 56:588–602

Jim F (1982) Global fission: the battle over nuclear power. Oxford University Press, Melbourne, pp 95–96. ISBN 978-0-19-554315-5

Li XJ, Zhang J, Xu LY (2015) An evaluation of ecological losses from hydropower development in Tibet. Ecol Eng 76:178–185

Lorenzin N, Abanades A (2016) A review on the application of liquid metals as heat transfer fluid in Concentrated Solar Power technologies. Int J Hydrogen Energy. doi:10.1016/j.ijhydene.2016.01.030

Mez L (2012) Nuclear energy – any solution for sustainability and climate protection? Energy Policy 48:56–63

Naik SN, Goud VV, Rout PK et al (2010) Production of first and second generation biofuels: a comprehensive review. Renew Sustain Energy Rev 14:578–597

Neubauer Y (2013) Biomass gasification. In: Rosendahl L (ed) Biomass combustion science, technology and engineering. Woodhead Publishing, pp 106–129

Ojovan MI, Lee WE (2005) An introduction to nuclear waste immobilisation. Elsevier Science Publishers, Amsterdam, p 315. ISBN 0-08-044462-8

Pribyl P, Cepák V, Zachleder V (2014) Oil overproduction by means of microalgae. In: Bajpai R, Prokop A, Zappi M (eds) Cultivation of cells and products, vol 1. Springer, Dordrecht, pp 241–273

Pryde EH (1983) Vegetable oil as diesel fuel: overview. J Am Oil Chem Soc 60:1557–1558

Rauch R, Hrbek J, Hofbauer H (2014) Biomass gasification for synthesis gas production and applications of the syngas. WIREs Energy Environ 3:343–362

Saxena RC, Adhikari DK, Goyal HB (2007) Biomass-based energy fuel through biochemical routes: a review. Renew Sustain Energy Rev 13:167–178

Wu N, Cai Q, Fohrer N (2012) Development and evaluation of a diatom-based index of biotic integrity (D-IBI) for rivers impacted by run-of-river dams. Ecol Indic 18:108–117

Xie W, Sheng P, Guo X (2015) Coal, oil, or clean energy: which contributes most to the low energy efficiency in China? Util Policy 35:67–71

Suggested Further Readings

Demirbas A (2008) Biodiesel: a realistic fuel alternative for diesel engines. ISBN: 978-1-84628-994-1 (Print) 978-1-84628-995-8 (Online). Springer-Verlag

Eksioglu SD, Rebennack S, Pardalos PM (2015) Handbook of bioenergy supply chain – models and applications. ISBN: 978-3-319-20091-0 (Print) 978-3-319-20092-7 (Online), Springer International Publishing Switzerland

Filho WL (2015) Handbook of renewable energy. ISBN: 978-3-642-39487-4 (Online). Springer-Verlag Berlin Heidelberg

Hakeem K, Jawaid M, Rashid U (2014) Biomass and bioenergy. ISBN: 978-3-319-07577-8 (Print) 978-3-319-07578-5 (Online). Springer International Publishing Switzerland

Luque R, Clark J (2016) Handbook of biofuels production, processes and technologies. Woodhead Publishing, Oxford. ISBN 9781845696795

Biodiversity and Its Conservation 10

Pradeep Kumar Singh, Rajveer Singh Chauhan, and Pankaj Singh

Contents

P.K. Singh (✉) • P. Singh
Department of Biochemistry, Dr. Ram Manohar Lohia Avadh University, Faizabad 224001, India
e-mail: pkbt99@gmail.com

R.S. Chauhan
Phycology Unit, Department of Botany, Lucknow University, Lucknow 226001, India

R.L. Singh (ed.), *Principles and Applications of Environmental Biotechnology for a Sustainable Future*, Applied Environmental Science and Engineering for a Sustainable Future, DOI 10.1007/978-981-10-1866-4_10

Abstract

Environmentalists might define biodiversity as the total sum of all plant and animal life on the Earth and air, water, and land that support animal and plant life. It is an attribute of an area and specifically refers to the variety within and among living organisms, assemblages of living organisms, biotic communities, and biotic processes, whether naturally occurring or modified by humans. The richness of biodiversity provides humans a food security, health care, and industrial commodities that have support to high standard of living in the modern world. This diversity of organisms makes a sustainable support system which is utilized by every society/nation for its growth, development, and betterment. Those that overused or misused it are decayed. Animal, plant, and marine biodiversity consists of the natural capital that keeps our ecosystems functional and economically productive. The real problem the world faces, however, is the conservation of biodiversity. If we utilize this biodiversity in a sustainable manner, we can develop new products/services for several generations. It is only possible when we treat biodiversity as a valuable resource and prevent extinction of species globally. But the world is facing a dramatic loss of biodiversity. The loss of biodiversity has adverse effects on living being, water supply, food security, and resilience to extreme events. It has consequences for 78 % of the world's extreme poor who live in rural areas and rely on ecosystems and the goods they produce to make a living. Conservation of biological diversity leads to conservation of crucial ecological diversity to preserve the continuity of food chains. Conservation of wildlife along with their natural habitats is the demand of the present scenario and the only way to moderate the self-destruction processes initiated by the mankind since the beginning of human civilization. The two conservation strategies are *ex situ* (outside natural habitat) and *in situ* (within natural habitat). Zoo, cryopreservation, and seed bank are the common examples of *ex situ* conservation, and protected areas like national park, sanctuary, biosphere reserve, conservation reserve, community reserve, etc. are examples of *in situ* conservation. Biological diversity has no regional/ national territories, and its conservation is therefore a combined responsibility of every society/country for the stable and healthy world.

Keywords

Diversity • National park • Species • Conservation • Keystone

10.1 Introduction

The term "biodiversity" was coined by Edward Wilson. Biodiversity refers to the number, variety, and variability of living organisms. In general, it refers to the variety of all forms of living being on our planet. It comprises two diverse ideas: one is the determination of number of living organisms present there, and the other is the determination of level of difference among them. Biodiversity is that component of nature which includes the differences in genes among the individuals of a species; the variety and richness of all the plant and animal species at different extents in space, locally, in a region, in the country, and in the world; and various types of ecosystems, both terrestrial and aquatic, within a given area.

According to the Convention on Biological Diversity, "*biological diversity* means the variability among living organisms from all sources including, *inter alia*, terrestrial, marine and other aquatic ecosystems and the ecological complexes of which they are part; this includes diversity within species, between species and of ecosystems."

Biodiversity can be measured in terms of genetic diversity, biotic communities, identity and number of different types of species, assemblages of species and biotic processes, and the amount (e.g., abundance, biomass, cover, rate) and structure of each. It is estimated that somewhere between 3 and 100 million species reside on the Earth (Heywood 1995). Any ecosystem that is overused or misused loses its productivity and gets degraded.

10.2 Level of Biodiversity

Biological diversity can be observed at three levels (Fig. 10.1).

10.2.1 Ecological Diversity

There are a large variety of different ecosystems on the Earth. Different natural ecosystems include grasslands, forests, deserts, mountains, etc. as well as aquatic ecosystems like rivers, lakes, and seas. Each of these also has human-modified areas such as grazing lands, farmlands, urban lands, etc. Ecosystems are either termed as

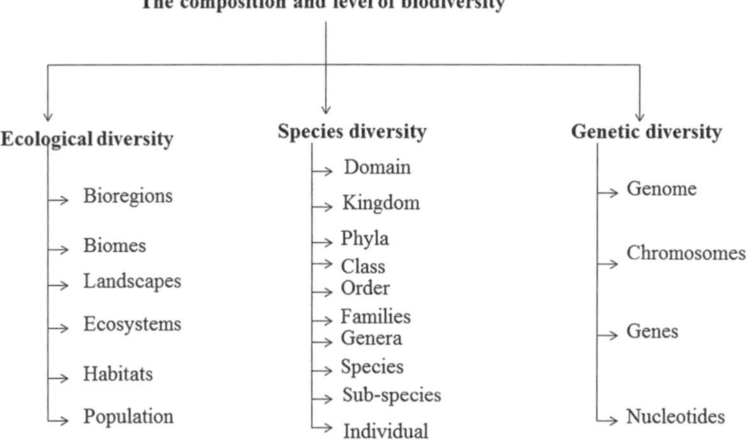

Fig. 10.1 The composition and level of biodiversity

natural or modified/artificial. Natural ecosystem is converted into artificial ecosystem such as farmland or urban areas by man-made action to fulfill their need for survival and development. There are several important ecosystems which are present on the globe that have their own importance and exist in different habitats. Ecosystem diversity may be used to depict a specific geographical region. Globally different types of classification systems exist to depict ecological diversity, for example, biogeographic zones, biomes, eco-regions, oceanic realms, etc. There is no unique definition or classification of ecosystems at the global level, and it is not possible in practice to review ecosystem diversity other than on a local or regional basis and then only largely in terms of vegetation.

10.2.2 Species Diversity

The species is the basic unit of classification in biology. Every natural and modified ecosystem is made up of a variety of animal and plant species. A number of ecosystems, for example, tropical rain forests, are very rich in the number of species as compared to other ecosystems such as the desert ecosystem. The total sum of animals and plants species in an area is said to be its species diversity which can be seen in both natural and artificial ecosystem. Natural undisturbed tropical forests have much greater species richness than plantations developed by the forest department for timber production. An area that is severely rich in species diversity is called "hotspots" of diversity. India is among the world's 15 nations that are remarkably rich in species diversity. A natural forest ecosystem is important in the sense that they are useful for neighborhood people. At present conservation scientists have been able to identify and categorize about 1.75 million species on the Earth (Table 10.1). However many new species are waiting to be identified, particularly in the flowering plants and insects. Although species may be described as a group of similar individuals that interbreed or share common lineage, there is no universally accepted definition of species. If we consider species as basic unit of classification, it symbolizes only one rank of phylogenetic classification. Further classification is genera consisting of related group of species, related genera grouped in families, order consisting of related families, and so on up to the highest level kingdom. Monera, Protista, Fungi, Plantae, and Animalia are the five kingdom systems of living organisms mainly described in biological classification. Worldwide, approximately 1.75 million of the estimated 13–14 million species have so far been described (Tables 10.2 and 10.3).

10.2.3 Genetic Diversity

It refers to the variety of genetic information present in all of the individual plants, animals, and microorganisms occurring within populations of species. In simple word it is the degree of variation of genes within species and populations. Each species from bacteria to higher plants and animals stores huge amount of genetic

Table 10.1 Global species diversity

Group	No. of described species
Mammals	4170
Birds	9198
Reptiles	6300
Amphibians	4184
Fishes (teleosts)	19,056
Starfish	6100
Molluscs	50,000
Insects	751,000
Other arthropods and minor invertebrates	132,461
Crustaceans	38,000
Roundworm and earthworm	24,000
Corals and jellyfish	9000
Sponges	5000
Protozoans	30,800
Angiosperms (flowering plants)	250,000
Gymnosperms (conifers)	750
Bryophytes (mosses and liverworts)	17,000
Algae	26,900
Fungi	46,983
Bacteria and blue-green algae	4760
Total	1,435,662 species

Source: Conserving the world's biological diversity WRI, IUCN, CI, WWF-US The World Bank

information, for example, 450–700 number of genes are present in mycoplasma, 4000 in bacteria (e.g., *Escherichia coli*), 13,000 in fruit fly (*Drosophila melanogaster*), 32,000–50,000 in rice (*Oryza sativa*), and 35,000–45,000 in human beings (*Homo sapiens sapiens*). The variation of genes is not only of numbers but also of structure which is of great value as it helps a population to acclimatize to its environment and to respond to the process of natural selection. If a species has more genetic variation, it can adapt better to the changed environmental conditions. New genetic variation arises in individuals by genetic and chromosome mutations, and it can be spread in population by recombination after sexual reproduction. Genetic variation makes possible to both natural evolutionary change and artificial selective breeding to occur. Global genetic diversity is extremely large and may be evaluated by various ways, for instance, by studying inherited character, by species specific karyotype, and by analyzing DNA structure with molecular tool. Presently some 109 genes are present in genome of various biological organisms. This large variation in genetic codes is the ultimate source for evolutionary change, adaptations to a changing environment, and the survival of species and process of speciation. In

Table 10.2 Comparative statement of recorded number of animal species in India and the world

	Species		
Taxa	India	World	% of India to the world
Protista	2577	31,250	8.24
Mollusca	5072	66,535	7.62
Arthropoda	69,903	970,670	7.20
Other invertebrates	8329	87,121	9.56
Protochordata	119	2106	5.65
Pisces	2546	21,723	11.72
Amphibia	240	5150	4.66
Reptilia	460	5817	7.91
Aves	1232	9026	13.66
Mammalia	397	4629	8.58

Source: MoEF (1999)

Table 10.3 Comparative statement of recorded number of plant species in India and the world

	Species		
Taxa	India	World	% of India to the world
Bacteria	850	4000	21.25
Viruses	Unknown	4000	–
Algae	6500	40,000	16.25
Fungi	14,500	72,000	20.14
Lichens	2000	17,000	11.80
Bryophyta	2850	16,000	17.80
Pteridophyta	1100	13,000	8.46
Gymnosperms	64	750	8.53
Angiosperms	17,500	250,000	7.00

Source: MoEF (1999)

present scenario biotechnology and crop or breeding programs depend on the identification of gene of interest and its successful integration into suitable host to produce desirable traits/products.

10.3 Threat to Biodiversity

The special feature of the Earth is the existence of life, and the most surprising feature of life is its diversity. Approximately nine million types of various plants, animals, protists, and fungi globally are living on the Earth, and the world population is seven billion people. Biodiversity losses can be attributed to the resource demands for support of our rapidly growing human population. In the first Earth Summit in the year 1992, several nations of the world pointed out that human activities were

Table 10.4 Summary of animal biodiversity threats

	No. of Indian species (% of the world's total)	% of Indian species evaluated	Species threatened in India as % of those evaluated	% of extinct those of evaluated
Freshwater fish	700	46 %	70 %	Unknown
Amphibians	207 (4 %)	79 %	57 %	Unknown
Reptiles	495	73 %	46 %	Unknown
Mammals	386 (7 %)	59 %	41 %	1.8 %
Birds	1219 (12 %)		7 %	Unknown

destroying the Earth's ecosystems, eliminating genes, species, and biological traits at a massive alarming rate. Since biodiversity is a fragile system, it is susceptible to all kinds of threats. For most of modern history, human actions have progressed without people giving much attention to the sustainability of ecosystems. Man started overusing and moreover misusing these natural resources which resulted in modification of economical forests and other natural ecosystems to deserts and wasteland throughout world, for example, mangrove vegetation destroyed for fuel-wood and prawn farming that leads to collapse of breeding places for marine fishes. Occasionally these unsustainable actions threaten biodiversity within a community or ecosystem and on the Earth as a whole. If human actions lead to the destruction of the entire ecosystems, such as wetlands or rain forests, biodiversity on the Earth must be decreased to greater extent. The process of speciation, species extinction, and some other ecological processes for shorter evolutionary time scale are the culprits for change in biological diversity. Species interactions, environmental fluctuation, and even cosmic disturbances have profound effect in shaping past and present biodiversity. It is estimated that the over 99 % of total species that have ever been present on our planet are now wiped out (Table 10.4).

Ecologists warn that if the present drift continues, nearly half of total species on our planet might be wiped out within the next 100 years. There are several reasons why species are threatened with extinction. Species losses and other declines in biodiversity are result from five main causes that lead to threats of natural biodiversity.

10.3.1 Habitat Destruction/Loss

Loss of habitat due to anthropogenic action is a major destructive threat to biodiversity (Pimm et al. 2001). Natural reasons of habitat destruction are actions such as volcanic eruptions, wildfires, droughts, and severe storms (e.g., hurricanes). Destruction of habitat occurs due to natural disasters or human activities which alter a terrestrial or aquatic ecosystem so severe that many species can no longer survive in it. If the organisms cannot drift somewhere else and if no other habitat is available, species may not survive, and biodiversity is threatened. Terrestrial ecosystem appears more prone to habitat loss due to various reasons as forest loss, urbanization, desertification, and artificial burns. Primary vegetation of Europe and North America has now been almost lost in the last few centuries, to raise crops to provide food for livestock which are used by humans as a food. The scenario is totally different in various tropic countries where most of the native flora and plants have been lost in the last century. In the tropical region where important essential nutrients reside in living biomass, these nutrients are lost due to loss of vegetation, and soils of these ecosystems are deprived of nutrients which accelerate the process of desertification. For instance, the use of fire by humans promoted the loss of previously dense forest of Madagascar to almost forestless region in the last century. Habitat loss results in fragmentation, which occurs when parts of a natural habitat become separated from one another because of changes in a landscape such as the construction of roads or from artificial creation. Although some habitat destruction is necessary to fulfill human needs, when natural habitats are changed or modified without fear for biodiversity, it imposes much negative impact on the structure of ecosystem. Habitat loss and degradation are the most prominent threat to birds, mammals, and plants affecting 89 % of all threatened birds, 83 % of the threatened mammals assessed, and 91 % of the threatened plants.

10.3.2 Exotic Species

Species that are by mistake or intentionally introduced from one ecosystem to another are a major root for global species extinction. Biological invasion by exotic species is second to habitat loss as a threat to biodiversity. Species that are introduced to new ecosystem from other parts have many names, for example, alien species, introduced species, nonnative species, invasive species, and exotic species. These exotic species destroy many indigenous (native) species as they increase in number and occupy habitats or compete for food, sometimes so entirely dominating an ecosystem that they exert pressure on native plants and animals for extinction. Exotic species are the significant threat affecting 350 threatened birds (30 %) and 361 threatened plant species (15 %). These species are harmful to native biodiversity in a number of ways, for example, as predators, vectors of disease, parasites, or direct competitors for shelter and food.

Only some exotic species become established in new environment, and out of them a few species cause huge change in local biodiversity. Few species are human friendly which includes house mouse, Norway rat, black rat, dog, pig, etc. Exotic plant species have disastrous remarkable effect on several ecosystems such as in North America; purple loosestrife and kudzu vine have captured numerous wetland areas and edge of forest, respectively. The most important plant exotic species are *Lantana camara*, *Parthenium* species, *Eupatorium glandulosum*, *Mimosa* species, *Eichhornia crassipes*, *Mikania micrantha*, *Cytisus scoparius*, *Ulex europaeus*, *Prosopis juliflora*, *Euphorbia royleana*, etc.

10.3.3 Pollution

Pollution is defined as contamination of the natural environment. Environmental pollution is a worldwide problem, and this arises with the unsustainable anthropogenic activities, resulting in a threat to biodiversity. Pollution can be in the form of solids, liquids, and gases or even forms of electromagnetic radiation that enters into air, water, or soil. From the time of industrial revolution, the release of organic and inorganic pollutants into the environment by humans has become a growing threat to biodiversity. Thousands of pollutants are released into the environment, and their unwanted presence threatens biodiversity and harms individual species or degrades the entire ecosystems. Pollutants may sometimes directly toxify the environment, such as lead or PCBs, whereas some are nontoxic but harmful to ecosystem, for example, fertilizer runoff causing excessive plant growth in aquatic ecosystem. We almost pour off our all types of waste products into oceans, rivers, and lakes as well as in soil. Pollutants enter the environment and move through the food web, thereby becoming incorporated into the tissues of the living organisms and leading to many lethal diseases. The absorption, transformation, and breakdown of large amount of pollutants we produce are due to finite capacity of the Earth for these types of processes. Especially some pollutants take thousands of years to decay (e.g., plastics) or may even become more toxic as its decomposition proceeds. Pollution may reduce or eliminate populations of sensitive species. Excessive use of pesticides and other agrochemicals has polluted both ground and surface water bodies due to which many sensitive species go extinct. Eutrophication of water bodies significantly reduces species diversity. Noise and light pollution threaten species by disturbing their behavior.

10.3.4 Overexploitation

Overexploitation or overharvesting of plants, animals, or natural resources in an unsustainable manner is another major threat to biodiversity. One of the most direct anthropogenic activities for the loss of diversity is the removal of living organisms from their natural habitat. Humans remove living organisms from their natural habitat to complete their need for food, energy, and other resources. Overexploitation or unsustainable use occurs when biodiversity is removed faster than it can regenerate or replenish, and over the long term, it can result in the extinction of particular species. Worldwide reduction in biodiversity arises due to major processes, for example, hunting, fishing, and logging. The killing of wildlife species by humans for food and other purposes constitutes a serious threat to biodiversity. Logging of forest is one of the most common threats to loss of biodiversity globally. Overexploitation can lead to dangerously low population numbers or extinction of a species. For example, the population of passenger pigeons was once about five billion, but fairly due to overhunting by early North American population, the last passenger pigeon died in the early 1900s. Large-scale fishing of yellow fish tuna and Atlantic cod during the past few decades has reduced the numbers of these species by 90%.

10.3.5 Global Climate Change

Climate change is a result of emission of greenhouse gases such as carbon dioxide, methane, etc. in the Earth's atmosphere, and it is also considered as a growing threat to biodiversity. Recent changes in climate such as warmer regional temperatures have significant impacts on biodiversity and ecosystems which causes change in population sizes and species distributions. There is significant effect of change in temperature and CO_2 concentration on growth, reproduction, and host-pathogen relationships in both plants and animals species. It is believed that the ecosystems with thriving species diversity and species with their genetic diversity integrity are able to cope up the impact of climate change. A report released by the Intergovernmental Panel on Climate Change in February, 2007, has indicated huge loss of biodiversity for biodiversity-rich mega-diverse countries such as India, due to emissions of large-scale greenhouse gas. Climate change alters the climate patterns and ecosystems in which species have evolved and survived. It is estimated that over the past 140 years, the global average surface temperature has increased 0.6 ± 0.2 °C. This rate of increase in temperature is sufficient to change the climate leading to more intense heat and cold in certain regions and more extreme weather events such as floods, droughts, and other weather fluctuations. Species are unable to adapt or disperse fast enough to cope up with these events of climate change and will go extinct with high rate, and loss of biodiversity occurs.

10.4 Hotspot of Biodiversity

The global biodiversity is distributed in specific ecological regions. There are over thousands of major ecological regions in the world. Of these, 200 are said to be the richest, rarest, and most distinctive natural regions. These regions are referred to as the Global 200. They are the regions especially rich in plant and animal species that are found nowhere else and are in great danger of extinction or serious ecological disruption. There are some 50,000 endemic plant species which are reported to consist of 20 % of the world's vegetation and are inhabitants of 18 hotspots of the globe. Nations that possess quite greater number of these hotspots of biodiversity are said to be "mega-diverse nations."

Ecologically hotspots are determined by four factors:

1. Number of species
2. Degree of exploitation
3. Degree of endemism
4. Degree of threat to habitat due to its degradation and fragmentation

Plant diversity is the biological basis for hotspot designation; to qualify as a hotspot, a region must support 1500 or more endemic plant species (0.5 % of the global total), and the region must have lost more than 70 % of its original habitat. The total 25 hotspots identified so far contain 44 % of all plant species and 35 % of all terrestrial vertebrate species and cover 1.4 % of the land area of the Earth and consist mostly of tropical forests. They have at least 60 % of the Earth's terrestrial biodiversity and are the only regions for more than one-third of the planet's known terrestrial plant and animal species. The vast proportion of the 121,000 potentially threatened species in the tropical regions is endemic to countries within the 25 so-called designated biodiversity hotspots, where high diversity and massive habitat loss occur. A global study on 2 February, 2005, identified nine new environmental hotspots of great ecological diversity. The finding brings to 34, the number of hotspots identified by leading scientists:

Number of Hotspots of Biodiversity Around the Globe

Asia-Pacific:

1. Himalayas
2. Indo-Burma
3. Japan
4. East Melanesian Islands
5. Mountains of Southwest China
6. New Caledonia
7. New Zealand
8. Philippines
9. Polynesia-Micronesia
10. Southwest Australia
11. Sundaland
12. Wallacea
13. Western Ghats and Sri Lanka

North and Central America:

14. California Floristic Province
15. Caribbean Islands
16. Madrean pine
17. Mesoamerica

South America:

18. Atlantic Forest
19. Cerrado
20. Chilean Winter Rainfall-Valdivian Forests
21. Tumbes-Chocó-Magdalena
22. Tropical Andes

Europe and Central Asia:

23. Caucasus
24. Irano-Anatolian
25. Mountains of Central Asia
26. Mediterranean Basin

Africa:

27. Cape Floristic Region
28. Coastal Forest of Eastern Africa
29. Horn of Africa

30. Succulent Karoo
31. Guinean forests of West Africa
32. Eastern Afromontane
33. Madagascar and the Indian Ocean Islands
34. Maputaland-Pondoland-Albany

The Eight Hottest Biodiversity Hotspots

1. Madagascar
2. Sundaland
3. Carribbean
4. Indo-Burma
5. Philippines
6. Brazil's Atlantic Forest
7. Western Ghats/Sri Lanka
8. Eastern Arc and Coastal Forest of Tanzania/Kenya

10.5 Keystone Species

The roles of some species in an ecosystem are much more important than their abundance or biomass suggests. Ecologists describe such species as keystone species. Such species play pivotal roles in the structure and function of an ecosystem because their strong interactions with other species affect the health and survival of these species. Of the various definitions, the most accepted is given by Power et al. (1996) as a species whose effect is large and disproportionately large relative to its abundance. Keystone species play critical ecological roles. The term has its origins in Robert Paine's studies on rocky shore communities in California; when a top predator (a starfish) was removed, the species assemblage collapsed (Paine 1996).

As used by Paine and other ecologists, there are two trademarks of keystone species: the first is their presence that is crucial for maintaining the organization and diversity of ecological communities, and second it is implicit that these species are exceptional relative to the rest of communities in their importance. The term keystone species is applied to many species at many trophic levels. For heuristic purposes we have collapsed the usages of keystone species into four types (Table 10.5).

The keystone species are grouped into four categories on the basis of their functional roles:

1. Keystone predators
2. Resource providers
3. Mutualists
4. Ecosystem engineers

These groups are not eventually exclusive, and individual keystone species may exhibit the characteristics of more than one functional type.

Table 10.5 Categories of presumed keystones and effect of their effective removal from ecosystem

Keystone category	Effect of removal
Prey	Other species more sensitive to predation may become extinct; predator populations may crash
Predator	Increase in one or several predators/consumers/competitors, which subsequently extirpates several prey/competitor species, e.g., starfish *Pisaster*, sea otters
Plants	Extirpation of dependent animals potentially including pollinators and seed disperser, e.g., palm nuts, figs
Modifier	Loss of structures/materials that affects habitat type and energy flow, e.g., American beaver (*Castor canadensis*)

Keystone predator species exerts a profound effect on their ecosystem by feeding on and helps to regulate the population of certain species. Among these the most famous is *Pisaster* starfish that keeps the mussels (*Mytilus californianus*) in check which occupy the entire rock surface. The removal of the mussels opens up enough rock surfaces to prevent competitive exclusion eliminating a range of less competitive species. Another example of this group is the triton gastropods *Charonia* which eats the starfish *Acanthaster planci*, thereby preventing overconsumption of corals by starfish. Tanner et al. (1994) found that high species diversity in coral assemblage in Great Barrier Reef was maintained not only by a single keystone species but also by the action of physical disturbances. In terrestrial ecosystem herbivores, grazers are the major keystone species. Disease-producing organisms can also function as keystone species, for example, *Bacillus anthracis*, fungal pathogens *Cryphonectria parasitica*, etc.

Resource providers are the group of keystone species which provide vital resources to a number of organisms at the time of scarcity. Interestingly these resource providers are not abundant in that ecosystem, but when removed, the dependent organisms are severely affected. The most familiar example involves the provision of keystone plant resources such as fruit during seasonal shortages (Terborgh 1986a; Van Schaik et al. 1993). Palm nuts, figs, and nectar are considered as keystone resource provider because they are vital to fruit-eating guilds, including primates, rodents, insects, and many birds. The fig genus (*Ficus*) has a central role as keystone species in many tropical forests, for example, in Malaysian lowland rain forest, at least 60 species of birds and 17 species of mammals use *Ficus* as food. Without the fruit trees, extinction of frugivores would occur.

When two species are mutually dependent on each other, the removal of one will have significant effect on the others. In this sense they act as keystone for each other. Common examples are in between fruit producers and seed-dispersing frugivores or between pollinators and group of plants. Fig and wasp are the interesting example of mutualistic keystone species. Another example is frugivores such as the cassowary, which spreads the seeds of many different trees, and some will not grow unless they have been through a cassowary.

Organism can also act as keystone species by modifying the physical environment in a way that releases resources for other species. Such organisms are known as ecosystem engineer because they are involved in structural alteration of habitat (Jones et al. 1994). A wide range of species included under this category are rabbits, moles, badgers, gopher, kangaroo rat, net building ants and termites, wallow creation by alligators, vegetation destruction by elephants, hole-drilling woodpeckers, etc. Gopher tortoise in Southeastern United States has been regarded as keystone species because their burrows provide homes for mice, possums, frogs, snakes, etc. Without tortoise' burrows many of these species would be unable to survive in sandy areas. Gopher tortoise acts as ecosystem engineer. African elephant may also act as ecosystem engineers through their browsing activity. Elephant destroys small trees and shrubs when browsing and converts woodland habitat into grassland. Termites are considered as true ecosystem engineers, and removal of termite from ecosystem has major effect on community structure and species diversity. Another well-known ecosystem engineer or keystone species is the beaver, which transforms its territory from a stream to a pond or swamp. Beavers affect the environment first altering the edges of riparian areas by cutting down older trees to use for their dams. This allows younger trees to take their place. Depending on topography, soils, and many factors, these dams change the riparian edges of streams and rivers into wetlands, meadows, or riverine forests. These dams have shown to be beneficial to myriad species including amphibians, salmon, and songbirds. The loss of a keystone species can lead to population crashes and extinction of other species that depend on it for certain reason, a ripple, or domino effect that spreads throughout an ecosystem.

10.6 Conservation of Biodiversity

Conservation is the protection, preservation, management, or restoration of wildlife and natural resources such as forests and water. By the process of biodiversity, conservation and the survival of many species and habitats which are threatened due to human activities can be ensured. Conservation involves the preservation, maintenance, sustainable use, and improvement of the components of biological diversity naturally as well as artificially.

Conservation biologists expect the extinction rate of species to get higher in this century, but are hopeful that people can significantly reduce this rate of biodiversity loss through well-planned systems of protected areas (Roberts 1997; Margules and Pressey 2000; Shaffer and Stein 2000). Our food and energy security strongly depends on biodiversity and so does our vulnerability to natural hazards such as fires and flooding. Biodiversity loss has negative effects on our health and material wealth and it largely limits our freedom of choice. Biodiversity is crucial to human well-being, sustainable development, and poverty reduction. But people particularly those in the developed world have become so far removed from nature that they have forgotten how much they and others depend on it. There is a crying need, not only to manage and conserve the biotic resources but also renovate the degraded ecosystems. Humans have been directly or indirectly dependent on biodiversity to a large extent for nutrients, fuel, etc. More than 70,000 plant species are used in traditional and modern medicines, and the value of ecosystem services is estimated at \$16 − 64 trillion. However, increasing demographic structure pressure and urbanization have led to large-scale depletion of the natural resources.

India ranks under 17 in mega-diverse nations of the globe. Having only 2.4 % of the world's soil cover, 16.7 % of the world's demographic structure, and 18 % of the world's livestock, it contributes about 8 % of the current world biodiversity. India is the reservoir of the world's largest wild tiger population and has got unique assemblage of globally important endangered species like Asian elephant, one-horned rhinoceros, Asiatic lion, Gangetic river dolphin, great Indian bustard, snow leopard, Kashmir stag, dugong, gharial, lion-tailed macaque, etc.

10.6.1 Conserving Biodiversity: What Should Be Done?

The Millennium Ecosystem Assessment (2005) underlines the following actions which may be useful for biodiversity conservation:

- There is a need to protect all the threatened species.
- Preservation of unique ecosystem on priority level.
- To develop public awareness.
- Reserve and protected areas must be developed.
- Wild relatives of all the economically important organisms be identified and conserved in reserve protected area.
- Introduction of exotic species must be stopped.
- Limiting the use of chemical pollutants, for example, pesticides, herbicides, etc.
- Ecosystem restoration.
- Establish funds and incentives to support protection of species.
- Removal of existing grant which supports unsustainable use of biodiversity resources.
- Sustainable growth of agriculture.
- Slowing and adapting to climate change.
- Correction of market failures and internalization of environmental externalities that lead to the degradation of ecosystem services.
- Integration of biodiversity conservation and development planning.
- Increased transparency and liability of government and private sector performance in actions that affect ecosystems, including through greater involvement of concerned stakeholders in decision making.
- Scientific findings and data need to be made available to all of society.

However, *ex situ* (off-site) conservation approach is used in case the need arises to save an endangered species.

10.6.2 *In Situ* Conservation

In situ (on-site) conservation includes the protection of plants and animals within their natural habitats or in protected areas. It refers to conservation of ecosystems and natural habitats including maintenance and recovery of viable populations of a species in their natural habitats. Protected areas are land or water bodies committed to protect and maintain biodiversity. All levels of biodiversity are protected when we conserve and protect the whole ecosystem, for example, we save the entire forest to save the tiger. This strategy is called *in situ* (on-site) conservation. *In situ* conservation is being done by asserting area as protected area. The natural surrounding or the complete ecosystem is protected and maintained so that all the native species, known or unknown to us, are conserved and benefited. The approach of *in situ* conservation revolves around defined small or large protected areas which are reserve exclusively for wildlife. Protected areas are ecological and biogeographical areas where biological diversity along with natural and cultural resources is protected and maintained through legal or other effective actions. *In situ* conservation includes biosphere reserves, national parks, wildlife sanctuaries, sacred forests, and lakes (Fig. 10.2).

10.6.2.1 Biosphere Reserves

The UNESCO has introduced the title "biosphere reserve" (BR) for natural regions to reduce clash between development and conservation. These regions are designated by national government which fulfill a minimal set of defined criteria and hold to minimal set of conditions for inclusion in the world network of biosphere reserves under

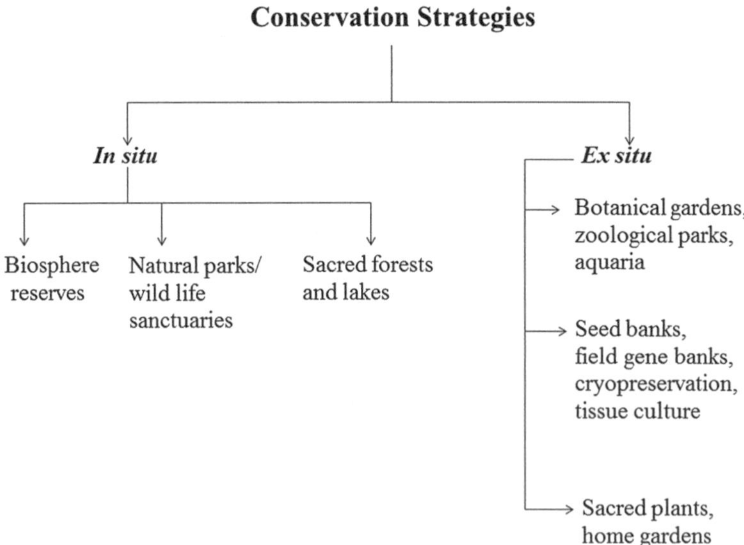

Fig. 10.2 Strategies for conserving biodiversity

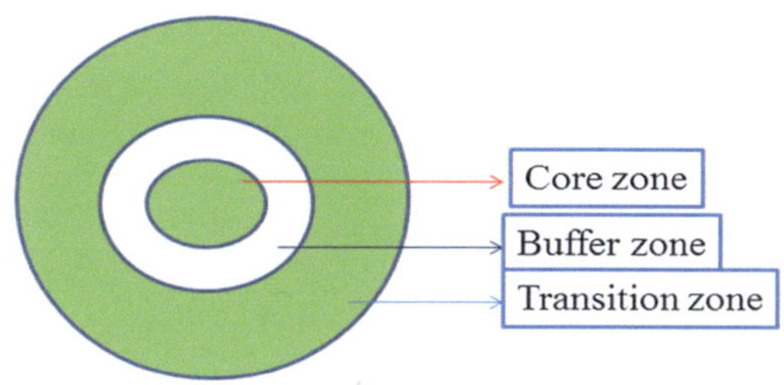

Fig. 10.3 Different zones of biosphere reserve

the Man and Biosphere Reserve Programme of UNESCO. They are the multipurpose protected areas which are used for conserved genetic diversity in representative ecosystem of different natural biomes and unique biological communities by protecting wild populations, traditional lifestyle of tribals, and domesticated plant and animal genetic resources. Biosphere reserves are a mean of conserving genetic resources, species, ecosystem, and landscapes without displacing the local people. They ensure culturally, ecologically, and socially sustainable economic development.

10.6.2.1.1 Structure and Functions of BR

Biosphere reserves consist of the following three interlinked zones (Fig. 10.3):

Core or Natural Zone

Core zone is a legally protected innermost region of biosphere reserve which includes appropriate habitat for various plant and animal species, with higher-order predators, and may include centers of endemism. Natural zone mainly protects the wild members of economic species and also represents important genetic resources having exceptional scientific interest. A natural zone is a national park or sanctuary/protected/regulated areas which are mostly covered under the Wildlife (Protection) Act, 1972. Here human activities are strictly not allowed.

Buffer or Manipulative Zone

This zone is presented outside next to natural zone, and activities are managed in this zone in a manner that supports in maintaining natural conditions of core area. These activities include restoration, demonstration sites for enhancing value addition to the resources, limited recreation, tourism, fishing, grazing, etc. which are allowed to minimize its effect on core zone. Research and educational activities are to be encouraged in this zone.

Transition Zone

Transition zone is the outmost or peripheral part of a BR. Human activities are allowed in this zone. The activities that are permitted in this zone are dwelling, agricultural lands, logging, and area for rigorous recreation, and other economic uses are characteristics of the region.

The program of biosphere reserve was initiated under the Man and Biosphere (MAB) program by UNESCO in 1971. The aim of the formation of the biosphere reserve is to conserve *in situ* all life forms, along with its support system in its totality. The first biosphere reserve of the world was established in 1979; since then the network of biosphere reserve has increased to 621 in 117 countries globally. Presently, there are 18 existing biosphere reserves, in India (Table 10.6).

10.6.2.2 National Park

National park is a vicinity having plenty of environmental, faunal, floral, geomorphological, natural, or zoological importance. The national parks are the site having the aim to conserve, propagate, or develop wildlife or its environment, similar to sanctuaries. In sanctuaries certain activities can be allowed but no activities are allowed in a national park. The term "national park" should indicate an area:

- Which is defined for the protection and conservation of important natural fauna, flora, geological formations, and natural scenic.
- Where also grazing is not permitted.
- In which hunting, killing, or capturing of fauna or removal of any wild animal of its habitat or damage and collection of flora and weapons are all banned except for the improvement and a better management of wildlife therein and on the condition that these issues are handled by, or are under the control of, the park authorities.
- No alteration of the boundaries of a national park shall be made except after the resolution passed by the legislature of the state.

There are 103 existing national parks in India covering an area of 40,500.13 km^2 (Tables 10.7, 10.8, and 10.9), which is 1.23 % of the geographical area of the country. In addition the 75 national parks covering an area of 16,608 km^2 are proposed in the Protected Area Network Report. The network of parks will go up to 178 after full implementation of the above report.

10.6.2.3 Wildlife Sanctuaries

There are 532 existing wildlife sanctuaries in India covering an area of 117,733 km^2, which is 3.58 % of the total geographical area of the country. Another 218 sanctuaries are proposed in the Protected Area Network Report covering an area of 16,829 km^2. The term "sanctuary" should denote an area:

- Which is defined for the conservation, protection, and management of wildlife and its habitat.
- Where human settlement and other human activities are restricted and prohibited.
- Within which destruction or damage, exploitation, weapons, hunting, and grazing are all prohibited, but activities like collection of forest products, private ownership of lands, harvesting of timber, tilling of land, etc. are allowed
- No change in boundaries of a sanctuary shall be made except after a resolution passed by the legislature of the state.

Table 10.6 List of biosphere reserves, their area, date of designation, location, type, and key fauna

Name of the biosphere reserve and total geographical area (km^2)	Date of designation	Location in the state (s)/union territory	Type	Key fauna
[a]Nilgiri (5520)	01.08.1986	Part of Wayanad, Nagarahole, Bandipur and Mudumalai, Nilambur, Silent Valley, and Siruvani Hills in Tamil Nadu, Kerala, and Karnataka	Western Ghats	Lion, tailed macaque
[a]Nanda Devi (5860.69)	18.01.1988	Part of Chamoli, Pithoragarh, and Almora districts in Uttarakhand		
[a]Nokrek (820)	01.09.1988	Part of East, West, and South Garo Hill districts in Meghalaya	East Himalayas	Red panda
Manas (2837)	14.03.1989	Part of Kokrajhar, Bongaigaon, Barpeta, Nalbari, Kamprup, and Darang districts in Assam	East Himalayas	Golden langur, red panda
[a]Sunderban (9630)	29.03.1989	Part of delta of Ganges and Brahamaputra river system in West Bengal	Gangetic delta	Royal Bengal tiger
[a]Gulf of Mannar (10,500)	18.02.1989	India part of Gulf of Mannar extending from Rameswaram Island in the North to Kanyakumari in the South of Tamil Nadu	Coasts	Dugong or sea cow
[a]Great Nicobar (885)	06.01.1989	Southernmost island of Andaman and Nicobar Islands	Islands	Saltwater crocodile
[a]Similipal (4374)	21.06.1994	Part of Mayurbhanj district in Orissa	Deccan Peninsula	Gaur, Royal Bengal tiger, elephant
Dibru-Saikhova (765)	28.07.1997	Part of Dibrugarh and Tinsukia districts in Assam	East Himalayas	Golden langur
Dihang-Dibang (5111.5)	02.09.1998	Part of Upper Siang, West Siang, and Dibang Valley districts in Arunachal Pradesh		

(continued)

Table 10.6 (continued)

Name of the biosphere reserve and total geographical area (km^2)	Date of designation	Location in the state (s)/union territory	Type	Key fauna
[a]Pachmarhi (4981.72)	03.03.1999	Part of Betul, Hoshangabad, and Chhindwara districts in Madhya Pradesh	Semiarid	Giant squirrel, flying squirrel
Khangchendzonga (2931.12)	07.02.2000	Part of North and West districts in Sikkim	East Himalayas	Snow leopard, red panda
Agasthyamalai (3500.36)	12.11.2001	Part of Tirunelveli and Kanyakumari districts in Tamil Nadu and Thiruvanthapuram, Kollam, and Pathanmthitta districts in Kerala	Western Ghats	Nilgiri tahr, elephants
[a]Achanakmar-Amarkantak (3835. 51)	30.03.2005	Part of Anuppur and Dindori districts of Madhya Pradesh and Bilaspur district of Chattisgarh	Aikala Hills	Four-horned antelope, Indian wild dog, sarus crane, white-rumped vulture, *Philautus sanctisilvaticus* (sacred grove bush frog)
Kachchh (12,454)	29.01.2008	Part of Kachchh, Rajkot, Surendranagar, and Patan districts in Gujarat	Desert	Indian wild ass
Cold Desert (7770)	28.08.2009	Pin Valley National Park and surroundings, Chandratal and Sarchu, and Kibber wildlife sanctuary in Himachal Pradesh	Western Himalayas	Snow leopard
Seshachalam (4755.997)	20.09.2010	Seshachalam Hill ranges in Eastern Ghats encompassing part of Chittoor and Kadapa districts in Andhra Pradesh	Eastern Ghats	
Panna (2998.98)	25.08.2011	Part of Panna and Chhatarpur districts in Madhya Pradesh	Catchment area of the Ken River	Tiger, sambhar, chital, chinkara, and sloth bear

Source: MoEF (2015b)

[a]Sites have been included in the world network of biosphere reserves of UNESCO

Table 10.7 List of national parks in India

Name of state/UTs	No. of national parks	Total area covered (km²)
Andaman and Nicobar Islands	9	1106.88
Andhra Pradesh	3	1368.88
Arunachal Pradesh	2	2286.82
Assam	5	1977.79
Bihar	1	335.65
Chhattisgarh	3	2899.08
Goa	1	107.00
Gujarat	4	480.12
Haryana	2	48.25
Himachal Pradesh	5	5031.40
Jammu and Kashmir	4	3925.00
Jharkhand	1	226.33
Karnataka	5	2795.76
Kerala	6	557.64
Madhya Pradesh	9	3656.36
Maharashtra	6	1273.60
Manipur	1	40.00
Meghalaya	2	267.48
Mizoram	2	150.00
Nagaland	1	202.02
Odisha	2	990.70
Rajasthan	5	3947.07
Sikkim	1	1784.00
Tamil Nadu	5	307.84
Telangana	3	19.62
Tripura	2	36.71
Uttar Pradesh	1	490.00
Uttarakhand	6	4915.02
West Bengal	6	1981.65
Total	**103**	**40,500.13 km2**

Source: National wildlife database November, MoEF (2015a)

10.6.2.4 Sacred Forest and Lake

Since the 1990s, sacred places have found its significant role as a new frontier for integrated research on their own merits and also for their actual or probable relevance for biodiversity conservation. Special regions or areas that have one or more attributes which distinguish them as somehow unusual, usually in a religious or spiritual sense, are called sacred places. Individuals may experience a sacred place in different ways as a site of fascination, healing, ritual, attraction, connectedness, danger, ordeal, meaning, identity, revelation, and/or transformation. Many sacred

Table 10.8 Numbers of threatened species of major groups of organisms (1996–2015), IUCN Red List 2015

Plants/animals	Number of described species	Number of species evaluated by 2015	Number of threatened species in 2015	Species evaluated in 2015, as % of species described
Vertebrates				
Mammals	5515	5502	1197	99.8
Birds	10,424	10,424	1375	100
Reptiles	10,272	4669	944	45
Amphibians	7448	6460	1994	87
Fishes	33,200	14,462	2271	44
Subtotal	**66,859**	**41,517**	**7781**	**62**
Invertebrates				
Insects	1,000,000	5573	1046	0.6
Molluscs	85,000	7216	1950	8
Crustaceans	47,000	3168	728	7
Corals	2175	862	237	40
Arachnids	102,248	210	164	0.21
Velvet worms	165	11	9	7
Horseshoe crabs	4	4	0	100
Others	64,658	472	67	0.69
Subtotal	**1,305,250**	**17,516**	**4201**	**1**
Plants				
Mosses	16,236	102	76	0.6
Fern and allies	12,000	365	197	3
Gymnosperms	1052	1011	400	96
Flowering plants	268,000	19,206	10,551	7
Green algae	6050	13	0	0.2
Red algae	7104	58	9	0.8
Subtotal	**310,442**	**20,755**	**11,233**	**7**
Fungi and protists				
Lichens	17,000	9	7	0.05
Mushrooms	31,496	25	22	0.079
Brown algae	3784	15	6	0.4
Subtotal	52,280	49	35	0.09
Total	**1,734,831**	**79,837**	**23,250**	**5**

places in nature are related to indigenous cultures. Although indigenes make up only about 15 % of the human population, estimates range from 200 to 600 million persons depending on definitions and sources, they covers a larger percentage of the land in the world, perhaps up to half. Indigenous people commonly use a wide array of natural resources for their survival, medicines, economy, rituals, and other purposes. Historical, spiritual, and cultural aspects of the ecology of indigenous people are grounded in the biodiversity, ecosystems, and landforms in their habitat. Thus,

Table 10.9 Numbers of species in the threatened categories (CR, EN, VU) (IUCN Red List version 2015) for the major taxonomic groups on the red list

Group	1996/1998	2015
Critically endangered (CR)		
Mammals	169	209
Birds	168	218
Reptiles	41	180
Amphibians	18	528
Fishes	157	446
Insects	44	176
Molluscs	257	576
Plants	909	2347
Fungi	0	5
Endangered (EN)		
Mammals	315	481
Birds	235	416
Reptiles	59	361
Amphibians	31	810
Fishes	134	614
Insects	116	305
Molluscs	212	503
Plants	1197	3510
Fungi	0	11
Vulnerable (VU)		
Mammals	612	507
Birds	704	741
Reptiles	153	403
Amphibians	75	656
Fishes	443	1211
Insects	377	565
Molluscs	451	871
Plants	3222	5376
Fungi	0	13

indigenes are the most important to be considered in exploring the relationships between sacred places, biodiversity, and conservation. In India several places in Maharashtra, Kerala, Meghalaya, and Karnataka are identified as sacred places in which the number of rare, endangered, and endemic species is thrived.

10.6.3 *Ex Situ* Conservation

It is a process of conservation of animals and plants outside their natural habitats. These include botanic gardens/arboretums, seed bank, zoos, gene banks, tissue culture, and cryopreservation. It involves in the conservation of important threatened and endangered species to circumvent their extinction and also known as captive conservation.

Cryopreservation or cryoconservation refers to the preservation in the frozen state. It is a process which is meant to be stored at very low temperatures. Generally cells, tissues, and other plant and animal origin materials are frozen and maintained at the temperature of liquid nitrogen (−196 °C or 77 K or −321 °F). At this temperature, the material to be preserved is maintained in a completely inactive state. The beauty of this technique is that the cell division and normal cellular metabolic reactions are totally paused at the very low temperature of liquid nitrogen which leads to high level of genetic stability and offers least cellular damage to preserved material by chemical reactions. This process is particularly useful for storage of any germplasm which needs to be maintained in a clonal form. In cryopreservation cryoprotectants are needed such as glycerol, dimethyl sulfoxide (DMSO), sucrose, sorbitol, glycols, glucose, mannose, proline, etc. for addition to the culture medium to protect the cells from cryoinjury. Most widely used cryoprotectants are DMSO, proline, sucrose, and glycerol. These cryoprotectants protect preserved materials by preventing the formation of large damaging ice crystals. Different types of tissues can be used for freezing such as the lateral and apical meristems, seeds, plant organs (embryos, ovules, endosperms, and anthers/pollens), protoplasts, calluses, somatic embryos, etc.

Gene banks are the institutions which preserved genetic materials and maintained stocks of viable seeds, live growing plants, and tissue culture with the whole range of genetic variability. For animals it is the process of preservation of sperms and eggs in zoological freezers. Types of gene bank are the seed bank, pollen bank, cryobank, tissue bank, and field gene bank.

The other *ex situ* conservation methods such as botanical gardens, zoological parks, aquaria, etc. are the valuable means for conservation and maintenance of our precious biodiversity.

10.7 International Status of Biodiversity

IUCN (International Union for Conservation of Nature and Natural Resources) carries out comprehensive research on the status of biodiversity. It runs projects to protect specific species and manage and restore national parks and other protected areas and promotes the sustainable use of natural resources. IUCN also provides the knowledge, standards, and tools for biodiversity conservation for governments, community organizations, and businesses. Loss of biodiversity, the variety of animals and plants, their habitats, and their genes, on which so much of human life depends, is one of the world's most pressing crises.

The entire basis of organic evolution is run by the appearance of some species and the disappearance of others; extinction is therefore a natural process. Extinction is the ultimate fate of all species just as a death is for all individual species. Biologist estimated that 99.9% of all the species that have ever existed are now extinct. Population traits which make species subject to extinction are:

- Large body size (e.g., elephant, lion, rhinoceros, Bengal tiger)
- Population with small size

- Higher position of trophic level
- Lack of genetic variability (e.g., Asiatic lion)
- Inability to switch over to alternate food (e.g., red panda)
- Fixed migratory route and habitat (e.g., blue whale)
- Narrow-range distribution (e.g., island species)

The IUCN (2015) has defined criteria to classify species into critically endangered, endangered, vulnerable, and lower risk. This classification is based on population biology principles given by Mace and Lande (1991).

The international historic Convention on Biological Diversity ("the Earth Summit") held in Rio de Janeiro in 1992 called upon all nations to take appropriate steps for conservation of biodiversity and sustainable exploitation of its benefits. In a follow-up, the World Summit on Sustainable Development held in 2002 in Johannesburg, South Africa, 190 countries promise their commitment to achieve by 2010 a significant decrease in the present rate of biodiversity loss at global, regional, and local levels.

References

Heywood VH (1995) Global biodiversity assessment. Cambridge University Press, Cambridge

IUCN (2015) Numbers of species in the threatened categories (CR, EN, VU) (IUCN Red List version 2015) for the major taxonomic groups on the Red List. http://www.iucnredlist.org/about/summary-statistics#TrendsInBiodiversityStatus

Jones CG, Lawton JH, Shachak M (1994) Organisms as ecosystem engineers. Oikos 69:373–386

Mace GM, Lande R (1991) Assessing extinction threats: towards a re-evaluation of IUCN threatened species categories. Conservation Biology 5:148–157

Margules CR, Pressey RL (2000) Systematic conservation planning. Nature (London) 405:243–253

Millennium Ecosystem Assessment (2005) Ecosystems and human well-being: synthesis. Island Press, Washington, DC

MoEF (1999) National policy and macro level action strategy on biodiversity. Ministry of Environment and Forests, Government of India, New Delhi

MoEF (2015a) List of national parks of India: national wildlife database. Ministry of Environment and Forests, Government of India, New Delhi

MoEF (2015b) List of biosphere reserve of India: National wildlife database. Ministry of Environment and Forests, Government of India, New Delhi. http://www.moef.nic.in/sites/default/files/BR%20List.pdf

Paine RT (1996) Food web complexity and species diversity. Am Nat 100:65–75

Pimm SL et al (2001) Can we defy nature's end? Science 293:2207–2208

Power ME, Tilman D, Estes JA, Menge BA, Bond WJ, Mills LS, Daily G, Castilla JC, Lubcher J, Paine RT (1996) Challenges in the quest for keystones. Bioscience 46:609–620

Roberts CM (1997) Ecological advice for the global fisheries crisis. Trends Ecol Evol 12:35–38

Shaffer ML, Stein BA (2000) Safe guarding our precious heritage. In: Stein BA, Kutner LS, Adams JS (eds) Precious heritage: the status of biodiversity in the United States. Oxford University Press, Oxford, pp 301–321

Tanner JE, Hughes TP, Connell JH (1994) Species coexistence, keystone species, and succession-a sensitivity analysis. Ecology 75:2204–2219

Terborgh J (1986) Community aspect of frugivory in tropical forests. In: Estrada A, Fleming TH (eds) Frugivores and seed dispersal. Junk, Dordrecht, pp p371–p384

Van Schaik CP, Terborgh JW, Wright SJ (1993) The phenology of tropical forests: adaptive significance and consequences for primary consumers. Annu Rev Ecol Evol Syst 24:353–377

Suggested Further Readings

Cain ML, Bowman WD, Hacker SD (2009) Ecology, 2nd edn. Sinauer Associates, Sunderland

Campbell NA, Reece JB (2005) Biology, 7th edn. Pearson Education, San Francisco

Chapman JL, Reiss MJ (2000) Ecology: principles and applications, 2nd edn. Cambridge University Press, Cambridge

Groombridge B, Jenkins MD (2002) World Atlas of biodiversity, Earth's living resources in the 21st century. University of California Press, Berkeley

Kormondy EJ (1996) Concept of ecology, 4th edn. Prentice-Hall, New York

Lovejoy TE, Lee H (2005) Climate change and biodiversity. Yale University Press, London

Molles J, Manuel C (2008) Ecology: concept & applications, 4th edn. Mc Graw-Hill Companies, New York

Odum EP, Barrett GW (2005) Fundamentals of ecology, 5th edn. Thompson Brooks/Cole, Belmont

Reaka-Kundla ML (1997) Biodiversity II: understanding and protecting our biological resources. Joseph Henry Press, Washington, DC

Roy S (2003) Environmental science. A comprehensive treatise on ecology and environment, vol 2, 1st edn. Publishing Syndicate, Kolkata

Smith TM, Leo R (2006) Elements of ecology, 6th edn. Pearson Education, Madrid

Biosensors 11

Minal Garg and Sudhir Mehrotra

Contents

M. Garg (✉) • S. Mehrotra
Department of Biochemistry, Lucknow University, Lucknow 226007, UP, India
e-mail: minal14@yahoo.com; sudhirankush@yahoo.com

R.L. Singh (ed.), *Principles and Applications of Environmental Biotechnology for a Sustainable Future*, Applied Environmental Science and Engineering for a Sustainable Future, DOI 10.1007/978-981-10-1866-4_11

Abstract

Biosensors appear as suitable, fast responsive, and cost-effective analytical tools that are extensively being used in monitoring programs including food quality control, agriculture, bioprocess control, environmental monitoring, military, and medical diagnostics. Biosensor is a self-contained integrated device which uses biological mechanism and provides specific quantitative and semiquantitative information about the analyte of interest. Biosensing systems and methods are being developed as environmental quality monitoring tools in the assessment of ecological/biological quality to determine the potentially harmful pollutants (organic and inorganic) and also provide information about their toxic effects. Detection of small amount of biological samples, requirement of minimal tissue damage for *in vivo* screening, on-site monitoring of clinical metabolites, and increased specificity and sensitivity in the order of ng/ml or pg/ml are some of the major concerns for the increasing need to develop biosensors as fast and economic methods for analysis in medical diagnostics. Mass production of molecular recognition elements with improved selectivity, affinity, and stability, immobilization techniques, miniaturization, multisensor array determinations, and operating conditions are some of the major potential areas of development that are expected to have an impact in biosensor technology. The future research for sustainable application of biosensors should rely on more efficient structure and function specificity of the biological components, noninvasive interfacing with the target molecule through mini-reactors, and improved digitization of the generated signal. Real-time parallel monitoring of multiple species is yet another driving force toward the development and commercialization of multichannel biosensors which are required for direct analysis in high-throughput screening systems.

Keywords

Biosensors • Environmental and clinical monitoring • Biorecognition element • Analyte • Transducer

11.1 Introduction

Over a period of recent years, analytes have been increased in number, and therefore, for better environmental control, more suitable analytical methods are required. They have been routinely used by regulatory authorities and industries for screening and testing large number of samples/analytes and provide enough information. Biosensor technology has gained enormous attention over the recent years in food and beverages; agricultural, environmental, and bioprocess control; clinical diagnostics; pharmaceutical industries; and many more.

Over the past many decades, discoveries at the frontiers of basic sciences (physics, chemistry, biology), engineering sciences (biotechnology, electronics and communication, biomedical engineering), and medical sciences (pathology, anatomy, and physiology) led to the development of new biosensors for various applications (Fig. 11.1). Clark in 1956 initiated research in the field of biosensors by publishing work on oxygen electrode (Clark 1956). Since then, biosensor research is one of the fast growing fields where billions of dollars are invested to develop them as potential analytical tools, many of them aimed at on-site analysis. Biosensors can provide fast, sensitive, reliable measurements at low cost as compared to conventional methods of detection which are time-consuming, are expensive, and require the use of highly trained personnel. Moreover, off-site analysis is possible with conventional methods where the samples are sent to laboratory. Nevertheless, field monitoring is more preferred which has driven the development of biosensors as new analytical tools which are easy to use.

The definition of biosensors has been appropriately chosen by Newman et al. (2004). Biosensors are compact analytical devices that incorporate recognition element of biological origin. Recognition element is either closely connected to or integrated with the second important component of biosensors called transducer. The property of selective interaction of biomolecules, utilized as biosensors with other molecules, is the basis for the designing of biosensors. This specific interaction of analyte of interest to the complementary biorecognition element immobilized on support matrix results in change in physicochemical properties including heat transfer, pH change, mass change, electron transfer, and uptake or release of

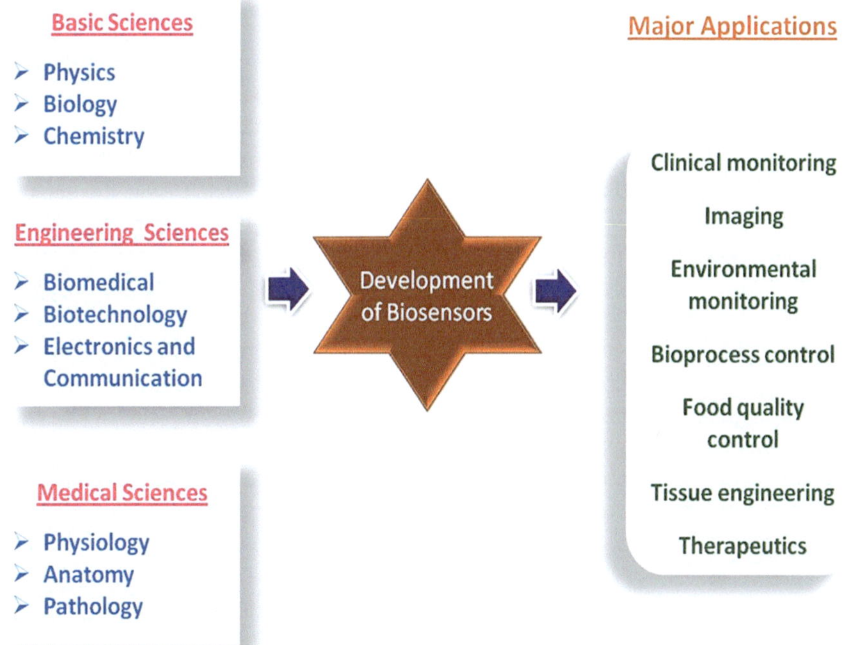

Fig. 11.1 Inventions at the frontiers of scientific fields contribute to the development of biosensors for wide range of applicants

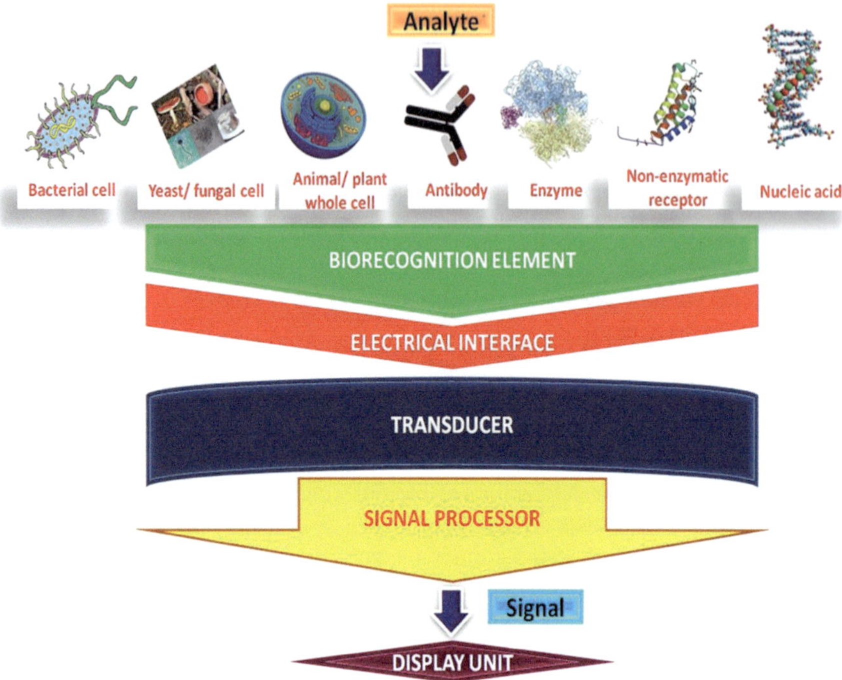

Fig. 11.2 Working principle of biosensor

gases or specific ions. These changes are detected and measured by transducer in the form of electronic signal which is proportional to concentration of bound analytes. Measuring biological outputs in the form of signals generated and monitoring the process is being utilized in biosensor technology. Figure 11.2 describes the general working principle of biosensors.

Most of the times, biosensors are case specific in its approach and use a specific bioactive component/inherent property of cells or tissues for the desired reaction/conversion to generate a signal which can be monitored. Sometimes enzyme-based biosensors generate the signal by catalytic conversion of substrate to form the product, disappearance of substrate, or coenzyme conversion. The biochemical reaction can either be measured to monitor the process or can be superimposed with other biochemical events like the use of kinetics or coupling with other reactions.

With the advent of genetic engineering technologies, more flexible and efficient biosensors can be constructed. Molecular tools make it possible to improve the analysis by:

1. Constructing microorganisms that produce surfactants to increase the bioavailability of pollutants
2. Constructing more efficient microorganisms in degrading pollutant

3. Constructing microorganisms which are capable of utilizing multiple types of compounds
4. Constructing microorganisms which can survive harsh environmental conditions like very high and very low temperature, high salinity, or areas where oxygen is in limited amounts
5. Selection of microorganisms with new enzymatic capabilities

Biosensors provide fast and accurate detection of contaminated sites and clinical metabolites for environmental and clinical monitoring, respectively. The other advantages that are being offered over other analytical tools include portability, working on-site, and ability to measure contaminants/metabolites in complex matrices with minimal sample preparation. Lots of other important information on samples including biological effects (toxicity, endocrine-disrupting effects, etc.) of the chemicals can also be accurately derived. This chapter presents a comprehensive overview of the fundamental principles for biosensor design, operating mechanisms, summarizes important recent applications of biosensors in environmental and clinical monitoring, and addresses the need for fundamental and continued research for further development of biosensor technology.

11.1.1 Components of Biosensor

Description on functions of a typical biosensor at five different levels is shown in Fig. 11.3. The five different components and their functions in a biosensor system include:

1. Biorecognition element – bioreceptors bind specifically to the analyte.
2. Electrical interface – specific biological processes occur at this interface which give rise to a signal.
3. Transducer element – specific biochemical reaction is being converted into electrical signal.
4. Signal processor – electronic signal is converted into a physical parameter.
5. Display unit – an interface to display the results to the operator.

Biosensor thus becomes an excellent analytical device to analyze, detect, and record the biological data for monitoring a number of biological materials and their conversion, verification of product content, and early detection of contaminants or hazardous chemicals or biological materials.

11.1.2 Biosensor Configurations

The general classification of biosensors based on recognition elements and transducers is described in Fig. 11.4. The type of recognition elements forms the basis of classification of biosensor. Enzymes, receptors [natural (proteins of non-catalytic,

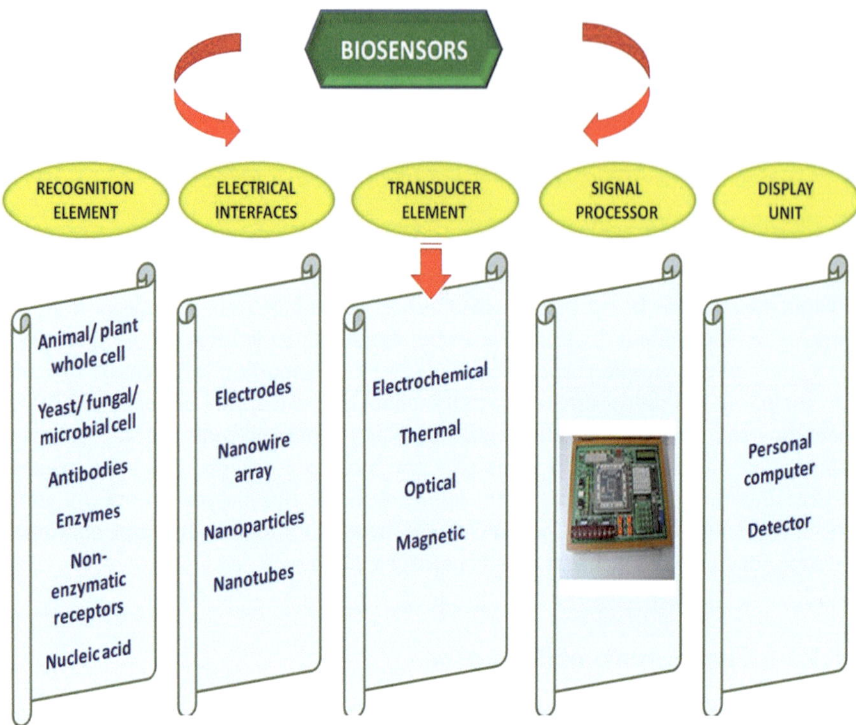

Fig. 11.3 Schematic representative of components of biosensor: a typical biosensor works at five different levels. (**a**) Recognition element; (**b**) electrical surfaces; (**c**) transducer element; (**d**) signal processor; and (**e**) display unit

or non-immunogenic origin) or synthetic], antigens/antibodies, nucleic acid/complementary sequences, animal or plant whole-cell organisms, yeast, fungi, or microbial cells and tissue slices are the examples of biological material that can be incorporated in a biosensing system. Hence, according to the biorecognition principle and the type of specific interaction between the analyte and biological element, biosensors are classified as immunosensors and enzymatic, nonenzymatic, whole-cell, and DNA biosensors. The support or matrix used to immobilize the type of biorecognition element and this immobilized molecule decides the specificity of a biosensor system. Immobilization imparts stability to the biomaterial and ensures proximity between the biomaterial and the transducer. Cross-linking between the molecules; physical adsorption at solid surface; entrapment within a membrane, polymer, or microcapsule and surfactant matrix; covalent binding to a surface; gel entrapment; self-assembled biomembranes; electropolymerization; and bulk modification are some of the popular methods of immobilization.

Depending on the type of transducing element and the method of signal transduction, biosensors are categorized as piezoelectric, electrochemical (amperometric and potentiometric), thermal, optical, mechanical, and magnetic biosensors.

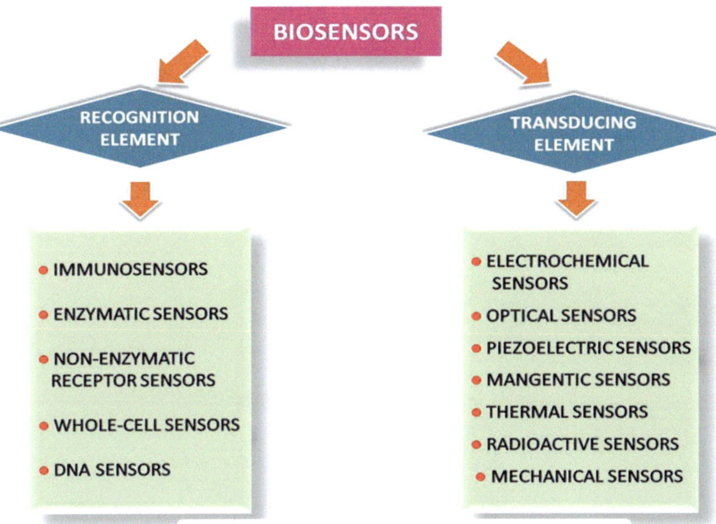

Fig. 11.4 Classification of biosensor

Light absorption, fluorescence/phosphorescence, reflectance, refractive index, bioluminescence, and chemiluminescence are the properties for detection that are being exploited by optical transducers. The characteristic property of bending of silicon cantilevers caused by the adsorption of target molecules onto the cantilever surface (where receptor molecules are immobilized) is being employed in cantilever biosensors.

A large amount of research has taken place recently to develop biosensors capable of efficiently determining several analytes alone or simultaneously and, therefore, represent an interesting tool in environmental monitoring and clinical screening (Fig. 11.5).

11.2 Biosensors for Environmental Monitoring

One of the first environmental biosensors was based on the enzyme acetylcholinesterase to detect nerve gas and developed for military in the late 1970s. This enzyme is involved in transmitting messages in the nervous system and produces an electrochemical reactive product. Presence of nerve gas inhibits the production of an enzyme and hence the electrochemical signal. Since then many biosensors are being developed to detect wide range of potentially harmful pollutants of environmental concern on the basis of specific recognition of biomolecule (Table 11.1). In the following section, the biosensors reported for different environmental applications are being described.

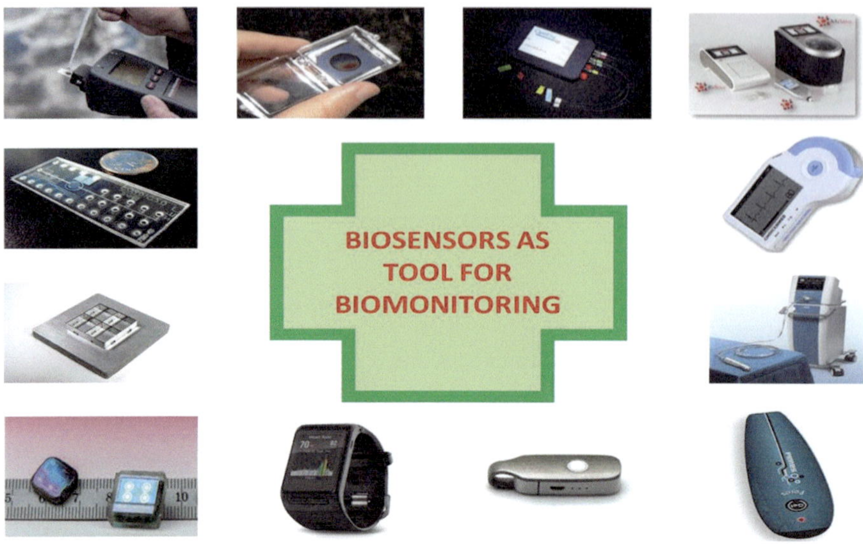

Fig. 11.5 Applications of biosensor in environmental monitoring and clinical screening

11.2.1 Toxicity

Bioluminescence assays, where genetic manipulation of lux gene in various organisms is done to allow controlled emission of light in response to metabolism of certain material, are used in toxicity sensors to determine toxicity in water, food samples, etc. Microtox® (Azure, Bucks, UK), or ToxAlert® (Merck, Darmstadt, Germany), and Cellsense are some of the reliable sensors for rapid ecotoxicity analysis. Among them, Cellsense gains wide popularity owing to multiple applications which include investigation of the toxicity of 3,5-dichlorophenol and other phenols in wastewater, analysis of wastewater treatment works (WWTW), influent and effluent evaluation of concentration of nonionic surfactants and benzene sulfonate compounds, and toxicity testing of wastewaters and sewage sludge (Farre et al. 2001; Aracic et al. 2015). This biosensor is an amperometric type which incorporates whole microbial cell, *Escherichia coli*. This sensor uses ferricyanide, a soluble electron mediator that allows the electrons to divert from the respiratory system of the immobilized bacteria which, therefore, generates current. The resultant current produced measures bacterial respiratory activity, and the change in the magnitude of current can be detected due to presence of pollutants.

11.2.2 Biochemical Oxygen Demand (BOD)

Biochemical oxygen demand (BOD) sensor is one of the commercially most successful environmental biosensors which uses immobilized microorganisms to measure assimilated carbons (amount of biodegradable organic matter) in wastewater.

Table 11.1 Biosensors designed and developed for various environmental applications

Type of biosensor	Analyte	Biorecognition element	Transducer element	Matrix	References
Toxicity					
Bacterial biosensor	Toxicity	*Escherichia coli*	Electrochemical (amperometric)	Wastewater	Farre et al. (2001)
Bacterial biosensor	Toxicity	Genetically engineered bioluminescent bacteria	Optical (bioluminescence)	No real sample	Choi and Gu (2002)
Biochemical oxygen demand (BOD)					
Bacterial biosensor	Low BOD	*Pseudomonas putida*	Optical (fiber optic)	River water	Chee et al. (2000)
Whole-cell biosensor	BOD	Multispecies culture	Electrochemical (amperometric)	Municipal and industrial water	Tan and Wu (1999)
Hormones and endocrine disruptors					
Immunosensor	Estradiol	Antibodies	Electrochemical (amperometric)	No real sample	Tiefenauer et al. (1997)
Immunosensor	Estrone	Antibodies	Optical	River water	Rodriguez-Mozaz et al. (2004)
Phenolic compounds					
Enzyme based biosensor	Phenols	Enzyme (tyrosinase)	Electrochemical/amperometric	Soil, sludge, and water after extraction	Parellada et al. (1998)
Enzyme-based biosensor	Phenols, m-cresol, p-cresol, catechols	Polyphenol oxidase	Amperometric	Wastewater	Chen et al. (2007)
DNA biosensor	m-cresol or catechol	DNA	Amperometric	Wastewater	Chen et al. (2007)
Heavy metals					
Bacterial biosensor	Zinc, copper, cadmium, nickel	*Pseudomonas fluorescens*	Optical	Soil	Mcgrath et al. (1999)

(continued)

Table 11.1 (continued)

Type of biosensor	Analyte	Biorecognition element	Transducer element	Matrix	References
Microalgae-based biosensors	Zinc, copper, cadmium, nickel, lead, iron, aluminum	*Chlorella vulgaris*	Electrochemical	Urban waters	Claude et al. (2007)
Bacterial biosensor	Nickel ions	*Bacillus sphaericus* strain MTCC 5100	Electrochemical	Industrial effluents and foods	Verma and Singh (2006)
DNA biosensor	Mercury (II) and lead (II) ions	DNA	Optical	Water	Knecht and Sethi (2009)
Pesticides and herbicides					
Enzyme-based biosensor	Organophosphorus compounds	Enzyme (acetylcholine esterase)	Optical	Water	Choi et al. (2001)
Immunosensor	Pesticides	Antibodies	Optical	River water	Rodriguez-Mozaz et al. (2004)
Enzyme-based biosensor	Paraoxon and carbofuran	Enzyme (acetylcholine esterase)	Electrochemical (amperometric)	Wastewater	Bachmann and Schmid (1999)
Microalgae-based biosensors	Atrazine and endrine (herbicides)	*Scenedesmus subspicatus* (*algal cells*)	Optical (fluorescence)	No real sample	Frense et al. 1998
Microbial organisms					
Immunosensor	*Salmonella enteritidis*, *Listeria monocytogenes*	Antibodies	Optical (SPR)	No real sample	Koubova et al. (2001)
Immunosensor	*Escherichia coli*	Antibodies	Electrochemical (potentiometric)	Drinking water	Ercole et al. (2002)
DNA biosensor	*Aeromonas hydrophila*	DNA (hybridization)	Piezoelectric	Mineral and drinking water	Tombelli et al. (2000)

BOD sensor was invented in Japan in the late 1970s, and the first commercial BOD sensor was produced by the Japanese company Nisshin Electric in 1983 (Rodriguez-Mozaz et al. 2004). Since then it had become a very popular and reliable device as an early warning system to detect possible contamination of waterways, and a number of other commercial BOD biosensors based on microbial cells have been marketed. Optical biosensor has been developed, for real-time analysis of BOD in multiple wastewater samples.

11.2.3 Hormones and Endocrine Disruptors

Endocrine-disrupting compounds (EDCs) are chemically ill-defined environmental contaminants that interfere with endogenous hormone homeostasis and impose adverse effects which include decrease of human sperm numbers and increased incidence of testicular, breast, and thyroid cancers. Hormones like estradiol, estrone and ethinyl estradiol, progesterone, and testosterone are discharged in the environment as a result of human/animal excretion or intensive farming and have been found to have endocrine-disrupting effects in aquatic or terrestrial organisms even at low concentrations (ng/l). An electrochemical biosensor to determine progesterone levels in cow's milk has been developed. The operating principle of this sensor is based on competitive binding between analyte and conjugate (alkaline-phosphatase-labeled progesterone) for the immobilized anti-progesterone monoclonal antibody (mAb) sites. The amperometric signal generated in the presence of *p*-nitrophenyl phosphate using either colorimetric assays or cyclic voltammetry measures the concentration of analyte. Surface plasmon resonance (SPR) biosensor BIAcore and piezoelectric biosensors utilize human estrogen receptor to determine the levels of estrogens and xenoestrogens.

11.2.4 Heavy Metals

Toxic heavy metals like chromium, zinc, mercury, cadmium, and copper are non-biodegradable and have been observed to be accumulated in the environment as contaminants and pose great threat to the environment and human health even in low concentrations. Need for trained personnel and high cost associated with traditional analytical methods, biosensors have become more popular to measure heavy metal concentration in environmental samples. Bacterial biosensors are developed which contain specific genes responsible for bacterial resistance to heavy metals fused with gene codes for bioluminescent proteins. Based on the inhibitory action of metal ions on urease activity, enzyme-based biosensors are employed to detect concentration of various heavy metal ions like Hg(II), Ag(I), Cu(II), Ni(II), Zn(II), Co(II), and Pb(II). An optical biosensor has been developed to determine lead and cadmium ions. These ions were shown to inhibit activity of alkaline phosphatase present on external membrane of *Chlorella vulgaris* microalgae as biological element (Durrieu and Tran-Minhw 2002).

11.2.5 Pesticides and Herbicides

An evitable use of toxic chemical pesticides for agricultural purposes has raised concerns for their persistence in atmosphere, food, water, soil, and plants. The limit of 0.1 ug/l for individual pesticide and 0.5 ug/l for total pesticides to check the quality of water for human consumption has been set by the European Community (Directive 98/83/EC) (Rodriguez-Mozaz et al. 2004).

Enzymatic sensors for faster on-site analysis of concentration of pesticides in samples have been developed which are based on inhibition of selected enzymes by the analyte. Choi et al. (2001) described the biosensors based on the inhibition of acetyl cholinesterase (AChE) and colin oxidase for examining the concentration of organophosphorus (paraoxon, parathion) and carbamate pesticides. Tyrosinase-based oxygen sensors can detect diazinon and dichlorvos at limits of 5 uM and 75 uM, respectively. Inhibition of the enzyme aldehyde dehydrogenase by dithiocarbamate fungicides can help in its detection. The lack of specificity of the enzymes to identify individual or class of pesticides can be overcome by genetic engineering of the existing enzymatic systems to produce new specific enzymes for desired analysis. Production of recombinant AChEs for various biosensor applications has been extensively reviewed.

The toxic effect of inhibition of photosynthetic electron flow by blocking the photosystem II (PSII) quinone-binding site and thus modification of chlorophyll fluorescence has been shown by the 30 % of herbicides including phenylurea, triazine, and phenolic herbicides. Biosensors are being developed that utilize membrane receptors of thylakoid and chloroplasts, photosystems and reaction centers, or complete cells like unicellular alga as biorecognition element and amperometric and optical transducers.

11.2.6 Nitrogen Compounds

Nitrites can react irreversibly with hemoglobin, and hence continuous consumption of these ions can cause serious health problems. Increased accumulation of nitrates in groundwater and surface water is of serious concern as they can harm the aquatic environment. The working principle of the biosensor designed to determine the nitrate/nitrite levels is based on the diffusion of nitrate/nitrite through a tip membrane into a dense mass of immobilized denitrifying bacteria which convert these ions into nitrous oxide (N_2O) followed by their electrochemical detection. These biosensors are commonly used for on-site determination in activated sludge systems, sewage treatment plants, and industry.

11.2.7 Inorganic Phosphorous Compounds

The degree of eutrophism, which is recognized as water pollution, depends upon the presence of inorganic phosphate mainly which enters in surface waters through detergents, fertilizers, or sewage. Increased biomass of toxic or inedible phytoplankton species, increased water turbidity, change in quality of water like its color and smell, depletion of dissolved oxygen, and increased incidence of fish kills and other aquatic animals are some of the major environmental pollution problems associated with eutrophication. Traditional methods for quantitative and qualitative determination of phosphate ions are available which include chromatographic methods, spectrophotometric analysis, and volumetric titration. However, these methods are time-consuming, labor-intensive, and not cost-effective; hence, there is a need to develop faster and cheaper ways to determine inorganic phosphate concentration in water samples.

Reagent less enzymatic phosphate sensors that are based on sequential action of enzymes have been recently developed and represent an effective alternative to conventional methods (Parellada et al. 1998). The sensitive enzyme electrode of the biosensor is constructed by co-immobilizing four enzymes, namely, maltose phosphorylase, phosphatase, mutarotase, and glucose oxidase. The concentration of co-substrate "maltose" is kept constant while measuring inorganic phosphate. Maltose phosphorylase in the presence of inorganic phosphate hydrolyzes maltose to glucose-1-phosphate and α-D-glucose. Mutarotase brings about the mutarotation of α-D-glucose to β-D-glucose followed by its conversion by glucose oxidase. The product of this catalytic reaction influences the sensor signal.

11.2.8 Polychlorinated Biphenyls (PCBs)

Polychlorinated biphenyls are considered as important environmental pollutants and are widely used in industries as dielectric fluids in electrical transformers and capacitors. In spite of strict ban on the production of PCBs in many countries since past many years, these pollutants are still present in the environment and pose threat to public health. Determining the presence of PCBs in the environment quantitatively can be possible with the use of various biosensor configurations. Few of them which are commonly employed in PCB detection include immunosensors with SPR, electrochemical or fluorescence detection principles, and DNA biosensor with chronopotentiometric detection principle.

11.2.9 Phenols

Paper and pulp industries, industries based on the production of drugs, dyes, and antioxidants, are the major contributors of the release of phenols especially chlorophenols as toxic pollutants which accumulate in the environment. Vasoconstriction, renal tube degeneration, decrease in liver function, cancer, and neurodegenerative diseases are some of the health issues observed when catechol, a phenolic derivative, is absorbed by the gastrointestinal tract. Determination of phenol index in environmental samples can be done using amperometric biosensor which consists of tyrosinase (polyphenol oxidase having wide selectivity for phenolic compounds) as a biological component, immobilized in a hidrogel on a graphite electrode. Chlorophenols or other substituted phenols can be detected using flow injection chemiluminescence fiber optic biosensors which exploit the ability of analytes to enhance the chemiluminescence reaction of luminol, catalyzed by horseradish peroxidases.

11.2.10 Surfactants

Surfactants can be anionic surfactants which are widely used and can be cationic surfactants which represent only 5 % of the total.

Pseudomonas rathonis T which bears a plasmid for surfactant degradation acts a biological component in an amperometric biosensor for detection of anionic surfactants. Biosensors based on *Achromobacter* have also been constructed for their ability to degrade anionic surfactants. Decrease in dissolved oxygen concentration and hence change in the oxygen electrode current due to degradation of surfactants by the bacteria can be monitored. Oxygen consumption not only acts as an indicator of cell metabolism but also provides information on the surfactant content in the sample. It is possible to achieve high sensitivity, selectivity, and reproducibility with microbial sensors. One of the examples of whole-cell biosensors consists of immobilized linear alkylbenzene sulfonates (LASs) degrading bacteria which are based on the detection of consumed dissolved oxygen concentration in the degradation of LASs.

Alkylphenol ethoxylates (APEs) are group of nonionic surfactants and are shown to be estrogenic both *in vitro* and *in vivo*. APEs are used in wide variety of industries including paper and pulp, textiles, paints, metals, rubber, resins, adhesives, plastics, and latex, but due to their endocrine-disrupting properties, their detection gains prominent importance. Degradation of APEs into alkylphenols (APs) pose more risk as they show grater estrogenic activity. Biosensors have been developed which utilize capillary-based immunoassay (CIA) where glucose dehydrogenase is used as label for the detection of APEs and APs.

11.2.11 Antibiotics

Release of antibiotics in the environment has recently been considered as a matter of great concern as they promote antibiotic resistance. Genetic selection of more harmful bacteria is one of the major limitations of the tremendous use of antibiotics for therapeutic purposes or as growth promoters in dairy cattle or as feed additives in fish farms or in livestock during the past few years. Due to widespread administration of antibiotics, antibiotic resistance can be transferred to humans via ingesting affected meat and milk products. This raises serious food safety issues, and hence biosensors are developed to determine their presence in biological or food samples. Detection of penicillin G or tetracyclines in milk, sulfamethazines (cause allergic reactions) using an optical immunosensor in animal urine and studying the cross-reactivity between two sulfonamides (sulfamethazine and furosemide) using a commercial biosensor, BIACORE 3000, are some of the potential applications of biosensors.

11.2.12 Microbial Organisms

Bacteria, viruses, and other microorganisms in polluted, treated, and untreated water pose great threat to public health worldwide and sometimes can also be used as biological warfare agents. Pathogens may enter rivers and streams through agricultural runoff and via domestic wastewater. These pathogens/pathogenic compounds may reach humans if such water is being consumed for recreation or sport, for irrigation of fruits and vegetables, and as drinking water. It is important to monitor water supply for the presence of pathogenic compounds or organisms and to prevent the transmission of diseases from these sources.

Current conventional analytical methods for detection of microorganisms are based on colony-forming unit (CFU) count; require selective culture, biochemical, and serological characterization; and thus are expensive and time-consuming (Ercole et al. 2002). Methods employed in biosensors to detect microbial content of the sample rely on identification of consumption of oxygen or the appearance/disappearance of electrochemically active metabolite, analysis of nucleic acid by polymerase chain reaction methods, and immunological techniques. One of the examples of commonly employed biosensors include determination of *Escherichia coli* (*E. coli*) in water samples by an immunochemical potentiometric biosensor by monitoring change in redox potential due to production of ammonia by urease-*E. coli* antibody conjugate linked with *E. coli* cells in wastewater. Another example of sensor capable of detecting 10^3–10^4 *E. coli* cells/ml after an enrichment step is based on tyrosinase-catalyzed oxidation of polyphenolic compounds, which are produced from salicylic acid microbiologically. SPR sensor based on antibodies immobilized on gold sensor surface has been designed to detect *Salmonella enteritidis* and *Listeria monocytogenes* (Koubova et al. 2001).

11.2.13 Bioremediation

Bioremediation is one of the important technologies which involve microorganisms for treating environmental pollution. It is useful in treating oil pollution and brings out an efficient mineralization of hydrocarbons into water and carbon dioxide. Microorganisms are also shown to curb heavy metal pollution by binding with heavy metals and removing them from the environment or change the valency of the metals like chromium or mercury and reduce their toxicity. This process is effective but it is time-consuming, and sometimes removal of pollutants is not possible by indigenous population of bacteria.

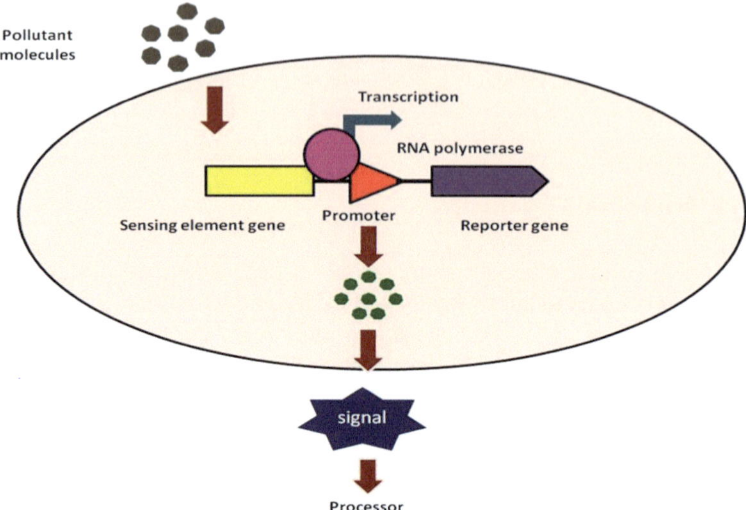

Fig. 11.6 Whole cell microbial bionsensor. Promoter selected from a generic operon and fused with a reporter system, can be made turn on or off based on the specific interaction of the biological component with the target molecule and generate signal

Biosensors are available molecular tools for monitoring pollution on site, *in situ*, and in a cost-effective manner. Biosensors consist of a biological component which is based on either a recombinant plasmid or a whole cell. It has got a reporter, a sensing element whose expression is sensitive to a target molecule (analyte), and a promoter which can be turned on or off in the presence or absence of target molecule. The reporter system is required to generate the signal where its intensity is directly proportional to the expression of promoter (Fig. 11.6).

Reporter systems code for specific gene is a part of expression vector and catalyzes biochemical reactions to generate a signal. Some of the commonly employed reporter systems include bacterial luciferase and green fluorescent protein (GFP) reporter system. The activity of bacterial luciferase reporter system depends upon the emission of light in the form of bioluminescence which is an enzymatic response of luciferase (coded by lux operon and fused with promoter and sensing element gene) activity. Emitted light can be received by a photomultiplier tube for signal analysis. GFP has got an internal chromophore, which confers its fluorescent property and, hence, emits bright green light when excited with ultraviolet or blue light.

Recent attempts have been made to construct biosensors/whole-cell biosensors to monitor the level of environmental pollutants such as toluene, octane, m-Xylene, and other aromatic hydrocarbons, heavy metals, etc. The efficacy of bioremediation can be determined by measuring the rate of elimination of pollutants from its site.

11.3 Biosensors for Clinical Monitoring

The complexity and diversity of human diseases have posed many challenges in the medical field, but owing to high selectivity and specificity toward the target analytes, inexpensive, integrated, and ready-to-use biosensor devices have been developed to improve the ability to detect pathogens or perform genetic analysis in hospitals and clinics or for point-of-care analysis.

More than 80 % of commercial devices based on biosensors are used in clinical monitoring. Yellow Spring Instruments produced first commercial biosensor device meant for glucose determination gained lot of popularity. Over the last decade, immunosensors and more recently DNA-based sensors have been extensively used to detect the presence of pathogenic species and to identify genetic polymorphisms and point mutations. Subsequent section focuses on applications of biosensors in clinical diagnostics for the appropriate interpretation of the identified and quantified biomarkers. Commercialization of biosensor technology in the field of clinical diagnostics can be explained by citing examples of its potential applications. Some of such examples are described in Table 11.2.

11.3.1 Cancer

Tumor biomarkers being the analytes can be detected by biosensors. Measuring the expression or secretion of proteins by tumor cells can help detect the presence or absence of tumor and its condition whether benign or cancerous or whether the treatment can be effective in eliminating or reducing cancerous cells. Detection of multiple tumor biomarkers by biosensors helps in diagnosis with improved sensitivity, specificity, and reproducibility.

No single oncogene or tumor suppressor gene has been found to be altered in all the cancers; hence, in the plethora of molecular biomarkers to decipher genome-related changes, protein under-/overexpression can be analyzed for tumor classification, diagnosis, monitoring treatment, and disease recurrence. Potential biomarkers can be of various molecular origins which include DNA (specific mutation, translocation, amplification, loss of heterozygosity), RNA, and protein (hormone, antibody, oncogene, tumor suppressor) and are typically detected in the serum, blood, cerebral spinal fluid, urine, and tumor tissues/cells.

Multi-analyte immunosensor has been constructed for the detection of seven tumor markers including α-fetoprotein (AFP), ferritin, β-human chorionic gonadotropin (β-hCG), carcinoembryonic antigen, cancer antigen 125 (CA 125), cancer antigen 15-3 (CA 15-3), and cancer antigen 19-9 (CA 19-9) by a competitive immunoassay with the detection limit of <2 ng/ml for all the markers. Capture antibodies are immobilized on electrodes which can capture specific target antigen. The detection is accomplished when signal transduction is realized via secondary antibody tagged with redox molecules or enzymes and oxidation current is generated simultaneously for all the electrodes.

Table 11.2 Potential clinical applications of biosensors

Type of diseases	Type of biosensor	Target/biomarker/analyte	Biorecognition element	Type of assay	Transducer element
Various cancers					
Cancer	Immunosensor	α-Fetoprotein (AFP)	Antibody	Antibody competitive assay	Electrochemical
Cancer	Immunosensor	Carcinoembryonic antigen (CEA)	Antibody	Antibody competitive assay	Electrochemical
Cancer	Immunosensor	Human chorionic gonadotropin (hCG)	Antibody	Antibody competitive assay	Electrochemical
Cancer	Immunosensor	Interleukin 6 (IL-6)	Antibody	Antibody direct assay	Acoustic (SAW)
Breast cancer	Immunosensor	Cancer antigen 15-3 (CA 15-3)	Antibody	Antibody competitive assay	Electrochemical
Breast cancer	DNA sensor	BRCA1 gene	Nucleic acid	Nucleic acid based assay	Electrochemical
Breast cancer	Immunosensor	Epidermal growth factor receptor 2 (HER-2)	Antibody	Antibody direct assay	Optical (SPR)
Ovarian cancer	Immunosensor	Cancer antigen 125 (CA 125)	Antibody	Antibody competitive assay	Electrochemical
Gastrointestinal tract carcinoma	Immunosensor	Cancer antigen 19-9 (CA 19-9)	Antibody	Antibody competitive assay	Electrochemical
Anemia cancer	Immunosensor	Ferritin	Antibody	Antibody competitive assay	Electrochemical
Oral cancer	Immunosensor	Interleukin 8 (IL-8)	Antibody	Antibody sandwich assay	Optical (fluorescence)
Prostate cancer	Immunosensor	Prostate-specific antigen (PSA)	Antibody	Antibody sandwich assay	Optical (SPR)

Chronic myelogenous leukemia	DNA sensor	BCR/ABL gene	Nucleic acid	Nucleic acid based assay	Electrochemical
Acute myocardial infarction					
	Immunosensor	Cardiac troponin T (cTnT)	Antibody	Antibody direct assay	Optical (SPR)
	Immunosensor	Cardiac troponin I (cTnI)	Antibody	Antibody sandwich assay	Electrochemical
				Antibody direct assay	Optical (FRET)
Cardiovascular diseases					
	Immunosensor	C-reactive protein (CRP)	Antibody	Antibody sandwich assay	Magnetic
	DNA sensor	Thrombin	Aptamer	Aptamer sandwich assay	Electrochemical
Inflammation					
	Immunosensor	C-reactive protein (CRP)	Antibody	Antibody sandwich assay	Magnetic
General physical stress					
	Immunosensor	Cortisol	Antibody	Antibody competitive assay	Optical (SPR)

Abbreviations: *SPR* surface plasmon resonance, *FRET* fluorescence resonance energy transfer, *SAW* surface acoustic wave, *BCR/ABL gene* breakpoint cluster region/Abelson gene

Prostate-specific antigen (PSA) is one of the first identified, reliable tumor marker for screening cancer at early stage and monitoring the recurrence of the disease after treatment. Conventional analytical methods to detect PSA have been replaced by innovative biosensors which are based on different transduction techniques, from electrochemical, piezoelectrico optical methods. Recently attention has been diverted to the use of electrochemical immunosensors based on carbon nanotubes with a detection limit of 4 pg/ml in low volumes of human serum and tissue samples. The design of biosensor is composed of 20–30 nm-long terminally carboxylated single-wallet carbon nanotubes (SWNTs) self-assembled on Nafion-iron oxide decorated conductive surfaces. The working principle of electrochemical immunosensor is based on specific interaction of primary antibodies attached to the SWNT forest with target antigen. The detection is achieved by monitoring the response of secondary antibody labeled with horseradish peroxidase (HRP) to hydrogen peroxide substrate. High sensitivity can be achieved by using high HRP/antibody ratio and linking secondary antibodies to wallet carbon nanotubes (CNT).

11.3.2 Hormones

Sex steroids regulate immune response and play important functions by modulating some inflammatory and autoimmune disorders. Radioimmunoassay (RIA) kits are commonly employed to determine progesterone, C21 (carbon 21) steroid levels in the serum or saliva. However, owning to problems associated with radioactivity, the use of immunosensor is an alternative approach. Immunosensors utilizes screen-printed electrodes as solid phase for a competitive immunoassay with estimated limit of detection of progesterone as 32 pg/ml.

Cortisol, another steroid hormone, is important for cardiovascular function and metabolic activities and considered as an indicator marker of stress and disease state. Detection of cortisol in saliva or serum using cortisol-specific monoclonal antibody is based on a competitive immunoassay with a six-channel portable SPR biosensor (Stevens et al. 2008). Human chorionic gonadotropin (hCG), an important diagnostic marker of pregnancy, has been considered as a target of electrochemical immunosensors which are based on the use of gold nanoparticles and ormosil sol-gel membranes.

11.3.3 Cardiovascular Diseases

Cardiac troponin I or T (cTnI/T), myoglobin, and natriuretic peptide (ANP), particularly of B type (B NP), are considered as potential biomarkers to diagnose heart infarction. Owning to slow response time and expensive nature of the conventional analytical methods including ELISA (enzyme-linked immunosorbent immunoassay), RIA, immune-chromatographic assays, and the use of several biosensors based on electrochemical and optical transduction are recommended. Immediate release of troponin T (TnT) in the bloodstream during heart infarction and its monitoring in short time by biosensors could improve patient care by allowing definitive diagnosis of myocardial infarction in real time (Cody Stringer et al. 2008).

C-reactive protein (CRP) has been recently used for conventional inflammation diagnosis. In routine clinical analysis, ELISA helps to determine CRP levels in the bloodstream (normal range in humans, 1–5 mg/l; protein levels, > 5 mg/l) with detection limits down to 0.2 mg/l as an indicator of inflammatory processes. It also serves as an important diagnostic marker to assist low-grade inflammation and risk in the patients for cardiovascular diseases. New methods in clinical diagnosis of CRP in cardiovascular diseases require its rapid quantification in native matrices such as saliva, urine and human serum. This is possible with the use of magnetic biosensors which utilize two CRP antibodies where one of them can be immobilized on polyethylene-sintered filters in ABICAP® plastic columns and acts as capturing antibody, while the other acts as secondary antibody, biotinylated and attached to streptavidin-coated magnetic beads. Interaction of this antibody-magnetic complex with the captured CRP on the primary antibody can be quantified by a magnetic reader. This highly sensitive system helps to determine CRP in native matrices with a very low detection limit of 0.025 mg/l (Meyer et al. 2007).

11.4 Future Perspectives

The primary aim of global research activities is to improve the quality of life and work for the welfare of the society. The better quality of life is closely related to medical diagnostics facilities, better control of diseases, drug and food quality control and safety, environmental monitoring, and pollution control. Sensitive, fast, continuous, and reliable monitoring is required to control important parameters, which is possible with the use of biosensors. Biosensors are the promising analytical tools which are simple to use, specific, cost-effective, and reliable and provide reproducible results.

Detailed study of biological processes in a biosensor for clinical and environmental analysis is required. Rendering artificial environment to the biomolecules in a biosensor can result in rapid loss of their activity, reduced stability, and low reproducibility of the response. The biosensor performance depends upon the nature and stability of biological element, method for immobilization of biomolecules, analyte specificity, type of transducer used, physiochemical properties of analyte, and operating conditions.

Detection of key substrate without prior separation, high selectivity, sensitivity in the order of ng/ml or pg/ml, short response times, quickness of data collection, and low cost-benefit ratio are some of the major advantages over traditional analytical tools which lead to their pronounced use in biomedical and environmental monitoring.

Although the concept of biosensor is simple and many of them with innovative working applications are at scientific stage. However, only few of them are commercialized and reached marketplace. Extensive research efforts need to be undertaken to produce new sensing elements with the capability to broaden the spectra of selectivities. Miniaturization of portable biosensors emerges as an important aspect of future bioelectronics as it allows on-field screening, handling of low-volume samples, reduction in reagent consumption and waste generation, and high-density information storage and increases sample throughput. The use of miniaturized flow cell and microsensor will have an impact not only on environment but also on economy and can become key technology of future times. Moreover, the vision of the future of biosensors is enormous and can be imagined to include chip-scale devices which when placed on human body, monitor vital signs for the disease, correcting abnormalities or even signaling a call for help after sensing an emergency situation.

References

Aracic S, Manna S, Petrovski S et al (2015) Innovative biological approaches for monitoring and improving water quality. Front Microbiol 6:826

Bachmann TT, Schmid RD (1999) A disposable, mutielectrode biosensor for rapid simultaneous detection of the insecticides paraoxon and carbofuran at high resolution. Anal Chim Acta 401:95–103

Chee GJ, Nomura Y, Ikebukuro K et al (2000) Optical fiber biosensor for the determination of low biochemical oxygen demand. Biosens Bioelectron 15:371–376

Chen H, Mousty C, Cosnier S et al (2007) Highly sensitive enitrite biosensor based on the electrical wiring of nitrite reductase by [ZnCr-AQS] LDH. Electrochem Commun 9:2240–2245

Choi SH, Gu MB (2002) A portable toxicity biosensor using freeze-dried recombinant bioluminescent bacteria. Biosens Bioelectron 17:433–440

Choi JW, Kim YK, Lee IH et al (2001) Optical organophosphorus biosensor consisting of acetylcholinesterase/viologen hetero Langmuir-Blodgett film. Biosens Bioelectron 16:937–943

Clark LC (1956) Monitor and control of blood and tissue oxygen tensions. Trans Am Soc Artif Intern Organs 2:41–48

Claude D, Houssemeddine G, Andriy B et al (2007) Whole cell algal biosensors forurban waters monitoring. Novatech 7:1507–1514

Cody Stringer R, Hoehn D, Grant SA (2008) Quantum dot-based biosensor for detection of human cardiac troponin I using a liquid-core waveguide. IEEE Sensors J 8:295–300

Durrieu C, Tran-Minhw C (2002) Optical algal biosensor using alkaline phosphatase for determination of heavy metals. Environ Res B 51:206–209

Ercole C, Gallo MD, Pantalone M et al (2002) A biosensor for *Escherichia coli* based on a potentiometric alternating biosensing (PAB) transducer. Sensors Actuators B83:48–52

Farre M, Pasini O, Carmen Alonso M et al (2001) Toxicity assessment of organic pollution in wastewaters using a bacterial biosensor. Anal Chim Acta 426:155–165

Frense D, Muller A, Beckmann D (1998) Detection of environmental pollutants using optical biosensor with immobilized algae cells. Sensors Actuators 51:256–260

Knecht MR, Sethi M (2009) Bio-inspired colorimetric detection of Hg^{2+} and Pb^{2+} heavy metal ionsusing Au nanoparticles. Anal Bioanal Chem 394:33–46

Koubova V, Brynda E, Karasova L et al (2001) Detection of foodborne pathogens using surface Plasmon resonance biosensors. Sensors Actuators B74:100–105

Mcgrath SP, Knight B, Killham K et al (1999) Assessment of the toxicity of metals in soils amended with sewage sludge using a chemical speciation technique and a lux-based biosensor. Environ Toxicol Chem 18:659–663

Meyer MHF, Hartmann M, Krause HJ et al (2007) CRP determination based on a novel magnetic biosensor. Biosens Bioelectron 22:973–979

Newman JD, Tigwell LJ, Turner APF et al (2004) Biosensors – a clearer view. Cranfield University Publication, Cranfield

Parellada J, Narvaez A, Lopez MA et al (1998) Amperometricimmunosensors and enzyme electrodes for environmental applications. Anal Chim Acta 362:47–57

Rodriguez-Mozaz S, Marco MP, Alda et al (2004) Biosensors for environmental applications: future development trends. Pure Appl Chem 76:723–752

Stevens RC, Soelber SD, Near et al (2008) Detection of cortisol in saliva with a flow-filtered portable surface plasmon resonance biosensor system. Anal Chem 80:6747–6751

Tan TC, Wu C (1999) BOD sensors using multi-species living or thermally killed cells of a BODSEED microbial culture. Sensors Actuators B54:252–260

Tiefenauer LX, Kossek S, Padeste C et al (1997) Towards amperometric immunosensor devices. Biosens Bioelectron 12:213–223

Tombelli S, Mascini M, Sacco (2000) A DNA piezoelectric biosensor assay coupled with a polymerase chain reaction for bacterial toxicity determination in environmental samples. Anal Chim Acta 418:1–9

Verma N, Singh MA (2006) *Bacillus sphaericus* based biosensor for monitoring nickel ions in industrial effluents and foods. J Autom Methods Manag Chem 2006:1–4

Suggested Further Readings

Berthuy OI, Blum LJ, Marquette CA (2016) Cancer-cells on chip for label-free detection of secreted molecules. Biosensors (Basel) 6(1):pii: E2. doi:10.3390/bios6010002

Dell'Atti D, Tombelli S, Minunni M et al (2006) Detection of clinically relevant point mutations by a novel piezoelectric biosensor. Biosens Bioelectron 21:1876–1879

Dzyadevych SV, Arkhypova VN, Soldatkin AP et al (2008) Amperometric enzyme biosensors: past, present and future. ITBM-RBM 29:17–180

Fernandez-sanchez C, McNeil CJ, Rawson K et al (2004) Disposable noncompetitive immunosensor for free and total prostate-specific antigen based on capacitance measurement. Anal Chem 76:5649–5656

Ivnitski D, Abdel-Hamid I, Atanasov P et al (1999) Biosensors for detection of pathogenic bacteria. Biosens Bioelectron 114:599–624

Minunni M, Tombelli S, Centi S et al (2007) Measurement of trace medical analytes. In: The handbook of biosensors and biochips. Wiley, Chichester, pp 1055–1067

Schulze H, Vorlova S, Villatte F et al (2003) Design of acetylcholinesterases for biosensor applications. Biosens Bioelectron 18:201–209

Soper SA, Brown K, Ellington AD et al (2006) Point-of-care biosensor systems for cancer diagnostics/prognostics. Biosens Bioelectron 21:1932–1942

Sun JZ, Peter Kingori G, Si RW et al (2015) Microbial fuel cell-based biosensors for environmental monitoring: a review. Water Sci Technol 71(6):801–809

Wilson MS, Nie W (2006) Multiplex measurement of seven tumor markers using an electrochemical protein chip. Anal Chem 78:6476–6483

Environmental Control of Biotechnology Industries

12

Rajesh Sharma, Rishi Srivastva, Kartikeya Shukla, and S.P. Tiwari

Contents

R. Sharma
Department of Biotechnology, VBS Purvanchal University, Jaunpur 222003, India

R. Srivastva • S.P. Tiwari (✉)
Department of Microbiology, VBS Purvanchal University, Jaunpur 222003, India
e-mail: sptiwarimicro@gmail.com

K. Shukla
Department of Environmental Sciences, VBS Purvanchal University, Jaunpur 222003, India

R.L. Singh (ed.), *Principles and Applications of Environmental Biotechnology for a Sustainable Future*, Applied Environmental Science and Engineering for a Sustainable Future, DOI 10.1007/978-981-10-1866-4_12

Abstract

Population explosion leads to deterioration and degradation of environment due to industrialization, urbanization, and agricultural practices. Industrial growth, economic development, urbanization, consumerization, etc. took place over last few decades to meet out the demand of growing population. All these activities result into generation of waste in enormous amount which is highly variable in nature. The nature of these wastes are simple organic compound to hazardous toxics materials using GMOs in industrial processing to produce desired products which involve different containment levels. Sustainable development includes the environment, economy, and community. It has become imperative to consider economic prosperity in such an integrated manner that social development is on one hand while environment protection on the other. There are various issues associated which greatly affect the sustainable development. These are regulation, planning, technological advancement assessment, globalization, and problems of developing countries. Environmental aspect of sustainable development and applications of technology must accept the recently implemented ambitious project CDM (Clean Development Mission) by the Government of India, wherein clean technology in general and green chemistry and white biotechnology in particular can make remarkable contribution toward the sustainable development. Wastes must be treated properly before disposing to the environment. Tools and techniques of biotechnology has given new impetus and opened new vistas in pollution control. Biosensors play critical role in detecting the pollutants even at very low concentration to assess the risk level.

Keywords

Environmental control • Industrial wastes • GMOs

12.1 Introduction

The Organization for Economic Cooperation and Development (OECD) defines biotechnology as "amalgamation of application of science and technology to living organisms, as well as parts, products and model thereof, to alter living or nonliving materials for the production of knowledge, goods and services." The pharmaceutical biotechnology reflecting the glamorous end of the market giving special emphasis on optimal production of desired products while less or no emphasis was given for the safe disposal of wastes/pollutant generated during the process of production. The prospects of a cure for many diseases and conditions currently promised by gene-based therapy and other biotech-based medical miracle affect us all. To cater to the needs of growing population in terms of food, feed, energy, pharmaceuticals, and clean environment forced to evolve technological advancement which revolutionized the industrial processing ultimately resulting into various problem across the globe. Industrial growth, economic development, consumerization, urbanization, and intensive farming (bad agricultural practices), all is leading to global problem of contamination of the whole components (primarily air, water, and soil) of the environment. On the other hand, there is continuous emergence of new diseases and resistance of pathogens to antibiotics such as Multidrug Resistance (MDR), Methicillin Resistance (MRSA), and Extended Beta Lactamase (ESBL) strains (Levy and Marshall 2004). In order to provide food, feed, energy, and pharmaceuticals to growing population, one has to search out for alternative sources and biotechnological processes that are being involved and play an important role in this context. The biotechnological processes are known from ancient times for the production of wine, bread, and vinegar using the microbes. "Sustainable development" has been introduced in 1987 in a report submitted to United Nations entitled *Our Common Future*. Development that meets the needs of the present without compromising needs of the future generations is the literal meaning of the sustainable development and by spirit too. The Earth Summit popularly known as UN Conference on Environment and Development (UNCED), 1992, and the World Summit on Sustainable Development, 2002, tried to work further on sustainable development of environment, economy, and community. Regulation, planning, technological advancement in the emerging era of LPG (Liberalization, privatization and Globalization), and problems of poor countries are several key and critical issues associated with sustainable development which are the matter of prime considerations. Industrial Ecology (a new term introduced in mid-1970s) is a growing concern for significant effects on global environment due to expansion of economic activities in which industries are supposed to optimize the cycle of materials and modify the process to reduce the negative impact on the environment. Similarly, industrial metabolism is a phrase introduced which deals industry as a one unit rather than assemblage of individual process.

New area of science emerged after the discovery of enzymes as tool for cutting and joining the DNA fragments from two different organisms and producing an organism expressing the gene from other never imagined before. The field was born with the initial effort of Paul Berg at Stanford University in 1971 (Berg et al. 1974) and Herbert W. Boyer (Univ. Calif. at San Francisco) and Cohen et al. (1973). The area is referred as biotechnology which is a combination of different branches of science for constructing the biological novel strains for the better service of mankind. Industrial application of biotechnology got its origin after the pioneer work of AM Chakraborty who constructed *Pseudomonas aeruginosa* strain for treatment of oil spills (Chakraborty et al. 1975).

There are three key points taken into accounts during industrial processing, namely, manufacturing process, waste management, and pollution control as shown in the following flow diagram. Accordingly, the range of business to which biotechnology has potential relevance is almost limitless. One area where this is most apparent is with regard to waste. Waste is a necessary outcome during industrial processing and their safe disposal leads to elevated cost of the product. Changes in

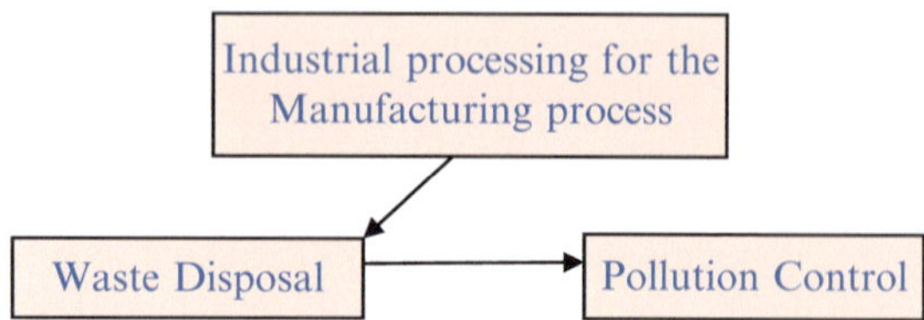

Fig. 12.1 The three intervention points

legislation throughout Europe, the USA, and other developed nations, combined with growing environmental awareness and burgeoning demand for reduced carbon footprint, have inevitably driven these issues higher up the political agenda, and biological methods of pollution control via waste treatment have gained greater acceptance. For those industries with particularly high biowaste production, the various available treatment technologies can offer considerable saving enabling the industries to grow in sustainable manner (Fig. 12.1).

Manufacturing industries can benefit from the applications of whole organisms or isolated biocomponents. Industrial processing generally involved harsh conditions such high pressure, temperature, and pH requiring high energy input, while microorganisms and enzymes typically catalyze the same process in mild conditions. This lower demand of energy greatly reduced the cost of the product on one hand and lesser generation of by-products on the other involving low cost for their waste management. There are two types of industrial products, one is the product of classical fermentation processes such as antibiotics and alcohol while recobinant product is another type such as insulin, interferons, and human growth hormones (Wurm 2004). Thus, mainly focuses on process technology, chemistry and classical chemical engineering aspects.

12.2 Colors of Biotechnology

The application of biotechnology in human welfare can be divided into different subarea.

Blue Biotechnology – is the use of marine and aquatic organisms for production of energy, new drugs, and active metabolites (for various purposes) or for production of sea food especially marine algae; natural pigments; and pharmaceuticals.

Red Biotechnology – is the application of biotechnology for the production of chemicals for health such as antibiotics and vaccines.

White Biotechnology – microbes are important role players in industrial process and have huge metabolic capability for production of valuable products such as enzymes, amino acids, flavoring compounds, biopolymers, etc.

Green and yellow Biotechnology – is related to the production of agricultural components like production of transgenic plants, e.g., for the production of bioinsecticides, biofertilizers, etc.

Black Biotechnology – is related to researches for use of microbes for biological warfare and bioterrorism.

Gray Biotechnology – is related to environment pertaining to biodiversity maintenance and decontamination aspects (DaSilva 2012).

12.3 Major Pollutant Released During Industrial Processing

Production can be improved by a shift in the original process and such a shift can be obtained by noting a process occurring in nature where there is interaction between the biotic and abiotic components. A term commonly used for such process is biomimicry (Volstad and Boks 2012). Industrial biotechnology employing set of technologies for production derives from adapting and modifying the natural properties of biological organisms and their metabolic capabilities for producing desired products at optimum level. These processes have evolved through evolutionary process over millions of years to combat existing environmental conditions with high efficiency.

Organisms preferred the route for their nutritive and metabolic processes though utilization of renewable resource, for example, sunlight for photosynthesis. Also, all bioorganic chemicals and materials occurring in nature are renewable and biodegradable (Rehm and Reed 1982). The waste product of one acts as nutrient source for other thus growing luxuriantly in nature by making interactions. There is nothing which can be considered as waste. In the last 25–30 years, biotechnology has evolved as powerful tools for developing bioprocesses and utilization of waste. This increased in efficiency and specificity for industrial sustainability. This has brought a drastic change in the industrial efficiency as use of microbial products for generating new products such as applications of microbial enzymes for generating pharmaceuticals earlier carried out by conventional chemical transformation process which cannot be considered as eco-friendly, e.g., synthesis of prednisolone and prednisone through biotransformation process (Vondrová and Čapek 1963).

As the biotechnological industries are growing at faster rate, it is responsible for producing waste and pollution which one has to curtail otherwise our existence will be in danger. Therefore, the target of future research will be more oriented toward sustainability of industries. Brundtland suggested the sustainability as the development and implications of plans to fulfill today's need altogether with future probabilities (Brundtland and Khalid 1987). An industry can be said as sustainable if it is sustainable economically, environmentally harmonious, and socially relevant, and a care should be taken before making plans for the economic as well as sustainable growth of industries. Major biotechnology-based industries and their wastes are summarized in Table 12.1 (Environmental Protection Agency Report 2015).

The biotechnology industries include the manufacture, extraction, processing, purification, and packaging of various biotechnological products. Biotechnological manufacturing can be divided into two major stages (Rowley 2010):

Table 12.1 Summary of waste type generating from industries

S. No.	Biotechnological industries	Type of waste
1.	Biometallurgical industries	Wastes from mineral metalliferous excavation
		Wastes from mineral non-metalliferous excavation
2.	Wastes from agriculture, horticulture, aquaculture, forestry, hunting, and fishing	Sludges from washing and cleaning, animal tissue waste, plant tissue waste, waste plastics (except packaging), animal feces, urine and manure (including spoiled straw), effluent, wastes from forestry
3.	Wastes from the preparation and processing of meat, fish, and other foods of animal origin	Sludges from washing and cleaning, animal tissue waste, materials unsuitable for consumption or processing, sludges from on-site effluent treatment, wastes not otherwise specified
4.	Wastes from fruit, vegetables, cereals, edible oils, cocoa, coffee, tea, and tobacco preparation and processing; conserve production; yeast and yeast extract production, molasses preparation, and fermentation	Sludges from washing, cleaning, peeling, centrifuging, and separation; wastes from preserving agents; wastes from solvent extraction; sludges from on-site effluent treatment
5.	Wastes from sugar processing	Soil from cleaning and washing beet, off-specification calcium carbonate, sludges from on-site effluent treatment
6.	Wastes from the dairy products industry	Materials unsuitable for consumption or processing, sludge from on-site effluent treatment
7.	Wastes from the baking and confectionery industry	Materials unsuitable for consumption or processing, wastes from preserving agents, sludge from on-site effluent treatment
8.	Wastes from beverages	Wastes from washing, cleaning, and mechanical reduction of raw materials; wastes from spirits distillation; wastes from chemical treatment and materials unsuitable for consumption or processing; sludges from on-site effluent treatment
9.	Wastes from pulp, paper, and cardboard production and processing	Wastes from sorting of paper and cardboard destined for recycling; lime mud waste; fiber rejects; fiber, filler, and coating sludges from mechanical separation; sludges from on-site effluent treatment
10.	Wastes from the leather and fur industry	Fleshings and lime split wastes, liming waste; degreasing wastes containing solvents without a liquid phase; tanning liquor containing chromium; tanning liquor free of chromium; sludge, in particular from on-site effluent treatment containing chromium; sludge, in particular from on-site effluent treatment free of chromium; waste tanned leather (blue sheetings, shavings, cuttings, buffing dust) containing chromium; wastes from dressing and finishing

(continued)

Table 12.1 (continued)

S. No.	Biotechnological industries	Type of waste
11.	Wastes from petroleum refining	Desalter sludges, tank bottom sludges, oil spills, oily sludges from maintenance operations of the plant or equipment, tars, sludges from on-site effluent treatment containing hazardous substances, sulfur-containing wastes from petroleum desulfurization
12.	Wastes from aerobic treatment of solid wastes	Non-composted fraction of municipal and similar wastes, non-composted fraction of animal and vegetable waste
13.	Wastes from anaerobic treatment of waste	Liquor from anaerobic treatment of municipal waste digestate from anaerobic treatment of municipal waste, liquor from anaerobic treatment of animal and vegetable waste, digestate from anaerobic treatment of animal and vegetable waste
14.	Landfill leachate	Landfill leachate containing hazardous substances

1. Primary/upstream processing – manufacture includes production of the active ingredient/drug.
2. Secondary/downstream processing – this involves conversion of the active ingredients/products in the form suitable for supply/administration.

Air emissions, wastewater, solid and hazardous wastes, threats to biodiversity, and bioethics are some potential environmental issues associated with pharmaceuticals and biotechnology-based manufacturing industries.

12.3.1 Volatile Organic Compounds (VOCs)

VOCs are emitted from various steps of upstream as well as downstream processing. Reactor vents from filtering systems during separation process; solvent vapors from purification tanks and dryers; and fugitive emissions from valves, tanks, pumps, and other equipment like centrifuges are the main source of VOCs emissions during upstream processing. Significant amount of VOCs are emitted also from various operations of downstream processing like mixing, compounding, granulation, and formulation where there is use of ethanol or isopropyl alcohol. Materials with high VOCs content should be reduced or substituted with the other products in order to reduce the VOCs emission. Equipment should be operated at lower temperatures, where possible. Cryogenic condensers should be used to reduce the gas stream temperature below dew point to achieve higher VOCs recovery and gas absorbers should be used to remove VOCs as well as other gaseous pollutants from a gas stream. Sometimes hypochlorite should be added in the gas absorbers for reducing emission of nuisance odors (Urashima 2000; Environmental Health and Safety Guidelines 2007).

12.3.2 Particulate Matter

Particulate matter can be emitted during milling, mixing, compounding, formulation, tablet making, and packaging. An abatement room should be also provided to remove the particulate from the air and High Efficiency Particulate Air (HEPA) filters should be used to control particulate matter emissions in the heating, ventilating, and air conditioning (Charlotte and Smith 2002).

12.3.3 Combustion Source Emissions

Exhaust gas emissions produced by combustion of fuels in turbines, boilers, compressors, pumps, and other machines for power and heat generation are the main source of air emissions from various biotechnology industries.

12.3.4 Odors

Odor emissions are typically associated with fermentation activities. Post-combustion of venting gases as well as appropriate height of exhaust stack should be used to reduce odor. Sometimes wet scrubbers are also used to remove odors with a high affinity to water (Chan et al. 2009).

12.4 Biological Waste Disposal

All biological, infectious, and chemotherapeutic waste that is generated should be disinfected and disposed properly. No infectious wastes are permitted to leave the place without first being disinfected or sterilized to ensure that they pose no harm to others or the environment. Infectious and chemotherapeutic waste requires special disposal procedures (Sadr et al. 2015).

12.4.1 Industrial Wastewater

Waste water streams in pharmaceuticals and biotechnology manufacturing depend on the specific process. The main conventional pollutants of concern in these waste water streams are added during fermentation, chemical synthesis, crystallization, purification, and biological/natural extraction. Various chemical compounds are present in the industrial wastewater including various solvents like methanol, ethanol, acetone, isopropanol, and methyl-ethyl ketone; organic acids like acetic acid and formic acid; organic halides; inorganic acids; ammonia; cyanide; toluene; and active pharmaceutical ingredients. These chemicals will affect various parameters of the water quality like Biochemical Oxygen Demand (BOD), Chemical Oxygen Demand (COD), Total Suspended Solids (TSS), ammonia, toxicity, biodegradability, and pH. The main aim of the liquid waste treatment is to reduce the BOD of the sewage, associated with suspended and dissolved organics, followed by the removal of inorganic nutrients, e.g., nitrogen and phosphorus, and recalcitrant organics before their disposal to water bodies (Chan et al. 2009).

Effluent treatment normally includes neutralization, flocculation, flotation, coagulation, filtration, settling, ion exchange, carbon adsorption, and detoxification of active ingredients by oxidation. These steps include grease traps, skimmers, and oil water separators for separation of oils and floatable solids which will reduce about 70–90 % of BOD depending upon the type of sewage. Filtration of suspended solids leads to flow and load equalization. Biological treatment is done by using trickling filters, anaerobic, activated sludge, and rotating biological contactors. Reverse osmosis or ultrafiltration is used for recovery of active ingredients. Exhausted carbon from adsorption processes may be sent for regeneration or combustion. Biological nutrient removal is done for reduction of nitrogen and phosphorus and chlorination of effluent is usually done when disinfection is required. Disposal of hazardous wastes is done for landfills. Sometimes, air or steam stripping and condensation processes are performed to remove volatile organics. Low-boiling compounds from waste water stream are removed by fractioned distillation. Toxic metals are precipitated and filtered out. Halogenated compounds are removed by solvent extraction to reduce the COD loads (Marrot et al. 2004; OECD report 1992).

12.4.2 Management of Unwanted/Expired Chemicals

Controlled substances that are expired or no longer used to research must be disposed of through a reverse distributor. Arrangements should be made to reverse the unwanted/expired products to the distributors to ensure the secured transport.

A little introduction for the treatment of wastes coming out from various industries in the form of solid, liquid, and gas are described below.

12.5 Treatment of Wastes

Bag filters, electrostatic precipitators, thermal precipitators, bioscrubbers, and activated carbon adsorber are used for the removal of air pollutants.

Waste water can be treated by conventional treatment methods involving primary, secondary, and tertiary treatments. In primary treatment, various physical and mechanical devices are used such as bar screen to remove the bigger objects with varying mesh sizes of the screen, communitor to reduce the particle size, hydrocyclones to remove the suspended matters, and fat trap to remove the oily and greasy materials. Secondary treatment involves biological treatment either aerobic or anaerobic mode involving microorganisms (bacteria, fungi, protozoans, worms, etc.). Aerobic treatment involves a set of microorganism with specific degradation capability, e.g., carbohydrates are attacked by cellulases, proteins by proteases, and lipid by lipases using specially designed process like activated sludge process, trickling filter, rotatory disk contactors, etc., while anaerobic processes involve anaerobic digester, Upflow Anaerobic Sludge Blanket Reactor (UASB), etc. involving sets of microorganisms. Tertiary treatment involves basically polishing of the water for reuse for various purposes. It involves chemical and biological methods. A summary of typical wastewater treatment plan is shown in Table 12.2.

Solid wastes are generally disposed into low-lying area as an open dumping site which involve low or no costs but putting the environment under stress condition. The two heat-involving methods of breaking down the solid wastes are:

Table 12.2 Summary of wastewater treatment

Primary treatment	Secondary treatment	Tertiary treatment
Using physical and mechanical devices	Using biological materials with various types of reactors. Categorized as:	Removal of specific pollutants and polishing of the water by chemical and biological methods
Bar screen – to remove bigger objects	i. Aerobic treatment	
Grit chambers – holds settleable mineral materials	Activated sludge process	Removal of nitrogen
Fat trap – to remove surface floating oily and greasy materials	Trickling filter	Removal of phosphorous
Hydrocyclone or centrifugal separators – to remove suspended solid material	Fluidized bed reactors	Removal of heavy metals
Flotation – another method to remove suspended solid materials using air bubble	Rotating biological contactors	Removal of biocides
	ii. Anaerobic treatment	Removal of pathogens using oxidizing/reducing agents
	Conventional method – septic tank	
	Anaerobic filters	
	Packed bed reactor	
	Anaerobic baffled digester	
	Contact digesters	
	Upflow anaerobic sludge blanket reactor	
	Anaerobic rotator biological contactors	
	Anaerobic fluidized bed reactor	
	iii. Specially designed	
	Membrane reactor	
	Sequential batch reactor	

(i) Pyrolysis is the destructive distillation in an oxygen-free environment. Hydrocarbons such as cellulose, plastic, and rubber (long-chain carbon) are the main components of solid wastes. On exposure at high temperature, these are broken down in to gases like CO_2, CO, H_2, C_2H_2, C_2H_4, and CH_4 and liquid like tar, light oil, and water-soluble distillate and solid tar. In this method, heterogeneous and complex nature of wastes is converted into homogeneous and simpler wastes.

(ii) Incineration: Muffle earth furnace and fluidized bed furnace is being used and of main types of incinerators. Almost 90 % volume is reduced after incineration but the main problem is associated with air pollution during burning process.

12.6 Recombinant DNA Technology

Recombinant DNA technology has tremendous potential and finds use in many novel experiments and applications. But it will be futile to run blindly after them. Application and release of genetically engineered organisms into the environment could lead to ecological consequences and potential risks unless necessary precautions are taken into account. Consequently several countries have formulated safety guidelines and regulations for research in the field of recombinant DNA technology. Considering the possible incremental risks associated with the use of recombinant strains and pathogenic microorganisms, the National Biotechnology Board issued a set of safety guidelines for India in 1983 to ensure the safety of workers in the laboratory environment. Remarkable developments have already been done in the last few years in the field of genetic manipulation and the scenario has shifted from the laboratories to the market. Recombinant organisms have several applications in large-scale fermentation operations as well as in the environment. The guidelines for the use of recombinant strains are prescribed in Good Industrial Large-scale Practices (GLSP). These criteria include measures such as proper engineering for containment, quality control, personnel protection, and medical surveillance. Nowadays most of the industrial processing shifted from chemical- to biological-based industries. Therefore, biological materials (recombinant or nonrecombinant) must be handled carefully and follow the recommended instructions as it may be potentially lethal. The following are the containment levels which must be taken in account while processing the biological materials.

12.7 Containment Levels

The term "containment" is used in describing safe methods for managing infectious materials in the laboratory environment where they are being handled or maintained. The purpose of containment is to reduce or eliminate exposure of laboratory workers, other persons, and the outside environment to potentially hazardous agents. *Primary containment*, the protection of personnel and the immediate laboratory environment from exposure to infectious agents, is provided by both good

Table 12.3 Risk group classification

Risk group	Individual risk	Community risk
1.	None or low	None or low
2.	Moderate	Low
3.	High	Low
4.	High	High

microbiological technique and the use of appropriate safety equipment. The use of vaccines may provide an increased level of personal protection. *Secondary containment*, the protection of the environment external to the laboratory from exposure to infectious materials, is provided by a combination of facility design and operational practices. Therefore, the three elements of containment include laboratory practice and technique, safety equipment, and facility design. The risk assessment of the work to be done with a specific agent will determine the appropriate combination of these elements. Pathogenic microorganisms are classified in different risk groups, in the increasing order of risk, based on the following parameters

1. Pathogenicity of the process organisms
2. Route of transmission and host range of the agent
3. Availability of effective preventive treatments or curative medicines
4. Capability to cause diseases to humans/animals/plants
5. Epidemic-causing strains
6. Volume and number of the process organisms

These parameters may be influenced by levels of immunity, density, and movement of host population, presence of vectors for transmission, and standards of environmental hygiene. Characterization of donor and recipient organisms, characterization of the modified organism, and expression and properties of the gene product are the important scientific considerations for assessment of potential risks. Containment facilities for different risk groups (Table 12.3) are as per the recommendations of World Health Organization (WHO).

12.7.1 Safety Equipment (Primary Barriers)

Safety equipment includes Biological Safety Cabinets (BSCs), enclosed containers, and other engineering controls designed to remove or minimize exposures to hazardous biological materials. The biological safety cabinet (BSC) is the principal device used to provide containment of infectious splashes or aerosols generated by many microbiological procedures. Open-fronted Class I and Class II biological safety cabinets are primary barriers which offer significant levels of protection to laboratory personnel and to the environment when used with good microbiological techniques. The Class II biological safety cabinet also provides protection from external contamination of the materials (e.g., cell cultures, microbiological stocks) being manipulated inside the cabinet. The gas-tight Class III biological safety cabinet provides the highest attainable level of protection to personnel and the environment. An example of another primary barrier is the safety centrifuge cup, an enclosed container designed to prevent aerosols from being released during centrifugation. To minimize this hazard, containment controls such as BSCs or centrifuge cups must be used when handling infectious agents that can be transmitted through the aerosol route of exposure.

12.7.2 Facility Design and Construction (Secondary Barriers)

The design and construction of the facility contributes to the laboratory workers' protection, provides a barrier to protect persons outside the laboratory, and protects persons or animals in the community from infectious agents which may be accidentally released from the laboratory. Laboratory management is responsible for providing facilities commensurate with the laboratory's function and the recommended biosafety level for the agents being manipulated. The recommended secondary barrier(s) will depend on the risk of transmission of specific agents. For example, the exposure risks for most laboratory work in biosafety level 1 and 2 facilities will be direct contact with the agents, or inadvertent contact exposures through contaminated work environments. Secondary barriers in these laboratories may include separation of the laboratory work area from public access, availability of a decontamination facility (e.g., autoclave), and hand washing facilities. When the risk of infection by exposure to an infectious aerosol is present, higher levels of primary containment and multiple secondary barriers may become necessary to prevent infectious agents from escaping into the environment. Such design features include specialized ventilation systems to ensure directional air flow, air treatment systems to decontaminate or remove agents from exhaust air, controlled access zones, airlocks as laboratory entrances, or separate buildings or modules to isolate the laboratory.

Elements of biosafety are important to reduce exposure of laboratory workers, other persons, and outside environment to potentially hazardous agents. It promotes a safe laboratory practice and procedures to ensure the proper use of containment equipment and facilities. It provides advice on laboratory design and risk assessment of experiments involving infectious agents/r-DNA *in vitro* and *in vivo*.

There are four different levels of lab containment protocols from BL-1 to BL-4 in rising order of danger, summarized in Table 12.4 (Benenson and Abram 1995). (Courtesy OECD Report 1992, rue André-Pascal, 75775 PARIS CEDEX 16, France).

Table 12.4 Recommended biosafety levels

Biosafety level	Agents	Practices	Safety equipment (primary barriers)	Facilities (secondary barriers)
1.	Not known to consistently cause disease in healthy adults	Standard microbiological practices	None required	Open bench top sink required
2.	Associated with human disease, hazard = percutaneous injury, ingestion, mucous membrane exposure	BSL-1 practice plus: Limited access Biohazard warning signs "Sharps" precautions Biosafety manual defining any needed waste decontamination or medical surveillance policies	Primary barriers = class I or II BSCs or other physical containment devices used for all manipulations of agents that cause splashes or aerosols of infectious materials; PPEs: laboratory coats; gloves; face protection as needed	BSL-1 plus autoclave available
3.	Indigenous or exotic agents with potential for aerosol transmission; disease may have serious or lethal consequences	BSL-2 practice plus Controlled access Decontamination of all waste Decontamination of lab clothing before laundering Baseline serum	Primary barriers = class I or II BCSs or other physical containment devices used for all open manipulations of agents; PPEs: protective lab clothing; gloves; respiratory protection as needed	BSL-2 plus Physical separation from access corridors Self-closing, double-door access Exhausted air notre-circulated Negative airflow into laboratory
4.	Dangerous/exotic agents which pose high risk of life-threatening disease, aerosol-transmitted lab infections, or related agents with unknown risk of transmission	BSL-3 practices plus Clothing change before entering Shower on exit All material decontaminated on exit from facility	Primary barriers = All procedures conducted in class III BSCs or class I or II BSCs in combination with full-body, air-supplied, positive pressure personnel suit	BSL-3 plus Separate building or isolated zone dedicated supply and exhaust, vacuum, and deconsystems Other requirements outlined in the text

12.8 Risk Assessment

"Risk" infers to the harm, injury, or incidence of disease to occur in context to biotechnological laboratories and the assessment of risk is to pay attention mainly on the prevention of laboratory-associated infections. Risk assessment is a critical and productive exercise while working on infectious or potentially infectious material. These consideration helps to allocate the biosafety levels (facilities, equipment, and practices) reducing the worker's as well as the environment's threat of exposure at minimum level. During qualitative risk assessment, one has to identify and explore all the risk factors and formulate according to the existing guidelines and manuals available across the globe (Collins 1983). The task becomes more complicated in those cases where complete information regarding these factors is not available. The factors of interest in a risk assessment include:

- The *pathogenicity* – generally, in biotechnological industries, the strain used is generally regarded as safe (GRAS), but in the case of the infectious or suspected infectious agent, one has to consider its ability of pathogenesis including disease incidence and severity. There is high incidence of risk if the organism has more severe potentiality for infection. For example, *Staphylococcus aureus* only rarely causes a life-threatening infection in laboratory conditions and is consigned to BSL-2. Some of the viruses, such as Ebola, Marburg, and Lassa fever that cause diseases with high mortality rates and for which there are no vaccines or treatment, are considered to be at BSL-4; however, other factors also play important role in regulating disease severity.
- The *route of transmission* – most of the laboratory infections spreads through the aerosol route. During the planning of experiments with an uncharacterized agent and its uncertain mode of transmission, one has to evaluate its potential for aerosol transmission, and if the aerosol potential is higher, there is high chance of risk.
- *Agent stability* – is a consideration that involves not only aerosol infectivity (e.g., from spore-forming bacteria) but also the agent's ability to survive over time in the environment. Factors such as desiccation, exposure to sunlight or ultraviolet light, or exposure to chemical disinfectants must be considered.
- The *infectious dose* can vary from one to hundreds of thousands of units. The complex nature of the interaction of microorganisms and the host presents a significant challenge even to the healthiest immunized laboratory worker and may pose a serious risk to those with lesser resistance.
- The *concentration* (number of infectious organisms per unit volume) will be important in determining the risk. Such a determination will include consideration of the milieu containing the organism (e.g., solid tissue, viscous blood or sputum, or liquid medium) and the laboratory activity planned (e.g., agent amplification, sonication, or centrifugation).
- The *origin* of the potentially infectious material is also critical in doing a risk assessment. Origin may refer to geographic location (e.g., domestic or foreign),

host (e.g., infected or uninfected human or animal), or nature of source (potential zoonotic or associated with a disease outbreak).

- The established *availability of an effective prophylaxis* or therapeutic intervention is another essential factor to be considered. The most common form of prophylaxis is immunization with an effective vaccine. However, important immunization only serves as an additional layer of protection beyond engineering controls, proper practices and procedures, and the use of personal protective equipment. Occasionally, immunization or therapeutic intervention (antibiotic or antiviral therapy) may be particularly important in field conditions.
- *Medical surveillance* is part of risk management. It may include serum banking, monitoring employee health status, and participating in postexposure management.

12.8.1 Materials Containing Known Infectious Agents

It is essential to recognize the characteristic features of infectious agent as it is worthwhile for risk assessment. This can be achieved through laboratory investigations of epidemiological studies and disease cycles. Several pathogens are associated with nosocomial as well as laboratory-related infections as mentioned in several guidelines such as *American Public Health Association's Manual, Control of Communicable Diseases* (Richmond et al. 1993). Existing literature on laboratory-acquired infections may also be helpful in formulating guidelines for risk exposure and assessment (Sewell 1995; Sulkin and Pike 1949, 1951; NIH Report Guidelines 1998; Sullivan et al. 1978; Pusztai and Bardocz 2006).

12.8.2 Materials Containing Unknown Infectious Agents

Most challenging area in biosafety consideration is on those specimens with limited information such as clinical specimens. In these circumstances, some interrogations may help in planning the risk assessment strategies to determine biosafety level such as studies on the infectious agent, epidemiological data available, its route of transmission, mortality and morbidity, and clinical data availability. In case of unavailability of data regarding the specimen, previously existing and reliable approach is advisable.

12.8.3 Materials Containing Recombinant DNA Molecules

This category includes those microbes which are genetically engineered containing chimeric DNA. Due to the rapid development of recombinant DNA technology, a lot of novel strains have been constructed using various organisms such as viruses, bacteria, yeast, and other microorganisms. Several strains have also been constructed having multiple antibiotic resistance genes and origin of replication from two different hosts (Shuttle vector), and in the future, it is likely that more complicated strains will be produced in future. Several guidelines and instructions are available from different organizations and will be useful for establishment of appropriate biosafety level for production and implementation of genetically modified organisms (GMOs).

While working on GMOs, one has to look for the following:

- What type of gene product is produced from inserted gene whether it is encoding a known toxin or unknown toxin?
- Whether the genetic alteration is likely to change the host range of the virus?
- Does the modification have the potential to increase the replication capacity of the virus?
- Does the inserted base sequence encode for a proto-oncogene?
- Does the inserted gene sequence have capability for altering the cell cycle?
- Does the viral DNA capable to integrate in host genome and undergoes lysogenic cycle?
- What is the possibility of generation of replication competent viruses?

All these questions must be addressed to judge whether a higher biosafety level is needed to work with GMOs (Abramowicz 1995).

12.9 Environmental Control

Industrial pollution not only poses the problem of effluents contaminating surface water bodies but it causes release of water bodies and it causes release of waste gases contaminating the air and contamination of land and ground water. Remediation of such contaminated land and groundwater is one of the major problems. Biotechnology has its contribution in the treatment of such sites using indigenous microflora or consortium of microorganisms that help in the degradation of pollutants and recover of land and groundwater. Cleanup of soil, groundwater, surface water, and wastewater coming out from various industries using biological agent is popularly termed as bioremediation.

Bioremediation is the use of biological materials to attain the environment in its original state making the environment pollution free. In broader sense, bioremediation comprised of enzymes, growth stimulators, bacteria, fungi, or plants to degrade, transform, sequester, mobilize, or contain contaminant organics, inorganics, or metals in soil, water, or air. Organism used in bioremediation may be characterized as either biostimulation, i.e., the addition of nutrients, or bioaugmentation, i.e., the addition of organisms, or processes that use both. Bioremediation operation may be made either on-site (*in situ*) or off-site (*ex situ*). Seeding contaminated site with competent microflora produced in the fermenters and used to speed up bioremediation is known as bioaugmentation (Frankenberger 2005). Constrains associated with bioremediations are as follows:

- The process of bioremediation is slow and large time is required in days or months.
- Heavy metals are not removed.
- *In situ* bioremediation for site cleanup is useful to biodegradable chemicals in soil with permeability.
- Cleanup of complex waste site will require application of several types of treatments.
- It does not remove all quantities of contaminants.
- For effective cleanup, rationally designed and optimized process should be introduced.
- Constant monitoring and evaluation should be done at regular time interval for which comprehensive pollutant audits are required on waste streams and contaminated sites. Decontamination of sites may be achieved either by natural microflora or by using intervention using remedial techniques. The following Fig. 12.2 shows principal types of bioremediations.

  The bioremediation of soil includes two processes:

(i) Pump and treat type process
This type of process is applied where soil is inaccessible (e.g., soil buried below the buildings). Groundwater is pumped to surface and treated under controlled condition in a reactor, and the cleaned water is returned to ground again. The process for surface soil and sludges are either solid-phase land farming or soil composting or liquid-solid process slurry.

(ii) Plume (saturated subsurface material) treatment
Plume is contaminant solutes in groundwater. Vinyl, trichloroethylene, and dichloroethylene are degraded using methane-oxidizing bacteria. Aromatic compounds, such as alkylbenzene, are also degraded by this technique.

12.9.1 Bioventing

Bioventing is also known as soil venting or soil vacuum extraction used for removal of oily contaminants above the water table. A bore is made near the point of contamination followed by application of vacuum from which extraction of volatile material is done. This removal of oily materials allowed oxygen, making the environment aerobic which helps in the degradation of organic materials, and appearance of carbon dioxide gas indicates the biodegradation process due to microbial activity.

12.9.2 Phytoremediation

Removal of contaminants in laboratory or field using various plant species (microscopic and macroscopic) is popularly known as phytoremediation. It generally

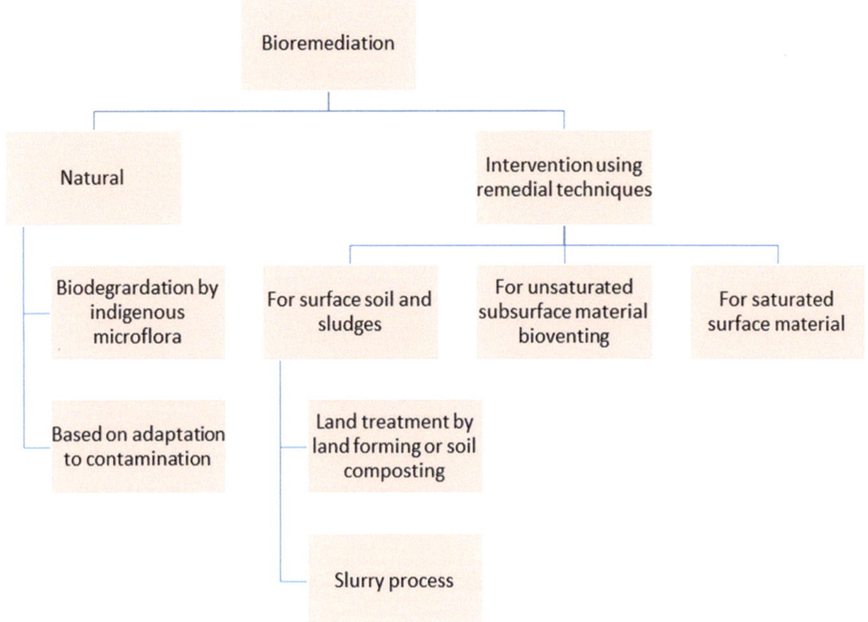

Fig. 12.2 Types of bioremediation

includes accumulation/degradation of heavy metal, radioactive material, chlorinated solvents, hydrocarbons, polyaromatic hydrocarbons, polychlorinated biphenyls, pesticides containing chlorine, military explosives and surfactants, etc. Phytoremediation may be (i) rhizofilteration, absorption, concentration, and precipitation of heavy metals by plant roots; (ii) phytoextraction, extraction and accumulation of pollutants from harvestable plant tissues; and (iii) phytotransformation, deals with conversion of complex organic material into simple forms. Phytoremediation is an alternative to other existing technologies such as incineration and chemical treatment including hazardous and radioactive waste. Phytoextraction gained attention for heavy metal uptake and can be enhanced by the addition of chelators. It has clear-cut two advantages over chemical treatments; one is capability of plants to accumulate pollutants from very dilute to highly concentrate solution, while the other is its selectivity and ability to adsorb target pollutant in a mixture of pollutants without any interference. Plants and algae like water hyacinth, *Chlorella*, *Scenedesmus*, etc. have been widely used for removal of phosphorus, nitrogen (nitrate, nitrite), and heavy metals for polishing of the water. This is a cheaper and eco-friendly method used for environmental cleanup, and the only limitation of phytoremediation is slow and time-taking (Moosavi and Mohamd 2013; Fantroussi and Agathos 2005; Fatima et al. 2005).

Biotechnological tools and techniques are being used to address some of the problems related to environmental concern and health hazard. The biological treatment of effluents is common practice for longer times throughout the globe.

However, conventional process now become obsolete because of highly variable nature of effluent which may overcome using biotechnological process such as degradation of xenobiotic compounds using co-metabolic properties of microbes via degradative plasmids (existing or reconstruct in laboratory).

Enzymatic detoxification (breaking down) of substances such as cyanides and also the by-products from the synthesis of S-triazine herbicides lignin degradation are used for the treatment of substances like polychlorinated biphenyls (PCBs) and dioxin effluents in several countries like USA and Europe. Microbial biotransformation of certain substances such as dibenzofurans, biarylketones halogenated dibenzodioxins, etc. is used to minimize the problem of pollution due to xenobiotic toxicity. The effluents released from paper and pulp industries contain chromophoric compounds and chlorogenated organic materials like chlorolignins, chlorosyringols, chloroaliphatics, catechols etc. which can affect the aquatic dynamics via mutagenic activities. Several white rot fungi (*Ganoderma lucidum*, *Trametes versicolor*, *P. chrysosporium*, *Coprinus macrorhizus*, *Hericium erinaceus*) and bacteria such as *Streptomycetes* etc. have been studied extensively for their capability to decolorize industrial dyes of potentially hazardous in nature. White rot fungi produces a variety of lignin-degrading enzymes (e.g., lignin-dependent peroxidases, manganese-dependent peroxidase, and lacasses commonly known as versatile peroxidases) that degrade phenolic, polyaromatic hydrocarbon, polychlorinated biphenyl, etc.

In order to solve the problem of soil contamination caused by extensive use of herbicides, pesticides, insecticides, biocides, and heavy metals in agricultural practices without any scientific consideration, decontamination of such soil has been carried out in past few decades using soil microorganisms.

12.10 Biosensors to Detect Environmental Pollutants

Biosensors are biophysical devices which may be an enzyme, an antibody, a nucleic acid, hormone, or an organelle/whole cell, which can detect the presence of specific substances, e.g., sugars, proteins, hormones, environmental pollutants, and a variety of toxic compounds capable of measuring the quantities of specific substances. The biosensors are being used in the area of medicine, industry, etc.; however, their use in environmental monitoring is of great benefit in biological monitoring program. The biosensor is basically comprised of the bioelement (enzyme, antibody, nucleic acid, tissue, microbial and polysaccharide, etc.) and sensor element (electric impedance, electric conductance, electric current, electric potential, intensity and phase of EM radiation mass, temperature, viscosity, etc.). Biosensors are classified into two broader categories on the basis of their applications: (i) clinical and (ii) nonclinical. Clinical biosensor may be used as *in vivo* or *in vitro* analysis such as artificial organs, glucose monitoring, anthrax, plague, cholera, and various other disease diagnosis. However, nonclinical biosensor may be used for pollution monitoring and pollutant detection, and industrial fermentation monitoring for real-time assessment.

Microbial biosensors are basically a unique combination of microorganisms and a transducer, which generate quantifiably the signal. The basic focus of this combination is the determination of a wide variety of substances in various fields related to biotechnology, pollutant monitoring, and pollution control. The emerging branch of biotechnology and immobilization technology play critical role not only for detecting the pollutants but also for controlling the pollution level such as fixed film bioreactors for wastewater treatment. The main advantages of microbial sensors are that they are inexpensive and easy to construct. Microbial biosensors based on bioluminescent techniques are also been used. However, the most recent focus on microbial biosensors is based on the use of recombinant microorganisms that recognize and report the presence of specific environmental pollutants. Environmental issues are most concerned nowadays as far as monitoring and pollution control is concerned. Pollution in drinking water from industry and agriculture is growing and of prime consideration and there is an immediate need for continuous online and real-time monitoring. Whole-cell biosensors gained attention during the past few years using living organisms as the sensitive agent to detection of pollutants and play important role in environmental monitoring and control. Microbial sensors are advantageous as they are tolerant to harsh conditions, repeated use, reproducibility, accuracy, robustness, and above all economically cost-effective. Ultrasensitive biological and chemical sensors are one of the great challenges before the scientific community. The next-generation biosensor requires significant improvements in sensitivity and specificity to meet out the requirements of a variety of fields including *in vitro* medical diagnostics, pharmaceutical discovery, and pathogen detection. Nanotechnology is an emerging branch of science that deals with the synthesis and transformation of materials to nanosize (10^{-9} m). The research in biosensor technology implemented either into transducers or receptors operation parts, so as to enhance their multi-detection capability and sensitivity. These nanomaterials are nanoparticles, nanotubes, etc. and contribute to either the bio-recognition element or the transducer or both and revolutionized the fields of analysis enabling rapidness with accuracy and specificity. Some of the important biosensors used in environmental pollution monitoring are:

(a) Gas biosensors – To detect gases such as sulfur dioxide, methane, and carbon dioxide etc, microbial biosensors have been developed. *Thiobacillus*-based biosensors can detect the pollutant sulfur dioxide, whereas methane can be detected by immobilized *Methylomonas*. A particular strain of *Pseudomonas* is used to monitor carbon dioxide levels (Dupont 1993).
(b) Immunoassay biosensors – Immunoelectrodes as biosensors are used to detect low concentrations of pollutants. Pesticide-specific antibodies can detect the presence of low concentrations of triazines, malathion, and carbamates, by using immunoassay methods (Pilon-Smits 2005).
(c) BOD biosensor – Biological oxygen demand is widely used to detect the levels of organic pollution load. This requires 5 days of incubation but a BOD biosen-

sor using the yeast *Trichosporon cutaneum* with oxygen probe takes only 15 min to detect organic pollution load (Moosavi and Mohamd 2013).

(d) Miscellaneous biosensors – A graphite electrode with Cynobacterium and *Synechococcus* has been developed to measure the degree of electron transport inhibition during the photosynthesis due to certain pollutants, e.g., herbicides, namely, 3-(3,4-dichlorophenyl)-1-dimethykl urea(DCMU), 2,4 diphenoxyacetic acid (2,4-D), and 2.4.5-trichloroacetic acid (2,4,5-T). To detect phenol, phenol oxidase enzyme obtained from white rot fungi (oyster mushrooms, *P. Chrysosporium*) is used as a biosensor. Biosensors for the detection of polychlorinated biphenyls (PCBs) and chlorinated hydrocarbons and certain other organic compounds have also been developed. Biosensors employing acetylcholine esterase which can be obtained from bovine RBC can be used for the detection of organophosphorus compounds in water.

12.11 Immobilization Technology in Environmental Control

The term immobilization is frequently used for confinement of enzymes/whole cells/part of the cells in inert support matrix in defined region of space with retention of its activity and can be used repeatedly or continuously. Presently immobilized systems are widely used in bioprocessing for the production of organic solvent, organic acids, drugs, artificial organs, and above all selective treatment of pollutants to solve the environmental problems. There are several methods available for immobilization as shown in Fig. 12.3.

(a) Carrier-binding method: This method is based on binding of cells/enzymes on insoluble matrix through different types of bonds. Carrier-binding method is of different types
   (i) Covalent Bonding: In this method the cell or enzyme is covalently attached to the support matrix and this technique is generally not suitable for cells because of toxic nature of inert matrices, e.g., cyanogen bromide, acid azide, etc.
   (ii) Ionic Bonding: Polysaccharide derivatives having an ion exchange group is generally used for immobilization of enzymes/cells.

(b) Physical Adsorption Method: This method is based on a physical interaction between the carrier matrix and enzyme/cells. It involves hydrogen bonding, hydrophobic interactions, van der Waal forces, or combination of these.

(c) Biospecific Binding Method: It is based on the biospecific interaction between enzyme/cells and other substances like coenzyme, inhibitor, effector, etc.

(d) Cross-linking: In this method a bi- or multifunctional compound is used for cross-linking which serves as the reagent for intermolecular cross-linking of the enzyme/whole cell.

(e) Entrapment Method: This method is itself categorized into five major types, i.e., lattice, microcapsule, liposome, membrane, and reverse micelle.

Immobilization technology offers several advantages such as repeated and continuous use of enzymes/cells which in turn enhances the overall efficiency; it does

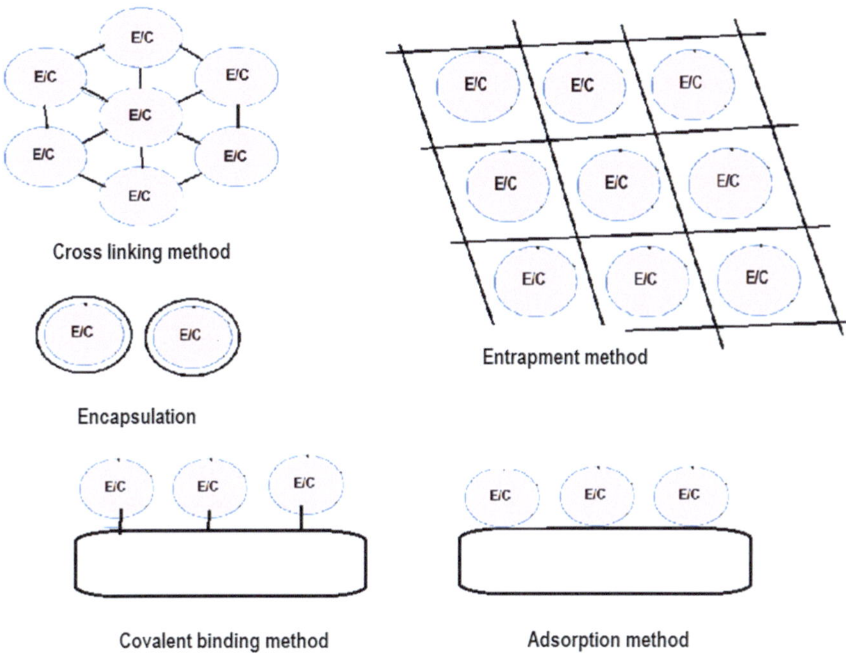

Fig. 12.3 Various methods of immobilization (*E* enzymes, *C* cells)

not contaminate the product; immobilized systems also offers ease of exploitation of kinetic features of continuous stirred tank and packed bed reactor. Some of the potential applications of the immobilization are as follows:

Hyphomicrobium species are grown on sand bed with added methanol causes nitrate reduction; *Micrococcus denitrificans* can be encapsulated in liquid membrane for reduction of nitrate to nitrite; immobilized amylase, lipase, protease, and pectinase are widely used for the wastewater treatment coming out from various industries like starch processing industries, detergent leather, and food and fruit processing industries. Similarly immobilized *Nitrosomonas europaea* is used for the oxidation of ammonia to nitrate and nitrite resulting into reduced BOD which prevents algal growth. Cyanide containing aqueous wastes may be detoxified with the help of immobilized *Stemphylium loti* by converting cyanide to formamide. Enzyme extract from consortium of microbial communities on immobilization can be used for hydrolysis of organophosphate insecticides. An activated sludge process, trickling filter, and rotatory disk contactor are basically a set of immobilized microbial population to reduce the BOD load by degrading the organic component.

Industrial revolution resulted into synthesis of various recalcitrant compounds commonly known as xenobiotic compounds, from Greek words “xenon” which means strange and “bios” which means life (stranger to existing microbial population), which is purely due to anthropogenic activity. Various factors make a compound recalcitrant such as unusual substitution of halogens (e.g., CFCs, polychlorinated biphenyl), nitro group (TNT, RDX, and other military explosives), and sulfonate group (surface active agents, commonly detergents) to aliphatic, aromatic, alicyclic, heterocyclic compounds. Recalcitrance is directly proportional to degree of branching of carbon chain, and when there is more length in the branching, resistivity to degradation is more. Xenobiotic compound itself is unable to induce the secretion of degradative enzymes. Insoluble nature and non-adsorption of these compounds at soil particles make it recalcitrant.

Partial degradation/mineralization may be achieved by degradative plasmids, an extrachromosomal material capable of autonomous replication. The genetics of degradative plasmid is similar to resistance plasmid may be high copy number or low copy number. Degradation of xenobiotic compounds involves either preexisting constitutive or newly synthesized induced enzymes of which part of the enzymes is encoded by chromosomal DNA, while the other part of the enzymes is encoded by plasmid DNA. In aromatic compounds, two types of ring opening have been reported either by ortho-cleavage (genomic DNA) or via meta-cleavage (plasmid DNA) as shown in Table 12.5. Pathway for benzene degradation using *Pseudomonas putida* containing degradative plasmid (TOL plasmid) is given below for better understanding of degradation of xenobiotic compounds. Beginning of degradation

Table 12.5 Bio-potentialities of microorganisms for xenobiotic compound degradation

S no.	Organism	Degradation of xenobiotic compounds
1.	*Pseudomonas* sp.	Benzene, anthracene, aliphatic and aromatic hydrocarbon, polychlorinated biphenyl, p-xylene, poly aromatic hydrocarbons, toluene, organophosphate pesticides like malathion, parathion, and rubber
2.	*Rhodotorula* sp.	Benzaldehyde
3.	*Arthrobacter* sp.	Benzene, hydrocarbons, pentachlorophenol, polyaromatic hydrocarbons, etc.
4.	*Azotobacter* and *Flavobacterium* sp.	Aromatic hydrocarbons
5.	*Alcaligenes* sp.	Halogenated hydrocarbons, alkyl benzene sulfonates
6.	*Corynebacterium* sp.	Halogenated hydrocarbons, phenoxyacetate
7.	*Bacillus* sp.	Aromatic hydrocarbon, long-chain alkanes, phenyl urea, phenol, cresol, salicylate
8.	*Rhodococcus* sp.	Naphthalene, polychlorinated biphenyl
9.	*Mycobacterium* sp.	Aromatic hydrocarbon, branched hydrocarbon, cycloparaffins
10.	*Nocardia* sp.	Hydrocarbons, alkyl benzene, naphthalenes, phenols, and substituted phenols
11.	*Lipomyces* sp.	Paraquet (herbicide)
12.	Methanogens	Aromatic hydrocarbons
13.	*Xanthomonas* sp.	Aromatic, aliphatic, polycyclic hydrocarbons
14.	*Candida* sp.	Polychlorinated biphenyl and formaldehyde
15.	*Streptomyces* sp.	Halogenated hydrocarbons, phenoxyacetate

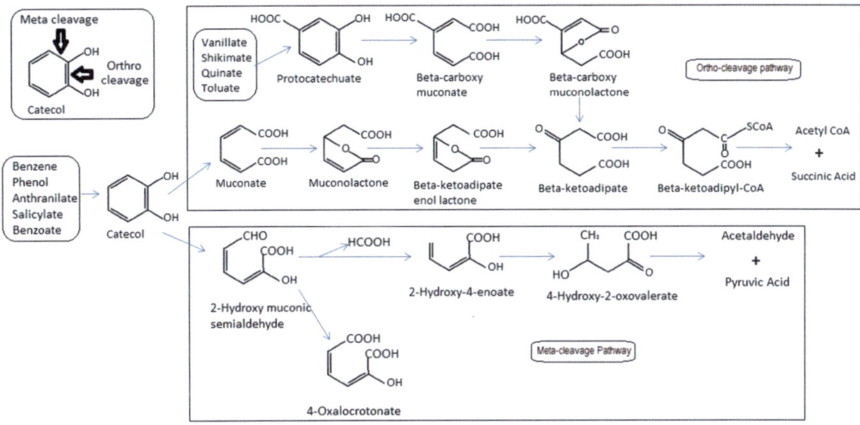

Fig. 12.4 Ortho and meta-cleavage

of aromatic compounds started with hydroxylation catalyzed by hydroxylases followed by oxidation resulting into generation of carboxyl group which ultimately degrade into carbon dioxide and via an essential intermediate protocatechuate and catechol as shown in Fig. 12.4.

References

Abramowicz DA (1995) Aerobic and anaerobic PCB biodegradation in the environment. Environ Health Prospect 103(suppl 5):97–99

Benenson AS (1995) In: Benenson AS (ed) Control of communicable diseases manual, 16th edn. American Public Health Association, Washington, DC, 20005

Berg P, Baltimore D, Boyer HW et al (1974) Letter: potential biohazards of recombinant DNA molecules. Science 185(148):303

Brundtland G, Khalid M (1987) Our common future ('Brundtland report')

Chakraborty AM, Mylroie JR, Friello DA, Vacca JG (1975) Transformation of *Pseudomonas putida* and *Escherichia coli* with plasmid-linked drug-resistance factor DNA. Proc Natl Acad Sci USA 72(9):3647–3651

Chan YL, Chong MF, Chung LL et al (2009) A review on anaerobic–aerobic treatment of industrial and municipal wastewater. Chem Eng J 155(1–2):1–18

Cohen SN, Chang AC, Boyer HW et al (1973) Construction of biologically functional bacterial plasmids in vitro. Proc Natl Acad Sci 70(11):3240–3244

Collins CH (1983) Laboratory-acquired infections, history, incidence, causes and prevention. Butterworths and Co Ltd, London

DaSilva EJ (2012) The colors of biotechnology: science, development and humankind. Electron J Biotechnol N Am 724:07

Dupont R (1993) Fundamentals of bioventing applied to fuel contaminated sites. Environ Prog Sustain Energy 12(1):45–53

Environmental Health and Safety Guidelines (2007) Pharmaceuticals and biotechnology manufacturing

Environmental Protection Agency Report (2015) Waste classification-list of waste and determining if waste is hazardous or non-hazardous, June, pp 1–55

Fantroussi SE, Agathos SN (2005) Is bioaugmentation a feasible strategy for pollutant removal and site remediation? Curr Opin Microbiol 8(3):268–275, ISSN 1369–5274
Fatima M, Bento A, Fla'vio AO et al (2005) Comparative bioremediation of soils contaminated with diesel oil by natural attenuation, biostimulation and bioaugmentation. Bioresour Technol 96:1049–1055
Levy SB, Marshall B (2004) Antibacterial resistance worldwide: causes, challenges and responses. Nat Med (Suppl) 10(12):122–129
Marrot B, Barrios-Martinez AP, Roche MN (2004) Industrial wastewater treatment in a membrane bioreactor: a review. Environ Prog Sustain Energy 23(1):59–68. doi:10.1002/ep.10001
Moosavi SG, Mohamd JS (2013) Phytoremediation: a review. Adv Agric Biol 1(1):5–11
National Institutes of Health (1998) Guidelines for research involving recombinant DNA molecules. GPO, Washington, DC, Federal Register. 59FR34496
Organization for Economic Cooperation and Development (OECD) Report (1992) Safety considerations for biotechnology, rue André-Pascal, 75775 Paris CEDEX 16, France
Pilon-Smits E (2005) Phytoremediation. Annu Rev Plant Biol 56:15–39
Pusztai A, Bardocz S (2006) GMO in animal nutrition: potential benefits and risks. Biol Nutr Growing Anim 4:513–540
Rehm HJ, Reed G (1982) Biomass, microorganisms for special applications, microbial products I. Energy from renewable resources. Verlag Chemie International, Inc., Deerfield
Richmond JY, McKinney et al. (1993) Biosafety in microbiological and biomedical laboratories. In: Richmond, Jonathan Y, McKinney (eds) Public Health Service, 3rd edn
Rowley JA (2010) Developing cell therapy bio manufacturing processes. Chem Eng Progr, SBE Stem Cell Eng, Nov Supplement 50–55
Sadr SMK, Saroj DP, Kouchaki S et al (2015) A group decision-making tool for the application of membrane technologies in different water reuse scenarios. J Environ Manag 156:97–105
Sewell DL (1995) Laboratory associated infections and biosafety. Clin Microbiol Rev 8:389–405
Smith, CA (2002) Managing pharmaceutical waste – what pharmacist should know. J Pharm Soc Wis: 17–22
Sulkin SE, Pike RM (1949) Viral infections contracted in the laboratory. N Engl J Med 241:205–213
Sulkin SE, Pike RM (1951) Survey of laboratory acquired infections. Am J Public Health 41:769–781
Sullivan JF, Songer JR, Estrem IE (1978) Laboratory acquired infections at the national animal disease center, 1960–1975. Health Lab Sci 15:58–64
Urashima K (2000) Removal of volatile organic compounds from air streams and industrial flue gases by non-thermal plasma technology. IEEE Trans Dielectr Electr Insul 7(5):602–614
Volstad NL, Boks C (2012) On the use of biomimicry as a useful tool for the industrial designer. Sustain Dev 20:189–199
Vondrová O, Capek A (1963) Microbial transformation of steroids. Folia Microbiol 8(2):117–128
Wurm FM (2004) Production of recombinant protein therapeutics in cultivated mammalian cells. Nat Biotechnol 11:1393–1398

Suggested Further Readings

Koziorowski B, Kucharski J (1972) Industrial waste disposal. Pergamon Press, New York, p 369
Scovazzo P, Chen WY, Wang LK et al (2005) Solvent extraction, leaching and supercritical extraction. In: Wang LK, Hung YT, Shammas NK (eds) Advanced physicochemical treatment technologies. The Humana Press, Totowa

Shammas NK (1997) Physico-chemically enhanced pollutants separation in wastewater treatment. In: Proceedings of the international conference: rehabilitation and development of civil engineering infrastructure systems- upgrading of water and wastewater treatment facilities, organized by The American University of Beirut and University of Michigan, Beirut, Lebanon, 9–11 June 1997

Vashon RD, Versteg DJ, McAvoy DC et al (2005) Aquatic environmental risk assessment of personal care product ingredients. In: Water Environment Federation 78th annual WEFTECH conference, Washington DC

Wang LK, Pereira NC, Hung YT (eds) (2005) Advanced air and noise pollution control. The Humana Press, Totowa, p 526

Bioleaching and Biomining

13

Surabhi Mahajan, Ankur Gupta, and Rajendra Sharma

Contents

S. Mahajan (✉) • A. Gupta
Department of Microbiology, School of Life Sciences, Dr. Bhim Rao Ambedkar University, Khandari Campus, Agra 282004, UP, India
e-mail: smahajanmb@gmail.com

R. Sharma
Department of Botany, School of Life Sciences, Dr. Bhim Rao Ambedkar University, Khandari Campus, Agra 282004, UP, India

R.L. Singh (ed.), *Principles and Applications of Environmental Biotechnology for a Sustainable Future*, Applied Environmental Science and Engineering for a Sustainable Future, DOI 10.1007/978-981-10-1866-4_13

Abstract

Universal reserves of high-grade ores are diminishing at an alarming rate due to the rapid increase in the demand for metals. Biomining is the extraction of specific metals from their ores through biological means, usually microorganism. Biomining is done in two steps often called bioleaching and biooxidation. Bioleaching commonly refers to biomining technology applied to base metals; whereas, biooxidation is normally applied to sulfidic-refractory gold ores and concentrates. Even though it's a new technique used by the mining enterprise to extract minerals equivalent to copper, uranium, and gold from their ores, however, nowadays, biomining occupies an increasingly primary place among the available mining applied sciences. The biomining methods are affordable, nontoxic, effective, and likewise environment pleasant. Utilizing biotechnology, efficiency of biomining can also be extended with the aid of genetically modified microorganisms.

Keywords

Acidithiobacillus • Heap leaching • *In situ* leaching • Commercial and technical challenges

13.1 Introduction

Minerals are assets of the nation which reflect in terms of strengthened values for economic growth of the country at large scale. Our ordinary mineral wealth has been exploited notably to a better extent for the past 50 years. With increasing industrialization coupled with population growth, the demand of metals has extended and is prone to go up additionally in the years yet to come. This has resulted in irreversible effects on diminishing high-grade ores with simultaneous generation of wastes and effluents containing metals. As a result, it leads to tackle the drawback for control of air pollution and healing of metallic values in a cost-potent procedure.

Worldwide reserves of excessive-grade ores are diminishing at an alarming rate as a result of the rapid expand within the demand for metals. Nonetheless there exist enormous stockpiles of low and lean grade ores yet to be mined. But the difficulty is that the restoration of metals from them using conventional systems could be very high priced due to excessive vigor and capital inputs required. A further main issue is environmental cost because of high level of pollution from these techniques. Environmental specifications proceed to stiffen, in particular related to poisonous wastes, so costs for making sure environmental security will continue to rise.

The use of microorganisms to facilitate the extraction and recovery of precious and base metals from primary ores and concentrates, referred to generically as "biomining," has developed into a successful and expanding area of biotechnology (Rawlings and Johnson 2007a, b). Biomining has application that works as an alternative to more traditional physical-chemical methods of mineral processing. The observe requires not one of the environmentally harmful strategies found in traditional refinement methods and as an alternative relies utterly on the ordinary interplay of organic organisms. Microorganisms are used to leach out the minerals, as a substitute than the common approaches of extreme heat or toxic chemicals, which have deleterious outcomes on the environment. Biomining procedure is a low-cost, secure, and efficient method of mineral recovery. This technology can be atmosphere pleasant because it generates minimal quantity of pollutants. It has a very low capital, low-operational fee, and it has a low power input method. It has the introduced advantage of mining low-grade ore and/or mine tailings. It is used to get better the various forms of minerals from an ore-utilizing microorganism. Utilizing biotechnology, effectively of biomining, may also be elevated with the help of genetically modified microorganisms.

Biomining is done in two steps referred to as bioleaching and biooxidation. Bioleaching usually refers to biomining technology applied to base metals, whereas mineral biooxidation is often utilized to biomining of sulfidic-refractory gold ores and concentrates. Nonetheless, within the technical literature, the terms are frequently used interchangeably. Physical-chemical procedures utilized in conventional mining technologies necessitate gigantic amounts of vigor for roasting/smelting and produce dangerous gaseous emissions similar to sulfur dioxide; biomining will aid to do away with these problems. Furthermore, the tailings generated by means of biomining operation are much less chemically lively. The biological exercise of those tailings is lowered to minimal as they have already been bioleached. The modest nutritional specifications and the irrigation needed to aid the preferred microbial existence in a heap or tank reactors are certainly much less expensive than the large cost associated with pyrometallurgical approaches.

13.2 Historical Past of Biomining

Biomining has an extended historical past, although the early miners did not know that microbes have been involved. Copper recovery from mine waters may also be dated to the fifteenth or sixteenth centuries. The usage of microorganisms to extract copper has its roots deep in antiquity. In his work on natural sciences, Plinius describes how copper minerals are obtained utilizing a leaching system. Georgius Agricola (1494–1555) a German medical professional and mineralogist describes the strategies for the healing of copper which might be founded on the leaching of copper-containing ores. In his book *De Re Metallica*, he illustrates the (guide) transport of metallic-containing leachates from mines and their evaporation in the sunlight.

On the Rio Tinto (Red River) mine in Seville, Spain, copper mine workings were rediscovered in 1556. Proof suggests that the mine used water from the Rio Tinto which comprises an extraordinarily excessive concentration of ferric ion as a result of microbial undertaking within the area. When the water from this river was irrigated into copper-containing deposits, the copper dissolved and later precipitated as smaller deposits. Although the persons at the moment likely believed this approach to be magic, we now understand that it was once the first recorded use of biomineralization.

The generation of bioleaching started with the discovery of the bacterium, *Thiobacillus ferrooxidans* (now *Acidithiobacillus ferrooxidans*), within the mid-1940s and the preliminary understanding of this microbe's involvement in copper extraction. In 1958, Kennecott Mining Enterprise patented the usage of *Thiobacillus ferrooxidans* for copper extraction and utilized the biohydrometallurgical approach to extract copper from run-of-mine (blasted, however uncrushed), low-grade copper ores from the Bingham Canyon Mine near Salt Lake City, Utah, USA. Development of the technology advanced rapidly during the 1980s, leading to the commissioning of the first commercial tank bioleaching plant at the Fairview Gold Mine near Barberton in South Africa.

13.3 Mechanisms of Biomining

Breaking down of minerals into constituent minerals is the basis of biomining. This provides energy to the microorganisms involved (Das et al. 1999). Originally, two types of mechanisms for the microbial metal solubilization of sulfide have been proposed (Ewart and Hughes 1991) to explain minerals (Figs. 13.1 and 13.2):

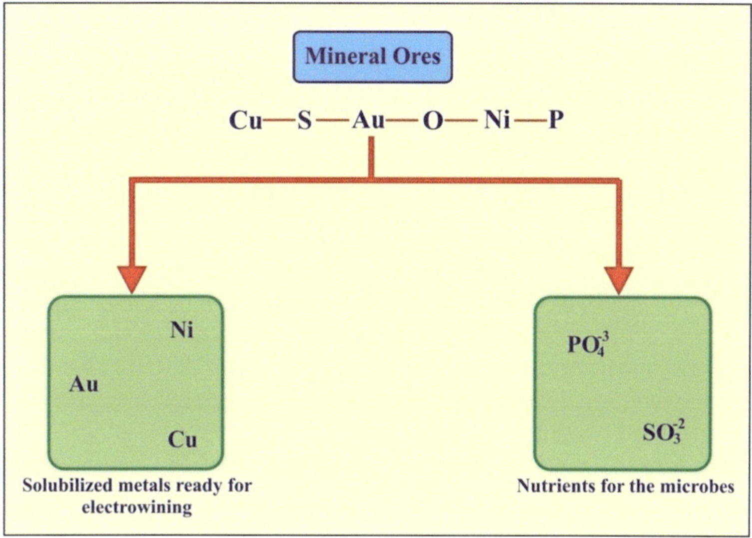

Fig. 13.1 Schematic diagram of biomining

Fig. 13.2 General mechanism of biomining

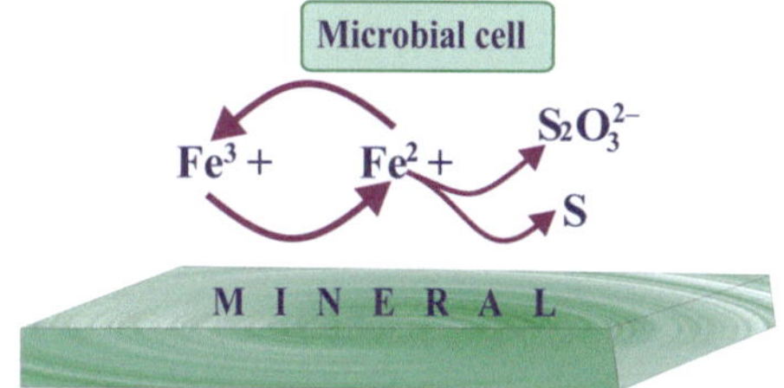

1. Direct mechanism
2. Indirect mechanism

13.3.1 Direct Mechanism

In this mechanism, microorganisms can oxidize metal sulfides obtaining electrons directly from the reduced minerals. Cells have to be attached to the mineral surface, and a close contact is needed. The adsorption of cells to suspended mineral particles takes place within minutes or hours. This has been demonstrated using either radioactively labeled *Thiobacillus ferrooxidans* cells grown on $NaH^{14}CO_3$ or the oxidative capacity of bacteria attached to the mineral surface. Cells adhere selectively to mineral surfaces occupying preferentially irregularities of the surface structure (Ewart and Hughes 1991). In addition, a chemotactic behavior to copper, iron, or nickel ions has been demonstrated for *Leptospirillum ferrooxidans*. Genes involved in the chemotaxis were also detected in *Thiobacillus ferrooxidans* and *Thiobacillus thiooxidans*.

13.3.2 Indirect Mechanism

The oxidation of reduced metals through the "indirect" mechanism is mediated by ferric iron (Fe^{3+}) originating from the microbial oxidation of ferrous iron (Fe^{2+}) compounds present in the minerals. Ferric iron can oxidize (e.g., metal sulfides) and is (chemically) reduced to ferrous iron which, in turn, can be microbially oxidized again (Ewart and Hughes 1991). In this case, iron has a role as electron carrier. It was proposed that no direct physical contact is needed for the oxidation of iron.

The following equations describe the "direct" and "indirect" mechanisms for the oxidation of pyrite:

Direct

$$2FeS_2 + 7O_2 + 2H_2O \rightarrow 2FeSO_4 + 2H_2SO_4$$

Indirect

$$4FeSO_4 + O_2 + 2H_2SO_4 \rightarrow 2Fe_2\left(SO_4\right)_3 + 2H_2O$$
$$FeS_2 + Fe_2\left(SO_4\right)_3 \rightarrow 3FeSO_4 + 2S$$
$$2S + 3O_2 + H_2O \rightarrow 2H_2SO_4$$

Now both the direct and indirect mechanisms have been combined, and a mechanism has been developed with the following characters:

1. Cells have to be attached to the minerals and in physical contact with the surface.
2. Cells form and excrete exopolymers.
3. The exopolymeric cell envelopes contain ferric iron compounds which are complexed to glucuronic acid residues. These are part of the primary attack mechanism.
4. Thiosulfate is formed as intermediate during the oxidation of sulfur compounds.
5. Sulfur or polythionate granules are formed in the periplasmic space or in the cell envelope.

There are two different reaction mechanisms that control the dissolution of metal sulfides. These are (1) acid-insoluble metal sulfides (thiosulfate pathway) and (2) acid-soluble metal sulfides (polysulfate pathway).

13.3.3 Thiosulfate Pathway

In thiosulfate mechanism, the attack of ferric ion on acid-insoluble metal sulfides like pyrite (FeS) and molybdenite (MoS) brings about solubilization via thiosulfate as an intermediate and sulfates as end product. In this group of metal sulfides, the chemical bonds between sulfur and metal moiety do not break until a total of six successive one-electron iron (III) hexahydrate oxidation steps have been conducted and thiosulfate is liberated. This mechanism is named after its first free sulfur compound (Fig. 13.3)

$$FeS_2 + 6Fe^{3+} + 3H_2O \rightarrow S_2O_3^{2-} + 7Fe^{2+} + 6H^+$$
$$S_2O_3^{2-} + 8Fe^{3+} + 5H_2O \rightarrow 2SO_4^{2-} + 8Fe^{2+} + 10H^+$$

13.3.4 Polysulfate Pathway

In polysulfide mechanism, a combined attack of ferric ion and protons on acid-soluble metal sulfides causes the solubilization with sulfur as intermediate in its elemental form which can be oxidized to sulfate by sulfur-oxidizing microbes (Fig. 13.3). Following are the reactions (Das et al. 1999; Schippers and Sand 1999):

$$MS + Fe^{3+} + H^+ \rightarrow M^{2+} + 0.5H_2Sn + Fe^{2+} \left(n \geq 2\right)$$
$$0.5H_2Sn + Fe^{3+} \rightarrow 0.125\,S8 + Fe^{2+} + H^+$$

The above reactions show the basic role of microorganisms in insolubilization, viz.:

1. To produce sulfuric acid for proton attack
2. To maintain iron in ferric state for oxidative attack on mineral

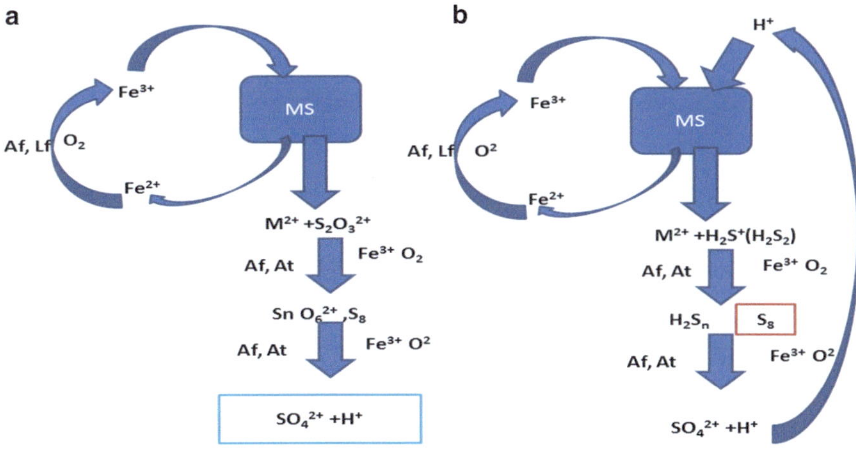

Fig. 13.3 Schematic diagram of the thiosulfate (**a**) and polysulfide (**b**) pathways

13.4 Techniques of Bioleaching and Biomining

According to the requirement, the following methods are used for the extraction of various metals:

1. *In situ* leaching
2. Dump leaching
3. Heap leaching
4. Vat leaching

13.4.1 *In Situ* Leaching

In this method, the mineral is extracted directly from the mine instead of collecting the ore and transferring to an extracting facility away from the site of the mine. *In situ* biomining is usually done to extract trace amounts of minerals present in the ores after a conventional extraction process is completed. The mine is blasted to reduce the ore size and to increase permeability and is then treated with water and acid solution with bacterial inoculums. Air supply is provided using pipes or shafts. Biooxidation takes place *in situ* due to growing bacteria and results in the extraction of the mineral from the ore (Fig. 13.4).

In situ leaching (ISL) is employed on underground ores to directly recover useful minerals and metals. The method is categorized under solution mining which involves drawing out the mineral from the earth by leaching and fluid recovery (Haque and Nargate 2013). It also comprises of drawing out minerals from the untouched ore. It is combined with mineral recovery operation time and again to

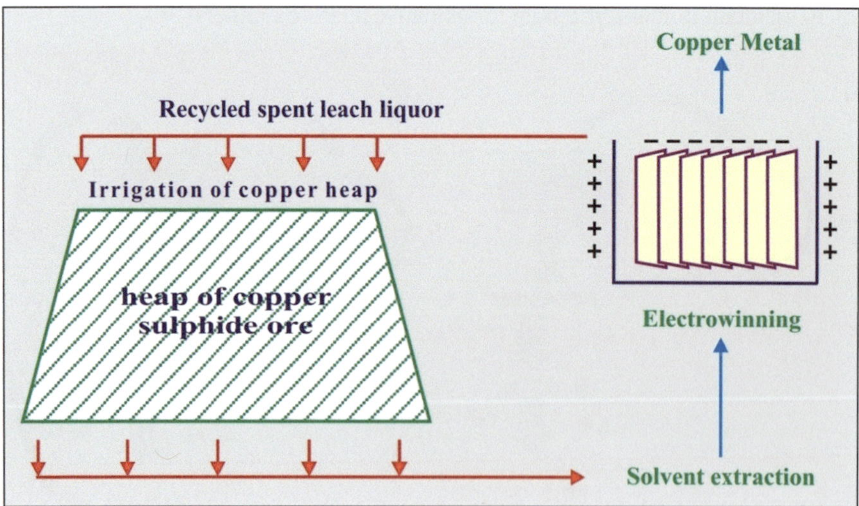

Fig. 13.4 Schematic representation of *in situ* leaching

pull out the minerals from recovered fluid solution or leachate. The large volume of solution is circulated with the aid of gravity flow and pumping. The extraction operation generates leachates known as "pregnant solutions," whereas the returning fluids to the extraction operation are known as "barren solution." The determining factor for this method is permeability of the ore body which can be increased by the fragmenting of ores in place called "rubblizing" (Fig. 13.5). *In situ* leaching was generally used for the extraction of uranium, copper, and gold (Haque and Nargate 2013).

13.4.2 Dump Leaching

Dump leaching is the oldest method. Dump leaching essentially includes the piling up of uncrushed waste rock in dumps (Siddiqui et al. 2009). It's used for mineral extraction which involves low-grade ores. The greater rocks are cracked via blasting in the pit and are carried as big fragments to dumps. These dumps incorporate run-of-mine ore in million tons. Acidified water is spread on the highest floor which percolates in the dump and creates required stipulations for the growth of microorganisms, with the intention to oxidize the mineral for extraction of the metal. Dump leaching was popularly used for the extraction of copper sulfide ores (Brierley 2008).

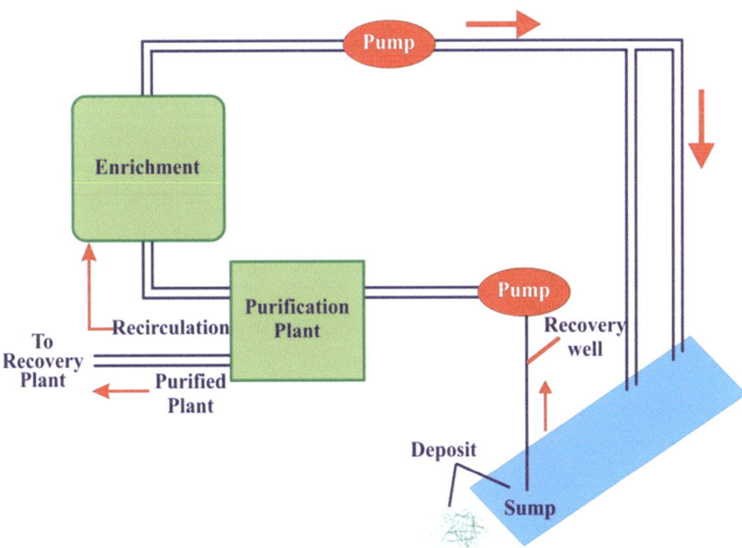

Fig. 13.5 Procedure flow for *In situ* Leaching

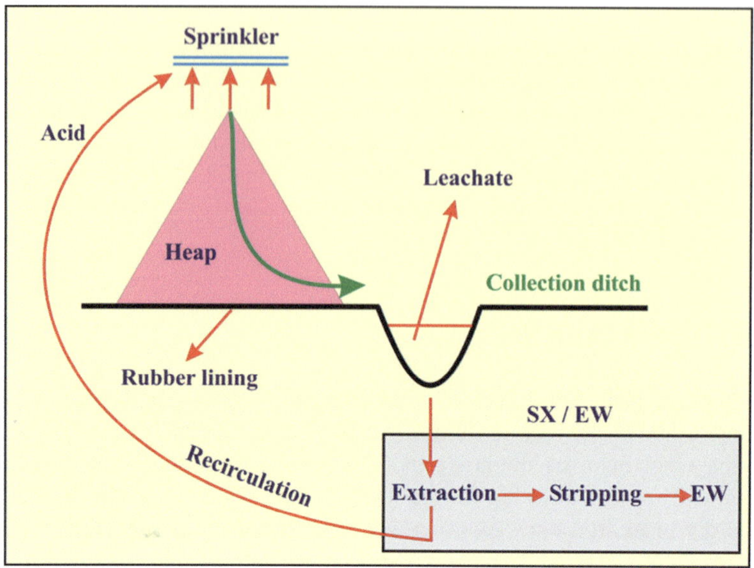

Fig. 13.6 Process flow for heap leaching

13.4.3 Heap Leaching

Heap bioleaching is a speedily rising method for the extraction of base metals from sulfide minerals. Bioheaps are a big amount of low-grade ores and effluents from extraction strategies that include trace quantities of minerals. In heap leaching, the overwhelmed secondary sulfidic ores are agglomerated with sulfuric acid adopted via stacking onto leach pads which can be aerated from the bottom of the heap. Then the ore is allowed to remedy for 1–6 weeks, and additional leached with acidic leach liquor for 400–600 days. Copper recovery of 75–95% is received in this period of time. Bioheaps are also known as biopiles, biomounds, and biocells. They're additionally used for biodegradation of petroleum and chemical wastes (Fig. 13.6).

As the development of heap reactors is affordable and easy to function, it's the favored cure of low-grade ores. Business utility of bioheap leaching designed to take advantage of microbial activity was pioneered in 1980 for copper leaching (Fig. 13.7).

13.4.4 Vat Leaching

Currently, vat leaching is operated to extract minerals from oxide ores which basically involves retaining of ore slurry and solvent for several hours in large tanks which are equipped with agitators (Fig. 13.8). It is used to carry out cyanidation for ores with high gold content and to extract precious metals from ores (Siddiqui et al. 2009).

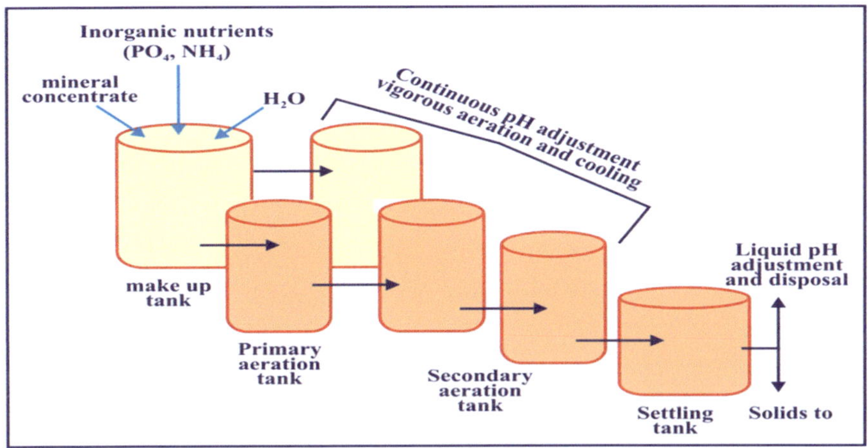

Fig. 13.7 Typical continuous-flow biooxidation approach for pretreatment of gold-bearing arsenopyrite concentrate

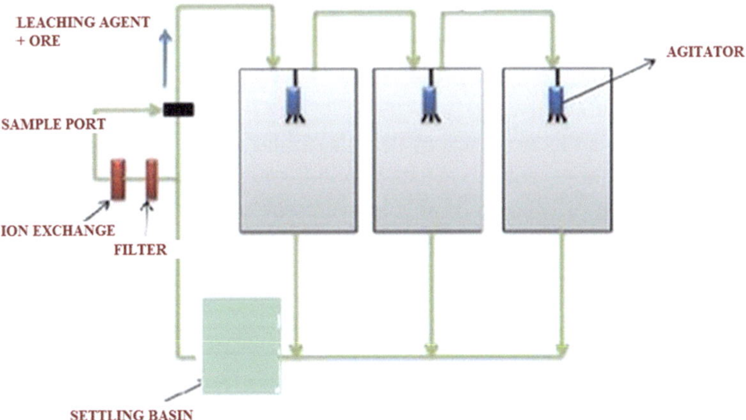

Fig. 13.8 Process flow for vat leaching

The techniques applied in biomining will be successful only when there is sufficient attachment of microorganism to the surface which is achieved by a successful biofilm development.

13.5 Biofilm Development

To initiate biofilm formation, firstly, properties of both the substratum (rock layer) and the cell surface are to be scrutinized. The attachment depends on the surface of the substratum such as smooth or rough, and also it depends on antimicrobial properties of the substratum. This feature of the substratum has a considerable effect on

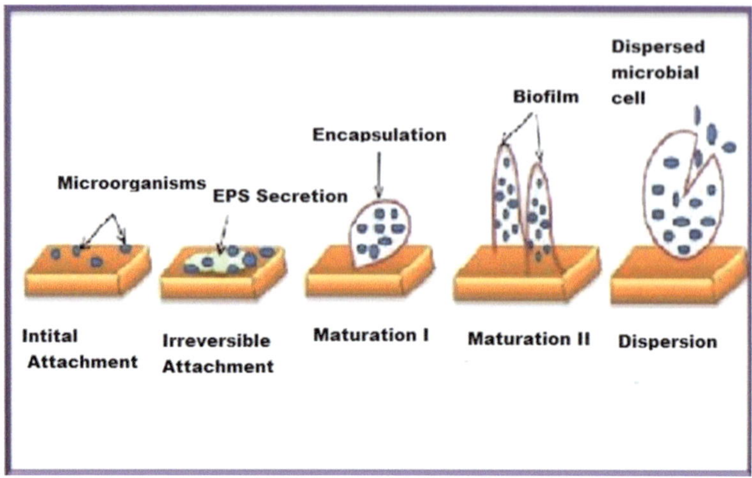

Fig. 13.9 Stages in biofilm development

the rate and degree of attachment by microorganisms. Basically, the rougher and highly hydrophobic materials (with some exceptions) will build up biofilm to be swifter (Donlan 2001; Fletcher and Loeb 1979; Pringle and Fletcher 1983; Characklis et al. 1990; Quirynen et al. 2000). The characteristics of cell surface also play a major role such as existence of flagella, fimbriae, pili, curli fibers, and outer membrane proteins or glycocalyx which may influence the rate of microbial attachment. The microbial cell should be able to overcome the repulsive forces, and these appendages which facilitate the cell continue the attachment until more permanent attachment procedures are in place (Donlan 2001). Biofilm development occurs in the following steps.

13.5.1 Initial Attachment

The initial attachment between any bacteria and nonliving surfaces is facilitated by nonspecific interactions such as hydrophobic interaction. But the attachment to the living surface is carried out through specific molecular docking mechanism (Dunne 2002). Further the specific adhesins such as polysaccharide intercellular adhesins (PIA) in biofilm on a nonliving surface (Dunne 2002; Rupp et al. 2001) (Fig. 13.9).

13.5.2 Irreversible Attachment

Cells attach to the surface irreversibly by secreting an extracellular polymeric substance (EPS) which comprises DNA, proteins, lipids, and lipopolysaccharides which smooth the progress of adhesion between cells and the substratum (Flemming and Wingender 2010).

13.5.3 Maturation I

The cells attached on the surfaces replicate and grow into microcolonies which have a diameter of tens or hundreds of microns. The encapsulation of bacteria in hydrogel formed due to the secretion of EPS generates a physical barrier between the community and the extracellular environment. Different microorganisms have different content of EPS which determines the growth conditions and chemical communication in cells in the community. Quorum sensing is important for chemical communication in bacteria. It adjusts the cellular functions, nutrient acquisition, motility, conjugation, pathogenesis, and secondary metabolite production.

13.5.4 Maturation II

The microbial community acquires a three-dimensional structure giving rise to a mature biofilm as a result of replication of cells and EPS accumulation. The EPS acts as glue for combining cells in an established biofilm to protect against mechanical stress and prevent the detachment from the surface.

13.5.5 Dispersion of Microbial Cell

Some microbial cells detach from the biofilm and disperse in bulk fluid forming new biofilms in different environmental niches. This step is important for propagation and self-renewal.

13.6 Microorganisms: Useful for Biomining

There are wide varieties of microorganism with various capabilities existing on earth. Hence, it's primary to establish exactly the forms that can participate in biooxidation/bioleaching effectively. A number of species of fungi can be used for biomining. Fungal strains *Aspergillus niger* and *Penicillium simplicissimum* had been competent to mobilize Cu, Sn, Al, Ni, Pb, and Zn. In a similar way, 'phytomining' is based on the tendency of some plant species to bioaccumulate immoderate quantities of metals from their host rock. The crops, referred to as hyperaccumulators, are grown on incredibly mineralized soils or post-mine lands, and their yield (bio-ore) is used as a pure metal supply. Compared to the bacterial mining, these technologies should not be so general particularly given that of the toughness of those approaches and so their unprofitability.

Biomining makes use of acidophilic microorganisms which might be autotrophic in nature and play a part in dissolution of metals from the sulfide ores. The acidophilic microorganisms actively participating are *Thiobacillus*, *Sulfolobus*, *Acidianus*, and *Leptospirillum* (Das et al. 1999; Buchanan and Gibbons 1974; Blake et al. 1993). The largest advantage of using these traces is that they can tolerate high concentrations of heavy metals (Asghari et al. 2013; Mousavi et al. 2007; Mishra et al. 2008). These microorganisms can also be classified as mesophiles, reasonable thermoacidophiles, and severe thermoacidophiles, based on their tolerance to temperature (Das et al. 1999).

Thiobacillus ferrooxidans is a chemophilic, moderately thermophilic microorganism which is able to produce vigor from oxidation of inorganic compounds like sulfur and iron. It is the mostly used microorganism in biomining. A number of other microorganisms such as *T. thiooxidans*, *Thermothrixthiopara*, *Sulfolobus acidocaldarius*, and *S. brierleyi* are additionally broadly used to extract quite a lot of minerals.

Thermothrixthiopara is a particularly thermophilic microorganism that can survive at very high temperatures between 60 and 75 °C and is used in extraction of sulfur. Strategies like genetic engineering and conjugation are used to provide microorganism with desired traits to expand the rate of biooxidation, for this reason increasing the mineral yield by means of biomining. It is usually foremost to identify biomining microorganism present in colonies of different bacteria.

13.6.1 Characteristics of Microorganism Used in Biomining

Bacteria are most suitable microorganisms for the extraction of minerals from low-grade ores. The traits of the microorganism used in biomining are:

1. Thermophilic bacteria which will continue to exist at high temperature are selected for biomining, as mineral extraction involves the excessive temperature strategies.
2. Microorganism that is chemophilic, as biomining uses both strong acid and alkalis.
3. Autotrophic bacteria which have the capability to provide vigor from inorganic compounds by means of photosynthesis or chemosynthesis are chosen for biomining.
4. Microorganism which has the capability to produce biofilms is selected for biomining purposes.

13.6.2 Microorganism Involvement in Biomining

13.6.2.1 Mesophiles

13.6.2.1.1 Acidithiobacillus

Biomining bacteria belonging to this genus were earlier incorporated in the genus *Thiobacillus*. Consequently of 16S rRNA gene sequence analysis, it is clear that the genus *Thiobacillus* included sulfur-oxidizing bacteria that belonged to α-, β-, and γ-divisions of the Proteobacteria. To resolve this anomaly, the genus *Thiobacillus* was subdivided and a new genus, *Acidithiobacillus*, was once created to accommodate the particularly acidophilic members of the former genus. These participants comprise *Acidithiobacillus ferrooxidans* (*A. ferrooxidans*) (previously *Thiobacillus ferrooxidans*), *A. thiooxidans* (previously *T. thiooxidans*), and *A. caldus* (previously *T. caldus*). These bacteria appear to be ubiquitous and have been isolated from sites that provide a compatible atmosphere for growth (such as sulfur springs and acid mine drainage) from many areas throughout the world.

Acidithiobacillus ferrooxidans

This was the first bacterium discovered that used to be competent of oxidizing minerals. Nutritionally, *A. ferrooxidans* isolates are obligate autotrophs. They are also in a position to develop on formic acid. For many years *A. ferrooxidans* was considered to be the important microorganism in biomining methods that operate at 40 °C or much less. It is currently understood that *A. ferrooxidans* just isn't preferred in circumstances wherein the ferric iron content material is very high than the ferrous iron.

Acidithiobacillus thiooxidans

A. thiooxidans is nutritionally very much like *A. ferrooxidans* besides that it's unable to oxidize ferrous iron and is as a consequence restricted to using reduced sulfur compounds as an electron donor. Like *A. ferrooxidans*, it is mesophilic and usually has a limit for growth of about 35 °C, but it's much more acid tolerant (pH 0.5–5.5). An investigation of copper-leaching columns making use of simulated recycled leach liquor with a high sulfate content material and pH 0.7 indicated that *A. thiooxidans* was the dominant sulfur oxidizer. The microorganism is exceptional in other respects, with *A. thiooxidans* having a genomic GC ratio of 53 %. The DNA-DNA similarity between *A. thiooxidans* and *A. ferrooxidans* is about 20 % or much less.

13.6.2.1.2 Leptospirillum

Bacteria of this genus are much like *Acidithiobacilli* in that they are also particularly an acid-tolerant (ideal pH 1.5–1.8), gram-negative, chemolithoautotrophic microorganism. On the basis of 16S rRNA gene sequence, they are not members of the Proteobacteria but belong to the division Nitrospira. The *Leptospirilli* are ready in utilizing ferrous iron as an electron donor. The *Leptospirilli* have an affinity for ferrous iron, and in contrast to *A. ferrooxidans*, their capacity to oxidize ferrous iron is not to be inhibited by ferric iron.

13.6.2.1.3 Acidiphilium

Acidiphilium are acid-tolerant, gram-negative heterotrophs as a substitute to iron- or sulfur-oxidizing autotrophs. As such, they are not specifically concerned in mineral decomposition. They are included considering the fact that they have been detected in a batch bioreactor and are commonly developing near bacteria similar to *A. ferrooxidans*, where they are believed to feed on the natural waste products produced via the iron and sulfur oxidizers. Certainly, species of *Acidiphilium* are so closely related to *A. ferrooxidans* that they have been difficult to separate.

Acidiphilium is ready to provide bacteriochlorophyll-a; however, they are not in a position to grow using light as their sole energy source. One species, *Acidiphilium acidophilum* (previously *Thiobacillus acidophilus*) is specific among members of the genus in that it's equipped to develop autotrophically utilizing diminished inorganic sulfur, heterotrophically using a style of carbon sources, or mixotrophically using each organic and inorganic carbon.

13.6.2.2 Moderate Thermophiles

13.6.2.2.1 Acidithiobacillus caldus

A. caldus is much like *A. thiooxidans* with regard to its potential to oxidize diminished sulfur compounds, but not ferric iron, and its capability to develop at very low pH. It is unusual differentiated from *A. thiooxidans* in that it is moderately thermophilic and has an optimum progress temperature of about 45 °C. It has a 16S rRNA gene sequence closely associated to *A. thiooxidans*, and for several years there used to be confusion as to whether new isolates of sulfur-oxidizing bacteria from mineral environments were *A. thiooxidans* or *A. caldus*. It was once observed that *A. caldus* is a dominant sulfur-oxidizing bacterium working within the temperature range 35–50 °C and is considered to be the "weed" of biomining bacteria underneath these conditions.

13.6.2.2.2 Sulfobacillus

Sulfobacilli are moderately thermophilic (40–60 °C), endospore-forming, gram-positive constructive bacteria that have been isolated from mineral waste and of biomining operations. These bacteria are able to grow autotrophically or heterotrophically. When growing autotrophically they use ferrous iron, reduced inorganic sulfur compounds, or sulfide minerals as electron donors.

Nonetheless, their potential to fix CO_2 is very low. To grow strongly, they require improved stages of CO_2 in the surroundings, small quantities of yeast extract, or close association with heterotrophic iron-oxidizing bacteria similar to *Acidimicrobium ferrooxidans*.

13.6.2.2.3 Ferroplasma

These organisms are pleomorphic in form and lack cell walls. *Ferroplasma acidiphilum* was once isolated from a pilot plant bioreactor treating arsenopyrite/pyrite in Kazakhstan. *Ferroplasma* oxidizes ferrous iron and appears to be obligatory aerobic. This is mesophilic, and optimum temperature required is 33 °C with an upper limit of 45 °C. pH for effective growth is 1.7 and a decrease limit of about 1.3. *Ferroplasma acidarmanus* was recovered from acid mine drainage. Same archaea have been isolated from industrial bioreactors additionally treating an arsenopyrite/pyrite concentrate at 40 °C at the Fairview Mine in Barberton, South Africa.

13.6.2.3 Thermophiles

13.6.2.3.1 Sulfolobus

These are obligatory autotrophic archaea and develop by using oxidizing ferrous iron, reduced inorganic sulfur compounds, or sulfide ores. *S. metallicus* is thermophilic. It has the capability of oxidizing minerals similar to arsenopyrite and chalcopyrite.

13.6.2.3.2 Metallosphaera

The species of *Metallosphaera* are aerobic iron- and sulfur-oxidizing chemolithotrophs which have the capability to develop on complex organics such as yeast extract or casamino acids but however no longer sugars. *M. sedula* has been reported to grow at pH 1.0–4.5 and is able to oxidize minerals at temperatures of 80–85 °C. *Metallosphaera*-like organisms have been suggested to be probably the most efficient at excessive temperature bioleaching of recalcitrant chalcopyrite ores.

13.6.2.3.3 Acidianus

Several species of this group oxidize minerals, although the industrial potential of this group is thought to be less promising than that of *Sulfolobus* and *Metallosphaera. Acidianus brierleyi* can grow autotrophically by oxidizing ferrous iron or sulfur or grow heterotrophically on complex organic substrates. The optimum temperature is 70 °C and the optimum pH is 1.5–2.0. *Acidianus infernus* and *A. ambivalens* are obligate chemolithotrophs that can grow either aerobically or anaerobically by the oxidation or reduction of inorganic sulfur compounds. *A. infernus* has an optimum temperature of 90 °C and an optimum pH of 2.0.

13.7 Metals Recovered in Biomining

For decades, biomining used to be a process for the recovery of metals from low-grade ores but at present is getting used as a major system for recovery of copper and as an essential pretreatment step for gold recuperation in their respective mining processes.

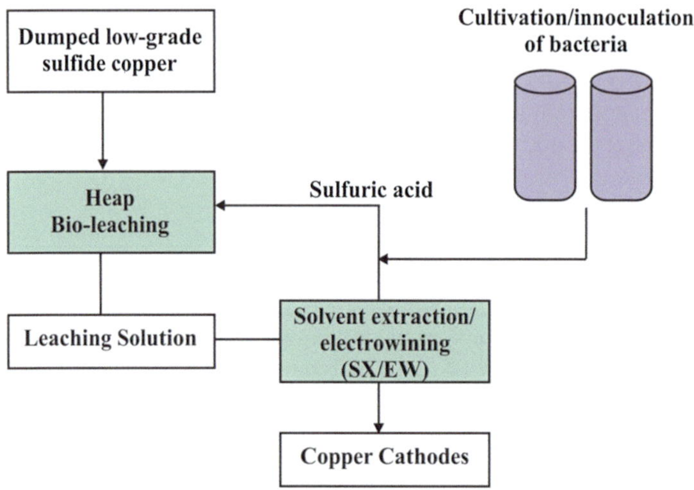

Fig. 13.10 Copper recovery process using biomining

13.7.1 Copper

Conversion of copper sulfides (water-insoluble) to copper sulfates (water soluble) is prerequisite for biomining of copper. Copper ores are crushed, acidified with sulfuric acid, and agglomerated in rotating drums to bind fine material to coarser particles before piling in heaps. The heaps are then irrigated with an iron which percolates through the heap and bacteria growing on the surface of the ore and in solution catalyze the release of copper (Fig. 13.10).

$$Cu_2S + 2Fe_2\left(SO_4\right)_3 \rightarrow 2CuSO_4 + 4FeSO_4 + S$$

$$CuS + Fe_2\left(SO_4\right)3 \rightarrow CuSO_4 + 2FeSO_4 + S$$

13.7.2 Gold and Silver

Bioleaching of precious metal ores to enhance gold and silver is the most promising application. Gold is recovered through bioleaching of arsenopyrite/pyrite ore and its cyanidation process. Silver is more readily solubilized than gold during microbial leaching of iron sulfide.

$$FeAsS + 13Fe^{3+} + H2O \rightarrow 14Fe^{2+} + AsO_4^{\ 3-} + SO_4^{\ 2-} + 16H^+$$

Biooxidation of refractory gold ores to extract gold is applied by means of an industrial method called BIOX developed with the aid of GENCOR SA Ltd. Johannesburg, South Africa, so that you could replace present tactics which posed severe pollution issues. The BIOX method had a number of benefits over current approaches together with low cost.

13.7.3 Uranium

Uranium recovery proceeds in a very similar way to copper recovery. Much like copper, uranium is recovered by the conversion of insoluble uranium oxides to soluble sulfates though the action of ferric iron and sulfuric acid produced by microbes. *In situ* uranium leaching from ore on a large scale is widely practiced in the USA, South Africa, Canada, and India.

$$UO_2 + Fe_2\left(SO_4\right)_3 \rightarrow UO_2SO_4 + 2FeSO_4$$

$$UO_3 + H_2SO_4 \rightarrow UO_2SO_4 + H_2O$$

13.7.4 Phosphate Solubilization

A large number of rock phosphates available in India are not being utilized owing to their low-grade rock phosphates, and our agricultural lands require phosphatic fertilizers for higher crop yield. Literature is available on microbial rock phosphate solubilization. The ability of microorganisms to utilize insoluble rock phosphates has a potential in agricultural applications. Therefore, opportunities do exist to isolate and identify such organisms and for optimization of process parameters for utilization of low-grade rock phosphates.

13.7.5 Silica

Magnesite, bauxite, dolomite, and basalt are the ores of silica. *Bacillus licheniformis* was isolated in India from magnesite ore deposits. Later, it was shown to be associated with bioleaching, concomitant mineralysis, and silicon uptake by the bacterium. Silicon uptake was restricted to adsorption of bacterial cell surface rather than internal uptake through the membrane.

13.7.6 Other Minerals

Recent technological developments have helped to make possible the recovery of oil. Using microorganisms is one such technique to improve the recovery process hence called "Microbially Enhanced Oil Recovery" (MEOR). It was discovered in

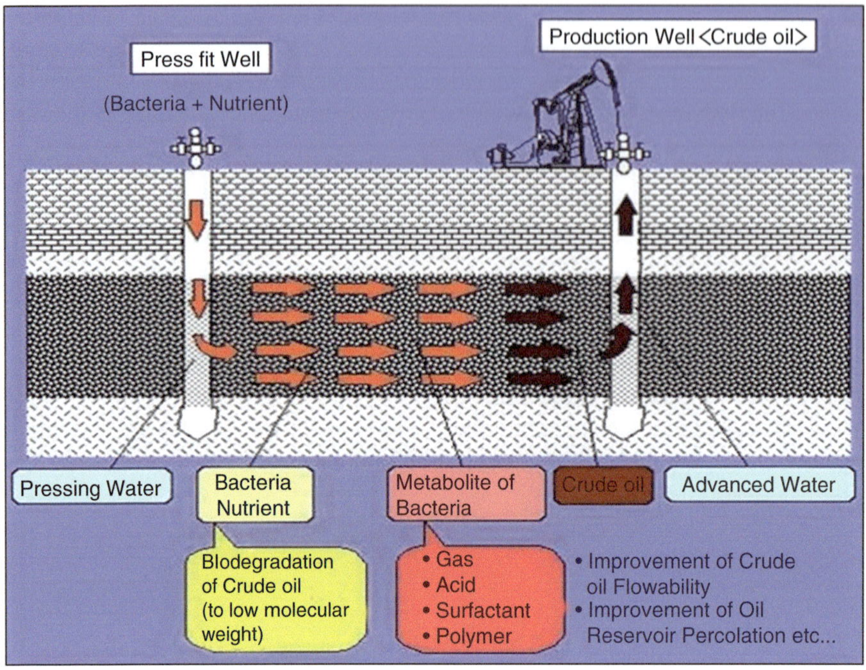

Fig. 13.11 Process of microbial-enhanced oil recovery

1926 that microorganisms can be used in the petroleum industry to enhance oil recovery, but the concept became popular only after the 1950s. Microbes can enhance the recovery of petroleum products directly or indirectly (Fig. 13.11).

13.8 Factors Affecting Bioleaching and Biomining

Success of biomining and efficiency in recovery of metals are dependent on mainly the following factors.

13.8.1 Choice of Microorganism

Suitable bacteria are the most important factor for the success of bioleaching. Selected bacteria should have the potential to survive at high temperatures, acid concentrations, and high concentrations of heavy metals and remain active under such circumstances.

13.8.2 Surface Area

Particle size of the ore is very important for the rate of oxidation of bacteria. The rate decreases with enhancement in size of the ore and vice versa.

13.8.3 Mineral Composition

Rate of bioleaching is also affected by the composition of ore such as concentration of sulfides, amount of mineral present, and the extent of contamination.

13.8.4 pH

The adjustment of correct pH is necessary for the growth of leaching bacteria and is decisive for the solubilization of the metals. Biomining requires a pH of 2.5–3.0 for maximum results. The rate of process decreases significantly if the pH is not in this range since the activity of acidophilic bacteria is reduced.

13.8.5 Temperature

The bacteria used in biomining are either mesophilic or thermophilic. Optimum temperature is required for bioleaching to proceed at a faster rate. Optimum temperature range for a given bacteria is between 25 and 35 °C depending on the type of ore being selected. The rate of biomining is reduced significantly if the temperature is above or below the optimum temperature. At higher temperature (50–60 °C) thermophilic bacteria can be used for the leaching process.

13.8.6 O_2 and CO_2

An adequate amount of oxygen is required for the good growth and high activity of leaching bacteria. Oxygen can be provided by aerators and pipes. Mechanical agitation is also an effective method to provide continuous air supply uniformly and also to mix the contents. CO_2 is the only carbon source required but there is no need for the addition of CO_2.

13.8.7 Solid-Liquid Ratio

The ratio of ore/sulfide to the leach solution (water + acid solution + bacterial inoculum) should be maintained at optimum level to ensure that biooxidation proceeds at maximum speed. The leach solution containing leached minerals should be removed periodically and replaced with a new solution.

13.8.8 Surfactants

Adding small amounts of surfactants like Tween 20 to the leaching process increases the rate of biooxidation of minerals from sulfide ores. The surfactants decrease the surface tension of the leach solution, thus, wetting the ore and resulting in increased bacterial contact which ultimately increases the rate of biooxidation.

13.9 Applications of Bioleaching and Biomining

13.9.1 Extraction of Precious Metals from Metal Waste

The modernization of human race and its increasing demand for a comfortable lifestyle has resulted in increased industrialization. This comfortable lifestyle includes usage of various machines which reduce the human workload. The increased usage of machines has resulted in generation of large amounts of metal waste in form of electronic waste, battery waste, the use of syringes in hospitals and its disposal by incineration forming fly ashes which contain a potential amount of metals, which may result in metal pollution. It includes wastes containing heavy metals such as nickel, copper, zinc, uranium, etc. which contaminate the industrial sites and are toxic to organisms particularly to human beings as well as to the environment. Therefore, the industries are obligated to enforce such environmental management system to prevent the environment strictly. Limited research has been done to show the use of bioleaching in extracting metals from wastes (Hoque and Philip 2011).

13.9.2 Extraction of Metal from Electronic Waste

Electronic goods like cell phone, laptops, desktop, television sets, etc. play a major role in human life. Therefore, world production of electronic waste is increasing persistently and posing a potential threat to the environment and mankind. For that reason, we need to carry out bioleaching of these electronic wastes which will not only decrease the hazardous nature but will help to recover the valuable metals (Hoque and Philip 2011). Microorganisms such as *Acidithiobacillus ferrooxidans*, *Acidithiobacillus thiooxidans*, and *Sulfobacillus thermosulfidooxidans* are involved in bioleaching of electronic waste. The electronic scrap being a mixture of various metals such as Cu, Al, Fe, Ni, and their alloys is an interesting kind of waste.

13.9.3 Extraction of Metal from Fly Ashes

Fly ashes produced from combustion are considered as a secondary source for extraction of heavy metals like Zn, Al, Cu, Ni, and Cr using a mixed culture of iron- and sulfur-oxidizing bacteria. The same mixed culture can be used to attain high metal-leaching ability and tolerability for extraction of Cu, Zn, Cr, and Cd from municipal solid waste incineration (MSWI) fly ash.

13.9.4 Extraction of Metal from Battery Wastes

Biomining techniques have been put into operation to extract toxic metals from hazardous spent batteries in developing countries. Leaching metals from Ni-Cd-spent batteries have been effectively done. With proper culture conditions provided, 95–98 % Co and nearly 1 % Ni can be recovered from spent lithium ion batteries. This was reported by Kang et al. wherein they recovered cobalt sulfate from crushed and screened prismatic type spent lithium ion batteries (Kang et al. 2010).

13.9.5 Extraction of Metal from Mine Wastes

Acid mine drainage (AMD) is the major form of waste from metal mines which has considerable amount of metal content which is a chief pollutant in water resource. The capability of bacteria to leach the metals can be exploited to retrieve metals from AMD. Iron- and sulfur-oxidizing bacteria show a promising iron oxidation potential in AMD sites. Zinc and aluminum have been successfully retrieved from AMD with the help of *A. ferrooxidans*. In addition, it has been reported that nickel can be recovered from aqueous waste with the help of polycrystalline hydrogen uranyl phosphate (HUP) bound to immobilized cells of *Citrobacter* sp.

13.9.6 Extraction of Metal from Contaminated Sediments

Contaminated sediments like anaerobic sludge from river sediments are a great source of heavy metal residues by biomining. It has been described that Cu, Co, Ni, Mn, and Fe have been retrieved from ocean manganese nodules in the pyrite and reducing agents. Iron- and sulfur-oxidizing bacteria have been shown to solubilize 90 % of Ni, Zn, Cu, and Cr at an optimum temperature of 37 °C.

13.10 Biomining: Environment-Friendly Mining

To sustain the environment, it is very important to return back all the components (wastes) in a recyclable way so that the waste becomes useful and helps to maintain the ecosystem. Biomining contributes to sustainable development in the same way as all microorganism-mediated processes do. It uses existing organisms and mechanisms in nature. Traditional mining is an especially toxic process involving the use of chemicals like cyanide. Although the process of biomining does not yet completely eliminate the use of harmful chemicals, it allows for a lessened use, resulting in lower production costs of cleaning up the mining processes (Fig. 13.12.)

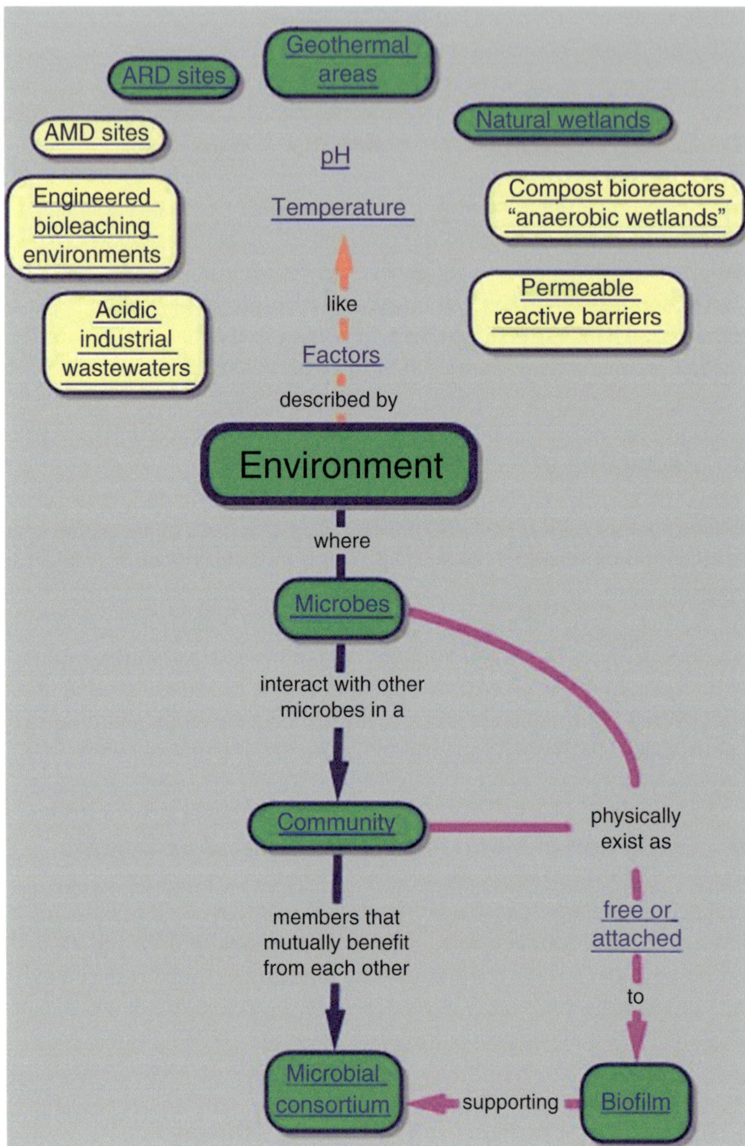

Fig. 13.12 Environment community interactions and microbes

13.11 Biotechnology and Biomining

Biomining makes use of natural microorganisms to leach and oxidate. It is important to learn more about microorganism genomes and to know how microorganism biology works. This will lead to the genetic engineering of organisms for foremost biomining results. Biotechnology or genetic engineering can be used to supply species of microorganism that are resistant against heavy metals like mercury, cadmium, and arsenic. These heavy metals lowered the system of biomining. Thus, these heavy metal-resistant microorganisms can also be effectually used in biomining process (Fig. 13.13).

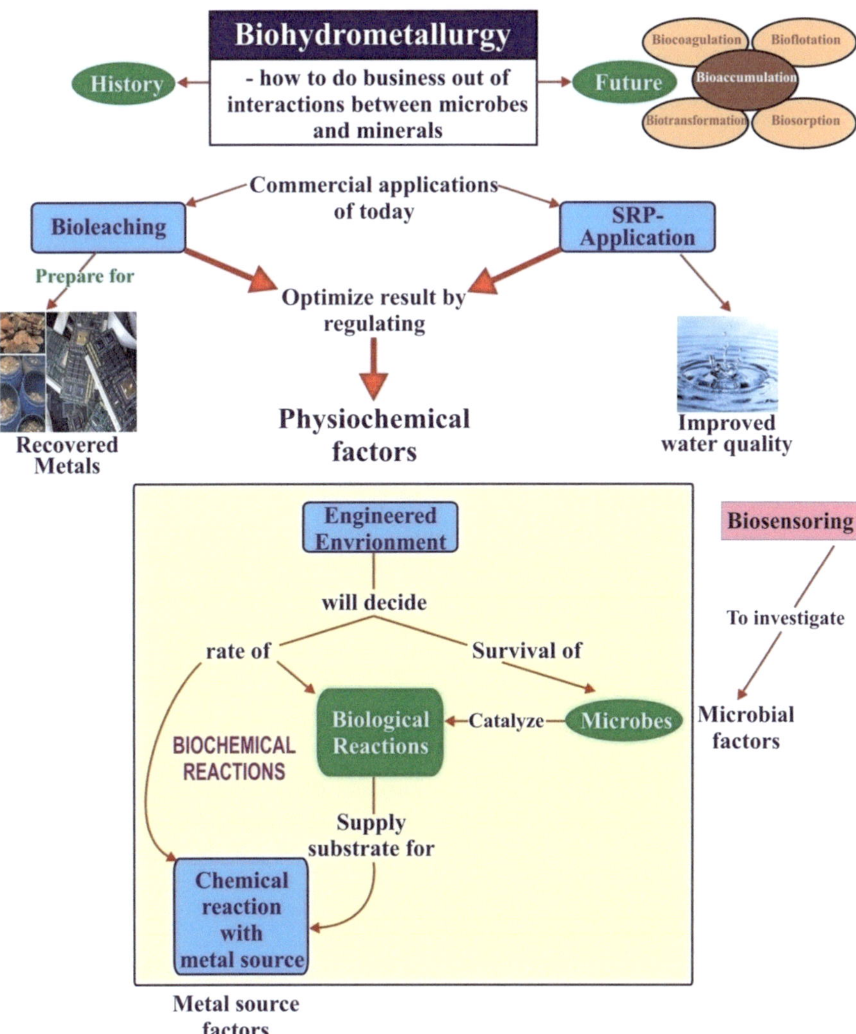

Fig. 13.13 Biomining and biotechnology

13.12 Current Research Areas in Bioleaching and Biomining

Biomining is used by the mining industry to extract copper, uranium, and gold from low-grade ores but not for low-grade manganese ore in industrial scale. The study of microbial genomes, metabolites, and regulatory pathways provide novel insights to the metabolism of bioleaching microorganisms and their synergistic action during bioleaching operations. This will promote understanding of the universal regulatory responses that the biomining microbial community uses to adapt to their changing environment leading to high metal recovery. Possibility exists of findings ways to imitate the intact process during industrial manganese biomining endeavor.

There are continuing efforts to fully understand the basic biology of microorganisms, such as the *Thiobacillus ferrooxidans*. Researchers are seeking enhanced microbial performance in biomining processes through the identification of better indigenous strains and also through genetic engineering. However, because genetically engineered microorganisms need careful control and monitoring, they will not likely be available for commercial use for several years to come, and then only for controlled processes, like those possible in reactors. Canadian researchers are also working on creating biomining conditions that will be optimal in the colder climates like Canada.

Currently, 25 % of all copper worldwide is produced through biomining. The process is used on a variety of other metals such as gold and uranium. Biomining is not yet a proven or profitable technology to apply to other metals such as zinc, nickel, and cobalt.

Some advantages of biomining over traditional methods include reduced noxious gas production and the elimination of toxic liquid waste produced as a result of chemical leaching. Biomining, however, is slower than traditional mining techniques and is not applicable to a wide variety of ores.

13.13 Challenges to Applying Biomining Technology

While biomining is gaining considerable success, both technical and commercial challenges remain that hinder biomining from achieving its full potential as an applied technology.

13.13.1 Technical Challenges

The technical challenges and opportunities faced by analysis and development units are the bioheap leach of primary sulfide minerals despite taking a few steps to perceive and progress with reference to mineral. However, a lot of attention has to be given to chalcopyrite biochip leaching to understand the problems faced during the leaching process such as diffusion barriers for the leachant to progress the leaching and finally how to get optimized conditions suitable for bacterial growth, recovery of copper minerals, and economic viability. Emphasis should also be given in understanding the fundamentals of the reducing conditions that can occur within the heap when thermophiles are used for dissolution of chalcopyrite and other primary copper sulfide minerals. The presence of silicate minerals bound to the complex polymetallic sulfides has been a big hindrance in the design of heap bioleaching. Bioheap leaching model development, integration, and validation developed via bioheap leaching aspects taking hydrology and heat balance into account need to be given a serious thought to understand mathematical modeling aspects to use the process in tough conditions. Other directions which could be looked into lies in better understanding of secondary copper sulfide heap leaching even though the crushed ore heap leaching of secondary copper sulfides has been widely used over a decade, lacking some information in production issues.

However, other than the all the aspects mentioned, a lot of questions are unanswered like the time taken by the microorganisms prevalent in the bioheap conditions where the source of inoculum is the raffinate together with the growth of microbes in the stacked ore. Profit of inoculation of microorganisms in the heap together with extent of aeration requirement in the heap to get better recovery and amount of aeration required. Is the temperature issue important within the heap bioleaching of secondary sulfides and also the reasons behind it as slow rate of oxidation of sulfides in the absence of pyrite, which needed to be solved? Finally, the microbe-mineral interactions followed by mechanisms involving in the heap with respect to galvanic interactions, oxidation-reduction potential, and pH and dissolved metals ion concentration lead to harmful effects on the microbes together with downstream process problems.

The use of *in situ* bioleaching is one of the growing considerations in each and every nick and corner around the world as urbanization have forced human environment nearby the mining operation which have made it very important to decrease mining foot prints. *In situ* biomining would significantly decline the impacts of mining on human habitation; however, this issue needs to be in retrospect and with efficiency accomplished and is of course a big challenge for tomorrow's mining organizations and intellectuals.

Advancement of technology is required to develop sustainable development to treat decommissioned cyanide-leached heaps by rotating biological contactors and by developing methods to treat cyanide-, thiocyanate-, and metal-contaminated waters resulting from gold treatment. However, the technology has not yet been successful and developed and needs to be further evaluated considering various environmental factors. The biotechnological approach can lead to huge cost savings. Another aspect to be looked into is the technology for stabilizing sulfide-bearing wastes, which tends to be a big challenge to future biomining professionals.

13.13.2 Commercial Challenges

Biomining and bioleaching technology is developed by mining companies, mining biotechnology companies, government laboratories, engineers, academic scientists, and mining consultants. The commercial application of biomining technologies developed by these various organizations faces very difficult challenges:

1. To research, develop, pilot, and commercialize the technology commitment by the Department of Research and Development, proper cooperation from financial terms and management issues is required. The biomining site is normally site-specific for which every technology developed needs an on-site pilot plant followed by display level, leading to a full-scale operation which can be a very costly affair.
2. Mining companies are motivated to develop biomining technology to process their own minerals deposits, which are technically and/or economically not comparable to conventional technology.
3. Many mining companies patent their technologies or copyright to publish their research in the public domain to protect the technology for their own use.
4. Mining companies mostly do not make available their technology for other mining companies.
5. Major mining companies also have good laboratory facilities and are staffed with efficient scientists and engineers who can resolve technical problems that arise during the complete process from testing to operation.

In response to the mining industry's interest in biomining, several biotechnology companies have developed their own technologies. They patent their technologies and develop strategies for its marketing. Some companies prefer to license their technologies to mining and/or companies in return for a licensing fee and a royalty. Other biomining companies opted to become mining companies themselves and use their technologies at their own mine sites. Unless the mining biotechnology company has ready access to a mining property at which the technology can be scaled-up and vetted, the biotechnology company must establish terms and conditions with mining companies that are willing to undertake the risk. Government laboratories (i.e., CSIRO in Australia, Mintek in South Africa, BGRIMM in China, and BRGM in France) contribute significant innovation to biomining. Innovations from government laboratories have entered the commercial monarchy through collaborations with mining companies and public domain publications.

University scientists throughout the world contribute elemental and practical biomining research. Much of this research is published in journals with broad, global readership. Laboratories are also funded through AMIRA International, an industry-funded association, and BioMinE – a European Union-funded consortium, to provide high-quality biomining technology. Independent researchers, specializing in biomining, may also transfer technology to the mining industry. Biomining research and development by universities, government laboratories, and consultants may be carried out in response to an industrial contract from a mining company.

The commercial application of biomining technology is very challenging, and most of the challenges are common to any new mining technology:

1. Biomining technologies are competing with alternative technologies like pressure oxidation, roasting/smelting, and developing chemical leach processes. There are so many factors that are considered when selecting a technology, and biomining may not always be a good option.
2. New technologies have a good chance of being successful. Failures are costly not only in financial terms, but also the reputations of mining, engineering, and biotechnology companies and individuals are at stake.
3. To bring new biomining technology from the abstract stage to commercialization, a lot of time, long-term financial, research, development, and managerial commitments are required.
4. New biomining technologies often require a considerable capital investment. For example, the 20,000 t/a BioCOP™ demonstration plant at the Chuquicamata Mine in Chile was estimated at $60 million.
5. Biomining technologies, like most hydrometallurgical processes, are site-specific. Therefore, for nearly every biomining technology, on-site piloting and large-scale demonstration may need to be conducted for every application of the technology. One exception is the BIOX™ technology, which has been applied at a large number of locations where an on-site demonstration is not necessary, but several months of laboratory research is a must. Site pilot trials and testing are costly and time-consuming.
6. Patents, which shield the intellectual property of each mining and biomining technology firms and are typically essential for the latter to lift funding, can smother industrial applications of biomining technologies. Mining companies might decide to choose an alternate method instead of pay licensing fees and royalties. Another unfortunate consequence of some patents is that technologies become inaccessible when homeowners of the patents have no interest in selling the technology or maybe using the method they developed.

13.14 Future of Biomining

Biomining or the "mining of the future" is cheaper and greener than traditional mining; there are a heap fewer dioxide emissions and carbon and water foot prints. The future of biomining is challenging, as it offers advantages of operational simplicity, low capital, and operating value and shorter construction times that no alternative different method will offer. In addition, minimum environmental impact and the use of this technology within the mining industry are set to extend. Once commercial-scale high-temperature processes have been designed, the variety of minerals can become obvious to biomining and will increase. Increased concern relating to the impact of mining on the atmosphere is probably to enhance the competitive advantage of microbic primarily based metal recovery processes.

1. While the demand for most metals has steady accumulated within the last decade, discoveries have declined and those deposits that are being discovered are declining in grade and quality. Processing choices for lower-grade ore deposits and

deposits of lower quality with advanced polymetallic mineral assemblages are restricted. Biomining technologies are notably adept at technically and economically processing these varieties of resources. Mining companies are aware of biomining's distinctive niche, and mineral heap bioleaching is already undergoing pilot- and demonstration-scale testing. There will, of course, continue to be opportunities for the commercial application of stirred-tank biooxidation of sulfidic-refractory gold concentrates, because that technology has been effectively marketed for over two decades and has competed well with pressure oxidization and cooking.

2. Because future biomining applications can probably be directed additionally on lower-grade, lower-quality complex ores, it is important that analysis and development specializes in the technical problems related to these biomining applications. Such studies should address those problems represented earlier: understanding however the totally different temperature groupings of microbes colonize and performance inside coarse ore plenty; engineering coarse ore and run-of-mine heaps to effectively exploit microbic development and activities as well as irrigation, aeration, and heat management.
3. For biomining technologies to be more wide applied commercially, they have to be incontestable at scale. To achieve this, there should be cooperation among the mining corporations' own UN agency and exploit the deposits and universities, government laboratories, biotechnology corporations and engineering companies that develop the technologies. Organizations such as AMIRA have succeeded to some extent in accomplishing this, and this cooperative concept has to be advanced and exploited.
4. Biomining patents are complicating the industrial development of biomining. Biotechnology companies have a cheap right to be financially rewarded for his or her innovations, yet the most pricey and most risky part of developing biomining technology is demonstrating the technology at scale and therefore the mining corporations each assume the price and therefore the risk for this. Mining companies, too, have an obligation to shield their rights to use technology that they need to develop.

References

Asghari I, Mousavi SM, Amiri F, Tavassoli S (2013) Bioleaching of spent refinery catalysts: a review. J Ind Eng Chem 19:1069–1081

Blake RC, Shute EA, Greenwood MM, Spencer GH, Ingledew WJ (1993) *Thiobacillus ferrooxidans*, *Leptospirillum ferrooxidans* and *Metallosphaera sedula* application in metal leaching: a review. FEMS Microbiol Rev 11(1–3):9–18

Brierley CL (2008) How will biomining be applied in future? Trans Nonferrous Met Soc China 18:1302–1310

Buchanan RE, Gibbons RE (1974) Bergey's manual of determinative bacteriology. Williams and Wilkins, Baltimore

Characklis WG, Mcfeters GA, Marshall KC (1990) Physiological ecology in biofilm systems. In: Characklis WG, Marshall KC (eds) Biofilms. Wiley, New York, pp 341–394
Das T, Ayyappan S, Chaudhury GR (1999) Factors affecting bioleaching kinetics of sulfide ores using acidophilic micro-organisms. Biometals 12:1–10
Donlan RM (2001) Biofilm formation: a clinically relevant microbiological process. Clin Infect Dis 33:1387–1392
Dunne WM Jr (2002) Bacterial adhesion: seen any good biofilms lately? Clin Microbiol Rev 15:155–166
Ewart DK, Hughes MN (1991) The extraction of metals from ores using bacteria. Adv Inorg Chem 36:103–135
Flemming HC, Wingender J (2010) Review: the biofilm matrix. Nat Rev Microbiol 8(9):623–633
Fletcher M, Loeb GI (1979) Influence of substratum characteristics on the attachment of a marine pseudomonad to solid surfaces. Appl Environ Microbiol 37:67–72
Haque N, Nargate T (2013) The greenhouse gas footprint of *in-situ* leaching of uranium, gold and copper in Australia. J Clean Prod 84:382–390
Hoque ME, Philip OJ (2011) Biotechnological recovery of heavy metals from secondary sources—an overview. Mater Sci Eng 31:57–66
Kang JG, Senanayake G, Sohn J, Shin SM (2010) Recovery of cobalt sulfate from spent lithium ion batteries by reductive leaching and solvent extraction with Cyanex 272. Hydrometallurgy 100:168–171
Mishra D, Kim DJ, Ralph DE, Ahn JG, Rhee YH (2008) Bioleaching of spent hydroprocessing catalyst using acidophilic bacteria and its kinetics aspect. J Hazard Mater 15:1082–1091
Mousavi SM, Yaghmaei S, Vossoughi M, Jafari A, Roostaazad R, Turunen I (2007) Bacterial leaching of low-grade Zns concentrate using indigenous mesophilic and thermophilic strains. Hydrometallurgy 85:59–65
Pringle JH, Fletcher M (1983) Influence of substratum wettability on attachment of freshwater bacteria to solid surfaces. Appl Environ Microbiol 45:811–817
Quirynen M, Brecx M, Van Steenberghe D (2000) Biofilms in the oral cavity: impact of surface characteristics. In: Evans LV (ed) Biofilms: recent advances in their study and control. Harwood Academic Publishers, Amsterdam, pp 167–187
Rupp ME, Fey PD, Heilmann C, Gotz F (2001) Characterization of the importance of *Staphylococcus epidermidis* autolysin and polysaccharide intercellular adhesion in the pathogenesis of intravascular catheter-associated infection in a rat model. J Infect Dis 183:1038–1042
Schippers A, Sand W (1999) Bacterial leaching of metal sulfides proceeds by two indirect mechanisms via thiosulfate or via polysulfides and sulfur. Appl Environ Microbiol 65:319–321
Siddiqui MD, Kumar H, KesariKavindra A, Arif JM (2009) Biomining – a useful approach toward metal extraction. American-Eurasian J Agron 2(2):84–88

Suggested Further Readings

Abhilash PBD, Nataranjan KA (2015) Microbiology for minerals, metals, materials and the environment. CRC Press Taylor and Francis Group, Boca Raton
Donadi ER, Wolfgang S (2010) Microbial processing of metal sulfides. Springer, Berlin
Ehrlich Henry L, Newman Dianne K (2009) Geomicrobiology. CRC Press Tyler and Francis Group, Boca Raton
Rawlings DE (1997) Biomining: theory, microbes and industrial process. Springer, Berlin
Rawlings DE, Johnson DB (2007) Biomining. Springer, Berlin

Genetically Modified Organisms (GMOs) and Environment

14

Rasna Gupta and Ram Lakhan Singh

Contents

R. Gupta
Department of Biochemistry, Dr. Ram Manohar Lohia Avadh University, Faizabad 224001, India
e-mail: rasna.gupta.biochem@gmail.com

R.L. Singh (✉)
Department of Biochemistry, Faculty of Science, Dr. Ram Manohar Lohia Avadh University, Faizabad 224001, India
e-mail: drrlsingh@rediffmail.com

R.L. Singh (ed.), *Principles and Applications of Environmental Biotechnology for a Sustainable Future*, Applied Environmental Science and Engineering for a Sustainable Future, DOI 10.1007/978-981-10-1866-4_14

Abstract

Genetically modified organisms (GMOs) or transgenic organisms are those whose genetic material has been altered for the production of desired biological products. GMOs are simply the most developed application of modern biotechnology in terms of research, commercialization, adoption, and regulation. Plants, animals, and microorganisms have all been genetically modified by various transformation methods for several purposes with medicinal, agricultural, environmental, and more recently industrial applications. Genetically modified (GM) plants are the predominant largest class of GMOs introduced into the environment for food and feed production.

These are theorized to reduce production costs due to reduced chemical and mechanical needs in planting, maintenance, and harvest. It is possible that this saving could be passed onto the consumer. Additional benefits to the consumers are the potential nutrition implications. GMO technology allows the creation of foods that are more nutrient dense. GMOs have many more applications but its total benefits have not yet been fully explored. So there is a need to increase the potential of researcher to generate the information to normalizing more benefits of GMOs in medical, agricultural, environmental, and industrial fields for sustainable future. This chapter summarizes GMOs, its types, labeling, applications, and health impact on human being.

Keywords

Genetic modification • Genetically modified organism • GMO regulation • GMO product • Health impact • GMO toxicity • GMO risk assessment

14.1 Introduction

Genetically modified organism (GMO) is an organism whose genetic material has been altered by using genetic engineering techniques. The term GMO is used to refer to any microorganism, plant, or animal, in which genetic engineering techniques have been used to introduce specific gene and remove or modify specific parts of its genome. GMOs produced through genetic technologies have become a part of everyday life, entering into society through agriculture, medicine, research, and environmental management. GMOs are the source of genetically modified foods and are also widely used in scientific research to produce goods other than food. GMOs show great promise in improving agriculture, health, and environment. Plant may be engineered to better tolerate temperature or weather extremes, to contain various vitamins, or to dispense medicines and vaccines. The production of foods with GMOs has risen rapidly over the past three decades comprising nearly 90% of crops grown in the USA today. Many researchers think that genetically modified foods have the potential to end world hunger. Monsanto®, USA, is the biggest company producing and releasing GM products in markets which owns patented seeds that are used for foods. GMOs are prepared by using genetic engineering techniques and were first developed by Herbert Boyer and Stanley Cohen in 1973 (Boyer and Cohen 1973).

In reproductive cloning, a nucleus is extracted from a cell of the individual to be cloned and is inserted into the enucleated cytoplasm of a host egg. The process results in the generation of an offspring that is genetically identical to the donor individual. RDT are employed to produce organisms whose genomes have been precisely altered at the molecular level, usually by the inclusion of genes from unrelated species of organisms that code for traits that would not be obtained easily through conventional selective breeding. The first animal produced by means of this cloning technique with a nucleus from an adult donor cell (as opposed to a donor embryo) was a sheep named Dolly, born in 1996. Since then a number of other animals, including pigs, horses, and dogs, have been generated by reproductive cloning technology. RDT, on the other hand, involves the insertion of one or more individual genes from an organism of one species into the DNA (deoxyribonucleic acid) of another (Fig. 14.1). Whole genome replacement involves the transplantation of one bacterial genome into the cell body or cytoplasm, of another microorganism.

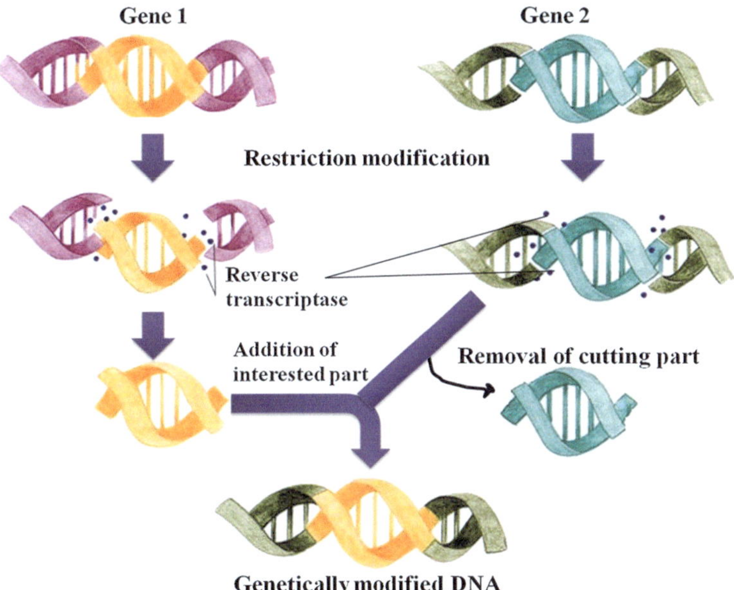

Fig. 14.1 Process of gene modification

Genetically modified (GM) foods were first approved for human consumption in the USA in 1994, and by 2007 about 90 % of the corn, cotton, and soybeans planted in the USA were GM. By the end of 2020, GM crops will cover more than 10 million square kilometers (3.86 million square miles) of land in 29 countries worldwide, one-tenth of the world's farmland. The majority of GM crops were grown in the Americas and top ten genetically modified crops are shown in Fig. 14.2.

Genetic transformation is a powerful tool and an important technique for the study of plant functional genomics. Plant transformation was first described in tobacco in 1984 (De Block et al. 1984; Horsch et al. 1984; Paszkowski et al. 1984). Since that time, rapid developments in transformation technology have resulted in the genetic modification of many plant species. Methods for introducing diverse genes into plant cells include either direct or indirect techniques (Fig. 14.3). Electroporation and particle bombardment are the most commonly used direct gene transformation methods in plant cells. In electroporation electric impulse of high field strength is used to reversibly permeabilized cell membrane to facilitate uptake of large molecules including DNA. This method is used for transformation of cells and protoplast of most of cereals (Fig. 14.3).

In particle bombardment method, gold and tungsten particles are coated with the DNA that is used to transform the plant tissue. The particles are propelled at high speed into the target plant material where the DNA is released within the cell and can integrate in the genome. Primary explants and proliferating embryonic culture, plant tissues are used for particle bombardment. Plant transformation mediated by *Agrobacterium tumefaciens*, a soil plant pathogenic bacterium, has become the most used indirect method for the introduction of foreign genes into plant cells and the subsequent regeneration of transgenic plants. *A. tumefaciens* has the exceptional

Fig. 14.2 Top ten genetically modified crops

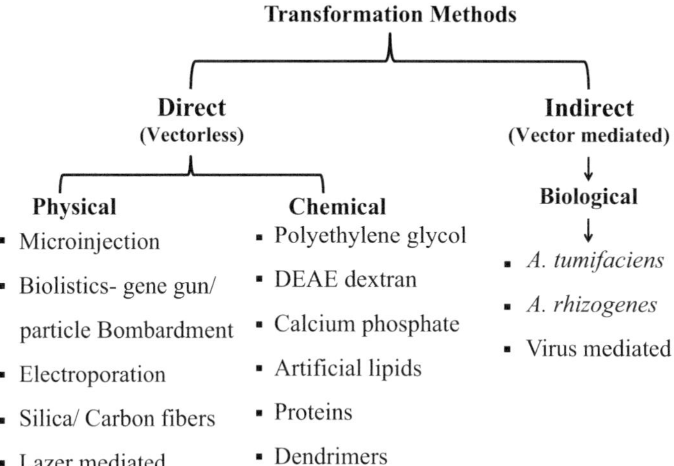

Fig. 14.3 Methods of DNA transformation

ability to transfer a particular DNA segment (T-DNA) of the tumor-inducing (Ti) plasmid into the nucleus of infected cells where it is then stably integrated into the host genome and transcribed, causing the crown gall disease. Tumorigenesis is a transformation process occurring by transfer, integration, and expression of T-DNA genes which can transfer any foreign DNA inserted in it (Fig. 14.4). This resulted in the construction of the first T-DNA vector to generate transgenic plants.

In the laboratory *Agrobacterium* can also transform non-plant species like fungi and yeasts; therefore, the yeast *Saccharomyces cerevisiae* can be used to investigate fundamental aspects of the transformation process. One method which takes advantage of natural forms of gene transfer is the ability of lentiviruses to transfer genes to animal cells. Genetic engineering techniques are employed to introduce the desired traits such as increased resistance to nonselective herbicides, improved nutritional contents, pest-resistant varieties, and enhanced fertility.

The GMOs came into light in 1980 for bulk production of medically useful proteins. The debate on safety of GMOs started just after that which partially slowed down the development and introduction of GM crops. A lot of deliberations have been going on among various stakeholders for safe use of GMOs. The group which is in favor of GM crops sees it as a potent option for alleviating poverty in the developing nations as almost 800 million people remain malnourished as per the United Nations Population Fund 2005 report. On the other hand, the group which is against it gives more weightage to the health safety and environmental concerns over the increased food production.

The present chapter focuses on the evaluation of GMOs, types, their health impact, and other applications along with environmental concerns.

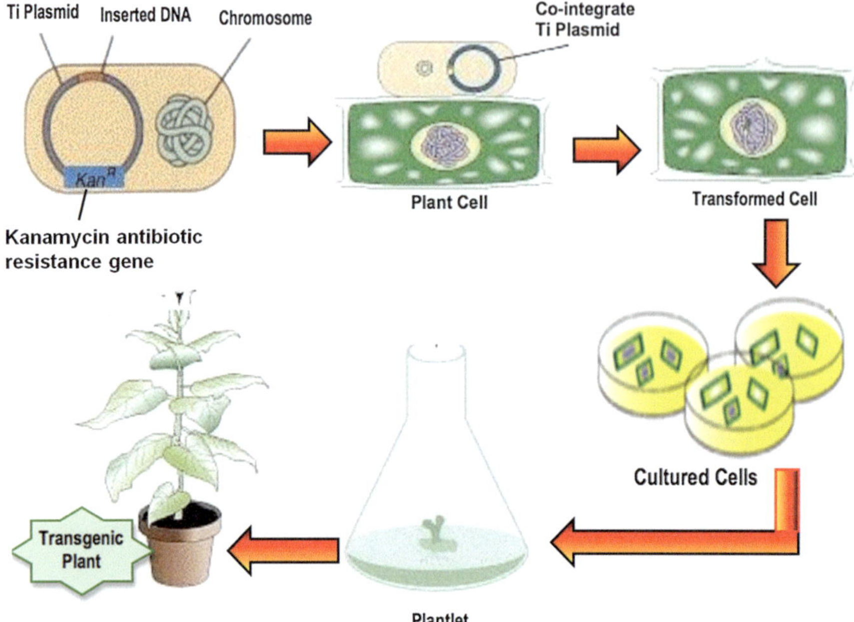

Fig. 14.4 *Agrobacterium*-mediated DNA transformation in plant cell

14.2 History of GMOs

Humans selected wild varieties of plants and animals with the trait they desired and began selective breeding to increase these desired traits, for example, to produce bigger, easier to harvest grains with greater yield and to breed animals that were the most docile and easy to handle. With increasing understanding of genetics, direct DNA alteration was possible which resulted in the creation of the first recombinant DNA molecule by Paul Berg in 1972 by combining DNA from a monkey virus with that of lambda virus.

Boyer and Cohen created the first GMO in 1973 by transferring antibiotic (kanamycin) resistance gene from resistant bacteria into nonresistant bacteria. The nonresistant bacteria were then able to survive in the presence of kanamycin (Boyer and Cohen 1973).

In 1973, Jaenisch and Mintz, from Fox Chase Cancer Center in Philadelphia, developed first transgenic mouse by introducing foreign DNA when mouse embryos were infected with simian virus (SV40) (Jaenisch and Mintz 1974). In 1981, the first transgenic mice were developed that passed the transgenic traits to their offspring (Gordon and Ruddle 1981; Costantini and Lacy 1981). In 1984, genetically modified mice were created with cloned oncogene, which subjected them to developing cancer. In 1985, Brophy developed first domestic transgenic animal (Brophy et al. 2003). In 1987, Gordon reported the first transgenic mice that synthesize human recombinant protein in their milk.

The year 1983 saw the development of the first transgenic plant, tobacco, by Bevan et al. (1983) (Fig. 14.4). It was transformed with *A. tumefaciens* containing antibiotic resistance gene followed by culture techniques to produce the resistant variety of tobacco. Also in 1994, the European Union approved tobacco which was engineered for resistance to the herbicide bromoxynil, making it the first genetically engineered crop commercialized in Europe. In 1995–1996, an insect-resistant (IR) potato was approved for release in the USA. Vitamin A-enriched golden rice was developed with increased nutrient value. A genetically modified tomato, Flavr Savr, was the first commercially grown genetically modified food granted a license for human consumption. It was produced by the Californian company Calgene.

In 1976, Genentech, the first genetic engineering company, was founded by Herbert Boyer and Robert Swanson, and within 1 year the company produced genetically engineered human somatostatin protein in *Escherichia coli* (*E. coli*). Genentech also announced the production of genetically engineered human insulin within 2 years (1978) known as humulin, and it was approved by the Food and Drug Administration for release in the market (1982).

The J. Craig Venter Institute (Maryland, USA) is a nonprofit genomic research institute founded by J. Craig Venter. Scientists at this institute created the first synthetic bacterial genome in the year 2010 and named it Synthia which was the world's first synthetic life form. The first genetically tailored animal to be commercialized was a zebra fish (GloFish) with a fluorescent gene protein added that allows it to glow in the dark under ultraviolet light.

14.3 Types of GMOs

14.3.1 Microorganism as GMO

14.3.1.1 Bacteria

Bacteria, because of their rapid replication rates, are used for manipulation of genes that are then introduced into plants or animals. In bacteria sexual reproduction occurs in the form of genetic recombination, involving conjugation, transduction, and transformation (Figs. 14.5a and 14.5b). They continue to be important model organisms for genetic manipulations:

1. Conjugation: direct physical interaction through conjugation tube (sex pili) between donor and recipient cell (Fig. 14.5a)
2. Transduction: when virus (bacteriophages or viruses infecting bacteria) infects a bacterium and transfers their genetic material to recipient cell (Fig. 14.5a)
3. Transformation: transfer of naked DNA from one bacterial cell to another by unknown mechanism (Fig. 14.5b)

Genetically identical population (clones) with gene of interest may be easily produced in a short period of time. The cells can then be lysed and DNA isolated

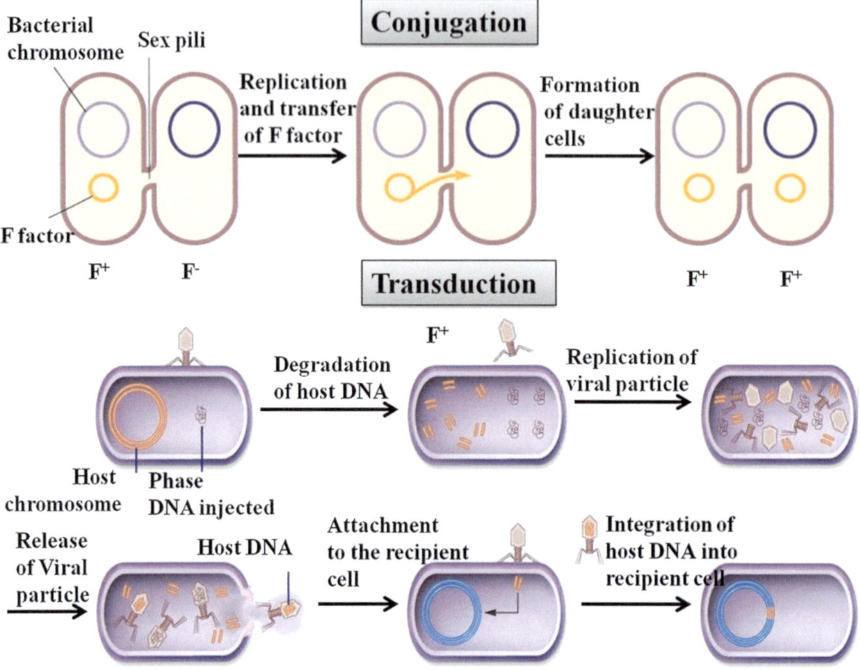

Fig. 14.5a Conjugation and transduction

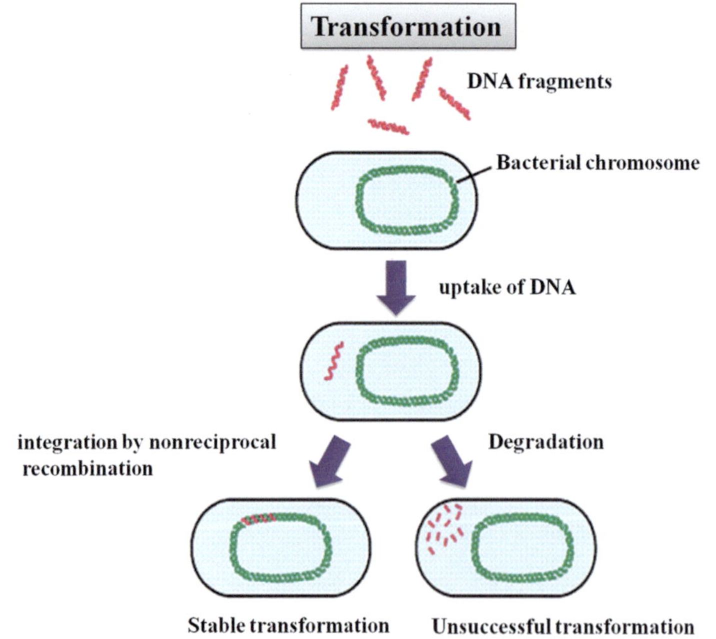

Fig. 14.5b Transformation

and insertion of isolated DNA in the target cells takes place in a very short period. Bacteria expression system is most common for regular production of desired non-bacterial proteins. An example is production of pure human insulin in large quantities for use in diabetic patients. Similarly, human growth hormones and clotting factors are also produced using bacterial systems.

In addition, various genetically engineered microorganisms are routinely used as sources of enzymes for manufacture of a variety of processed foods. These includes alpha-amylase from bacteria which converts starch to simple sugars, chymosin from bacteria or fungi which clots milk protein for cheese making, and pectin esterase from fungi which improves fruit juice clarity. Plasmid is a small, circular, self-replicating part of bacterial DNA that can be genetically manipulated for the production of proteins. The proteins encoded by plasmid always offer selective advantages for its host. For example, antibiotics encoded by bacterial plasmids allow their host bacteria to destroy competing microbes. Alternatively, some resistant antibiotics are encoded by bacterial plasmids that may lose its ability to effectively control or kill the bacterial growth.

Some genetically engineered bacteria and fungi may enhance the rate of biodegradation of environmental wastes and keep our environment clear from toxic compounds. Their most common genetically modified bacterial name, their modified genes, and their encoded gene products and function have been summarized in Table 14.1.

14.3.1.2 Fungi

The transformation results in either knock-in or knockout strain where the wild-type gene has been replaced by a foreign or mutated one. Homologous recombination is an alternative method of DNA-mediated transformation in bacteria, fungi, and higher eukaryotes. However, DNA-mediated transformation experiments in fungi differ from bacteria and eukaryotes in two ways. First, self-replication vectors are rare and integration of transferred DNA occurs ectopically into genomic DNA. As a consequence, the recombinant DNA is not easy to recover from transgenic fungal strains. Second, DNA integration is driven by nonhomologous end. Homologous recombination is necessary in order to allow effective analysis of fungal gene function. In principle, two types of transformants can be distinguished: knock-in and knockout. In knock-in strain foreign DNA sequence integrates in the target gene, resulting in abnormal transcripts. In knockout transformants the target gene is replaced by a marker gene (resistance gene) and the target gene is therefore no longer expressed.

A recent alternative technique for manipulating fungal gene expression is RNA interference (RNAi), a mechanism that is also known as gene silencing. This technique is mainly important when target genes are present as multi-copy genes and deletion of target genes may cause lethal effect to the recipient. RNAi leads to reduction of the transcript level in knockdown transformants in the recipient. Homologous recombination and RNAi are powerful and versatile used methods of genetic engineering that can be applied for functional genomic approaches to uncover gene function but also will have a major impact on the use of fungi for the

Table 14.1 List of genetically modified bacteria

S.N.	Name of bacteria	Gene	Protein product	Function
1	*E. coli*	aad	Adenylyltransferases	Resistance to spectinomycin and streptomycin antibiotics
		aph4	Hygromycin B phosphotransferase	Resistance to the hygromycin B antibiotic
		bla	Beta-lactamase	Detoxification of beta-lactam antibiotics such as penicillin and ampicillin
		nptII	Neomycin phosphotransferase II	Metabolization of antibiotics neomycin and kanamycin
		spc	Spectinomycin adenyltransferase	Resistance to spectinomycin/streptomycin antibiotics
		ecbeta	*E. coli* choline dehydrogenase	Tolerance to water stress
		dam	DNA adenine methylase	Promote male sterility
		pmi	Phosphomannose isomerase	Metabolization mannose for selection
		uidA	β-D glucuronidase	Production of blue stain on treated transformed tissue, used for selection

2	*Bacillus thuringiensis strain PS149B1*	cry34Ab1	δ-Endotoxin	Resistance to insecticide against coleopteran insects
		cry35Ab1	Cry35Ab1 δ-endotoxin	Resistance to insecticide against coleopteran insects
		cry3A	Cry3A δ-endotoxin	Resistance to insecticide against coleopteran insects
		cry3Bb1	Cry3Bb1 δ-endotoxin	Resistance to insecticide against coleopteran insects
		mcry3A	Cry3A δ-endotoxin	Resistance to insecticide against coleopteran insects
		cry1A	Cry1A δ-endotoxin	Resistance to insecticide against lepidopteran insects
		cry1Ab-Ac	Cry1Ab-Ac δ-endotoxin	Resistance to insecticide against lepidopteran insects
		cry1C	Cry1C δ-endotoxin	Resistance to insecticide against lepidopteran insects
		ecry3.1Ab	Cry3A-Cry1Ab δ-endotoxin	Resistance to insecticide against coleopteran and lepidopteran insects
3.	*E. coli bacteriophage T3*	sam-k	S-Adenosylmethionine hydrolase	Delay ripening
4.	*Stenotrophomonas maltophilia strain DI-6*	dmo	Dicamba monooxygenase	Tolerance to the herbicide dicamba (2-methoxy-3,6-dichlorobenzoic acid)
5	*Bacillus subtilis*	5azazz	Cold shock protein B	Tolerance to water stress

(continued)

Table 14.1 (continued)

S.N.	Name of bacteria	Gene	Protein product	Function
6	*Rhizobium meliloti*	rmbeta	Choline dehydrogenase	Tolerance to water stress
7.	*Bacillus amyloliquefaciens*	barstar	Barnase ribonuclease inhibitor	Restore fertility by inhibiting barnase on tapetum cells of the anther
		barnase	Barnase ribonuclease (RNAse)	Induction of male sterility by interfering with RNA production in the tapetum cells of the anther
8.	*Streptomyces hygroscopicus*	bar	Phosphinothricin N-acetyltransferase	Resistance to bialaphos herbicide
9	*Streptomyces viridochromeogenes*	pat	Phosphinothricin N-acetyltransferase	Resistance to phosphinothricin herbicides
10	*Agrobacterium tumefaciens strain CP4*	epsps	5-Enolpyruvylshikimate-3-phosphate synthase	Acetylation of glufosinate
		nos	Nopaline synthase	Synthesize nopaline, which permits the identification of transformed plant embryos
11.	*Arthrobacter globiformis*	epsps	5-Enolpyruvylshikimate-3-phosphate synthase	Resistance to glyphosate herbicides
12.	*Bacillus licheniformis*	gat4601 and gat4621	Glyphosate N-acetyltransferase	Resistance to glyphosate herbicides
13.	*Ochrobactrum anthropi strain LBAA*	gox	Glyphosate oxidase	Resistance to glyphosate herbicides
14.	*Pseudomonas fluorescens strain A32*	hppd	p-Hydroxyphenylpyruvate dioxygenase	Resistance to isoxaflutole herbicide
15.	*Bacillus thuringiensis* subsp. *kumamotoensis*	cry1A.105	Cry1A.105 protein which comprises the Cry1Ab, Cry1F, and Cry1Ac proteins	Resistance to insecticide against lepidopteran insects
		cry1Ab	Cry1Ab δ-endotoxin	Resistance to insecticide against lepidopteran insects

16.	*Bacillus thuringiensis* var. *aizawai*	cry1F	Cry1F δ-endotoxin	Resistance to insecticide against lepidopteran insects
		cry1Fa2	Cry1F protein	Resistance to insecticide against lepidopteran insects
17.	*Bacillus thuringiensis* subsp. *kumamotoensis*	cry2Ab2	Cry2Ab δ-endotoxin	Resistance to insecticide against lepidopteran insects
18.	*Bacillus thuringiensis* subsp. *dakota*	cry2Ae	Cry2Ae δ-endotoxin	Resistance to insecticide against lepidopteran insects
19.	*Bacillus thuringiensis* subsp. *tolworthi strain BTS02618A*	cry9C	Cry9C δ-endotoxin	Resistance to insecticide against lepidopteran insects
20.	*Bacillus thuringiensis* var. *aizawai*	mocry1F	Modified Cry1F protein	Resistance to insecticide against lepidopteran insects
21.	*Bacillus thuringiensis strain AB88*	vip3A(a)	VIP3A vegetative insecticidal protein	Resistance to insecticide against lepidopteran insects
22.	*Corynebacterium glutamicum*	dhdps	Dihydrodipicolinate synthase	Increase production of lysine amino acid
23.	*Klebsiella pneumoniae* subsp. *ozaenae*	bxn	Nitrilase specific	Eliminate herbicidal activity of oxynil herbicides such as bromoxynil

Table 14.2 Some genetically modified fungi

S.N.	Name of fungus	Gene	Process	Function
1	*Neurospora crassa*	Nc.Fad3	δ-15 desaturase protein	Enhance production of unsaturated fatty acid
2	*Aspergillus niger* var. *van Tieghem*	Phya3	3-Phytase	Promote degradation of phytates
3	*Aspergillus niger strain 963*	phya2	2-Phytase	Promote degradation of phytates

production of medicinal and industrial products. For example, production of insulin by GM fungi is benefited for diabetic patients.

One of the medical benefits of GM fungi is that they are used to produce certain medicine such as insulin. GM fungi have a great approach for bioprocess engineering, remediation of contaminated site, and production of many secondary metabolites. Some GM fungi with their name, modified gene, encoded products, and their function have been listed in Table 14.2.

14.3.1.3 Viruses

The development of vaccines against infectious diseases represents one of the most important contributions of viruses in medical science. However, vaccine-preventable diseases still cause millions of deaths each year due to the thermal instability and poor efficacy of vaccines. Human enterovirus type 71 vaccine strain has been used as a model for rational design approach to improve the thermostability and immunogenicity of live vaccines by self-biomineralization.

Some stabilizers such as deuterium oxide, proteins, $MgCl_2$, and nonreducing sugars have been introduced to produce stabilized formulations of vaccines. In nature, biomineralization is adopted by many organisms to improve their performance in harsh environments. A natural virus cannot induce biomineralization by itself during its normal life cycle. By using reverse genetic technology, the viral capsid can be rationally genetically engineered to improve its physicochemical properties. A combination of genetic engineering and biomineralization techniques is used to produce a thermostable vaccine. The self-directed biomineralized vaccine can be used efficiently after storage at ordinary temperatures which significantly increases the efficacy of immunization systems and lowers the cost of vaccine delivery and storage.

Viruses that have been genetically altered to prevent them from causing disease such as retrovirus or adenovirus are often used as the vehicle for delivering the gene into certain human cell types, in much the same way as ordinary viruses infect cells. This delivery method is fairly imprecise and limited to the specific types of human cells. For example, some viruses commonly used as gene delivery vehicles can only infect cells that are actively dividing. This limits their usefulness in treating diseases of the heart or brain, because these organs are largely composed of nondividing cells.

Retroviruses (Moloney murine leukemia virus) and adenovirus were mainly used for manipulation of eukaryotic genes. Retroviruses were used for knock-in and adenoviruses for knockdown strain production and their gene expression. However, there has been little success in gene transfer with such virus vectors because they may degrade the nuclear membrane of target cell for integration of their gene into host chromosome. However, scientists soon realized that members of the subfamily lentivirus, such as the retrovirus human immunodeficiency virus (HIV), would have the same ability to transfer genetic material into the genomes of cells, but could do this with nondividing dormant cells *in vivo* and growth-arrested cells *in vitro*. Exploring this new method of gene therapy has been the work of many laboratories in the past few years.

Lentiviral vectors can infect both dividing and nondividing cells because of their shell integration complex which helps in the integration of virus into host cell through nuclear membrane. Lentiviruses are effectively used in gene therapy as they can change the expression of target gene for up to 6 months. These are also used specifically for gene manipulation in terminally differentiated cells such as neurons, macrophages, hematopoietic stem cells, retinal photoreceptors, and muscle and liver cells. Lentiviral vector is very effective because it has evolved to infect and express its genes in human helper T cells and other macrophages (Fig. 14.6).

Lentiviruses are the only type of virus that is diploid; they have two strands of RNA. Thus, HIV contains a diploid single-stranded positive-sense RNA genome that is approximately 10 kb long. The ends are flanked with long terminal repeats (LTRs). A Psi-sequence is found near the 5′ end of the RNA genome which is necessary for packaging viral RNA into virus capsids to continue the infection of HIV in its host. However, the HIV's genetic information is integrated into the DNA of host cell, so its RNA must be converted into DNA inside the host by reverse transcription for viral replication to be successful. Reverse transcriptase synthesizes the first

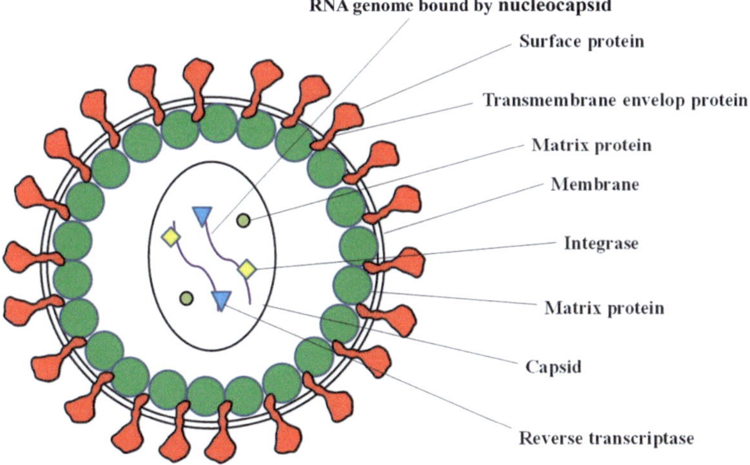

Fig. 14.6 Structure of lentivirus

strand of DNA from the RNA template, and the host DNA polymerase synthesizes the second strand to produce dsDNA. The DNA copy just made, which contains the genes *gag*, *env*, and *pol*, is inserted by integrase into the host genome. LTRs are also necessary for integration of the dsDNA into the host chromosome.

It has been shown that the lentivirus vector has higher rate of expression in its host cells than other retroviruses. These vectors do not trigger immune reactions in the host making them very attractive delivery systems in gene therapy. As it has been proved that lentiviruses are successful gene delivery vehicles, the researchers has now turned their attention to produce GM lentivirus vectors with built-in safety features to prevent the development of replication competent lentiviruses (RCL). However, even the earliest studies with HIV lentiviral vectors did not generate RCL *in vitro* or *in vivo* (Amado and Chen 1999), but precautions are still very important. Thus some genetically modified HIV lentiviral vectors are being produced without HIV genes in their capsids. GM vector is a great increase in safety because potential RCLs cannot use the HIV genes necessary for replication of HIV in humans. One of the drawbacks to use lentivirus vectors is that they cannot transduce macrophages because of the absence of accessory genes necessary for HIV replication. Thus, scientists have shown that the type of lentiviral vector is dependent on the type of target cell.

To complete the key goal of management of neuronal disorders (Parkinson's disease, Alzheimer's disease, Huntington's disease, motor neuron diseases, lysosomal storage diseases, and spinal injury), vectors based on lentiviruses have shown particularly useful features. Lentiviral vectors can deliver 8 kb of sequence, and after delivery into the nervous system, they induce no significant immune responses, there are no unwanted side effects of the vectors per se to date, and manufacturing and safety testing for clinical applications are well advanced. Some genetically modified viruses with their name and function are listed in Table 14.3.

14.3.2 Plants as GMO

The new traits introduced to crop plants by genetic engineering have the potential to increase crop yields, improve agricultural practices, or add nutritional quality to products (Table 14.4). For example, transgenic crop plants capable of degrading weed killers allow farmers to spray weeds without affecting yield. The use of herbicide-tolerant crops may also allow farmers to move away from preemergent herbicides and reduce crop cultivation, thereby decreasing soil erosion and water loss. Transgenic plants that express insecticidal toxins resist attacks from insects. Crops engineered to resist insects are an alternative to sprays, which may not reach all parts of the plant. They are also cost-effective, reducing the use of synthetic insecticides. Genetic engineering has also been used to increase the nutritional value of food; "golden rice" is engineered to produce beta-carotene, for example. Edible vaccines, introduced in the plants we eat, may be on the prescription list of doctors. The new traits expressed in such transgenic plants are derived from a variety of other organisms (Table 14.4). Researchers have isolated a gene from the bacterium *Salmonella* and inserted it into a variety of crops such as soybeans, corn,

Table 14.3 List of genetically modified plant and animal viruses

S.N.	Name of virus	Gene	Viral components	Function
1.	Bean golden mosaic virus (BGMV)	ac1	Actin-1 (sense and antisense RNA of viral replication protein	Inhibit synthesis of viral replication protein of BGMV, conferring resistance to the BGMV
2.	Plum pox virus (PPV)	ppv_cp	Coat protein of plum pox virus	Resistance to plum pox virus (PPV)
3.	Papaya ring spot virus (PRSV)	prsv_cp	Coat protein of papaya ring spot virus	Resistance to papaya ring spot virus (PRSV)
4.	Papaya ring spot virus (PRSV)	prsv_rep	Replicase domain of papaya ring spot virus	Resistance to papaya ring spot virus (PRSV)
5	Potato virus Y (PVY)	pvy_cp	Coat protein of potato virus Y	Resistance to potato virus Y (PVY)
6.	Watermelon mosaic potyvirus 2 (WMV2)	wmv_cp	Coat protein of watermelon mosaic potyvirus 2	Resistance to watermelon mosaic potyvirus 2
7.	Zucchini yellow mosaic potyvirus (ZYMV)	zymv_cp	Coat protein of zucchini yellow mosaic potyvirus	Resistance to zucchini yellow mosaic potyvirus (ZYMV)
8.	HIV (lentivirus)	tat gene	Encode tat protein	Decrease level of chloramphenicol acetyltransferase w, induce immune response
9.	HSV	UL39	Encode the large subunit of ribonucleotide reductase	Inhibit viral transcription

canola, and cotton to make them insect resistant against glyphosate (Roundup™) pesticide. In transgenic crops (cotton, potato, and corn), the insecticidal gene comes from the bacterium *Bacillus thuringiensis* (Bt) (Fig. 14.7). A gene *crtI* from the soil bacterium *Erwinia uredovora* and the other gene *psy* from plant *Narcissus pseudonarcissus* (daffodil) were inserted for vitamin A productions in golden rice. Synthesis of beta-carotene occurs in the chloroplast of rice plant and not in the endosperm of seed. Geranyl geranyl diphosphate (GGPP), a precursor of carotenoid production, is present in rice seed endosperm. It is converted into beta-carotene by an enzyme expressed in genetically engineered golden rice plant. Conversion of GGPP into beta-carotene requires four biochemical reactions, each catalyzed by different enzymes.

Agrobacterium tumefaciens may be used in the transfer of gene into rice plant for the production of beta-carotene. Bacteria contain three plasmids, each encode an enzyme for complete biochemical pathway for beta-carotene production. Bacterium uses three enzymes instead of four by plant system for this conversion (Fig. 14.7).

Edible vaccines are designed to induce immune response without causing any side effect. These vaccines are produced from different crops and could help people in developing countries where safe transportation of medical supplies is limited.

Table 14.4 List of genetically modified plants

S.N.	GM plant (common name)	GM plant (scientific name)	Gene	Product	Function
1	Alfalfa	*Medicago sativa*	ppo	Polyphenol oxidases encode dsrna	Induce silencing of suppresser genes
			ccomt	Caffeoyl-CoA O-methyltransferase	Induce silencing of suppresser genes
2	Apple	*Malus domestica*	ppo	Polyphenol oxidases encode dsrna	Induce silencing of suppresser genes
3	Bean	*Phaseolus vulgaris*	hva1 gene	Encode *Hordeum vulgaris* abundant protein (late embryogenic protein III)	Tolerance to abiotic stresses
4	Cotton	*Gossypium hirsutum* L.	Bar	Phosphinothricin N-acetyltransferase	Resistance to l-phosphinothricin herbicides
			dmo	Encode dicamba monooxygenase	Resistance to dicamba herbicide

5	Eucalyptus	*Eucalyptus* sp.	cbd and cel1	Cellulose-binding domain and endo-1,4-β-glucanase	Increase cellulose content and provide strength to cell wall
			echb1	Encoding for a hd-zip class II protein	Xylem cell wall biosynthesis
			egrtub1	Encode β-tubulin gene	Influence the microfibrillar orientation in cellulose secondary cell wall
			fra1	Encode fragile fiber 1 (kinesin-like protein)	Influence the microfibrillar orientation in cellulose secondary cell wall
			ccr	Cinnamoyl-CoA reductase	Enhance lignin biosynthesis
			dreb1	Dehydration-responsive element-binding 1 (encode transcription factor)	Resistance to drought stress
			egucbf1a/b	Eucalyptus c-repeat binding factors	Resistance to cold stress
			p5cs	Pyrroline-5-carboxylate synthase (encode transcription factor)	Resistance to cold stress
6	Maize	*Zea mays* L.	zm-hra	Herbicide-tolerant acetolactate synthase (als) enzyme	Resistance to imidazolinone herbicides such as sulfonylurea and
			rpi-vnt1	Late blight resistance protein	Resistance to potato late blight
7	Papaya	*Carica papaya*	prsvcp	Encode papaya ring spot virus coat protein	Resistance to papaya ring spot virus
			npti	Neomycin phosphotransferase	Resistance to neomycin antibiotics

(continued)

Table 14.4 (continued)

S.N.	GM plant (common name)	GM plant (scientific name)	Gene	Product	Function
8	Petunia	*Petunia hybrida*	cytb5	Encode cytochrome b5	Induce synthesis cyt b5 protein that acts as an electron donor to the cyt p450
			dfr	Dihydroflavonol-4-reductase (dfr) hydroxylase	Catalyze the production of anthocyanin pigment
			hfl	Flavonoid 3′,5′-hydroxylase enzyme	Catalyze the production of anthocyanin pigment
9	Potato	*Solanum tuberosum*	Gbss	Granule-bound starch synthase (gbss)	Reduces the levels of amylose and increases levels of amylopectin
			pphl	Phosphorylase-l gene	Decrease formation of reducing sugars through starch degradation
			pr1	Pathogenesis-related protein 1 (encode ds rna)	Decrease formation of reducing sugars through starch degradation
			asn1	Asparagine synthetase [(encode ds rna)	Decrease asparagine synthesis
10	Rice	*Oryza sativa* L.	osa1	Encode *Oryza sativa* protein kinases	Resistance to syncytial virus infection
			cbf1	Encode inducible crt/dre binding factor	Resistance to water stress
			pma1959 pma80	Physiological mechanisms and adaptation factors encode late embryogenic proteins	Induce salt tolerance

11	Golden rice	(*Oryza sativa*)	Psy	Encode enzyme	Involved in biosynthesis of carotenoids
				Phytoene synthase	
			crtI	Encode enzyme carotene desaturase	Involved in biosynthesis of carotenoids
11	Rose	*Rosa hybrida*	p(sag12)-ipt gene	Encode ipt gene	Delay leaf senescence and enhance resistance to exogenous ethylene
12	Soybean	*Glycine max*	fad2-1a	Delta-12 fatty acid desaturase	Reduce desaturation of unsaturated fatty acids
			fatb1-a	Acyl-acyl carrier protein thioesterases	Reduce desaturation of unsaturated fatty acids
			gm-fad2-1	Glycine max-delta-12 fatty acid desaturase	Reduce desaturation of unsaturated fatty acids
			gm-hra	Glycine max modified acetolactate synthase	Tolerance to sulfonylurea-based herbicides
13	Sugar beet	*Beta vulgaris*	bv_22240_ksro	Encode viral coat protein	Induce defense against viruses
			bnyvv	Encode beet necrotic yellow vein virus coat protein	Resistance to beet necrotic yellow vein virus
14	Sweet pepper	*Capsicum annuum*	rf gene	Encode restore fertility	Restore fertility
				Protein	
			orf456	Encode orf protein	Enhance male sterility

(continued)

Table 14.4 (continued)

S.N.	GM plant (common name)	GM plant (scientific name)	Gene	Product	Function
15	Tobacco	*Nicotiana tabacum* L.	qpt1	Quinolinic acid phosphoribosyltransferase (enhance siRNA)	Suppression of QPTase gene reduces production of nicotinic acid
			als	Acetolactate synthase	Resistance to sulfonylurea herbicides
16	Tomato	*Lycopersicon esculentum*	acc	1-Amino-cyclopropane-1-carboxylic acid (acc) synthase gene (enhance) siRNA)	Reduce ethylene production and delay fruit ripening
			anti-efe	1-Amino-cyclopropane -1-carboxylate oxidase (aco) gene (enhance siRNA)	Reduce ethylene production and delay fruit ripening
			pg	Polygalacturonase (enhance siRNA)	Induce breakdown of pectin molecules in the cell wall and thus cause delayed softening of the fruit
17	Wheat	*Triticum aestivum*	lr	Encodes anti-leaf rust protein	Resistance to leaf rust disease

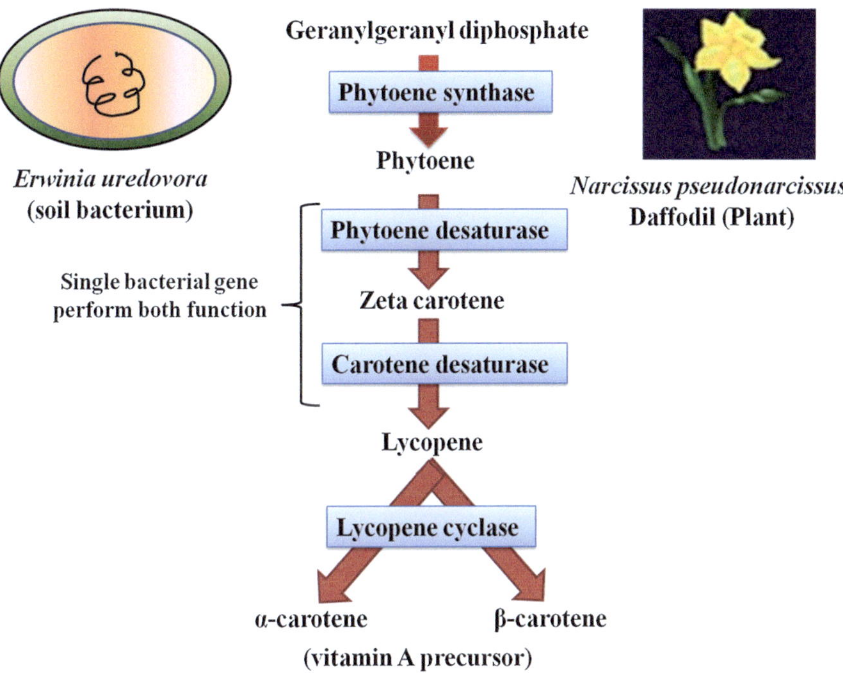

Fig. 14.7 Comparison of beta-carotene production pathway in plant and bacteria cell

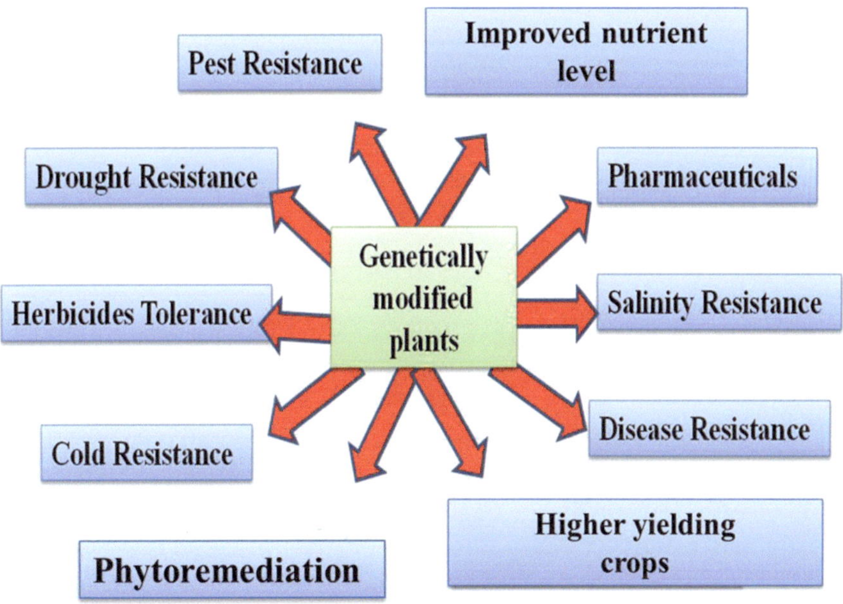

Fig. 14.8 Benefits of genetically modified plants

Overall, transgenic plants have many applications to make environment eco-friendly and producing natural and economically efficient vaccines (Fig. 14.8).

14.3.2.1 Methods of Plant Trait Introduction

14.3.2.1.1 *Agrobacterium* Mediated

A. tumefaciens and *A. rhizogenes* are virulent bacterial strains that induce crown gall and hairy root disease when interacting with susceptible dicotyledonous plant cells. Both bacterial strains contain a large megaplasmid (more than 200 kb) named as Ti and Ri plasmid, which plays a key role in tumor induction and hairy root formation (Fig. 14.9).

Classification of Ti plasmids is done on the basis of type of opines which are produced and excreted by the induced tumor cells. During T-DNA infection, a mobile segment of Ti or Ri plasmid is transferred to the plant cell nucleus and integrated into the plant cell chromosome. The T-DNA fragment is flanked by 25-bp direct repeats, which act as signal for T-DNA transfer. Proteins encoded by genes present in virulence region (vir) of Ti plasmid play an important role in T-DNA transfer (Fig. 14.10). The Ti plasmid also contains the genes for opine catabolism produced by the cancerous cells for its own integrity and stability. The 30-kb vir region is a regulon organized in six operons that are essential for the T-DNA transfer (virA, virB, virD, and virG) or for the increasing efficiency of T-DNA transfer (virC and virE).

Different chromosomal-determined genetic elements have shown their functional role in the attachment of *A. tumefaciens* to plant cell and bacterial colonization. The loci chvA and chvB are involved in the synthesis and excretion of the b-1,2 glucan. The chvE is required for the sugar enhancement of vir gene induction

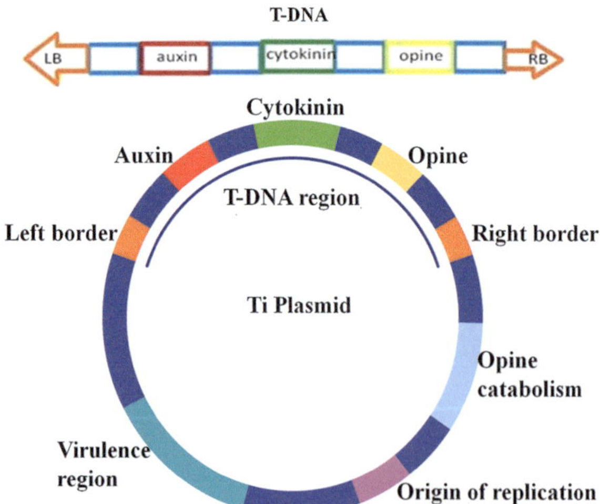

Fig. 14.9 Ti plasmid

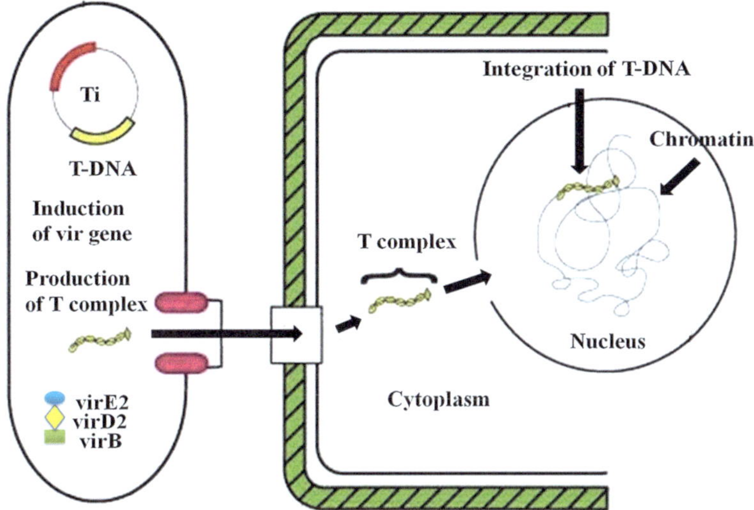

Fig. 14.10 Molecular mechanism of gene transfer through *Agrobacterium*

and bacterial chemotaxis, the cel locus for the synthesis of cellulose fibrils, the pscA (exoC) locus for the synthesis of both cyclic glucan and acid succinoglycan, and the att locus involved in the cell surface proteins.

The *Agrobacterium* system delivers a single-stranded (ss) DNA molecule with, at the 5′-end, the pilot protein VirD2 into host cells through a type 4 secretion system (TFSS). The nuclear localization sequence in VirD2 guarantees a rapid translocation into the nucleus of the host cells. Besides the T-strand-VirD2 complex, the TFSS is also used to secrete separately a set of virulence proteins into the host cells. The latter are effector proteins that aid in the transformation process. For instance, the VirE2 protein is an ssDNA-binding protein which coats (and thus protects) the T strand on its way to the nucleus and VirF, an F-box protein, that is thought to help in the integration process by uncoating the T-DNA.

Host factors also play an important role in *Agrobacterium*-mediated DNA transfer and involve several steps: bacterial colonization, induction of bacterial virulence system, generation of T-DNA transfer complex, T-DNA transfer, and integration of T-DNA into plant genome. However, *A. tumefaciens* naturally infects only dicotyledonous plants. For these regions, alternative direct transformation methods have been developed for the transfer of genes in monocot plants such as rice, banana, corn, wheat, and sugarcane (Shillito et al. 1985; Potrykus 1991) such as electroporation, microinjection, particle gun bombardment and polyethylene glycol-mediated DNA transfer, protoplast and intact cell electroporation, and gene gun technology.

Agrobacterium-mediated transformation has stunning advantages over direct transformation methods. It reduces the copy number of the transgene, potentially leading to fewer problems with transgene cosuppression and instability. In addition, it is a single-cell transformation method not forming mosaic plants, which are more frequent when direct transformation is used.

14.3.2.1.2 Electroporation

It is the process whereby electrical impulses of high field strength are used to reversibly permeabilized cell membrane to facilitate uptake of large molecules, including DNA. It has been used for transient and integrative transformation of protoplasts. The genes of interest require plant regulatory sequence. Plant materials are incubated in a buffer solution containing DNA and subjected to high-voltage electric pulse. The DNA then migrates through high-voltage-induced pores in the plasma membrane and integrates into the genome (Fig. 14.11). It is used to transform all the major cereals particularly rice, wheat, maize, etc.

14.3.2.1.3 Particle Bombardment

It is the most powerful method for introducing nucleic acids into plants, because the helium pressure can drive microcarriers through cell walls. It is a much easier and less time-consuming process. It requires less DNA and fewer cells than other methods and can be used for either transients or stable transformation. In this process the gold or tungsten particles are coated with the DNA that is used in the transformation of plant tissue. The particles are propelled at high speed into the target plant material where the DNA is released within the cell and can integrate in the genome (Fig. 14.12). Two types of plant tissue are used for particle bombardment: (1) primary explants that are bombarded and then induced to become embryogenic and (2) proliferating embryogenic cultures that are bombarded and then allowed to proliferate further and subsequently regenerate. This process also allows the transformation of animal cells that have unique growth requirements and that are not permeable to gene transfer using any other method.

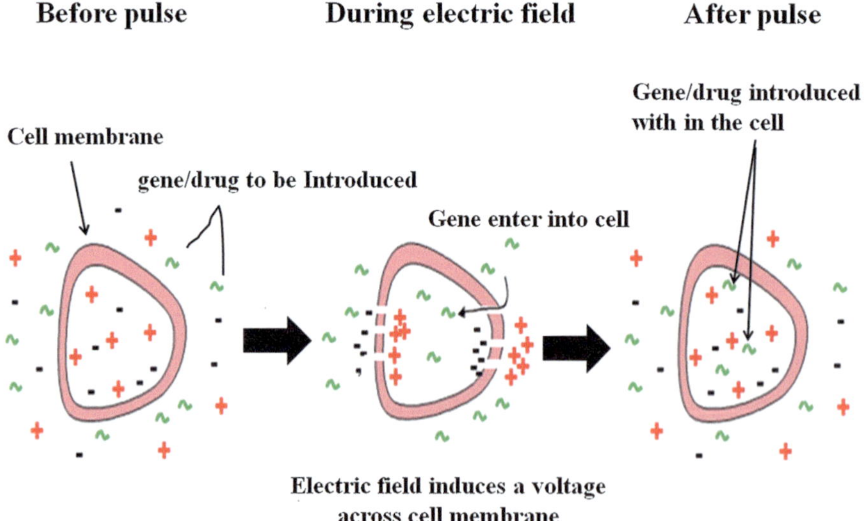

Fig. 14.11 Electroporation method of gene transfer

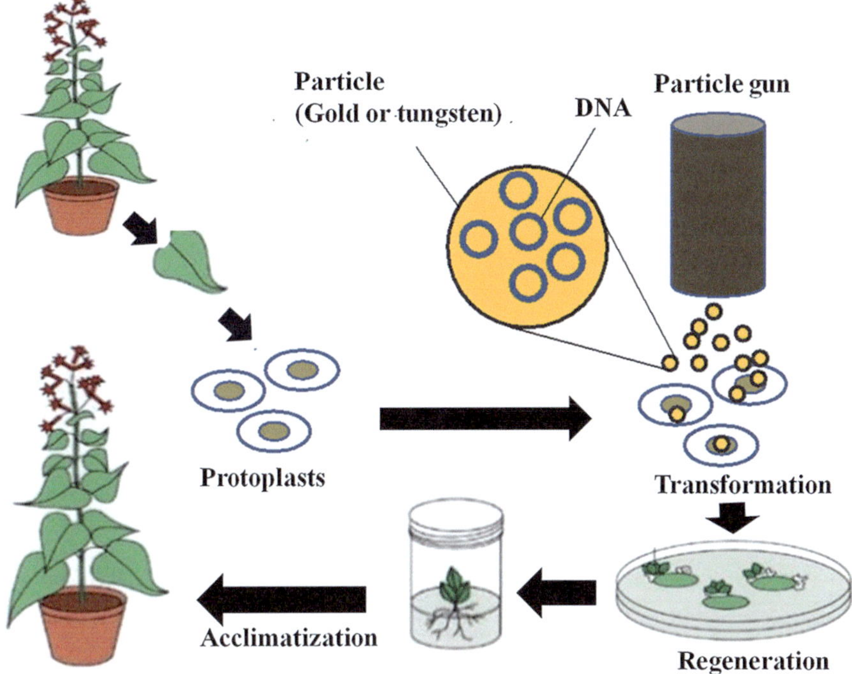

Fig. 14.12 Particle gun bombardment method of DNA transformation

14.3.2.1.4 Microinjection

In this method the DNA is directly injected into plant protoplasts or cells (specifically into the nucleus or cytoplasm) using fine-tipped (0.5–1.0 μm diameter) glass needle or micropipette. This method of gene transfer is used to introduce DNA into large cells such as oocytes, eggs, and the cells of early embryo (Fig. 14.13).

14.3.2.1.5 Lipofection

Liposomes are circular lipid molecules with an aqueous interior that can carry nucleic acids. Liposomes encapsulate the DNA fragments and then adhere to the cell membranes and fuse with them to transfer DNA fragments. Thus, the DNA enters enter into the target cell (Fig. 14.14).

14.3.2.1.6 Polyethylene Glycol Mediated

Plant protoplasts treated with polyethylene glycol more readily take up DNA from their surrounding medium, and this DNA can be stably integrated into the plant's chromosomal DNA. Protoplasts are then cultured under conditions that allowed them to grow cell walls, start dividing to form a callus, develop shoots and roots, and regenerate whole plant.

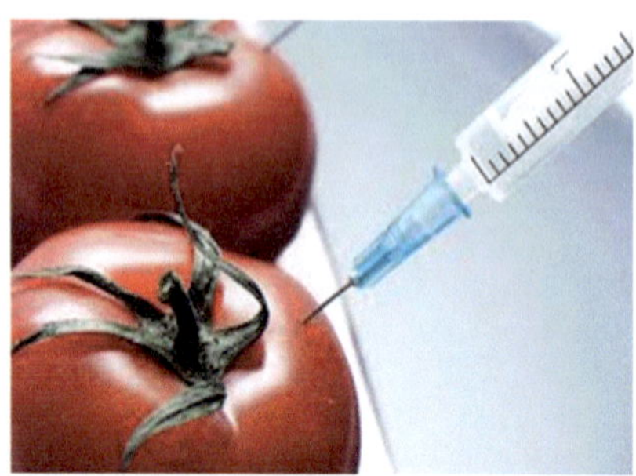

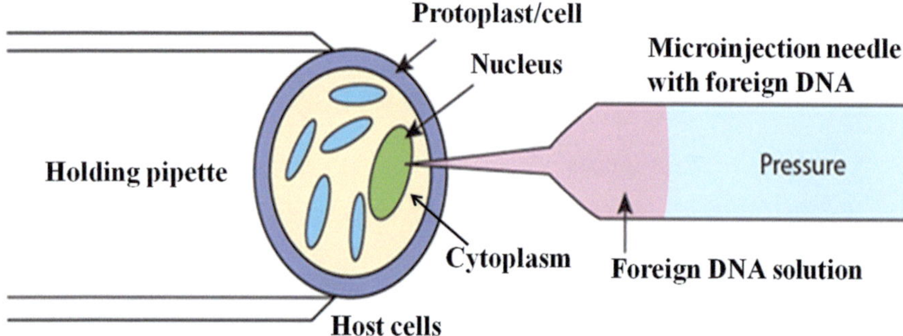

Fig. 14.13 Microinjection-mediated DNA transfer

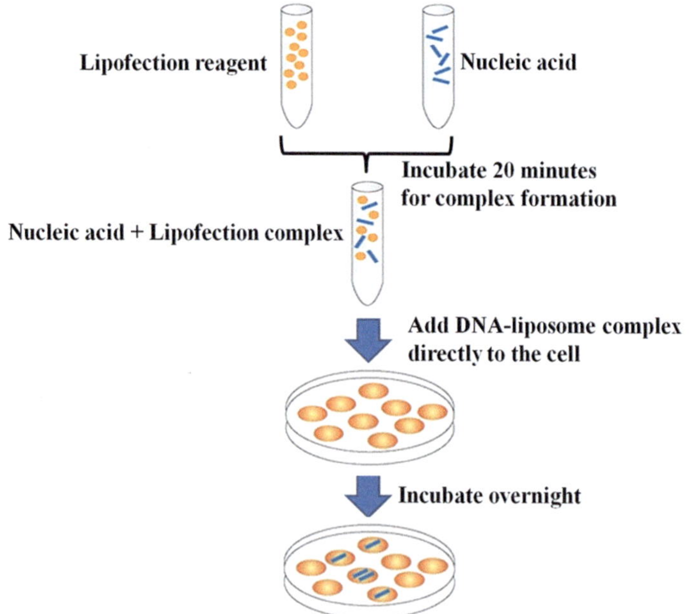

Fig. 14.14 Liposome-mediated DNA transformation

14.3.2.2 Health Concerns

Several health concerns have been raised about gene manipulation/transgenic crops by various groups of researchers and lobbyists. Some of these are listed here:

(a) Altered genes could affect the environment and human health by expressing unanticipated allergens. For example, some researchers of Nebraska University showed that on the introduction of allergen from Brazil nuts into soybeans, it makes some specific IgE antibodies against certain proteins in nuts which may result into hypersensitive reactions (Nordlee et al. 1996).
(b) Development of insecticide-resistant plants of target pests is a concern, as few possibilities that nontarget beneficial insects might be affected by transgenic-resistant plants. For example, hundreds of known subspecies of *Bacillus thuringiensis* produced insecticidal toxins against certain species of insects. Some of these insect species may be beneficial for pollination and other pest control.
(c) Manipulated genes in engineered crops could move to other organisms in the environment and may create health issues.
(d) GM foods may cause antibiotic resistance in human beings as some or the other antibiotic-resistant gene is used as marker in genetic manipulation techniques.
(e) GM foods have more quantity of a single nutrient, but other nutrients may be lost in this process which may result into overall decrease in the nutritional quality of this food.
(f) Some groups also claim that there may be immune suppression after consumption of GM foods.
(g) The most important concern about GMOs is its unpredictability due to unavailability of long-term exposure data. The foreign DNA in a host cell may induce other genetic material in that host to behave erratically and out of this genes may be overexpressed or underexpressed creating numerous health problems.

14.3.3 Animal as GMO

Transgenic animals can be created by manipulating embryonic stem (ES) cells. ES cells are obtained from the inner cell mass of a blastocyst. Transgene is incorporated into ES cells by microinjection, by retrovirus (lentivirus), or by electroporation. Transgenic ES cells are grown *in vitro* and then they are inserted into blastocyst which is implanted into a host's uterus to grow normally. Somatic cell nuclear transfer-mediated cloning in animals depends on the availability of donor cells and appropriate genetic information. Fibroblast, mammary epithelia, and ovarian and muscles cells are grown in cell culture and are genetically altered by fusion with the enucleated egg cells (Fig. 14.15). Virus is a vector of choice to transfer DNA in animal cells.

Lentiviral vectors have become a promising new tool for the establishment of transgenic animals (Fig. 14.16) and the manipulation of the mammalian genome. While conventional microinjection-based methods for transgenesis have been

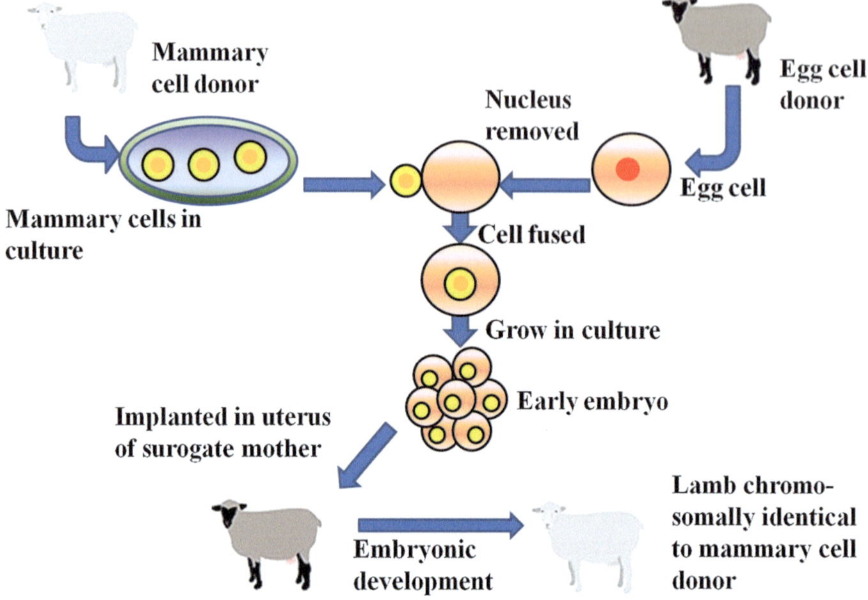

Fig. 14.15 Gene transfer in animals

Fig. 14.16 Lentivirus-mediated gene transfer in animals

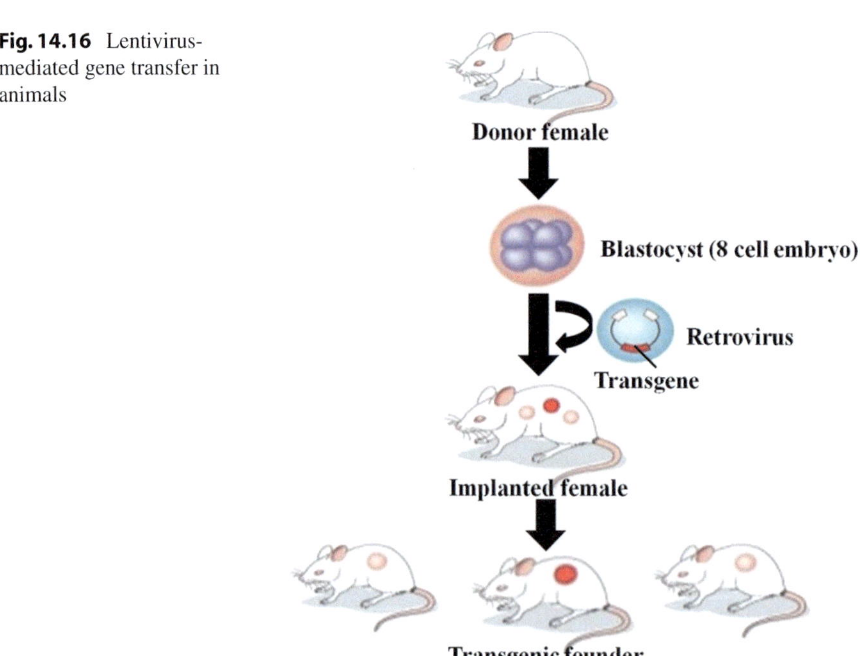

successful in generating small and large transgenic animals, their relatively low transgenic efficiency has opened the door for alternative approaches, including lentiviral vectors. Lentiviral vectors are an appealing tool for transgenesis in part because of their ability to incorporate into genomic DNA with high efficiency, especially in cells that are not actively dividing. Lentiviral vector-mediated transgene expression can also be maintained for long periods of time. Recent studies have documented high efficiencies for lentiviral transgenesis, even in animal species and strains, such as NOD/SCID and C57Bl/6 mouse that are very difficult to manipulate using the standard transgenic techniques. These advantages of the lentiviral vector system have broadened its use as a gene therapy vector to additional applications that include transgenesis and knockdown functional genetics.

An alternate and reliable method for gene transfer in animals is microinjection. In this method, a gene construct is characterized in culture system and sufficient quantity of the visual and desired DNA is obtained. Before the starting of cell division, the DNA is injected into fertilized ova. This increases the probability that all of the cells of the organisms will harbor the gene. After fertilization the injection is done soon, before the female and male pronuclei have fused. Surrogate mothers are implanted with the injected eggs after they are made pseudopregnant with the help of hormones. Tissue samples of the offspring are assessed for the presence of the required gene, after birth. Nuclear gene transfer in germ line cells is given special attention. If the novel gene is present in germ line cells, the animal may be used as a founder for breeding.

The use of transgenic animals may sometimes create ethical issues such as genetic diversity in livestock which may increase susceptibility to disease.

14.3.4 Human as GMO

In humans, embryonic stem cell-mediated gene transfer occurs which involve isolation of totipotent stem cells from embryos. Then the desired gene is inserted into these cells. Recombinant cells are incorporated into the host embryo (Fig. 14.17). Gene therapy is an approach that uses genes to treat or prevent disease. In future, this technique will be used to treat a disorder by inserting a healthy gene into a patient's cells instead of using drugs or surgery. Gene therapy using genetically modified stem cells offers several unique advantages over direct gene transfer into the body and over cell therapy, which involves administration of cells that have not been genetically modified. The major reason for using stem cells in cell-based gene therapies is that they are a self-renewing population of cells and thus may reduce or eliminate the need for repeated administrations of the gene therapy. For GM stem cell-based gene therapy trials that have had a therapeutic goal, approximately one-third have focused on cancers (ovarian, brain, breast, myeloma, leukemia, and lymphoma), one-third on HIV-1, and one-third on so-called single-gene diseases (e.g., Gaucher's disease, severe combined immunodeficiency, Fanconi anemia, Fabry disease, and leukocyte adherence deficiency).

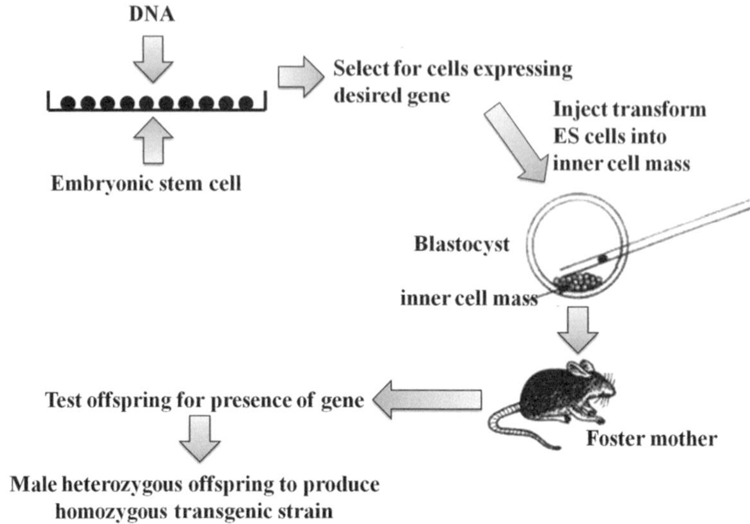

Fig. 14.17 Stem cell-mediated gene transfer

Brain metastasis is the most common malignant tumor of the central nervous system. Primary brain tumors are known as glioma and glioblastoma. Brain metastasis most commonly arises from lung, breast, and skin cancers. In the treatment of lung cancer metastasis to the brain, neural stem cells (NSCs) derived from human fetal telencephalon were used for a genetically engineered stem cell (GESTEC)-based therapy. Human stem cells can be modified by different techniques including electroporation, viral transduction, and nucleofection. Genetically modified hematopoietic stem cells are used as therapy against viruses (Fig. 14.18).

14.4 GMO Production

A transgene is a unit of gene or genetic material that has been transferred by any transformation method into host cells to change the phenotype of an organism means production of GMO. Transgene must contain a promoter, coding sequence, and a terminator region (Fig. 14.19). A promoter is a regulatory sequence of transgene which gives signal for starting the transcription, coding sequence determines the amino acid sequence of the protein, and a terminator gives the signal for ending the transcription.

GMOs are produced either by natural or artificial gene modification systems. Natural genetic modification is the change in phenotypic expression of an organism themselves and not by any genetic engineering technique. In contrast, artificially created GM constructs have to be either transferred into recipient cells by pathogenic microorganism or otherwise forced into the cells by gene guns or electric shock-mediated transformation. Natural modification is very precise and

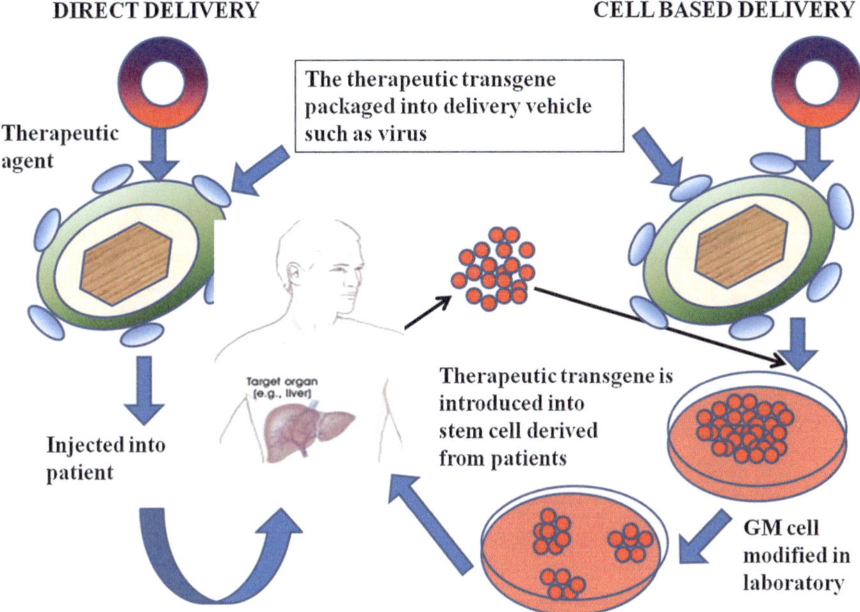

Fig. 14.18 Direct and indirect gene transfer in human

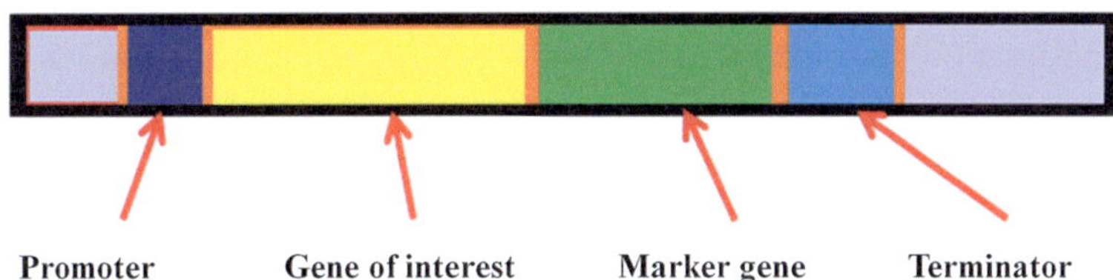

Fig. 14.19 Transgene construct

predictable and happens in the right place at the right time without damaging the genome, but it is time-consuming, whereas artificial constructs could land anywhere in the genome, scrambling and damaging the genome in the process. GMO production technique has been summarized in Fig. 14.20.

14.5 GMO Food Labeling

Labeling of foods is the best way of providing information about the food to consumers by producers/manufacturers. The main aim of any food labeling is to make the consumer aware of the contents and quality of food so that the consumer can choose the best out of available products for their good health. Various policies of food labeling differ in their nature, scope, coverage, and rate of execution leading to

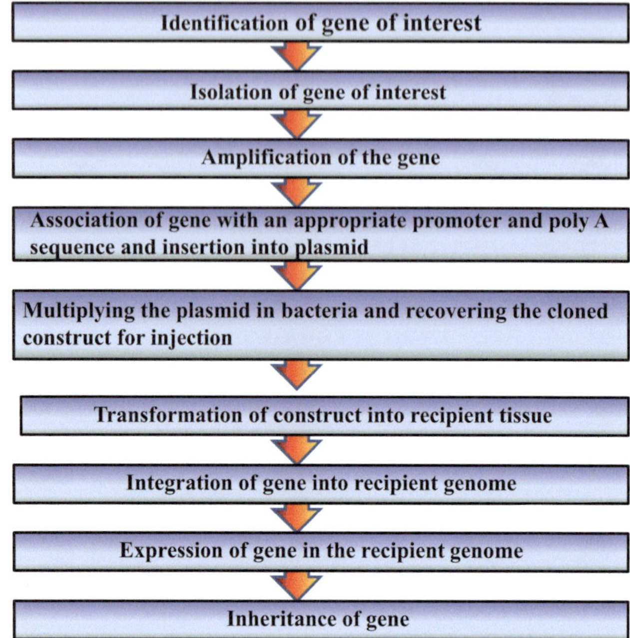

Fig. 14.20 Steps of GMO production

varying degrees of information to consumers such as nutritional facts; fresh meat and vegetable origin; date of production and expiry; allergen information, preventing misleading practices; and clear indication of defrosted products. It varies from country to country.

Labeling of GMO foods creates an impression and even acts as warning that it is different from or less safer than non-GMO counterparts. Without labeling information, the consumer is said to lack control, leading to irrational decisions. Such gap between the producer and consumer categorizes GMO products as authorized good as the consumer cannot assess the long-term attributes of GMO. Thus, labeling the credibility qualities can improve asymmetric information to consumers.

14.6 Economic Motivation

Genetic engineering and biotechnology continue to provide major environmental benefits and allow farmers to grow more, using fewer resources. A majority of these benefits are in developing countries in the field of crop development. "In the 17th year of widespread adoption, crops developed through genetic modification delivered more environmental friendly farming practices while providing clear improvements to farmer productivity and income," said Graham Brookes (2014).

Along with genetic engineering, crop biotechnology has contributed to significantly reducing the release of greenhouse gas emissions from agricultural practices and reduced pesticide spraying. Genetic engineering that has been used to develop insect-resistant (IR) technology in cotton and corn has consistently delivered gains from reduced pest damage. The herbicide-tolerant technology used in soybeans and canola has also contributed to increased production in some countries by helping farmers. Genetic engineering continues to be a good investment for farmers around the world coupled with higher average levels of benefits in developing countries.

14.7 GMO Irregularity

More than half a century ago, before the onset of genetic engineering, horizontal gene transfer (HGT) was already known to occur in nature, via mechanisms such as conjugation, phage transduction, and transformation. HGT is the movement of gene information between cells in unicellular or multicellular organism; it is also known as lateral gene transfer. It has been shown to be a confounding factor in the evolution of many organisms. GM DNA may be transferred at high frequency to bacteria, fungi, and plants. Plant wounds and rhizosphere are hot spots for HGT. Higher organisms including humans are more susceptible to HGT than bacteria due to sequence homology in higher organism.

HGT is a natural process, normally under the control of the organism itself, which is why GM DNA is such a threat. Due to high intensity of HGT, GM DNA may silence the transgene expression in successive generations. Instability of one transgene would create another transgenic plant with different unstable characters due to cell-to-cell transformation. Unstable transgenic lines are not good and their use should be reduced. Recent evidence showed that GM DNA of bacteria cannot transfer stably through HGT because of the majority of bacteria present in the environment and the human gut cannot be cultured.

The worse thing is that most living cells of the body including germ cell lines secrete DNA and RNA intercommunication system, which integrates into other cells and changes the expression of normal gene. This nucleic acid intercom can free the cells to take GM DNA and RNA, which may alter the cell structure and function, for example, conversion of normal cell into cancerous cells. As time pass, cancer cells spread cancer all around the body by HGT.

HGT is the primary reason for the spread of antibiotic resistance in bacteria. Through HGT, genes that are responsible for antibiotic resistance in one species of bacteria can be transferred to another species of bacteria. HGT plays an important role in the evolution of bacteria that can degrade novel compounds such as recalcitrant xenobiotics and in the evolution, maintenance, and transmission of virulence.

14.8 GM Controversies

Growing controversies about the use of GM foods are a major challenge to consumers. There are claims that genetically modified foods are nontoxic and nutritious and help save money and time. Generally, the genetically modified food benefits are greater than its risk. According to the Food and Drug Administration (FDA), foods from GM plant varieties marketed to date are as safe as non-GM foods. However, GM foods are still controversial. Allergenicity, gene transfer, and outcrossing are the main points of these controversies. Various government organizations have begun generating guidelines and recommendations regarding foods derived from transgenic organisms. The World Health Organization (WHO) and Food and Agriculture Organization of the United Nations (FAO) have developed protocols for proper testing of GM foods. In the USA the Food and Drug Administration (FDA) regulates the safety of food for humans and animals. The FDA made it mandatory for GM plant developers to consult the agency before marketing their products. During this process, the food developer is asked to conduct a safety assessment study by certified laboratory, and the report is submitted to the FDA for evaluation. After thorough evaluation the permission to release the food is granted with proper labeling.

14.9 GMO Toxicity

GMOs are a broad group of plants, animals, and bacteria that are engineered for a wide variety of applications ranging from agricultural production to scientific research. The types of potential hazards posed by GMOs vary according to the type of organism being modified and its intended application. Most of the concern surrounding GMOs relates to their potential for negative effects on the environment and human health. Food allergy affects approximately 5 % of children and 2 % of adults in the USA and is a significant public health threat (Bakshi 2003). Allergic reactions in humans occur when a normally harmless protein enters the body and stimulates an immune response (Bernstein et al. 2003). If the novel protein in a GM food comes from a source that is known to cause allergies in humans or a source that has never been consumed as human food, the concern that the protein could elicit an immune response in human increases. GM crops have inserted genes for pest resistance, nutritional quality, or other characteristics. Sometimes addition and deletion of genes create hundreds to thousands of mutations that disrupt normal gene function in GMO plants. Among herbicide-tolerant GM crops, the first to be grown commercially were soybeans which were modified to tolerate glyphosate. Glyphosate [N-(phosphonomethyl) glycine] is a nonselective, postemergence herbicide used for the control of a wide range of weeds. The target of glyphosate is 5-enolpyruvoylshikimate-3-phosphate synthase (EPSPS), an enzyme in the shikimate pathway that is required for the synthesis of many aromatic plant metabolites, including some amino acids. Glyphosate is believed to be less toxic than other pesticides. However, several recent studies showed its potential adverse health effects to humans as it may be an endocrine disruptor. A strain of genetically modified soybean produced lower levels of phytoestrogen compounds, believed to protect against heart disease and cancer, than traditional soybeans (Thongprakaisang et al. 2013). GM versus non-GM maize feeding to rats for 1 month increased the rate of tumor formation in the GM-fed animals (Séralini et al. 2012). GM gene is inserted in the plant genome efficiently, which may disrupt normal plant genes, further increasing the potential for toxicity. Hundreds to thousands of mutations in normal genes are caused by the genetic manipulation required to introduce the transgene to plants. This results in the development of plants that are dissimilar to native or non-GM plants and may have potential toxicity risk.

14.10 GMO Benefit-Risk Assessment

14.10.1 Increased Food Production

GMOs may play a very important role in alleviating hunger and poverty around the globe by way of increased quality food production. Transgenic plants of more than 20 species that are resistant to over 30 different viral diseases have been produced using variations of the pathogen-derived resistance strategy. δ-Endotoxin of *B. thuringiensis* has been used by many plant varieties such as tobacco, tomato, potato, cotton, walnut, maize, sugarcane, and rice to make them insect resistant. Genetically manipulated plants may delay ripening; it has a potential application for tropical fruit crops, which suffer severe losses from rapid ripening, lack of appropriate storage conditions, and efficient transport systems to reach to consumer. GM seeds theoretically provide the means for increased food production, which may accelerate the availability of food for close to one billion people in the world. These benefits have the potential to increase agricultural productivity and help to minimize the worldwide hunger burden.

14.10.2 Human Antibiotic Resistance

Ingestion of transgenic edible vaccines may develop resistance in humans. Excessive use of oral antibiotics is a universally accepted and well-known source of bacterial resistance. Immunotoxicity of transgene products and their unintended disruption of normal gene functions are additional potential hazards. Proper labeling to GM products may help prevent some of these concerns (Liu et al. 1999).

14.10.3 GMO-Resistant Weeds

Increased application of glyphosate confers tolerance of crops to this pesticide. The long-term consequences with increased pesticide application and resistance may increase cancer rates among exposed humans and animals. Glyphosate-resistant weeds increased rapidly since the introduction of GMOs in agriculture, medicine, industries, and other fields (Benbrook 2012).

14.10.4 Increased Cancer Rates

Glyphosate is an active ingredient of the most widely used herbicide and it is believed to be less toxic than other pesticides. Glyphosate induces human breast cancer cell growth via estrogen receptors only in human hormone-dependent breast cancer, T47D cells, but not in hormone-independent breast cancer (Thongprakaisang et al. 2013). High pesticide exposure may be a cause of development of non-Hodgkin's lymphoma (Hardell and Eriksson 1999).

14.11 GMO Detection Techniques

14.11.1 Real-Time (RT) PCR

RT-PCR is a laboratory technique of molecular biology, based on the polymerase chain reaction (PCR), which is used to amplify and simultaneously detect or quantify a targeted DNA molecule. In this method, genomic DNA extracted from an agricultural product is analyzed using various qualitative real-time PCR assays on a 96-well PCR plate, targeting for individual GM events, recombinant DNA (r-DNA) segments, taxon-specific DNAs, and donor organisms of the respective r-DNAs.

14.11.2 Differential Quantitative PCR

This method is based on the presence of several common elements (e.g., promoter, genes of interest) in different GMOs. A statistical model was developed to study the difference between the number of molecules of such a common sequence and the number of molecules identifying the approved GMO (as determined by border-fragment-based PCR) and the donor organism of the common sequence.

14.11.3 Multiplex PCR

In order to cope with the multiplicity of new GMOs released in the market and to save money and time, analytical strategies should be based on multi-detection systems that search for a high number of possible modifications in a single step. In addition, the systems must be capable of adjusting rapidly to new GM foods. As previously highlighted, an interesting feature of PCR techniques is the possibility of multiplexing. Multiplex PCR is a variant of PCR in which two or more target sequences are simultaneously amplified in a single reaction. A number of multiple simultaneous detections through the use of multiplex PCR have been reported.

14.11.4 Loop-Mediated Isothermal Amplification (LAMP)

It is a rapid and specific detection technique of GMO. The LAMP method is capable of auto-cycling strand displacement synthesis which amplifies DNA with high efficiency and specificity very rapidly under isothermal conditions. In a typical LAMP assay, a set of four specially designed primers including the inner primers (FIP/BIP) and the outer primers (F3/B3) are used to recognize six independent target regions.

14.11.5 Pyrosequencing

Pyrosequencing is a sequence-based detection technology that enables rapid and accurate quantification of sequence variation. High-throughput 16S rRNA gene-targeted pyrosequencing was used with commonly used risk assessment techniques to evaluate the potential microbial risk in soil after inoculating genetically modified (GM) *Corynebacterium glutamicum*. To verify the risk, reference experiments were conducted in parallel using well-defined and frequently used GM *E. coli* and wild-type strains. The viable cell count showed that the number of GM bacteria in the soil was reduced to below detection limit within 10 days, while the molecular indicator for GM plasmids was detected throughout the experiment by using quantitative real-time polymerase chain reactions.

14.11.6 Microarray

A multiplex DNA microarray chip was developed for simultaneous identification of nine GMOs. The on-chip detection was performed directly with PCR-amplified products. Particular emphasis was placed on the reduction of the number of PCR reactions required and on the number of primers present per amplification tube. The targets were biotin labeled and the arrays were detected using a colorimetric methodology. Specificity was provided by specific capture probes designed for each GMO and for the common screening elements. The chip detection system complies with the requirements of current EU regulations and other countries where thresholds are established for the labeling of GMO.

14.11.7 DNA Biosensor

DNA biosensors have the potential to overcome the limits of DNA microarrays by offering rapid and high sensitive analytical tools for genetic detection. The most important challenges are (1) the integration of microelectronics to microchip-based nucleic acid technologies in a high scalable process, (2) the automation of the detection step, and (3) the ability to perform direct signal transduction avoiding the image processing and statistical analysis, necessary in canonical DNA microarray workflow. Potential applications of DNA biosensors include molecular diagnostics, pharmacogenomics, drug screening, GMO analysis, medical diagnosis, food analysis, bioterrorism, and pollution or environmental monitoring. Recently, new generations of chips that can perform DNA sequencing have been developed accelerating biological and biomedical research in the genetics.

14.12 Future of GMO

Genetic engineering introduces organisms with novel phenotypes into existing ecology networks. If these traits escape the space and spread elsewhere, the effects on surrounding ecosystems could be novel and complex. GMOs have proven to be a very unique and relatively new wave in science. As the need for foods that can not only taste better but also withstand obstacle such as droughts, the future development in this area of science becomes a necessity. With these needs on the rise and the opportunity to help developing countries produce large and more advanced crops, there is more attention given to this area of science. Only a handful of studies have been done to predict the future of this important wave in science. This is primarily due to the fact that most of the findings thus far have been more accidental than controlled. The two major expert groups in regard to the advancement of GM food are the Food and Agriculture Organization and the World Health Organization. Both organizations have released statements defending the need for GM research as for future development. While criticism have been released about GMO research and the health risks linked to them, most major research group still urge the need for future research and development. The importance of development in this field is pushed by scientists that highlight the need to develop foods that can withstand the changing environment of current times. We hope that deadlock in the scientific sector will not stop the continuation of study in this field.

References

Amado RG, Chen YIS (1999) Lentiviral vectors-the promise of gene therapy within reach? Science 285:674–676

Bakshi A (2003) Potential adverse health effects of genetically modified crops. J Toxicol Environ Health B Crit Rev 6:211–225

Benbrook C (2012) Impacts of genetically engineered crops on pesticide use in the US-the first sixteen years. Environ Sci Eur 24:1–13

Bevan MW, Flavell RB, Chilton MD (1983) A chimaeric antibiotic resistance gene as a selectable marker for plant cell transformation. Nature 304:184–187

Bernstein JA, Bernstein IL, Bucchini L, Goldman LR, Hamilton RG, Lehrer S, Rubin C, Sampson HA (2003) Clinical and laboratory investigation of allergy to genetically modified foods. Environ Health Perspect 111:1114–1121

Boyer H, Cohen SN (1973) Develop recombinant DNA technology, showing that genetically engineered DNA molecules may be cloned in foreign cells. Gen Genomics Timeline

Brookes G (2014) What are the real economic and environmental impacts of GM crop use? Bus Pract Impact Environ Futur GMO 13:14

Brophy B, Smolenski G, Wheeler T et al (2003) Cloned transgenic cattle produce milk with higher levels of β-casein and k-casein. Nat Biotechnol 221:157–162

Costantini F, Lacy E (1981) Introduction of a rabbit β-globin gene into the mouse germ line. Nature 294:92–94

De Block M, Herrera-Estrella L, van Montagu M, Schell J, Zambryski P (1984) Expression of foreign genes in regenerated plants and their progeny. EMBO J 3:1681–1689

Gordon J, Ruddle F (1981) Integration and stable germ line transmission of genes injected into mouse pronuclei. Science 214:1244–1246

Gordon K, Lee E, Vitale J et al (1987) Production of human tissue plasminogen activator in transgenic mouse milk. Biotechnology 5:1183–1187

Hardell L, Eriksson MA (1999) Case-control study of non-Hodgkin lymphoma and exposure to pesticides. Cancer 85:1353–1360

Horsch RB, Fraley RT, Rogers SG, Sanders PR, Lloyd A, Hoffmann N (1984) Inheritance of functional foreign genes in plants. Science 223:496–498

Jaenisch R, Mintz B (1974) Simian virus 40 DNA sequences in DNA of healthy adult mice derived from preimplantation blastocysts injected with viral DNA. Proc Natl Acad 71:1250–1254

Liu YB, Tabashnik BE, Dennehy TJ et al (1999) Development time and resistance to Bt crops. Nature 400:487–598

Nordlee JA, Taylor SL, Townsend JA, Thomas LA, Bush RK (1996) Identification of a Brazil-nut allergen in transgenic soybeans. N Engl J Med 334:688–692

Paszkowski J, Shillito RD, Saul M, Mandak V, Hohn T, Potrykus I (1984) Direct gene transfer to plants. EMBO J 3:2717–2722

Potrykus I (1991) Gene transfer to plants: assessment of published approaches and results. Annu Rev Plant Physiol Plant Mol Biol 42:205–225

Séralini GE, Clair E, Mesnage R et al (2012) Long term toxicity of a Roundup herbicide and a Roundup-tolerant genetically modified maize. Food Chem Toxicol 50:4221–4231

Shillito R, Saul M, Paszkowski J, Muller M, Potrykus I (1985) High-efficiency direct transfer to plants. Biotechnology 3:1099–1103

Thongprakaisang S, Thiantanawat A, Rangkadilok N et al (2013) Glyphosate induces human breast cancer cells growth via estrogen receptors. Food Chem Toxicol 59:129–136

Suggested Further Readings

Ahmed F (2004) Testing of genetically modified organisms in foods. CRC Press, New York 317. ISBN: 9781560222743

Giacca M (2010) Gene therapy. Springer, Dordrecht, 303. ISBN: 978-88-470-1642-2 (Print) 978-88-470-1643-9 (Online)

Josef T, Richard W, Sandra S (2012) Gene vaccines. Springer, New York, 330. ISBN: 978-3-7091-0439-2

Mundembe R, Allison RF, Niangin IS (2012) The book "Genetic engineering-basics, new applications and responsibilities". InTech Publisher, Rijeka, 350. ISBN: 978-953-307-790-1

Piguet P (Basel), Poindron P (Strasbourg) (2012) Genetically modified organisms and genetic engineering in research and therapy. Karger, Medical and Scientific Publisher, Basel, 412. ISBN:978-3-8055-9066-2

Sarad RP (2004) Genetically modified animals, microbes and plants in biotechnology. Springer 374. ISBN: 978-1-61737-482-1 (Print) 978-1-59259-801-4 (Online)

Setlow JK (2002) Genetic engineering principles and methods. Springer, Boston, 350. ISBN: 978-1-4615-0721-5

William FG (2008) Biotechnology and bioengineering. Nova Science Publisher, New York, 272. ISBN: 978-1-60456-067-1

Degradable Polymers and Plastics of the Future: Steps Toward Environmental Sustainability, Regulations, and Safety Aspects

15

V.P. Sharma, Ram Lakhan Singh, and R.P. Singh

Contents

V.P. Sharma (✉)
CSIR-Indian Institute of Toxicology Research, Lucknow 226001, India
e-mail: vpsitrc1@rediffmail.com

R.L. Singh
Department of Biochemistry, Faculty of Science, Dr. Ram Manohar Lohia Avadh University, Faizabad 224001, India
e-mail: drrlsingh@rediffmail.com

R.P. Singh
Department of Biochemistry, Dr. Ram Manohar Lohia Avadh University, Faizabad 224001, India
e-mail: rajat2330@gmail.com

R.L. Singh (ed.), *Principles and Applications of Environmental Biotechnology for a Sustainable Future*, Applied Environmental Science and Engineering for a Sustainable Future, DOI 10.1007/978-981-10-1866-4_15

Abstract

Plastics are an integral part of our life, and the usage is increasing due to its salient characteristics. Both synthetic and natural plastics have great application and a ubiquitous role in our modern lifestyle due to the broad range of properties. They range from newly designed degradable or synthetic plastics to natural biopolymers such as DNA and proteins that are fundamental to biological structure and function. Polymers are synthesized via polymerization of many small molecules, known as monomers. Plastics are ubiquitous material of choice in the modern economy – combining unrivaled functional properties with low cost and wide acceptability by consumers. While delivering many benefits, the current plastic economy has drawbacks that are becoming more apparent day by day. Polylactic acid and polyhydroxyalkanoates are considered to be two main polymers that have a future role in biodegradable mulches. In order to overcome drawbacks of nondegradable and toxic plastic products, efforts are being done. Challenges range from enhancing system effectiveness to achieve better economic and environmental outcomes to continuing to harness several benefits of plastic packaging. Biodegradable polymers are ensured for compliance to claims, mechanical properties, customer demands, and environmental sustainability.

Keywords

Plastics • Degradable • Sustainability • Environmental • Regulatory

15.1 Introduction

The packaging waste is quite high in concentration and quantity when we visualize the municipal solid waste of any developing city. According to an estimate, of the 300 million tons of plastics produced annually, about 3 % is recycled. It has raised environmental concerns, resulting in strengthening of various regulations aimed at reducing the amounts generated. Moreover, wide ranges of oil-based polymers are also being used in formulation of new finished products for packaging applications. These are generally nonbiodegradable, and some are difficult to recycle or reuse due to complex composites having varying levels of contamination. Commonly, the finished plastic may be inert and nontoxic, but the monomers used in manufacture of parent polymers may be toxic (Casarett and Doull's 2013; Hayes and Kruger 2014). In some cases, small amounts of those chemicals may remain trapped in the product unless suitable processing methods are applied. Biobased polymers are replacing existing polymers in a number of applications and also provide new combinations of properties for new usages. The new environmental regulations, societal concerns, and developing ecological mindfulness have raised the interest multifold for new items and procedures that are sustainable and biocompatible with the Earth and its surrounding environment.

The ASTM/OECD/ISO guidelines mention clause-wise details toward classification, sampling procedure, preparation, application, analysis, quality assurance as per protocols, and end-user criterion fulfillments or requirements beside the need for periodical review and updating. Several degradable products do not approach the minimum criteria of biodegradability, and thus, they cannot be categorized as biodegradable as per ASTM D 5338/6400. Recently, BIS has notified IS/ISO: 17088:2008, which characterizes the criteria for biodegradable plastics under compostable condition similar to ASTM D 6400. It is beyond doubt that biodegradable packaging materials are most suitable for single-use disposable applications where the postconsumer waste may be locally composted. Few studies conducted to assess realities and potential innovative scenarios from life cycle assessment of nonrenewable energy use and greenhouse gas emissions reveal innovative concepts for minimization of CO_2 emissions using biodegradable concepts in farms. As Organization for Economic Co-operation and Development (OECD) countries emerge from the global financial crisis, several countries have their plans for the development of a future bioeconomy, an economy in which biobased materials and production techniques will significantly contribute toward economic and environmental sustainability (OECD 2014). Futuristic plans involve building a biobased production industry in which fuels, energy, and materials are generated. Under green economy concepts, minimum use of toxic chemicals and additives generated from fossil resources such as oil and natural gas is incrementally replaced by equivalent or novel products generated from renewable resources. The realization of this strategy will require sustainably harnessing the vast biomass resource. The new types of polymers known as oxo-degradable polymers are becoming popular due to the intrinsic features. In case of oxo-degradable polymers, the metal salts (mainly transition metals instead of heavy metals) are used to initially speed up the process of degradation followed by action through bacteria or other living organisms. Degradation is oxidative and a cell-mediated formula as per CEN/TR 1535–2006 and ASTM D 6954. It may be simultaneous or successive in process and dependent on resins, thickness, antioxidants, temperature, etc.

15.2 Bioresorbable Polymers

Bioresorbable polymers are polymer materials which may be safely absorbed by the body so that the materials from which a construction is made disappear over time. Various sources of degradable polymers are shown in Fig. 15.1. The common bioresorbable polymer is polylactic acid (PLA), also known as polylactide, and is made from a lactide monomer. Generally speaking, PLA is the main building block for bioresorbable polymer materials. The derivatives of PLA are poly-L-lactide (PLLA), poly-D-lactide (PDLA), and poly-DL-lactide (PDLLA). When in the body, PLA degrades into lactic acid, a nontoxic chemical appearing in the body. Polyglycolic acid (PGA), or polyglycolide (PG), is another type of bioresorbable polymer usually used for bioresorbable sutures. The material may be copolymerized with lactic acid to form poly(lactic-co-glycolic acid), or PLGA, with e-caprolactone to form poly(glycolide-co-caprolactone), or PGCL, and with trimethylene carbonate to form poly(glycolide-co-trimethylene carbonate), or (PGA-co-TMC). PGA generally degrades to form glycolic acid. The mechanical properties and degradation time of a bioresorbable device may be designed to a specific application by adjusting the molecular weight, crystallinity, and hydrophilicity of the polymer. Resorbable polymer scaffolds are printed using stereolithography and then injected, or seeded, with stem cells from the target organ. Additive manufacturing is crucial to the process because it allows an individual product to be made from a single data file. The data file is taken from a CT scan of the patient's organ, providing the required geometry to give a perfect fit. The objective of printing biological materials is to produce functional cells, tissues, and organs to repair, replace, or enhance biological function which has been lost by disease or injury. The method is seen by many as the

Fig. 15.1 Varied sources of degradable polymers

most promising solution to meeting high demand for suitable organs for transplantation or engineering of the tissues.

15.3 Biodegradable Plastics and Sustainability

Marvelous developments have been made in the fabrication or designing of biodegradable plastics, largely from renewable natural resources, to produce biodegradable materials with similar functionality to that of oil-based polymers (ATSDR). The expansion to biobased materials has several potential benefits for greenhouse gas balances and other environmental impacts over whole life cycles and in the use of renewable, rather than finite, resources. It is intended that use of biodegradable materials will contribute to sustainability and reduction in the environmental impact associated with disposal of oil-based polymers. Nowadays, the limited fossil resources, increased cost of fossil resources, public concern about climate changes, and important technology breakthroughs in white biotechnology are significant drivers to move from fossil-based polymers to biobased polymers in both low- and high-value polymer categories and markets (Narayan et al. 1999a, b). The structures of biodegradable polymers are critical in their specific properties. Biodegradable polymers are of potential interest to multiple fields including medicine, bioimplants, tissue engineering, agriculture, packaging, etc. An active niche area of research in biodegradable polymer is in controlled drug delivery and release.

The large molecular mass relative to small-molecule compound produces unique physical properties, including toughness and tendency to form glasses and semicrystalline structure rather than crystals. Biodegradable plastics may degrade upon exposure to sunlight, humidity, bacteria, enzymes, wind abrasion, and in some instances pests or extrinsic factors that may alter forms of biodegradation or lead to environmental degradation. Some modes of degradation require that the plastic be exposed at the surface, whereas other modes will only be effective if certain conditions exist in landfill or composting systems. Corn or starch powder has been mixed with plastic as a filler to allow it to degrade more easily, but it still does not lead to complete breakdown of the plastic. In the medical field, extensive research activities are underway to develop new carbon nanotube (CNT) biomaterials for use in the treatment and diagnosis of disease, viz., application of CNTs to cancer treatment and diagnosis, such as in drug delivery systems (DDSs) for treatment of cancer, *in vivo* imaging or for regenerative medicines, scaffold materials for nerve and bone tissue regeneration, etc. Researchers are doing innovative R&D work to improve the mechanical strength and durability of implants by combining CNTs with existing biomaterials. The field of microscale and nanoscale biomedical device design will continue to revolutionize medicine with varied applications in all areas of medicine, ranging from drug delivery to sensing to regenerative medicine. In addition to prosthetics, stereolithography of tissue engineering scaffolds is a proven technique for ceramic-polymer composites, biodegradable polymers, and polymer-cell solutions. Bioplastics are a family of materials, and they are not just one single substance; they comprise of a whole family of materials with differing properties and applications.

It may be either biobased or biodegradable or may feature with both properties. The term degradable biobased polymer means that the material or product is degradable in nature and derived from biomass, viz., corn, sugarcane, or cellulose. Bioplastics may provide biodegradability to deal with the increasing problems of litter and blockage of sewer lines and adversely effecting sanitation. Long-lasting plant-based bioplastics and their traditional counterparts can be recycled helping the development of a more sustainable world economy.

Biobased products are the new developments in view of environmental sustainability. There is consistent push to supplant materials got from petrochemicals with those that can be produced using biodegradable components. These polymers have the particular point of interest that over time they will degrade. Along these lines, biodegradable polymer coatings may sequester, as well as keep away from further creation of CO_2 in countless metric tons in few years.

15.4 Biodegradation

During biodegradation, microorganisms that are available in the environment convert materials into natural substances, viz., water, carbon dioxide, and compost. The process depends on the surrounding environmental conditions with reference to location, temperature, and moisture content. The property of biodegradation is affected by chemical structure and is not dependent only on the basic resource. Degradable polymers are naturally occurring as well as artificially synthesized and chiefly comprise of ester, amide, and ether functional groups. Their properties and breakdown system are decided by their definite structure. The reactions used for the synthesis of these degradable polymers are generally condensation reactions, ring-opening polymerization, and metal catalysis. Otherwise stated, 100% biobased plastics might be nonbiodegradable, and 100% fossil-based plastics can biodegrade (ISO, BIS).

They are classified as under Fig. 15.2:

- Biobased or partly biobased nonbiodegradable plastics such as biobased polyethylene (PE), polypropylene (PP), or polyethylene terephthalate (PET) (so-called drop-ins) and biobased technical performance polymers such as polytrimethylene terephthalate (PTT) or TPC-ET 2
- Plastics that are both biobased and biodegradable, viz., polylactic acid (PLA) and polyhydroxyalkanoate (PHA) or polybutylene succinate-3 (PBS 3)
- Plastics that are based on fossil resources and are biodegradable, such as polybutylene adipate-co-terephthalate (PBAT)

With reference to claims of compostability of degradable plastics (ISO 17088, EN 13432/14995, or ASTM 6400 or 6868), guidelines need to be ensured. Products may be categorized as biobased or biocomposites by determining its biobased mass content. Plastics are derived from renewable biomass sources ranging from oils, corn, sugar beet, banana, starch, pea starch, or microbiota. Low-density

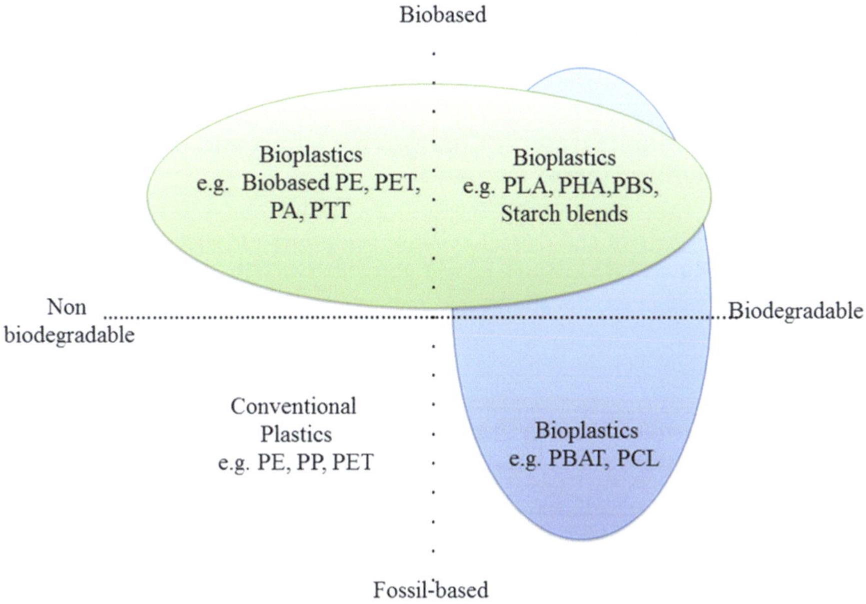

Fig. 15.2 Material coordinate system of bioplastics (Bioplastics coordinate system © European Bioplastics 2016)

polyethylene may degrade by ultraviolet (UV) light and oxygen, and to prevent this manufacturers add stabilizing chemicals for preventing breakdown or oxidation disintegration process. Such variety of polymers may be termed as degradable plastic or oxo-degradable plastic or photodegradable plastic as the process is not initiated by microbial action. Oxo-biodegradable plastics are conventional petroleum-based products with some additives that initiate degradation.

15.5 Risk Assessments

Qualitative and quantitative risk assessment is essential to evaluate risk posed to human health and or environment by the actual or potential presence and use of specific pollutants. Sustainable development urges that cities must coexist with nature and that they must become resource pools by reusing their own energy and resources, both natural and human. Water-sensitive design could improve the quantity and quality of water in cities, the alternative being increased water pollution, especially river pollution and drainage. According to reports, few cities may be covered with smog during low temperature which is air pollution typically associated with oxidants. The applications of polymers in indoor environment of the buildings and modern lifestyle comforts have a direct relationship with the concentration and matrix of the pollutants. It is not only the paints, fixtures, upholstery, and soft toys, but other items are responsible for pulmonary implications on health. The appropriate hygiene, ventilations, and preventive measures are vital for healthy living conditions and improving quality of life.

15.5.1 Polymers: Strategic Changes for Smart and Intelligent Polymer Market Demands

Humans are progressive and demand change with time for upcoming-generation requirements. Polytetrafluoroethylene and glass fiber reinforced plastics are products with tailor-made specification for dental applications to use in shipping, aircrafts, railway, etc.

15.5.2 Polymers: Health Problems Related to Synthetic Polymers

Natural or synthetic polymers may be produced with co-mixing with fillers; polytetrafluoroethylene for properties of stiffness, strength, heat resistance, density, and conductance; or nonflammability for curtains or liners in aircrafts or railways. With continued research into the science and applications of polymers, they are playing an ever-increasing role in society. Plasticizers such as DEHP, DBP, and DEP may mimic or antagonize the actions of naturally occurring estrogens [Interagency Coordinating Committee on the Validation of Alternative Methods (ICCVAM) 2003, 2006; National Research Council, 1999]. They may interact with one or more of the classical nuclear estrogen receptor (ER) subtypes: ERα, ERβ, or nonclassical membrane or ER-related subtypes (Hewitt et al. 2005; Matsushima et al. 2008; National Research Council 1999). These chemicals cause early puberty in females, reduced sperm counts, altered functions of reproductive organs, obesity, altered sex-specific behaviors, and increased rates of some breast, ovarian, testicular, and prostate cancers (Della Seta et al. 2006; Patisaul et al. 2006, Kavlock et al. 1996; National Research Council 1999).

The plastic products leach in to its content upon exposure to stresses such as ultraviolet (UV) radiation in sunlight, microwave radiation, and/or moist heat via boiling or dishwashing.

Polymers are studied in the fields of biophysics and macromolecular science and polymer science. Historically, products arising from the linkage of repeating units by covalent chemical bonds have been the primary focus of polymers; emerging important areas of science now focus on non-covalent links. Toxicity and concerns regarding furans and dioxins arise from degradation particles and gases emitted on combustion or accidental fires in multistoried apartments or business settings which have used inferior-quality commercial products of polyvinyl chlorides (PVCs) or old wires and building items. The emissions have a warming potential as they persist in the atmosphere, contributing to the greenhouse effect.

15.6 Plastic Debris/Litter in Marine Atmosphere: A Threat for Biodiversity

Carbon sequestration process is vital to estimate whether carbon is taken from the atmosphere and stored in a carbon sink using underground reservoirs. Examining the effect of plastics in the marine environment, raising public awareness, preventing litter by means of improved collection, and recycling plans are the need of the present time. Commercially available biodegradable plastic varieties of mulch are films produced using plant starch. These are prepared by utilizing traditional plastic processing technology. However, because of substandard mechanical properties and brittleness of starch, it must be blended with different polymers and plasticizers.

Positive drivers in this sector are:

- Use of renewable and biobased raw material
- Biodegradability
- Positive attitude of government toward green procurement policies
- Shift in consumer preferences toward eco-friendly packaging

15.7 Degradable Plastics from Renewable Resources

Plastic materials are the dominant materials of use – from agriculture to electronics to medical devices to packaging. Starting from a mere 1.65 million tons in 1950, they reached 280 million tons in 2011 and continue to show explosive growth as populous countries like India and China become more industrialized and consume more plastic materials (ISO TC/61, FAO, CFR). Due to the ubiquitous use of petroleum-based plastics, their persistence in the environment, and their fossil fuel derivation, alternatives to these traditional plastics are being explored. Issues surrounding waste management of traditional and biodegradable polymers are being emphasized in the context of reducing environmental pressures and carbon footprints. There is a negative viewpoint connected with transgenic plants, particularly food crops, plant-based biodegradable polymers, produced as value added coproducts which have the potential to become viable options to petroleum-based plastics and an environmentally benign and carbon eco-friendly source for production.

Polyhydroxybutanoate (PHB) is prepared by fermentation of glucose solutions to which propanoic acid has been added and bacterium is *Ralstonia metallidurans*. Biopol is a well-known copolymer with blocks of PHB and blocks of polyhydroxypentanoate. The film of biopol is used as coatings for paper and cardboard, food wrapping, kitchenware, and different medical uses.

15.7.1 Degradable Plastics from Fossil Fuels

Few degradable polymers are not made by biological processes, but still degrade naturally. For example, synthetic polymers, viz., polyethene which have starch granules encapsulated in variable amounts ranging from 5 to 50 %. The bacterial system degrades the finer granules and produce tiny fragments of polyethene which degrade at a quicker rate than polyethene without the starch. Synthetic polymers that might be treated with starch granules include polypropene and polyphenylethene.

Polycaprolactone which is prepared using benzene and cyclohexanone degrade readily in composts. Some polymers hydrolyze slowly in acid conditions and are used for stitching internal wounds and for drug delivery in the target region. The gradual dissolution of coating permits the drug to be delivered over a measured time frame.

15.8 Properties of PLA and Other Copolymers

The properties of PLA, for example, its strength, can be enhanced by utilizing added substances without decreasing its optical properties and is in this way utilized for food packaging. Copolymers of PLA with polyhydroxyethanoic acid (polyglycolic acid) have been found in significant application for surgery. Strings of the copolymer are block copolymers and utilized for stitching internal wounds and are degraded back to their parent monomers inside 3 months or prior after surgery.

15.9 Photodegradable Polymers

Photodegradable plastics are generally made up of oil-based polymers. These plastics are especially designed in order to control their degradability when exposed to sunlight in the environment. Photodegradable plastics have bonds (mainly carbonyl groups) in its structure that can be broken in the presence of sunlight, in this way deteriorating the polymer chain into smaller fragments rather than decomposing completely. These smaller fragments will then biodegrade. Photodegradable plastics contain an additive for their decomposition which absorbs light and afterward attacks the polymer resulting in breakage of some bonds. Relatively, these polymers are degraded in a compost facility in a couple of months and have preferable mechanical properties over the starch-filled polymers.

15.10 How Plastics Are Energy Efficient

Plastics are taken into account as energy-efficient materials when the whole life cycle of the material is considered. They are easy to install; versatile in texture, color, and designs; and cost-effective, with easy maintenance. Compared to traditional building materials, plastics are good building blocks composed of predesigned blocks, laminates, and composites. Moreover, they are lightweight, require

85 % less energy to manufacture, generate 95 % less CO_2, and use minimal water during the production process.

15.10.1 Economics and Recent Developments

High energy prices and increased awareness of environmental issues have encouraged regulators and authorities to initiate legislation to encourage a gradual shift from conventional to renewable energy sources. Modern polymer economy is fragmented in terms of composition and formulations. Sporadic availability of standards for each product range in the fast-developing scenario is still a concern for the regulatory bodies. The usage of polymers is increasing in other fields such as appearance of solar panels and windmills in coastal regions. Plastics are used in the construction of solar system accessories like blades of glass fiber reinforced plastics (GFRP). The consumption of biodegradable plastics may help in reducing the carbon footprint, pollution risks, and greenhouse gas emissions. The economy is related to consumption pattern and volume of plastics and polymeric products. The easy availability, installations, and cost are the critical criteria during acceptance of the technology and procurement by the consumers. Periodical feedback and quick improvements may help in increasing the market trends.

15.11 Growth and Future of Plastics

Growth of plastics is not limited due to extreme range of composition and increasing innovation in the developing global scenario. The plastic materials and the processes to shape them are developing faster than in the past. The challenges and opportunities are, therefore, multifold, and scientists or engineers are taking steps to imbibe new concepts of engineering with advanced knowledge of biotechnology, nanotechnology, and framing strategies or guidelines for exploiting plastic materials from orthopedic applications to biomedical device designing. Medical devices and instruments are in a constant state of evolution, responding to trends within and outside the hospital environment to achieve better care at lower system costs. Modern plastics have an important role to play in infrastructural changes to facilitate the acceptance by consumers and innovators.

Efforts are being made for improving repellitizing for reuse of recycled scrap-based resin. Replacement of materials with plastic has made many of our possessions cheaper, lighter, safer, and stronger. Since it is clear that plastics have a valuable place in our lives, some scientists are attempting to make plastics safer and more sustainable. Some innovators are developing bioplastics, which are made from plant crops instead of fossil fuels, to create substances that are more environmentally friendly than conventional plastics. Others are working to make plastics that are truly biodegradable and make recycling more efficient and possible to convert plastics back into the fossil fuels from which they were derived. The preparation of biodegradable and thermoresponsive enzyme-polymer bioconjugates with controllable enzymatic activity via reversible addition fragmentation chain transfer polymerization and amidation conjugation reaction is being attempted with better properties. Moreover, novel and intelligent polymers are also being fabricated to meet customers' expectations.

15.12 Plastics in the Food Packaging Sector

In order to meet the targets of safe and healthy food for all, we foresee greater usage of good-quality, nontoxic plastics for food delivery and packaging. The global market share for the packaging sector is expected to grow at a faster rate during the next few decades in spite of challenges related to degradability and disposal issues. The market requirements are regulated by demand, increased environmental awareness, and implementation of stringent environmental regulations. The growth is also boosted by the stringent policies and regulations implemented by governments in the region. Few countries have restricted the use of petroleum-based plastic bags, whereas there is a tax levied on the use of petroleum-based plastic bags in specific regions. Researchers are working to find ways to extend the shelf life of packaged food and making packing which is comfortable to open and recap for later use. In the future, plastic packaging can incorporate antifungal compounds in its polymeric matrix. The new coatings will provide 30 times better security for beverages or barriers that diminish oxygen transmission to close nonexistent levels. Drinking water should also be safely transported with almost zero wastage. Plastic storage tanks and pipes contribute significantly for supply of water through distribution system or transporting potable water to unreached areas while preventing bacteriological contamination. Most of the water filters are made from components made up of attractive and safe plastics using the latest technology-based cartridges or membranes.

15.13 Plastics: Agriculture and Impact on Soil

Filling of barren land or low-lying areas with municipal solid waste is one common practice in most of the adjoining cities of the world. Landfilling has certain advantages and disadvantages and thus needs to be viewed in terms of local and environmental sustainability without adverse effects on aquifers. Plastic lining or sheeting is an effective erosion control method adopted in some parts of the globe. Plastic nets linked to the concrete blocks fitted into soil close to riverbanks may prevent erosion. Biobased plastics or biobased natural fibers may also play a role in coastal erosion control.

Convenient and environmentally friendly solutions are the need of the day. Some biodegradable polymers which are utilized in agricultural practices degrade rapidly, while others may take a longer time. The measurement of degradation method is not standardized for degradable polymers. The rate of degradation of some biodegradable polymers is more rapid in compost soil than in landfills or in marine situations. The synthetic plastic sheets are not advisable for the lining in disposal sites for waste disposal due to varied reasons. In the longer period of time, there might be leaching from these liners, and unknown moieties may contaminate soil and water sources.

15.14 Wastes to Wealth: An Innovative Concept for Environmental Sustainability

We may generate wealth from waste depending on the chemical characterization and ease of segregation and extraction. It is a process by which plastic materials that would otherwise become solid waste are collected, separated, or processed and reused in the form of raw materials or finished goods. Industrialists and researchers are taking an active role in helping to minimize the amount of plastic materials that end up in landfill with initiatives across the globe aimed at stimulating progress in all types of recycling and recovery options. According to World Bank estimates, 1.4 billion tons of trash is generated globally each year, and approximately 10 % of it is plastic. The dumping of plastic waste at sea has been banned already in several parts of the globe, but still incidences appear without strict legislation. Thus, great content of plastic, metallic, and electronic waste generated each year escapes into the environment instead of being landfilled, incinerated, or recycled. The plastic industry also believes that an optimum waste collection scheme including comingled plastic collection may lead to better results and altogether increase recycling rates if proper recycling framework is set up.

In few countries, segregation units are being constructed that may divert nonrecyclable waste from landfill for the production of renewable energy. In the future, it is expected that the plant will likewise generate a renewable source of hydrogen for commercial purpose. An open-access facility is initiating new directions for concerting products and materials, for example, biomass into high-grade fuels and energy. The new facility may increase the amount of high-quality bottle grade plastics.

In spite of the fact that the technology of carbon capture reduces CO_2 emissions considerably, there is efficiency loss in power plants and increase in fuel consumption, capital cost, and overall costs of electricity generation. The main technologies that are currently used to generate power from fossil fuels are pulverized coal-fired (PC) steam cycles and natural gas combined cycles. Integrated gasification combined cycle (IGCC) technologies are being developed although it is not considered to be economically competitive. Reducing primary carbon footprints may be achieved through relatively straightforward actions. There is a philosophy that states and policymakers should not wait for scientific proof of harmful effects before taking steps to limit harmful environmental and human health impacts from new products or activities. Specific areas of application include genetically modified food products and chemicals that may have harmful developmental effects in low doses.

15.15 Guidelines Related to Degradable Plastics

The ASTM D 6400 specification encompasses several ASTM standardized tests for biodegradability, and they need to be consulted in light of recent improvements in guidelines and requirements of the country. The referred ASTMs of importance are ASTM D 5209, D 5338, D 6002, D 5988-03, D 6954, etc. A standard that is embedded in ASTM D 6400 specifies that 90 % of C atoms must be mineralized, that is, converted to CO_2 within 180 days by microorganisms (ASTM 2003, BIS, ISO).

Regulatory reforms may improve the effectiveness of regulation and reduce pollution. The use of third-party auditing to monitor compliance of organizations with regulations is ubiquitous. Consumers and commodity markets use third-party audit reports to monitor standards for food safety and health care and other interrelated goods. The dissemination strategy plan must be through both policy and academic channels for true implementation of guidelines which are developed with extreme efforts of experts.

Biodegradability-related standards followed by various countries are mentioned in Table 15.1. The Bureau of Indian Standards (BIS) has also framed few guidelines related to degradable plastics in harmonization with International Organization for Standardization (ISO) specifications, viz., 14851:1999; 14852:1999; 14853, 14855-1:2005; 15885:2004; 16929:2002; 17088: 2008; 20200:2004, European Standard EN 13432, etc. In terms of industrial standard, the definitions of ASTM, EPA, and ISO are well-set standards for biodegradability. International methods cover the test for biodegradable plastic, anaerobically and aerobically, as well as in marine environments.

Table 15.1 Salient guidelines on plastics[a]

S. No.	Title	Guidelines
1.	Standard specification for labeling of plastics designed to be aerobically composted in municipal or industrial facilities	ASTM D 6400
2.	Determination of the ultimate aerobic biodegradability of plastic materials in aqueous medium – method by measuring oxygen demand in a closed respirometer	DIN/EN/ISO 14851
3.	Determination of the ultimate anaerobic biodegradability of plastic materials in aqueous medium – method by analysis of evolved carbon dioxide	DI/EN/ISO 14852
4.	Packaging requirements for packaging recoverable through composting and biodegradation	BS/EN 13432
5.	OECD 301/310: Ready biodegradability tests	OECD 301 C
6.	Determination of the ultimate aerobic biodegradability of plastic materials in an aqueous medium – method by measuring the oxygen demand in a closed respirometer (Japanese guidelines)	JIS K 6950
7.	Specifications for compostable plastics (PCD 12: plastics)	BIS 17088:2008
8.	Guide on suitability of plastic for food packaging	IS 10171
9.	Aerobic biodegradation under controlled conditions	ISO 14855
10.	Anaerobic biodegradation in a high-solid sewerage environment	ISO 15985

[a]Updated guidelines/specifications are being framed globally and harmonized to have state-of-the-art and uniform acceptable guidelines

15.15.1 Bioplastics and Multifunctionality

Bioplastics are plastics derived from renewable biomass sources, such as vegetable fats and oils, corn starch, or microbiota. Bioplastic can be made from agricultural by-products and also from used plastic bottles and other containers using microorganisms. Common plastics, such as fossil fuel plastics, are derived from petroleum or natural gas. Production of such plastics tends to require more fossil fuels and to produce more greenhouse gases than the production of biobased polymers. Few bioplastics are designed to biodegrade. Biodegradable bioplastics can break down in either anaerobic or aerobic environments, depending on how they are manufactured. These may be composed of starches, cellulose, biopolymers, and a variety of other materials. The environmental impact of bioplastics is often debated due to different metrics for greenness, viz., water use, energy use, deforestation, biodegradation, etc.

15.15.2 Selection of Polymers for Specific Purpose and Designing

During selection of materials and packaging technologies, it is necessary to keep in mind the product which needs to be packaged and the duration. In order to retain freshness of vegetables, cling may be sufficient, while for packing pickles or composites food items, the situation is quite different. Moreover, duration from packaging date to commercialization and consumption may vary in different parts of the country where the product has been manufactured. Appropriate plastic packaging offers excellent advantages with multifunctional applications and versatility for intended purpose. Polymeric packaging with bar coding and intelligent sensor or signal-based indicative plastics offers several advantages over shelf-life indications, reducing wastage of food, beverages, flexibility in thermal and mechanical properties, lightness, aesthetic appeal, etc. at an economical cost. The recent food packaging systems like retort packing are responsive to consumer preferences toward mildly preserved fresh, taste retention, healthier, and convenient food product availability with increased shelf life to the maximum extent. The novel packaging technologies may be further improved for preventing shortcomings in the packaging design and control the fizz, barrier properties, mechanical strength, oxygen content, and water or CO_2 levels in the headspace. They have become the ubiquitous material of the modern economy combining unrivaled functional properties with minimum cost and thus compete with other packaging solutions. Their use has increased 20-fold in the past half century and is expected to increase exponentially. In spite of delivering several advantages, the plastic industry feels sometimes over-commenting by other groups of packagers due to competitive attitudes. With an explicit systemic and collaborative approach, the industry aims to overcome the limitations with initiatives of green chemistry and economical paradigm shift. Thus, it may be anticipated to kindle wave of innovation and to move the plastic value chain into a positive spiral of value capture, value addition in features for improved economics, and better environmental satisfaction in tune to regulatory requirements.

15.15.3 Innovations

Innovative R&D with efforts involving predictive modeling, simulation studies, engineering upliftment through machinery, and trained manpower innovations are clearly visible in the market. Novel polymers serve well in therapeutics for purpose of guided tissue regeneration, tissue engineering, drug delivery systems, gene transfection, and with properties of biocompatibility as well as serve as functional polymers for medicine and dentistry. Introducing reactive chemical functions within or along polymer backbones is an attractive route to generate functional polymers for medicine. There have been development and changes in packaging materials including paper or metallic liners for effective distribution chains reaching to the remotest places with high temperature or freezing locations in transboundary situations for the defense personnel. Polymeric materials have a great potential for exciting new applications in the foreseeable future. They are being developed in varied diverse areas for conduction and storage of electricity, heat, UV-protected light, molecular-based information storage, processing, molecular composites, unique separation, food processing, health, housing, packing, and transportation with an effective and vital role in all aspects of life including under illness by helping to reach the food by Ryle's tube and medication through nontoxic tubes. Recent studies indicate that the exposure to certain engineered nanomaterials (ENMs) has potential to lead to health complications, and thus need arises for detailed investigations to unravel the biological outcomes of nanofood consumption. Plants naturally produce numerous polymers, including rubber, starch, cellulose, and storage proteins, and most of them are exploited in quest for safe and degradable plastic manufacture. Bacterial bioreactors fed with renewable resources are successful in producing biodegradable polymers.

15.15.4 Polyethylene: Usage and Recycling

Plastics are man-made, lightweight, long-chain polymeric molecules that find use in an extensive variety of applications. Most types of plastics are durable and endure for long time in nature since they are not biodegradable. Likewise, due to the durability of the polymers involved in production of plastics, significant quantities of disposed plastics are aggregating as debris in landfills and in natural habitats. Recycling can accordingly give chances to reduce energy, materials, carbon dioxide emissions, and the quantity of waste needed to dispose with enhanced eco-efficiency. The complexities of composition adjunct with contamination during use often render recycling uneconomic compared with disposal in landfill. Due to lack of biodegradability, plastics have attracted criticism from the environmentalists as they take about 100–1000 years to degrade when used in landfills besides polluting the air and water around. Mainly plastics of the waste stream are dealt by incineration, landfilling, recycling, or composting. Polyethylene plastic mulch is widely used for crop production as it controls weeds, conserves soil moisture, increases soil temperature, and improves crop yield and quality. Most mulch films are generally produced from petroleum-based plastics, e.g., polyethylene, which leads to considerable waste disposal problem. Despite the fact that recycling might be a choice, polyethylene mulches utilized as a part of farming are contaminated with an excess of dirt and debris to be recycled specifically from the field. The abnormal state of contamination might be the explanation for low amount of agricultural plastic film recycling. In addition, burning of polyethylene mulch releases persistent organic pollutants such as dioxins and furans which have an undesirable effect on the environment. Polyethylene mulch is utilized in less quantity by soil microorganisms in landfills with the microbial conversion and abiotic oxidation products consisting of harmful chemicals such as aldehydes and ketones. An alternative solution for decreasing waste from polyethylene mulches is to develop biodegradable mulch which is easily degraded by microorganisms. The use of biodegradable mulches may save labor and disposal costs, conserve resources, and decrease pollution if they are in line with the extended producer's responsibility. Accumulations of plastic debris or litter in the environment or coastal regions with associated consequences are largely preventable. Minimization of waste entering the natural environment might be accomplished by better waste disposal practices and process management. Littering is a behavioral mindset issue comprised of waste items that have been disposed improperly at an inappropriate location. Litter of plastic is presumably the most noticeable part of the production of such high volumes of plastics.

Recovering, reutilizing, and recycling may help to reverse this trend such that we start to regard end-of-life materials as valuable feedstocks for new production rather than waste. Improved mass dissemination and education to the students when they are young would be a better idea to see the concepts growing with the children as a habit. In fact, most of the products are expected to degrade in natural habitats, and oil-based polymers may merely disintegrate into small pieces. To pick up the most extreme advantage from degradable, biodegradable, and compostable materials, it is important to distinguish particular uses that offer clear preferences and to refine national and international norms (e.g., EN 13432, ASTM D 6400-99) and associated product labeling to indicate appropriate usage and appropriate disposal. Plastics offer considerable benefits for the future, but it is evident that our current approaches to production, use, and disposal are not sustainable and present concerns for wildlife and human health.

To achieve appropriate use, disposal, and recycling of plastics, we require executing a comprehensive waste disposal and litter management as well as recycling strategy through education and enforcement for environmental sustainability. Recycling, recovery, or disposal of plastic waste should be carried in such a way that it minimizes the environmental contamination issues. There is an urgent need for industries to develop waste management and recycling programs and execute demonstrated waste minimization schemes that exemplify the best in environmental safety. There is additional requirement for governments and policymakers by setting norms and focuses which can reduce the amount of waste and integrate environmentally sustainable waste disposal and management plans. These measures might be considered inside a structure of life cycle investigation, and it must fuse all key stages in plastic generation, including synthesis of the chemicals that are utilized as a part of production, together with utilization and disposal (Palmer 2003; OECD 2013).

15.16 Environmental Regulations and Guidelines for Monitoring or Management

The recent environmental regulations, societal concerns, and developing environmental mindfulness have started the search throughout the world for new products and processes that are cost-effective and compatible with the environment. Globally, strengthening of standardization activity may continue to be the main thrust as it is the core activity of regulatory organizations to keep in trends to the rapid growth being witnessed internationally along with fast-paced industrialization and technological advancements. In formulating standards, the key factors of importance are harmonization with international standards, consideration of new and emerging technology areas, and environmental concerns and other regional or national priorities for meeting requirements of consumer needs. Few researchers are of the opinion that another greenhouse gas, methane, might be released when any biodegradable material degrades in an anaerobic landfill environment. Methane production from managed landfill environments may be captured and used for energy. Disposing of nonbiodegradable plastics made from natural materials in anaerobic/landfill environments may result in the plastic lasting for the next few decades. Attempts are taken into perfecting the processes for cellulosic biofuels. Producing biofuels at the scale required to influence markets has proved extremely challenging. Plastics and polymeric industries have to grow amid challenges and meet the new dimensions of society. Environmental dilemmas with long-term implications are climate change and consequently on life-related issues. Reducing the effect on the climate is a key component in sustainable development and of specific importance for industries. The quick and profound greenhouse gas emission minimization can be achieved with industrial biotechnology applications. Compostable plastics are a new generation of plastics capable of undergoing biological decomposition in a compost site and do not produce any visual or toxic residue.

Commercial composting for selected plastics is economically viable using state-of-the-art waste disposal techniques with prescreening and selection thereafter on a pilot scale. Municipalities may divert appropriate quantities of waste from overburdened landfills and devise local need-based units or plants in coordination with local governing authorities and NGOs. There are paradigm shifts in our requirements and expectations. The pattern of production is shifting from the true biodegradable plastics to the biobased plastics, and that trend is likely to persist in future. The carbon footprint of a bioplastic is dependent on the carbon extracted from the air by the growing plant and on CO_2 captured by the plant in the photosynthesis process and physiological processes. Bioplastics are degraded back into carbon dioxide and water, giving back all the sequestered carbon to the atmosphere. The production of most bioplastics results in decreased carbon dioxide emissions compared to their conventional alternatives. Efforts have been made to develop environmentally sustainable plastic products by using renewable polymers as an alternative to petroleum-based plastics. Under appropriate conditions, microorganisms can completely degrade the biodegradable plastics, and subsequent basic components might be utilized again. As compared to bioplastics, petrochemical-based plastics are stubborn because they cannot be degraded by microorganisms and persist in the environment for a long time.

Hydrophobic, high-surface-area plastic residues may migrate into the water table and other compartments of the ecosystem causing irreparable harm to the environment. The end products of the biodegradation of bioplastics have insignificant environmental impacts. Bioplastics can be integrated directly into the soil where microflora completely degrade them into environmentally compatible end products such as carbon dioxide or methane, water, and biomass. Another type of degradable plastics is photodegradable plastic. The photodegradable plastics consist usually of polyethylene with additives that enhance degradation in the presence of sunlight. However, the oxo-degradable plastic bags may take a couple of years to biodegrade depending on exposure to sunlight, oxygen, and temperature and produce tiny fragments of plastic that do not continue to degrade. Plastic inertness and resistance to microbial attack were reduced by incorporating starch as an additive. Polyethylene plastic bags that are produced with starch additives likewise incompletely degrade over time because microorganisms digest the starch, but leave the polyethylene fragments. Many traditional plastics may contain added substances which facilitate their processing and enhance the physical qualities of the products fabricated from them.

It is expected that plastics made in the future using biotechnological methods will contribute to a massive reduction in the carbon footprint by the exploitation, in industrial productive processes, of renewable monomers such as polyols and dicarboxylic acids.

15.17 Engineered Films and Polyesters

Engineered film manufacturers are designing films to create the environment-friendly polyesters for day-to-day requirements with efficient packaging by using and producing the least waste with minimal depletion of renewable resources. Through state-of-the-art extrusion technologies, an array of renewable building blocks, ultrahigh-efficiency films ensuring excellent quality with significantly reduced material costs, and lowered carbon footprint are in the exploitable stage for process in developed countries. Polyhydroxyalkanoates (PHA) are a group of biologically synthesized linear polyesters. It is produced during bacterial fermentation of sugars and lipids. PHA is more ductile and less elastic than other plastics, and it is widely used in the medical industry. PHA is an eco-efficient biobased plastic which is recyclable and biodegradable.

Polylactic acid is a thermoplastic polymer made from lactic acid and has mainly been used for biodegradable products, such as plastic bags and planting cups, but in principle, PLA may be used as a matrix material in composites.

15.18 Inference

Most of the degradable polymers undergo decomposition into CO_2, CH_4, water, inorganic compounds, or biomass with the predominant mechanism of enzymatic action through microorganisms, which may be estimated by standardized tests, in a specified period of time. Plastics have a great future with skill development, innovative polymeric developments for smart cities, and intelligent marketing novelties for the next generation. Developed countries are viewing the product to meet challenging customer needs and compliance as per ISO 9001:2015 and ISO 17025:2015 and OECD requirements with respect to quality. Dedicated design and improved process engineering enable plastic recycling processes to be constantly improved, with new technologies being developed and patents registered. Packaging materials are becoming more colorful and chemically sophisticated. Films are often fully printed and in some cases even feature multiple layers. They are also thinner, reducing weight and helping protect resources. Earlier recycling technologies struggled to recycle these waste plastics with binding agents and other additives remaining in the result, making them no longer suitable for use in high-quality end products.

Safe food packaging does quite more than simply holding a product as they maintain food safe and fresh and also explicit how to safely store and prepare it, facilitate purchasing, provide nutritional information, and protect products during transport, delivery, and storage. Packaging also fills trash containers and landfills, lasting far longer than the products it were made to contain. It consumes natural resources and may transfer chemicals into our packaged food with unknown health effects of varied nature due to poor disposal methods and improper segregation for treatment or recycling. New technology is now enabling the recycling of such materials in a single step, turning them into high-quality recycled pellets. An economically viable market is being created for recycled plastic film, suitable for all applications. Thus, in fact, plastics have revolutionized the packaging industry because they are highly moldable into infinite shapes, lightweight, inexpensive, easy to seal, and durable.

References

Agency for toxic substances and disease registry ATSDR for plasticizers DEHP, DBP and other chemicals related to plastics. http://www.atsdr.cdc.gov/toxprofiles/tp9-c1-b.pdf

ASTM D5988-03 (2003) Standard test method for determining aerobic biodegradation in soil of plastic materials or residual plastic materials after composting. ASTM International, West Conshohocken

Biodegradable polymers, website. https://en.wikipedia.org/wiki/Oxo_Biodegradable

Casarett CD, Doull's LJ (2013) Toxicology: the basic science of poisons, 8th edn. Curtis Klaassen, New York

Della Seta D, Minder I, Belloni V, Aloisi AM, Dessi-Fulgheri F, Farabollini F (2006) Pubertal exposure to estrogenic chemicals affects behavior in juvenile and adult male rats. Horm Behav 50:301–307

Hayes AW, Kruger CL (2014) Hayes' principles and methods of toxicology, 6th edn. CRC Press, Boca Raton

Hewitt SC, Deroo BJ, Korach KS (2005) Signal transduction: a new mediator for an old hormone? Science 307:1572–1573

Kavlock RJ, Daston GP, DeRosa C, Fenner-Crisp P, Gray LE, Kaattari S et al (1996) Research needs for the risk assessment of health and environmental effects of endocrine disruptors: a U.S. EPA-sponsored workshop. Environ Health Perspect 104(Suppl 4):715–740

Matsushima A, Teramoto T, Okada H, Liu X, Tokunaga T, Kakuta Y et al (2008) ERRγ tethers strongly bisphenol A and 4-α-cumylphenol in an induced-fit manner. Biochem Biophys Res Commun 373(3):408–413

Narayan R, Krishnan M, Dubois P (1999a) Polymer 40:3091–3100

Narayan R, Krishnan M, Snook JB et al (1999b) US Patent# 5,969,089

National Research Council (1999) Hormonally active agents in the environment. National Academies Press, Washington, DC

OECD (2013) Guidelines for testing of chemicals and policies for bioplastics in the context of a bioeconomy. Paris

OECD (2014) Biobased chemicals and bioplastics: finding the right policy balance, OECD science, technology and industry policy papers, No. 17, OECD Publishing. http://dx.doi.org/10.1787/5jxwwfjx0djf-en

Patisaul HB, Fortino AE, Polston EK (2006) Neonatal genistein or bisphenol-A exposure alters sexual differentiation of the AVPV. Neurotoxicol Teratol 28:111–118

Suggested Further Readings

Andersson SR, Hakkarainen M, Inkinen S et al (2010) Leads to higher hydrolytic stability but more acidic hydrolysis product pattern. Biomacromolecules 11:1067–1073

Brehmer B (2013) Polyamides from biomass derived monomers. In: Kabasci S (ed) Bio-based plastics: materials and applications. Wiley, Chichester, pp 275–293

Briassoulis D, Dejean C (2010) Critical review of norms and standards for biodegradable agricultural plastics part I: biodegradation in soil. J Poly Env 18(3):384–400

Bureau of Indian Standards (2016) Guidelines on plastics- PCD 12 and PCD 21, 2016. New Delhi

European Bioplastics (2016) What are bioplastics? http://en.european-bioplastics.org/wp-content/uploads/2011/04/fs/Bioplastics_eng.pdf

Kawasaki N, Nakayama A, Yamano N et al (2005) Synthesis, thermal and mechanical properties and biodegradation of branched polyamides 4. Polymer 46:987–993. doi:10.1016/j.polymer

Lee MW, Tan HT, Chandrasekaran M et al (2005) Synthesis and characterization of PLLA by melt polycondensation using binary catalyst system. SIMTech Tech Rep 6:40–45

Palmer RJ (2003) Polyamides, plastics. In: Mark H (ed) Encyclopedia of polymer science and technology, 3rd edn. Wiley, New York, pp 618–642

Thielen M (2010) Basics of bio-polyamaides. Bioplastics Mag 5:50–53

Vert M (2005) Aliphatic polyesters: great degradable polymers that cannot do everything. Biomacromolecules 6:538–546

Yamano N, Kawasaki N, Oshima et al (2014) Polyamide with long-chain fatty acid groups suppressing the biodegradability of biodegradable polymers. Polym Degrad Stable 108:116–122

Yao K, Tang C (2013) Controlled polymerization of next generation renewable monomers and beyond. Macromolecules 46:1689–1712

MIX
Papier aus verantwortungsvollen Quellen
Paper from responsible sources
FSC® C105338

If you have any concerns about our products,
you can contact us on
ProductSafety@springernature.com

In case Publisher is established outside the EU,
the EU authorized representative is:
Springer Nature Customer Service Center GmbH
Europaplatz 3, 69115 Heidelberg, Germany

Printed by Libri Plureos GmbH
in Hamburg, Germany